Lost
Enlightenment

❀

Lost
Enlightenment

CENTRAL ASIA'S GOLDEN AGE

FROM THE

ARAB CONQUEST TO TAMERLANE

❀

S. Frederick Starr

PRINCETON UNIVERSITY PRESS

PRINCETON AND OXFORD

press.princeton.edu

Jacket art: Detail of ms. Elliott 339, fol. 95v, courtesy of
the Bodleian Library, University of Oxford

Library of Congress Cataloging-in-Publication Data
Starr, S. Frederick.
Lost enlightenment : Central Asia's golden age from the Arab conquest
to Tamerlane / S. Frederick Starr.
pages   cm
Includes bibliographical references and index.
ISBN 978-0-691-15773-3 (hardcover)
1. Asia, Central—History—To 1500. I. Title.
DS288.3.S73 2013
958'.02—dc23
2013013684

British Library Cataloging-in-Publication Data is available

This book has been composed in Minion Pro text with Archer display

Printed on acid-free paper. ∞

Printed in the United States of America

1 3 5 7 9 10 8 6 4 2

To strive for knowledge is the duty of every Muslim.
—Saying or Hadith of the Prophet Muhammad,
recorded in the tenth century by the scholar
Abu Isa Muhammad Tirmidhi (824–892)
from Tirmidh (Termez), and inscribed at the
entrance to the madrasa of Ulughbeg, ruler and
astronomer, Samarkand, ca 1420

❁

Wisdom is the principal thing; therefore get wisdom.
And with all thy getting, get understanding.
—Bible, Proverbs 4.7, King James Version

Contents

❁

Illustrations

❀

MAPS

PLATES

Following page 292

Figures

Preface

❁

This book was written not because I knew the answers to the questions it poses, or even because I had any particular knowledge of the many subjects and fields it touches upon, but because I myself wanted to read such a book. It is a book I would have preferred someone else to have written so I could enjoy reading it without the work of authorship. But no one else took up the assignment. Central Asia as yet has no chronicler comparable to Joseph Needham, the great historian from Clare College, Cambridge, whose magisterial, twenty-seven-volume *Science and Civilization in China* has no equal for any other people or world region. And so I backed into the task, in the hope that my work might inspire some future Needham from the region or from among scholars abroad.

The questions raised in this book became my constant companions for nearly two decades and over several scores of trips through every corner of the region—trips that included scorching treks in the Karakum Desert of Turkmenistan and being snowbound for nearly a week in the Pamirs at minus 40 degrees. Enormous, predigital piles of notes made entrance to my office a challenge that few chose to face. Now, with the volume done, I find myself saying, with Edward Gibbon in the preface to his *Decline and Fall of the Roman Empire*, that "I have ventured, perhaps too hastily, to commit to the press a work which in every sense of the word deserves the epithet of imperfect."[1] And, by the way, I know all too well that I am no Gibbon.

It would be more than a stretch to say that I am qualified to have undertaken this book. But at least I can claim a long-term interest in the subject. The Persian world first opened to me when, at age eighteen, I met my freshman roommate at Yale, Hooshang Nasr, whose father was mayor of Tehran under the shah. "Hoosh" went on to become a dedicated medical doctor who loyally served his country. My first contact with the Turkic world began through archaeological work at Gordium in Turkey, where Alexander the Great cut the Gordian knot, and eventually extended over several seasons spent mapping ancient roads in Anatolia. Neither of these links qualified me as an expert on anything, but from these early contacts to the present it has been natural for me to view both the Persian and the Turkish worlds as places inhabited by exceptionally interesting people, among whom are many good friends of mine.

The number of scholars and experts who have plowed the separate furrows of this book is staggering. It is fashionable in some quarters to fault Western and Russian scholars of the past two centuries for their "orientalism." But without their painstaking research, the larger story of the intellectual effervescence of the Islamic East would never have become known to the world. This has been a thoroughly an international effort. Among the many participants are French savants like Jean Pierre Abel-Remusat, Farid Jabre, Étienne de la Vaissière, and Frantz Grenet, not to mention the many authors of the publications, since 1922, of the Délégation archéologique française en Afghanistan. In Germany Heinrich Suter, Adam Mez, and others founded a tradition that continues today in the likes of Josef van Ess, Gotthard Strohmaier, and a host of younger scholars from both the former East and West, while the Czech Republic claims the great literary scholar Jan Ripka.

Across the English Channel, adventurers Armenius Vambery and Sir Aurel Stein, both of them immigrants from Hungary, sparked the imagination of the English-speaking world and of all Europe with the accounts of their explorations in Greater Central Asia. Then came linguists like Edward Granville Browne and translator Edward Fitzgerald, who together did much to bring the treasures of regional literature to broader notice. In the twentieth century the awesomely prolific Clifford Edmund Bosworth from Manchester wrote with insight on scores of topics essential to a book like this, while Georgina Herrmann and her colleagues extended this tradition into archaeology. Patricia Crone and other British scholars have advanced the study of many philosophers from the region, while E. S. Kennedy did authoritative work on the scientists. American scholars should also be noted, especially Richard N. Frye and Richard W. Bulliet, whose research on Nishapur, Bukhara, and the broader region inspired a generation of historians. Such gifted linguists and translators as Robert Dankoff and Dick Davis have opened windows on unknown or underappreciated masterpieces. Dimitri

Gutas and other distinguished scholars have analyzed the writings of Farabi and other Central Asian thinkers who wrote in Arabic. Raphael Pumpelly and Fredrik Hiebert should also be saluted for their pioneering archaeological research that traced the first grain for bread to a site in what is now Turkmenistan. In addition to all these, a host of younger scholars, especially in Europe and the United States, are on the lip of transforming our understanding of the region and time.

Iranian scholarship also continues to make important contributions. Tehran scholars have undertaken the monumental task of locating, editing, and publishing the complete works of Ibn Sina and several other major thinkers of the Age of Enlightenment. They are also conducting important research on the various traditions of Sufism. Persian scholarship also thrives in emigration, where it has given rise to such valuable productions as the New York-based *Encyclopedia Iranica*, not to mention distinguished luminaries like Seyyed Hossein Nasr of George Washington University, also in the United States. The Indian subcontinent, with its deep cultural ties with Central Asia, has produced important editions and English translations from the Arabic works of Central Asian authors and has given rise to valuable studies on specific figures, notably Biruni, who spent time in Kashmir. Japanese scholars, meanwhile, have developed a strong base in language and linguistic studies and have been among the world leaders in the extent and depth of their recent research on Buddhism in Central Asia.

Russian research did not begin with Vasilii Bartold (Wilhelm Barthold), but he brought it to a high level from which it has only rarely descended since. A superior linguist with a passion for detailed chronology, the austere and tireless Bartold spent a lifetime poring through forgotten texts in medieval Arabic and Persian and reconstructing the outlines of a lost history. His research remains today the gold standard for the region. After his death in 1930 his students not only carried on his work but extended it into new areas, notably archaeology and the history of science.

Few mourn the passing of the Soviet Union, but a monument should nonetheless be raised to the research on Central Asia that its Academy of Sciences supported. The special strengths of this research lay in the history of science, literature, and archaeology. Multiyear projects collected neglected manuscripts, surveyed whole regions for archaeological

remains, and reconstructed the outlines of the lives and work of great figures from the past. Names like Mikhail E. Masson, Galina A. Pugachenkova, P. G. Bulgakov, and Iurii A. Zavadovskii are prominent among
those Soviet scholars who rescued Central Asia from oblivion. Their
successors within the region continue to advance important research
on many fronts. Ashraf Akhmedov, B. A. Abdukhalimov, Edvard Rtveladze, and Otanazar Matyakubov in Uzbekistan; Gurtnyyaz Hanmyradov in Turkmenistan, and K. Olimov and N. N. Negmatov in Tajikistan
all carry on this tradition of high scholarship and, increasingly, make its
fruits available in the languages of the region. Thanks to them, and to
many others, a younger generation of highly qualified scholars is emerging across the region. Trained by leaders of the last Soviet generation
of scholars and in regular contact with their counterparts in Europe,
America, Iran, and the Middle East, these talented young researchers
are raising fresh questions and arriving at unexpected answers. Doubtless, the story that follows will look very different as it is deepened and
corrected by the fruits of their investigations over the coming decades.

This book has benefited from comments and assistance from a number of colleagues and friends. Among these are Anna Akasoy of Ruhr-
Universität, Bochum; Christopher Beckwith of Indiana University;
Jed Z. Buchwald of California Institute of Technology; Farhad Daftary
and Hakim Elnazarov of the Institute of Ismaili Studies, London; Gurtnyyaz Hanmyradov, rector of Turkmenistan's National University; Deborah Klimburg-Salter of the Universität Wien; Azim Nanji of Stanford
University; Morris Rossabi of Columbia University; Edvard Rtveladze of
the Academy of Sciences of Uzbekistan; Pulat Shozimov of the Academy
of Sciences of Tajikistan; Nathan Camillo Sidoli of Waseda University in
Japan; and Sassan Tababatai of Boston University. Each of these people
offered generous personal advice and assistance, often taking time to
instruct me on matters I should have known about in the first place. In
many cases their keen-eyed reading of the manuscript led to the correc-
tion of errors of fact and interpretation. Many others doubtless remain,
but these are solely my responsibility as author, not theirs.

This list could be infinitely extended, but the point is clear: that whatever strengths this book may possess trace to the research of scores of
dedicated scholars in many countries. Together they have been my pro-
fessors, and I am profoundly grateful to them.

The history of American publishing at its best is a history of great editors. Among the genuine stars of that firmament is Peter Dougherty of Princeton University Press, who gently but firmly encouraged this project beginning in 2006. His colleague and my editor, Rob Tempio, brought a priceless combination of creative imagination and high professionalism to the final revision, design, and production of the book. He turned what could have been a chore into a thoroughly pleasant process. Marja Oksajärvi Snyder, a professional bibliographer specializing on Inner Asia, handled with skill and a good eye the time-consuming task of locating or commissioning, selecting, and assembling images for the illustrations and gaining the necessary reproduction rights. Anita O'Brien undertook the copyediting with patience, tenacity, and precision.

Very few of the works on which this books draws are available electronically. Thanks are therefore due to the staff of the library at Johns Hopkins University's School of Advanced International Studies, especially Barbara Prophet and Kate Picard of the Inter-Library Loan department, who found and assembled hundreds of sources from all over the world. Over many years the staff of the Central Asia-Caucasus Institute, notably Katarina Lesandric and Paulette Fung, provided timely support to the project.

Above all, my wife Christina, children Anna and Elizabeth, their spouses Patrick Townsend and Holger Scharfenberg, and their children—my grandchildren—patiently tolerated the entire undertaking for longer than anyone should have had to do. This book is dedicated to them all, with deep love and gratitude.

A Note on Names, Spellings, and Transliterations

This book is rooted, in the first instance, in the Persian, Turkic, and Arabic worlds. All three languages present challenges for anyone wishing to render their proper names and specific terms into English. This is vastly complicated by the fact that many relevant names and terms have become known mainly through the works of scholars writing in English, French, German, or Russian, who have standardized them in accordance with the particular rules of their own languages, in the process often distorting the originals. Thus, to take just one example, is the region of northeastern Iran, southern Turkmenistan, and western Afghanistan to be called Khwarism, Khwarizm, or Khwarasm, after Persian usage, or Khorezm, as transliterated from Russian? Any of the first three will be useful when scouring the indexes of books by Persianists, while Khorezm must be used for Russian indexes. Since sources in regional languages or by scholars using regional languages are increasingly important, I have chosen the form Khwarazm. And should we refer to the Prophet's grandson as Hosain, Hossein, Husain, Husayin, Husein, Huseyin, Hussain, Hussayin, Huseyn, Hussein, Husseyin, or Husseyn? In this and other cases, I have sought to use whatever version is most familiar to an English-speaking audience.

Personal names like this pose a particular problem. The great St. Petersburg orientalist was baptized Wilhelm Barthold, the name used in the English translations of his works, but he spent his career in a Russian environment and signed his works Vasilii Bartold. I have chosen to use Bartold, except when citing English editions that use the original German spelling of his name. The name of the dread Mongol khan and conqueror appears variously in the literature as Gengis, Genghis, Genghiz, Gengiz, Chinggis, and Chingiz. An informal poll of experts left me with Chinggis. For nearly every language there exists an accepted system for transliterating into English, but these often result in spellings that only a linguist could decipher. I have solved all these diverse problems in favor of whatever is most familiar to an English-speaking audience.

Arabic and Persian names pose a particular challenge. By the time he has been identified in terms of his father, his son, and his place of origin (nisba), a man can end up with a name with six or more elements. Thus we encounter Abu Ali al-Ḥusayn ibn Abd Allah ibn Sīna, and Ghiyath al-Din Abu'l-Fath ʿUmar ibn Ibrahim Al-Nishapuri al-Khayyami. But contemporary nonspecialist readers of English demand shortened names, which leaves the reader with Ibn Sina, and Omar Khayyam. Similarly, contemporaries doubtless referred to the author of the famed Algebra as Muhammad ibn Musa, but for several hundred years scholars have chosen to refer to him by his nisba, hence "Khwarazmi"; I have followed their practice. Latin versions of names are cited but not widely used in the text. When referring to figures who are well-known around the world, I have followed common English usage, hence Ibn Sina and not Ibn Sino.

Whole books can be, and have been, written about the transliteration or transcription of Arabic or Persian proper nouns. Even though the formal systems of transliteration to various languages devised by linguists would produce results that would satisfy most scholars, they force general readers to confront letters, markings, and spellings that are all but incomprehensible to them. Hence Romanization is here achieved through simple transcription, which has been modified as necessary to conform to the rules of English orthography or conventional English usage. Thus the reader will encounter the thinker Omar Khayyam but not Umar Hayyam or Omar Chajjam.

Linguists and other specialists will doubtless rue the absence of diacritical markings on most words and names throughout the text. Such markings as the breve, accent mark, circumflex, hacek, and diaeresis can be useful aids to pronunciation. But they can also put off readers while giving little or no benefit in return. And so diacritical markings have been deleted for Central Asian names. Those readers who long for them are free to add them by hand, as a medieval copyist might have done.

Abul-Wafa BUZJANI (940–998). Afghan-born pioneering researcher at Baghdad and Gurganj. His method of developing sine and tangent tables produced results accurate to the eighth decimal point. By applying sine theorems to spherical triangles, Buzjani opened the way to new methods of navigating on open water.

Abu Mansur Muhammad DAQIQI. An ardent patriot from Balkh, champion of the Zoroastrian past, and author of versified sections of the Persian epics that Ferdowsi incorporated into his *Shahnameh*. At Daqiqi's death in 976, Ferdowsi took over the project.

DEWASHTICH (r. 721–722). The last pre-Islamic ruler of Panjikent in present Tajikistan; fleeing before Arab armies in the early eighth century, he hid a collection of official documents in a large pot and buried them at Mount Mug. Rediscovered by a shepherd in 1933, the Mug documents enabled scholars to reconstruct details of Sogdian government and society.

Abu Nasr Muhammad al-FARABI (870–950). A native of Otrar in modern Kazakhstan; known in the West as Alfarabius and revered in the East as "The Second Teacher," after Aristotle. A great expounder of logic, Farabi set out the foundations of every sphere of knowledge.

Ahmad al-FARGHANI (ca. 797–860). An astronomer who hailed from the Ferghana Valley in present-day Uzbekistan. Farghani's *The Elements* was among the earliest works on astronomy to be written in Arabic. In the West "Alfraganus," as he was known, became the "Arab" astronomer with the widest readership; among his readers was Columbus.

Abul Hasan ibn Julugh FARUKHI. Eleventh-century poet and musician from Sistan at the court of Mahmud of Ghazni and the author of lucid but complex poems built around the symbolic image of the garden. His verse on the death of Mahmud is one of the finest elegies in Persian.

Abolqasem FERDOWSI (ca. 934–1020). Author from Tus in Khurasan (now Iran) who toiled for thirty years—happily under the patronage of the Samanids of Bukhara and unhappily under the patronage of Mahmud of Ghazni—to produce the Persian epic *Shahnameh*. Combining legend with historical fact and spanning fifty reigns, his epic was a ringing affirmation of Persian values after the Arab conquest.

Abu Hamid Muhammad al-GHAZALI (1058–1111). Theologian and philosopher from Tus in what is now Iranian Khurasan, and author of *The Incoherence of the Philosophers*, which threw down the gauntlet to rationalism. After undergoing a nervous breakdown following the death of his chief patrons, he adopted Sufism and, in a series of brilliant works, integrated his views on faith into the mainstream of Islam, eventually influencing Christianity as well.

GHOSAKA. A deeply respected Buddhist theologian and author from Balkh who played an important role in the deliberations at the Fourth Buddhist Council in Kashmir in the first century AD.

HABASH al-Marwazi (769–869). Astronomer and mathematician from Merv who led a team at Baghdad to calculate a degree of terrestrial meridian and hence Earth's circumference, and whose tables plotted planetary motion.

Ahmad ibn HANBAL (780–855). An Arab collector of Hadiths from Merv who refused to succumb to Caliph Mamun's rationalist inquisition, thereby establishing himself as an early martyr of Sharia-based traditionalism in Islam.

HIWI al-Balkhi. Late ninth-century skeptic and polemicist from Khurasan who launched blistering assaults on the Old Testament but spared neither Christian nor Islamic holy writ from his scathing criticism.

Abu Ali al-Husayn IBN SINA (980–1037). Philosopher, theologian, polymath, and author of the *Canon of Medicine*, which remained for half a millennium the classic medical text throughout the Muslim world and Europe. The impact of his *Book of Healing* and *Book of Deliverance* on theology in the Muslim world and Christian Europe was equally powerful owing to his intricate affirmation of both reason and faith. Ghazali frontally challenged his legacy in theology.

Abu Nasr Mansur IRAQ (960–1036). A prince of the Khwarazm royal house, mathematician, and astronomer who did pioneering work in spherical geometry and applied it to finding solutions to problems of astronomy.

Nuradin JAMI (1414–1492). Leader of the Naqshbandiyya Sufi order in Timurid Herat, poet, and author of complex mystical allegories that are rich with Sufi symbolism.

Abu Abdallah al-JAYHANI. Geographer and Samanid vizier from 914 to 918; author of a massive *Book of Roads and Kingdoms* that was prized for its scope and detail.

Zayn al-Din JURJANI (1040–1136). Author in Gurganj of a massive compendium of medical knowledge, the *Khwarazm Shah's Treasure*, which focused on the needs of the practicing doctor.

KANISHKA I. Powerful second-century AD Kushan ruler of much of Central Asia whose synthesis of Buddhism, the Greek pantheon, and Zoroastrianism was manifest at his capital at Begram and other sites in Afghanistan.

Mahmud al-KASHGARI. Eleventh-century author of *A Compendium of the Turkic Dialects*, a comprehensive guide to the Turkic languages and their oral literature. A masterful treasure of linguistic, anthropological, and social information, Kashgari's work was designed to claim for Turkic culture the same status as Arabic and Persian in the Muslim world.

CHINGGIS KHAN. Mongol ruler whose devastation of Central Asia between 1218 and 1221 has been called an "attempted genocide," but who opened both China and Persia to new waves of intellectual influence from Central Asia.

Omar KHAYYAM (1048–1131). Mathematician, astronomer, philosopher, engineer, and poet from Nishapur whose landmark *Treatise on the Demonstration of Problems of Algebra* first conceived a general theory of cubic equations. His new solar calendar was introduced in 1079.

Abu al-Rahman al-KHAZINI (d. ca. 1130). Astronomer and polymath whose *Book of the Balance of Wisdom,* written in Merv, has been called "the most comprehensive work on [weighing] in the Middle Ages, from any cultural area."

Abu Mahmud KHUJANDI (945–1000). A native of Khujand, Tajikistan, and designer of astronomical instruments who reached conclusions on Earth's axial tilt that were more precise than those of anyone before him.

Nasir KHUSRAW (1004–1088). A Seljuk civil servant turned Ismaili missionary and poet. This native of Balkh province in Afghanistan left works of travel and philosophical poetry of unsurpassed beauty.

Abu Abdallah Muhammad al-KHWARAZMI (780–850). From Khwarazm; worked in Baghdad. He systematized and named algebra, contributed to Arabic and Western understanding of spherical trigonometry, championed the decimal system, compiled data on the locations of 2,402 places on earth, and gave his name to algorithms.

MAHMUD OF GHAZNI (971–1030). Born a Turkic slave, founder of an orthodox Sunni empire stretching from India to Iran, and patron of Biruni, Ferdowsi, and four hundred poets. Mahmud was at the same time the enemy of all heterodoxy in religion.

Caliph Abu Jafar Abdullah MAMUN (786–833). Worked initially from his capital at Merv and then shifted to Baghdad, where he promoted science and philosophy and carried out an unsuccessful inquisition against Muslim traditionalists.

MANAS. Legendary or, to some, historical Kyrgyz leader who became the main subject of the huge oral epic of the Kyrgyz people, *Manas.* The government of the Kyrgyz Republic celebrated the thousand-year anniversary of Manas in 1995.

Muhammad Abu Mansur al-MATURIDI (853–944). A truculent and influential defender of literalist and traditionalist Islam from Samarkand, author of many combative "Refutations" of rationalism and other errors.

Bahaudin al-Din NAQSHBAND Bukhari (1318–1389). Founder of a major Sufi order who helped bring about a reunion between Sufism, traditionalist Islam, and the state.

NAVAI, pen name of Nizam al-Din Alisher Harawi (1441–1501). Timurid official, art patron, and poet who singlehandedly elevated his native Turkic language, Chaghatay, to the same high level as Persian.

Al-Hakim al-NAYSABURI (821–875). An Asharite traditionalist in theology from Nishapur who collected and issued two thousand Hadiths and quarreled with Bukhari and others over questions of authenticity.

NESTORIUS. Archbishop of Constantinople (428–431) and founder of a branch of Syrian Christianity that long dominated Christian life and learning in Central Asia.

NIZAM AL-MULK, or "Order of the Realm" (1018–1092). Honorific title of Abu Ali al-Hasan ibn Ali, powerful Seljuk vizier from Tus who railed against the Ismailis in his *Book of Government* and championed Ghazali against perceived threats to Muslim orthodoxy.

Ali QUSHJI (1402–1474). Son of Ulughbeg's falconer and later a renowned astronomer, founder of Ottoman astronomy, and author of a ringing defense of astronomy's autonomy from philosophy.

RABIA Balkhi. A tenth-century poetess and friend of Rudaki from Balkh, now Afghanistan, whose brother killed her on learning of her love for a Turkic slave.

Abu Hasan Ahmad Ibn al-RAWANDI (820– 911). Prolific thinker from Afghanistan who abandoned Judaism and Islam to become a thoroughgoing atheist and champion of unfettered reason.

Muhammad ibn Zakariya al-RAZI (865–925). From Rayy near modern Tehran, but educated in Merv by Central Asian teachers; his principal intellectual heirs were also from Central Asia. Razi was the first true experimentalist in medicine and the most learned medical practitioner before Ibn Sina. He was a thoroughgoing skeptic in religion.

RUMI (ca. 1207–1273). Common name of the hugely popular poet Jalaluddin (Jalal al-Din) Muhammad Balkhi, from Balkh, Afghanistan.

Ismail Ibn Ahmad SAMANI (849–907). Founder of the Samanid state, which for a century gathered Central Asia's cultural resources to Bukhara.

Ahmad SANJAR ibn Malikshah (1085–1157). Sultan who moved the Seljuk capital back to Central Asia and oversaw a last, albeit limited, period of flowering, symbolized by his massive double-domed mausoleum at Merv.

Abu Sulayman al-SIJISTANI (932–1000). Moved from his native Khurasan to Baghdad, where he led a humanist seminar and advocated a strict separation of science/humanities from religion.

Abdallah ibn TAHIR. Mid-ninth-century Tahirid ruler of Khurasan and all Central Asia who advocated universal education on the grounds that the welfare of society depends on the welfare of the common people.

TAHIR ibn Husayn (d. 822). Founder of the Tahirid dynasty, which ruled Central Asia virtually as a sovereign state between 821 and 873, and supporter of intellectual life at its capital, Nishapur.

TAMERLANE (TIMUR) (1336–1405). Turkic marauder who conquered territory from the Mediterranean to India, founded a century-long dynasty, and assembled artists and craftspeople at his capital at Samarkand.

THEODORE. Appointed Nestorian Christian archbishop at Merv in 540. A linguist and expert on Aristotle in general and on his *Logic* in particular.

Abu Isa Muhammad TIRMIDHI (824–892). Hadith collector from Tirmidh, now Termez in Uzbekistan, where Buddhist monks earlier carried out similar work on religious texts.

Nasir al-Din al-TUSI (1201–1274). Polymath native of Tus in Khurasan and founder of the Maragha observatory under the

Mongols. He challenged Aristotle's notion that all motion is either linear or circular.

ULUGHBEG (1394–1449). Honorific name of Mirza Muhammad Taraghay. Ulughbeg, a grandson of Timur, briefly ruled Central Asia and was an educator and astronomer. His tables of the movements of stars were long unsurpassed for accuracy, while his encouragement of mathematical and scientific studies was the Islamic world's last great push in these fields.

Abul Qasim UNSURI (968–1039). Native of Balkh and the prodigiously prolific "King of Poets" at Mahmud's court at Ghazni, Afghanistan.

YAKUB ibn Laith, "The Coppersmith" (840–879). Founder of a short-lived dynasty from Sistan on the border of Iran and Afghanistan that frontally challenged the hegemony of Arab rule and the Arabic language in Central Asia and Iran.

Ahmad YASAWI (1093–1166). Sufi mystic and poet from Isfijab, now Sayram, in southern Kazakhstan. His Turkic quatrains carried a message of private prayer and contemplation of God to large numbers of heretofore unconverted Turkic nomads.

ZOROASTER (ca. 1100–1000 BC). Founder, probably in the eleventh century BC, of the monotheistic system that became the core religion of urban Central Asia down to the rise of Islam. Its doctrine of an individual judgment, Heaven and Hell, and bodily resurrection were later reflected in both Christianity and Islam.

Chronology

❁

Note: all dates are approximate.

3500– 3000 BC	Lapis lazuli mined in Afghanistan is exported to India and Egypt. Complex urban centers appear in Central Asia. Bronze Age Central Asians become first to raise grain for bread.
1100–1000 BC	Most commonly accepted dates for Zoroaster, founder of Zoroastrianism.
	Bactrian camel becomes the backbone of regional and continental transport in Central Asia.
563 BC	Birth of Gautama Buddha in India.
329–326 BC	After destroying Persepolis, Alexander the Great wages a three-year war in Central Asia and Afghanistan.
180 BC	Armies of Greek Bactria invade India, the apogee of the Hellenistic Greek kingdom of Bactria in Central Asia.
127 BC	Approximate start of reign of Kushan ruler and champion of Buddhism, Kanishka I.
114 BC	Start of export of Chinese silk westward and eventual rise of indigenous sericulture in Central Asia.
53 BC	Parthian army from Central Asia defeats the Roman general Crassus to take control of the Levant.
540	Theodore, a Syrian Christian linguist, scholar, and philosopher, is named archbishop at Merv.
570	Birth of Muhammad, prophet of Islam.

651	The last Sassanian ruler, Yazdegerd III, is murdered at Merv.
660	Arab conquest of Central Asia begins.
743	Umayyad caliph Abu Hashim dies, unleashing the Abbasid revolt or "civil war of Islam."
	Abbasid armies of Abu Muslim defeat Tang Chinese army at Battle of Taraz (Talas).
762	Caliph Mansur founds Baghdad as capital of the Abbasid Caliphate, basing his plan on Merv.
780	Birth of Abu Abdallah Muhammad al-Khwarazmi (Khwarazmi).
797	Birth of Central Asian astronomer Ahmad al-Farghani.
810	Birth of Muhammad al-Bukhari, the collector of the most widely accepted compendium of Hadiths of the Prophet.
810–819	Merv in Central Asia is capital of caliphate.
813–833	Abu Jafar Abdullah Mamun, son of Caliph Harun al-Rashid, reigns as caliph, after serving as governor of Khurasan, 809–813.
ca. 820	Birth of Abu Hasan Ahmad Ibn al-Rawandi, a radical critic of Islam and other revealed religions.
822	Tahir ibn Husayn, governor of Khurasan, asserts his independence by purging the caliph's name from coinage and Friday sermons.
824	Birth of Abu Isa Muhammad Tirmidhi, prominent collector of the Hadiths of Muhammad, at Tirmidh (Termez) in Uzbekistan.

833–848	Caliph Mamun's rationalist inquisition takes place in Baghdad.
849	Birth of Ismail Ibn Ahmad Samani, founder of the Samanid dynasty.
850	Birth of Abu Zayd al-Balkhi, a student of al-Kindi, who pioneered an innovative school of terrestrial mapping.
865	Birth of the medical authority and polymath Muhammad ibn Zakariya al-Razi (Rhazes).
ca. 870	Birth of philosopher Abu Nasr Muhammad al-Farabi.
875–876	Ismail Samani establishes his dynasty at Bukhara.
ca. 900–934	Construction of Tomb of Ismail Samani, Bukhara.
	Onset of the so-called Shiite century (900–1000), epitomized by Fatimids in Egypt and Buyids in Iran.
ca. 934	Birth of Abolqasem Ferdowsi, author of *Shahnameh*.
940	Birth of Afghan mathematician Abul-Wafa Buzjani.
960	Birth of Abu Nasr Mansur Iraq, student of Buzjani, astronomer and mathematician.
977–978	Construction of the mausoleum at Tim.
980	Birth of Abu Ali al-Husayn Ibn Sina (Avicenna).
986	Mahmud of Ghazni's first raid into the Indus Valley.
993	Mahmud of Ghazni drives the Samanis from Khurasan.
995	Birth of Abolfazi Beyhaqi, historian of the Ghazni dynasty.
	Abu Hasan, son of Mamun I, becomes the shah of Khwarazm.

998	Mahmud of Ghazni gives himself the title of sultan.
998–999	Biruni and Ibn Sina enter into correspondence.
ca. 1000	Al-Hakim al-Naysaburi writes eight volumes on the lives of local scholars of law and religion and other learned men from Nishapur.
1002	Fall of the Samanid dynasty.
1004	Birth of Ismaili scholar and poet Nasir Khusraw.
1005	Ibn Sina leaves Bukhara for Gurganj to work in an administrative post.
1010	Ferdowsi completes the *Shahnameh*.
1017	Officers and patricians of Abu Abbas, shah of Khwarazm, kill him in a coup, leading to Mahmud of Ghazni's conquest and destruction of Gurganj and the end of the Mamun Academy.
	Mahmud gains control of the entire Indus Valley.
1018	Birth of the Seljuk vizier Abu Ali al-Hasan Ibn Ali, known as Nizam al-Mulk.
	Ibn Sina completes his *Canon of Medicine*.
ca. 1027	Birth of Mahmud al-Kashgari.
1037	Biruni completes his compendium of astronomy and mathematics, the *Canon Masudicus*.
1040	Masud, the son of Mahmud of Ghazni, is defeated by Seljuks in battle at Dandanqan near Balkh and escapes to India.
1048	Birth of mathematician, astronomer, and poet Omar Khayyam.
1055	Seljuks sweep to power in all Central Asia.

1065	Abu Hamid Muhammad al-Ghazali (Algazel) sets up the first Nizamiyya madrasas at Baghdad, Nishapur, Khargird, and, later, Shiraz.
1066	Norman invasion of England.
1069	Yusuf Balasaguni completes his *Wisdom of Royal Glory*.
1072	Normans reestablish European rule in Sicily.
1077	Mahmud al-Kashgari completes his *Compendium of the Turkic Dialects*.
1079	Sultan Malikshah and his vizier, Nizam al-Mulk, introduce Khayyam's new solar calendar.
1092	Nizam al-Mulk is murdered by an Ismaili assassin on October 14.
1093	Birth of Sufi Ahmad Yasawi at Isfijab (Sayram) on the southern border of modern Kazakhstan east of Chimkent.
1095	Ghazali, at the peak of his popularity, falls severely ill.
1118	Ahmad Sanjar assumes power in Merv after the death of Sultan Malikshah.
1127	The Kalyan minaret, also known as the Minaret of Death, is built in Bukhara.
1135	Birth of Ibero-Egyptian thinker Maimonides.
1141	Karakhitai utterly rout the entire Seljuk army on the wasteland northwest of Samarkand.
1145	Birth of Sufi poet Farid al-Din Attar in Nishapur.
ca. 1180	Ibn Sina's *Canon of Medicine* is translated into Latin by Gerard of Cremona.
1201	Birth of astronomer Nasir al-Din al-Tusi.

1207	Traditional date of birth of the great Sufi poet Jalal al-Din Rumi.
1219	Chinggis Khan launches the Mongol invasion after the ruler of Otrar destroys a Mongol caravan and murders Chinggis's ambassador.
1270s	Marco Polo travels twice from Venice to Beijing.
ca. 1300	Printing with inscribed blocks and movable type in East Turkestan.
1308–1321	Dante Alighieri writes his *Divine Comedy*.
1318	Birth of the Sufi Bahaudin al-Din Naqshband Bukhari, founder of the Naqshbandiyya order.
1336	Birth of Tamerlane (Timur the Lame).
1350s	Black Death reduces populations across Persia and the Caucasus.
1380	Muscovy defeats the Mongols of the Golden Horde.
1389	On death of Bahaudin Naqshband, the ruler of Bukhara permanently endows his school and mosque.
1402	Timur attacks the Ottoman Turks and captures Sultan Bayazit.
	Birth of astronomer Ali Qushji at Samarkand.
1414	Birth of the Persian poet Nuradin Jami.
1417	Ulughbeg founds his madrasa in Samarkand.
1436	Filippo Brunelleschi completes the Basilica of Santa Maria del Fiore in Florence.
1439	Johannes Gutenberg, working in Strasbourg and Mainz, uses movable type to print.
1441	Birth of Nizam al-Din Alisher Harawi, known as "Navai," who played a great role in elevating his native

Turkic language, Chaghatay, to the same high level as Persian.

1449   Murder of the Timurid ruler and astronomer Ulughbeg.

1450   Birth of the painter Kamoliddin Bihzad.

1470   Husayn Bayqara becomes the ruler of Herat and stays in power until his death in 1506.

1483   Birth of the conqueror of India and founder of the Mughal dynasty, Babur.

1493   The Ottomans allow Sephardic Jews to print a volume of Jewish laws in Hebrew.

1506   End of Timurid dynasty in Central Asia.

1556   Jesuits introduce movable type printing in India.

1575   Mughal emperor Akbar is shown type fonts for printing books in Persian but does not pursue the new technology.

1576   Damascus-born Taqi al-Din persuades the sultan to fund an observatory in Constantinople patterned after Ulughbeg's.

1600   Giordano Bruno is burned at the stake for championing plurality of worlds.

ca. 1620s   The Vatican sends a printing press with Arabic letters to Isfahan.

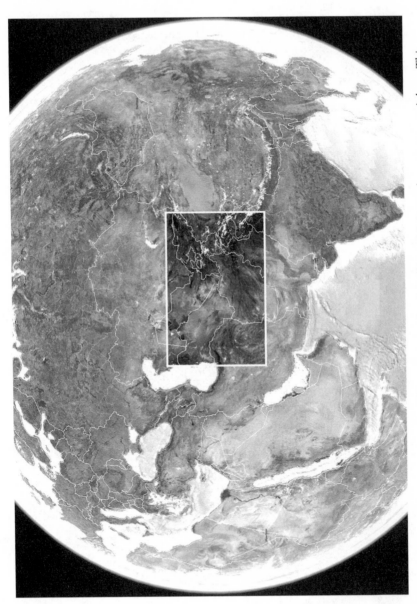

Map 1. Maps that focus on Europe or Asia chronically place Central Asia on the remotest periphery. This satellite photograph presents it instead as it was perceived for nearly two millennia: at the very center of the Eurasian land mass and the pivot of communications in every direction. ▪ © 2013 Google.

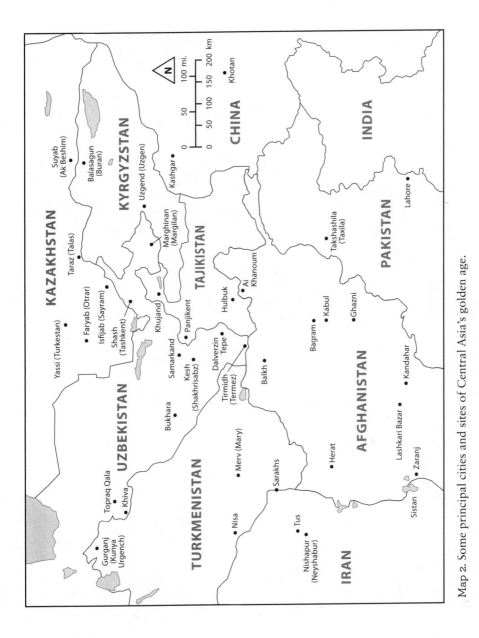

Map 2. Some principal cities and sites of Central Asia's golden age.

Lost
Enlightenment

CHAPTER 1

❧

The Center of the World

In the year 999 two young men living over 250 miles apart, in present-day Uzbekistan and Turkmenistan, entered into a correspondence. They could have sent their messages by pigeon, as was often done then, but the letters were all too long and hence too heavy. The exchange opened when the older of the two—he was twenty-eight—sent his eighteen-year-old acquaintance a list of questions on diverse subjects pertaining to science and philosophy. Nearly all his questions still resonate strongly today. This opened a round of verbal jousting that, through at least four long messages on each side, reads like a scholarly feud waged today on the Internet.

Are there other solar systems out there among the stars, they asked, or are we alone in the universe? Six hundred years later, Giordano Bruno (1548–1600) was burned at the stake for championing the plurality of worlds (the actual charge was pantheism), but to these two men it seemed clear that we are not alone; unique, probably, but not alone. They also asked if the earth had been created whole and complete, or if it had evolved over time. Here they accepted the notion of Creation but emphatically agreed that the earth had undergone profound changes since then. This blunt affirmation of geological evolution was as heretical to the Muslim faith they both professed as it would have been to medieval Christianity. This bothered one of the two young scientists but not the other, so the first—Ibn Sina—hastened to add an intricate corrective that would be more theologically acceptable. But at bottom both anticipated evolutionary geology and even key points of Darwinism by eight centuries.

Few exchanges in the history of science have leaped so boldly into the future as this one, which occurred a thousand years ago in a region now often dismissed as a backwater and valued mainly for its natural resources, not its intellectual achievements. We know of it because copies survived in manuscript and were published almost a millennium later. Twenty-eight-year-old Abu Rayhan al-Biruni, or simply Biruni (973–1048), hailed from near the Aral Sea and went on to distinguish himself in geography, mathematics, trigonometry, comparative religion, astronomy, physics, geology, psychology, mineralogy, and pharmacology. His younger counterpart, Abu Ali al-Husayn ibn Sina, or just Ibn Sina (ca. 980–1037), grew up in the stately city of Bukhara, the great seat of learning in what is now Uzbekistan. He was to make his mark in medicine, philosophy, physics, chemistry, astronomy, theology, clinical pharmacology, physiology, ethics, and music theory. When eventually Ibn Sina's magisterial *Canon of Medicine* was translated into Latin, it triggered the start of modern medicine in the West and became its Bible: a dozen editions were printed before 1500. Indians used Ibn Sina's *Canon* to develop a whole school of medicine that continues today. Many regard Biruni and Ibn Sina together as the greatest scientific minds between antiquity and the Renaissance, if not the modern age.

In due course it will be necessary to return to this correspondence, which left a residue of bad blood between the two giants. But one detail concerning it warrants particular note. At one point Ibn Sina threateningly reported that he would check Biruni's claims with authorities elsewhere to see if they concurred or not. This was a pathbreaking acknowledgment of the existence of separate fields of knowledge, each with its own body of expertise, and that he, as a philosopher and medical expert, was not necessarily qualified to pass judgment on every field. No less important, he was demanding what today we would call peer review—a clear sign of the existence of a large, competent, and interconnected community of scientists and thinkers. Ibn Sina and Biruni were by no means alone in their scientific passions. Both had honed their skills at intellectual jousting with learned colleagues. This imparted a direct and feisty tone to their exchange, which was festooned with frequent charges like "How dare you . . . ?" But neither side stooped to appeal to authority. Evidence, not authority, is what counted.

It was precisely the authority of one writer, Aristotle, that was most at issue throughout the debate. Syrian Christians in Baghdad had only recently translated his *On the Heavens* into Arabic. Both correspondents had read the translation and were now arguing over whether the observable evidence proved or challenged its claims. It fell to Biruni to point out the discrepancies between Aristotle's observations and his own. Far from brushing these concerns aside, Ibn Sina tried to account for them within the framework of Aristotelian theory, even as he showed himself open to questioning it.

Both Biruni and Ibn Sina were, in fact, engaged in the very essence of scientific discovery. As Thomas Kuhn pointed out in his magisterial analysis *The Structure of Scientific Revolutions*, scientific breakthroughs are rarely, if ever, a matter of "Eureka!" moments. Rather, Kuhn explained, science is a cumulative process, in which discrepancies between observed reality and accepted theory (what he famously called the "paradigm") slowly pile up. Breakthroughs occur when the accumulation of such discrepancies or "anomalies" leads to the development of a new theory or paradigm. Under the new paradigm, what had formerly been considered anomalous becomes what is expected. Ibn Sina and Biruni were identifying, sifting, and testing anomalies. Their efforts, and those of scores of their colleagues in Central Asia, led directly to the great breakthroughs that occurred much later, and they were an essential part of the process that created those breakthroughs. Medieval Central Asians produced more than a few genuine breakthroughs of their own. But in looking for achievements by these scientists and scholars, we should be equally attuned to this science-making process, at which they were masters, and not just to their "Eureka!" moments.

What is most astonishing about our correspondents is that they were but two—admittedly a very distinguished two—of a pleiad of great scientists and thinkers who worked in the region a millennium ago. Many other instances of learned exchanges involving Central Asians could be cited. Some were friendly and even fraternal: collaborative research was by no means unknown, especially in astronomy and geography, where teams of a dozen or more investigators were assembled. Some collaborations lasted a lifetime. Others were filled with abuse and nasty ad hominem attacks. But whatever the tone, across Central Asia there existed hundreds of learned people who delighted in disputations such as that between Ibn Sina and Biruni, and who expected them to be resolved, so far as possible, on the basis of reason.

This phalanx of scientists and thinkers did not work in a vacuum. Philosophers and religious scholars fleshed out the implications of the latest ideas, sometimes cheering on the innovators and at other times digging in their heels against them. Rigorous and demanding, these learned men continually asked not only what could be known through reason but also what could not. It was an intellectual and philosophical free-for-all. Adding yet more yeast to the environment was a bevy of talented poets, musicians, and artists, who were creating immortal works at the same time and in the same places. No less than the scientists and scholars, these creative folk left masterpieces that are still revered and admired today.

This was truly an Age of Enlightenment, several centuries of cultural flowering during which Central Asia was the intellectual hub of the world. India, China, the Middle East, and Europe all boasted rich traditions in the realm of ideas, but during the four or five centuries around AD 1000 it was Central Asia, the one world region that touched all these other centers, that surged to the fore. It bridged time as well as geography, in the process becoming the great link between antiquity and the modern world. To a far greater extent than today's Europeans, Chinese, Indians, or Middle Easterners realize, they are all the heirs of the remarkable cultural and intellectual effervescence in Central Asia that peaked in the era of Ibn Sina and Biruni.

Time and Place

Neither the beginning nor the end of this great era of creativity can be fixed precisely in time. It is customary to link its beginning with the Arab conquest of the region, which began in AD 670 but was not really completed until 750. It would be more accurate to date the start of Central Asia's Age of Enlightenment to 750, when forces based in Central Asia overwhelmed the Arabs and their Umayyad Caliphate in Damascus and established a new capital at Baghdad. This event, followed by the installation in 819 of a caliph whose power base was in Central Asia, was akin to a reconquest of the Islamic world from the East. As such, it released enormous cultural energies.

Where did these energies come from? We know disappointingly few details of the intellectual life of pre-Islamic Central Asia. But the fragmentary evidence all points in one direction: that Central Asia entered its golden age with a rich accumulation of cultural and intellectual experience in both the secular and religious spheres. As we shall see, the process of Islamization in the region proceeded very slowly, with many other intellectual traditions thriving side-by-side with Islamic thought down to the year 1000 and beyond. This allowed ample time for cross-fertilization in every direction.

There is no more vexing question regarding the flowering of intellectual and cultural life in the era of Ibn Sina and Biruni than the date of its end. The most commonly accepted terminus point is the Mongol invasion, which Chinggis Khan launched in the spring of 1219. But this turns out to be both too early and too late. It is too early because of the several bursts of cultural brilliance that occurred thereafter; and it is too late because the cultural and religious crisis that threw the entire enterprise of rational enquiry, logic, and Muslim humanism into question occurred over a century prior to the Mongol invasion, when a Central Asian theologian named Ghazali placed strict limits on the exercise of logic and reason, demolished received assumptions about cause and effect, and ruthlessly attacked what he considered "the incoherence of the philosophers."[1] That he himself was at the same time a subtle and nuanced thinker and a genuine champion of the life of piety made his attack all the more effective.

Taking these reservations into account, it is fair to fix the start and finish of this great intellectual effervescence as 750 and 1150, with important developments occurring both prior to this period and thereafter, but of a different scale and character.

It is important also to fix the geography of this cultural flowering. This, too, turns out to be no easy matter. Those who look at the region through the lens of Arabic religious and political history see Central Asia as nothing more than a vague "Islamic East" that starts somewhere in eastern Iran and fades into nothingness the further east, or south, one goes.[2] This approach defines most scholarship that has arisen in the Mediterranean world, whether Arabic or European, and has spread from there to other parts of the globe. Champions of the approach now write of the existence in Central Asia of "a network of cities and their hinterlands"[3] but do not acknowledge a broader identity that might set off those cities and their hinterlands from other settled zones further west.

Meanwhile, during three and a half generations of Soviet rule we became accustomed to think of the region as a Middle (Srednaia) Asia that included only Kazakhstan, Kyrgyzstan, Tajikistan, Turkmenistan, and Uzbekistan, that is, the five former Soviet republics that became independent states in 1991–1992. But over the previous two thousand years, most observers recognized the existence of a much larger cultural zone at the heart of Asia, one that included what are now the new states of Central Asia, but much more besides.

Afghanistan was considered to be a central component of this broader cultural zone, as were adjacent regions of what is now northern Pakistan. So, too, was the Chinese province of Xinjiang, which remained overwhelmingly Turkic and Muslim down to the Communist takeover in 1949. No less intimate a part of this Central Asian cultural sphere was the ancient region of Khurasan, or "Land of the Rising Sun." Now reduced to the modest status of a dusty province in the far northeast of Iran, Khurasan once embraced large parts of western Afghanistan and southern Turkmenistan as well. Culturally, Khurasan is inseparable from the regions that later came under Soviet rule, from Afghanistan, and from traditional Xinjiang. In spite of differences of language, ethnicity, nationality, and geography, the inhabitants of all these areas belonged to a single, albeit highly pluralistic, cultural zone.

Anyone seeking a simple and uniform explanation for the burst of intellectual life in medieval Central Asia will be confounded by the diversity of land types within the region.[4] A northern band of grassy steppes stretches practically to the Syr Darya; a central band of deserts and irrigated oases extends with interruptions nearly to Afghanistan and then picks up again in Afghanistan's Helmand Valley. A third band of mountains covers the region's South, with a single massif jutting northward along what is now China's western border.

Beyond these radical contrasts, it is worth noting that each geographical zone represents the ultimate of its type globally. The steppes of Kazakhstan are part of the largest grassland on earth; the middle desert zone includes the Taklamakan, a desert so dry that its sands preserve apple cores for three thousand years; while the mountain chains embrace the Pamirs and Karakorams, both of which are far higher than the Alps. One Karakoram peak in Pakistan-ruled Kashmir, K2, is a mere 777 feet shorter than Everest, and it is only one of many neighboring peaks over five miles high, the largest concentration of peaks of this scale on earth.

To put it mildly, this is not a forgiving landscape. The total area of the cultural sphere of "Greater Central Asia"[5] is smaller than the eastern United States or western Europe, and large parts of all three of its constituent zones are nearly uninhabitable. Its three main rivers, the Syr Darya, Amu Darya, and Helmand, were all formerly used for transport, but none of them provides a direct water route to the outside world. Worse, the open terrain and location of the mountain chains expose the entire region to invasion from outside, which has occurred as frequently as armies from the region have invaded others.

One of the three geographical zones—the irrigated deserts—was host to the greatest amount of intellectual activity, but it did not function in isolation from the other two. Indeed, without the constant economic and social interaction between desert and steppe, the entire intellectual adventure would never have unfolded as it did.

Introducing the Players

Before plunging into the long and winding drama of Central Asia's Age of Enlightenment, let us follow Goethe in his drama *Faust* and allow some of the key players, the dramatis personae, to take an initial bow on stage. Rather than have them come out together, let us summon them according to the fields of learning in which they made their more noteworthy contributions. In doing so, however, two important caveats must be borne in mind.

First, this was an age of polymaths, of individual thinkers who accumulated truly encyclopedic bodies of knowledge and then went on to make original contributions to as many as six or more different fields. Indeed, the very notion of discrete disciplines was alien to their thinking, the product of a later and more specialized age. During the centuries under study, there was a prevailing interest in assembling the full range of known things into encyclopedias and organizing them into rational categories. A similar impulse had led one ancient writer, Pliny the Elder, in AD 77, to pen an encyclopedia, which was followed up by the Sicilian Cassiodorus in the mid-sixth century. But for the sheer number and variety of their products, no one surpassed the Central Asians as encyclopedists. Their passion for producing meticulous compilations and analyses extended to all nature and to human life as well. Indeed, Central Asia's Age of Enlightenment anticipated the eighteenth-century Enlightenment in Europe, when the Frenchman Diderot issued his famous *Encyclopédie* and the Swede Linnaeus organized all plants into neat categories.

The second caveat is that our judgments of individual thinkers and scientists will be deeply distorted by the fact that a mere fraction of their known writing has come down to us. Whole bodies of works by scientists and thinkers who were considered stars of the intellectual world have been lost or are known only through an occasional quotation buried in the works of others. We know that Ibn Sina wrote over 400 books and treatises ranging in length from a few pages to multiple volumes, but only 240 survive in any form, of which only a fraction have been edited and published. Biruni is known to have written 180 works, of which only 22 have survived.[6] The problem does not end there. Large numbers of surviving manuscripts still languish in archives, having never been transcribed, edited, and issued in the original Arabic or Persian, let alone translated into today's scientific languages. Judging by the quality of works that have recently seen the light of day, what remains to be edited and translated is no less important than what has been. Only thirteen works by Biruni have been published, 7 percent of his total oeuvre. Dedicated scholars in many countries have made progress at this immense task, but much remains to be done. And so we are constrained by what by chance has survived and appeared in modern editions.

Let us, then, proceed with our brief introductions. In astronomy, we might start with Khwarazmi, from what is now western Uzbekistan, who was among several Central Asian astronomers who organized a major project to measure the length of a terrestrial degree and developed tables for constructing horizontal sundials that were precisely adjusted to latitude.[7] He also devised an instrument that used sine quadrants to derive numerical solutions to problems of spherical astronomy. Biruni's astronomical research led him to conclude that planetary orbits could be elliptical, not circular, and that the sun's apogee varied in predictable ways. In a bold move against Aristotle and his followers who used "natural philosophy" to solve scientific problems, he argued that such issues could be resolved only by mathematical astronomy. Recently the existence of elliptical orbits among planets circulating other suns in our galaxy has drastically shrunk the estimated number of such "exoplanets" that might be inhabitable. Biruni's teacher and close friend, Abu Nasr Mansur Iraq, was said to have been the "second after Ptolemy," but next to nothing of his voluminous astronomical work survives.

Khujandi, from northwest Tajikistan, built a large mural sextant and produced several measurements of unprecedented precision on the obliquity of the ecliptic, such as the angle formed by the plane that is perpendicular to Earth's axis, and the angle in which Earth and the sun moved in relation to each other. Of course, he still assumed that the sun orbits Earth, but his measurements represented an important step forward in the study of this relationship. He also developed an instrument that applied spherical trigonometry to astronomical problems.

Farghani, from the Ferghana Valley in present-day Uzbekistan, wrote a treatise on the main astronomical instrument of the Middle Ages, the astrolabe, that later gained a wide readership in Europe. He also penned a study on astronomy that became the best-known "Arab" work in that field in Europe. Among his many readers was Christopher Columbus, who, working half a millennium after Farghani, seized on the Central Asian's calculation of a degree of Earth's circumference as 56 2/3 miles. Farghani's calculation was in Arab miles, however, while Columbus, eager to reduce the distance between Europe and China to the greatest extent possible, jumped to the conclusion that he had meant Roman miles. This, along with several other computational errors, reduced the distance Columbus would have to travel by 25 percent. Because of this miscalculation, the "Admiral of the Ocean Sea" fully expected to find "Cipango," or Japan, at about the same meridian as the Virgin Islands. The shorter measurement was woefully inaccurate, but it conveniently provided Columbus with a powerful argument when he began to present his case for funding to the king of Portugal, and later to the Spanish court.[8]

Several Central Asians prepared astronomical tables of stunning accuracy. Ulughbeg, a ruler in Samarkand who pursued a lifelong passion for astronomy, determined the length of the sidereal year more accurately than Copernicus and measured Earth's axial tilt so precisely that his figure is still accepted today. A student of Ulughbeg's, Ali Kushji, considered that the motion of comets provided empirical evidence for the possibility of Earth's rotation and was the first to declare astronomy's full independence from "natural philosophy."

In mathematics, Khwarazmi was the first to elaborate a theory of equations solvable through radicals, which can be applied to the solution of a variety of arithmetical and geometrical problems. The result was a book, *Algebra*, that gave its name to the field; the term *algorithm* is a corrupted form of his own name. Khwarazmi advanced the field of spherical astronomy and did more than anyone else to popularize the decimal system that had been invented in India. His friend Marwazi from Merv in

Turkmenistan did pioneering work on tangents and cotangents. Biruni was one of several Central Asians who championed the importation of the concepts of zero and negative numbers from India and broke new ground in their use. Several Central Asians competed for priority in the development of trigonometry and its establishment as an independent field, which was to be reinvented in Italy in the seventeenth century.[9]

The construction by poet Omar Khayyam (yes, the poet!) of a geometrical theory of cubic equations was a genuine breakthrough, as was his extension of arithmetical language to ratios. Khayyam was the first to identify and classify the fourteen types of third-degree equations and to propose geometric proofs for many that had previously confounded experts. He was also among the first, if not *the* first, to accept irrational numbers as numbers. In attempting to prove Euclid's axiom that parallel lines cannot meet, he produced a new theory of parallels. Two Soviet historians of science concluded that some of the propositions derived by Khayyam in his theory of parallels were "essentially the same as the first theorems of the non-Euclidean geometries of Lobachevski and Riemann,"[10] both of whom, it should be noted, lived seven hundred years after Khayyam.

In optics, Ibn Sahl from what is now the border area between Turkmenistan and Iran wrote an important treatise on the use of curved mirrors to focus light. Building on the work of his predecessors, he also solved the problem of using lenses to focus light to a point, which no ancient scientist had accurately addressed. In the process he discovered the law of refraction. In medicine, Ibn Sina's *Canon* contains powerful passages on the impact of the environment on health, and also stunningly prescient passages on what we today call preventive medicine. He considered the principles of treatment for hundreds of maladies, including psychosomatic illnesses of all types. Besides Ibn Sina, several other Central Asians wrote massive compendiums of practical and theoretical medicine. One, Central Asia–trained Muhammad ibn Zakariya al-Razi, was the boldest diagnostician and surgeon of the Middle Ages. Pharmacology, too, attracted many pioneering scientists from the region, including some who had no connection with medical practice.[11] In biology, Biruni, in a book on India, directly anticipated Malthus in predicting the proliferation and collapse of species.[12]

The large-scale effort to analyze the medicinal effects of plants was paralleled by research in chemistry to identify the hardness and other properties of minerals. Building on the work of Archimedes, Biruni was a leader in this. He was also the first anywhere to measure the hardness of minerals and their specific gravity. A Persian disciple of Central Asian scientists was the first to identify reverse reactions. Large-scale mining throughout the region also encouraged pioneering research in chemistry, which was successfully pursued by numerous regional investigators, whose existence is known only from occasional mentions in the literature.

Geology and the earth sciences also advanced strongly during these marvelous centuries, with Ibn Sina and Biruni being credited with the first theory of sediments, a theory on the formation of mountain ranges that has gained acceptance only in recent centuries, and important hypotheses on the process by which continents emerged from seabeds.

Geography also flourished, with Mahmud Kashgari's issuing of the earliest map showing Japan, and numerous astronomers and experts on trigonometry combining their skills to pinpoint the latitude and longitudes of hundreds of locations from India to the Mediterranean. A whole school of geographers was born when a researcher at the Afghan city of Balkh came up with an innovative way of mapping the earth based on the application of spherical geometry and mathematics.

Beyond doubt, the era's greatest achievement in geography was the work of our friend Biruni, who used astronomical data to postulate the existence of an inhabited land mass somewhere between the Atlantic and Pacific Oceans. The astonishing process, which led to this earliest "discovery of America," is described in detail in chapter 11. It represents the triumph of the mathematician and geometrician over the metaphysicians and theologians. Equally, it demonstrated that a cloistered scientist could be as bold an explorer as an intrepid mariner who hoped against hope that land would appear on the horizon, that rational analysis could be even more effective as a tool for discovery than seafaring.

Central Asia produced many talented historians. Beyhaqi from Khurasan wrote a highly intelligent history of one the many mega-states to arise in the region, that of Mahmud of Ghazni, whose realm stretched from India to the Middle East. And later a descendant of Tamerlane (Timur), Babur, wrote an extraordinary history of his own rise in Central Asia, his conquest and rule in Afghanistan, and his eventual creation of the Mughal empire in India. But the main focus of most Central

Asian historians was, tellingly, on their native cities, where cultural life thrived, and on the great leaders who changed the fate of the region. Consequently it fell to a Persian from Hamadan, beyond the borders of Central Asia, Rashid al-Din, to write the world's first universal history.[13]

Central Asia's most astute student of societies was Biruni, who founded the field of anthropology and pioneered the field of intercultural studies and comparative religion.[14] It is no exaggeration to say that Biruni was the greatest social scientist between Thucydides and modern times. By comparison, Hugo Grotius (1583–1645), Thomas Hobbes (1588–1679), Samuel von Pufendorf (1632–1694), and John Locke (1632–1704) were all more interested in theorizing about the nature of society than in studying it as it actually is. Mahmud of Kashgar, now in Xinjiang, China, was an accomplished Turkologist and ethnographer who virtually invented the field of comparative linguistics, while Yusuf Balasaguni, from Balasgun in present-day Kyrgyzstan, and Nizam al-Mulk, from Nishapur in Khurasan, were adept at combining political reality and philosophic principles in their manuals for leaders.[15] The polymath Farabi, from Otrar in southern Kazakhstan, penned an important theoretical treatise on the ideal city, in which he warned that any society that fails to make use of the thinkers in its midst has only itself to blame.[16]

One of the glories of the Central Asian intellect was philosophy, which natives of the region pursued with an innovating passion that far surpassed that of everyone else in their era; their writings were to exercise a decisive influence on Muslims everywhere and on the Christian West as well. Cosmopolitan, individualistic, and profoundly humanistic, Central Asian philosophy was also, in the eyes of its critics, skeptical, irreverent, and profane. It reached its apogee with Farabi, who was called "the second teacher" after Aristotle and who, with Ibn Sina, achieved what many considered to be a harmonious blend of reason and revelation, logic and metaphysics, Aristotle and the Neoplatonists.[17] The great German scholar Adam Mez declared that the humanism of the European Renaissance would have been impossible without this earlier explosion of philosophical inquiry in Central Asia.[18]

Logic, largely ignored today, was an essential tool of all those who fought on the intellectual barricades in the Middle Ages. Thanks to Farabi and other Central Asian logicians, its austere principles became established, or reestablished, as a prime tool for attaining truth. Ibn Sina and others proceeded to show how logic could be applied in the mathematical sciences as well. In the West, meanwhile, Aristotelian logic lost out to Scholasticism.

Libraries, the essential tool of all these scientists and scholars, abounded in Central Asia. Ibn Sina's work took off when he gained access to the royal library at Bukhara, and the Middle Eastern scholar Yaqut traveled halfway across Eurasia to avail himself of the dozen libraries at Merv, now in Turkmenistan. Other great collections existed at Gurganj in the North, Balkh in the South, Nishapur in the West, and Samarkand. Indeed, it is all but certain that every major Central Asian city at the time boasted one or more libraries, some of them governmental and others private. Central Asians were also among the main users at the caliph's library at Baghdad, which was built up mainly by learned men from the region. The West could also claim libraries by this time, especially after 780, when Charlemagne sent out an appeal for copies of remarkable or rare books,[19] but their number and wealth in classical texts left then incomparably weaker. Nor could the West claim anything to compare with the countless book dealers in every major Central Asian city, or the well-attended auctions of books and manuscripts that attracted well-heeled buyers.

What can be said of these assembled thinkers as a group? Central Asian intellectuals of this golden age affirmed that there are not one but many means of reaching scientific truths, including deduction, logical argumentation, intuition, experimentation, and observation. By so doing, they enormously broadened and deepened the scientific enterprise.[20] Equally important, they held that the rules they set down in each of these areas applied equally to the simple and the bewilderingly complex, to domestic objects and the movement of the heavens. This notion of *universality* has often been seen as a signal achievement of the scientific revolution and the age of Isaac Newton.[21] But it was accepted as a fact by most of the leading figures of Central Asia's Age of Enlightenment.

Theology, too, reached a high peak in Central Asia during the Age of Enlightenment. Ibn Sina was but one of many thinkers from the region who explored the rational basis for religion while acknowledging the mysteries of revelation and faith. Some pushed the first part of this equation, notably the so-called Mutazilites, who favored the most uncompromising application of reason to Muslim theology. While not founded by Central Asians, this important and controversial school of thought found its most ardent supporters in Central Asia. Beyond these, the region was also home to Hiwi al-Balkhi, Abu Bakr al-Razi, and Ibn al-Rawandi, all outspoken skeptics of religion or outright atheists.

On the other side of the equation, those who would found their faith on revelation alone or the words of the Prophet that had been passed down through the centuries also had their most effective champions in Central Asia. Islam's second most hallowed book, the collected Hadiths, or Sayings, of the Prophet Muhammad, was the work of a Central Asian, Bukhari; beyond this, of the six collections of the Hadiths considered canonic by Sunni Muslims (and most Shiites), fully five were the work of Central Asians.[22] One of the four schools of Sunni Islamic jurisprudence was founded by a Central Asian, and a second found its most congenial home there. Also, the greatest official defender of Sunni orthodoxy was Nizam al-Mulk from Khurasan, who also gave the madrasa the purpose and form it retains today. In sharp contrast to both the rationalists and the traditionalists were those who adhered to the mystical current known as Sufism. This movement, too, found early expression and attained its greatest influence in Central Asia, where several of the major worldwide Sufi orders were founded by the likes of Najmuddin Kubra, Ahmad Yasawi, and Bahaudin al-Din Naqshband Bukhari.

The Age of Enlightenment produced consummate achievements in the arts and letters. Sufi poets like Rumi from Balkh in Afghanistan and Omar Khayyam have large audiences worldwide even today. Earlier poets like Rudaki and Asjadi stand at the source of the great Persian literary tradition. Ferdowsi, whose immense panorama of the civilization of Iranic peoples, the *Shahnameh*, set a world standard for other national epics, was a native of Khorasan, and most of his epic was set in Central Asia, not the lands that now constitute the state of Iran. Nearly all the scientists, including Ibn Sina, wrote at least part of their works in verse.

The building arts and painting flourished. The stunning multicolumnar minaret of baked brick constructed in 1108–1109 at Jarkurgan, Uzbekistan, still stands, as does the now extensively restored tomb of the Seljuk sultan Sanjar, designed in 1157. Both were designed by architects from Sarakhs on the Turkmenistan-Iran border, Ali of Sarakhs and

Muhammad Ibn Atsiz al-Sarakhsi. Roman architects had employed a double dome at the Pantheon in the early second century AD, but knowledge of this innovative technique seems to have died out soon thereafter. However, it emerged again at the immense eleventh-century tomb of Sanjar at Merv and at smaller structures in Khurasan. Eventually the technique found its way across Iran and the Middle East to Brunelleschi's dome at the Duomo in Florence, and other double domes across Europe and the Americas. Similarly, the diamond pattern in the brickwork on the exterior of the twelfth-century Kalyan minaret in Bukhara was later imitated on the walls of the Doge's Palace in Venice.[23]

Painting had deep roots in pre-Muslim cultures across the region. Despite Muslim prohibitions against depicting the human form, it lived on in Muslim times and even staged a great revival at the end of our period. Kamoliddin Bihzad from Herat in Afghanistan stands as one of the great painters of the late Middle Ages, and his exquisite book illustrations and miniatures are now recognized as one of the highest achievement of Islamic art. Meanwhile, craftspeople do not often sign their products, but in Central Asia many silversmiths and casters of bronze did so, revealing their justified pride in even their most utilitarian wares. Finely woven fabrics from the region were so prized in the West that they found their way into the treasuries of many European cathedrals, where they remain today.

The people who made these and other seminal contributions to science, thought, and the arts were not anonymous toilers or withdrawn ascetics. On the contrary, they were activists who traveled widely, wheeled and dealt with patrons, and engaged in sharp polemics with colleagues. In spite of Muslim dictates to the contrary, many, if not most, drank wine, and one poet, Anvari, was so earthy that a modern Muslim editor declared that "a large part of his writings are unfit for translation."[24] In short, they were energetic, worldly, and resourceful—the kind of people who naturally impose their personalities on their work and those around them.

The great Swiss historian Jacob Burckhardt argued that what he called "the discovery of the individual" was the very hallmark of the Italian Renaissance, separating that dynamic era from the Middle Ages that had gone before. Even if one modifies this to acknowledge the Greeks and

Romans, the Central Asian thinkers and artists assembled on the stage before us must still be credited with the same discovery, or rediscovery, half a millennium before the age of da Vinci. This may have been their greatest innovation of all.[25]

THE INTELLECTUAL CLASS

Even this superficial and incomplete list of names and achievements confirms that these medieval Central Asians were not mere transmitters of the achievements of the ancient Greek past but were also, in diverse fields, the creators of important new knowledge. The scale and range of their achievement prompts one to ask, "Who were these people?" They fit no single stereotype, but a few generalizations are in order, beginning with their ethnic identity.

Many, if not most, Western writings down to the present day identify Ibn Sina, Biruni, Khwarazmi, Farabi, Ghazali, and the others as Arabs. This important misidentification is to be found even in some of the most authoritative European and American histories of philosophy and science.[26] It is true that most, but not all, of Central Asia's thinkers in this era *wrote* in Arabic. Indeed, the adoption of Arabic as a single lingua franca for intellectual interchange throughout the Islamic world was of huge importance to the creation of an international marketplace of ideas. The speed with which the Arabic language absorbed unfamiliar concepts and adapted to the needs of scientific and technical communication is impressive. But it was Central Asians, by their prolific writings, who were at the forefront of this process of enriching Arabic with new concepts and terms. This occurred at almost exactly the same time that Latin in the West was shrinking from the status of a universal tongue to the language mainly of religion and ideas. As the French scholar Jacques Boussard put it, "Latin gradually became deformed and simplified, and finally gave place to a new and extremely rough and uncivilized language—Vulgar Latin."[27]

A Central Asian who wrote in Arabic a millennium ago was no more an Arab than a Japanese who writes a book in English is an Englishman. Most of the writers and thinkers mentioned above may have passed their professional lives in an Arabic-speaking professional milieu, but

Arabic was not their native language, nor were they Arabs. As Harvard's Richard N. Frye archly observed, "It is a remarkable fact that, with few exceptions, most Muslim scholars both in the religious and intellectual sciences [were] non-Arabs."[28] When a learned Arab of the eleventh century compiled a list of all the "praiseworthy peoples of the age" who wrote in Arabic, a third of the total of 415 he enumerated were from Central Asia.[29] Of the remaining two-thirds, more than half were Persians from what is now Iran. The hegemony of Central Asians was more overwhelming in the sciences, philosophy, and mathematics, in which fields they constituted up to 90 percent of the total.[30] Most were of some Iranian stock and spoke diverse Iranian languages, but increasing numbers were Turkic as well. Their many native tongues belonged to either the Iranian or the Turkic language groups.

Were they, then, what we think of today as Iranians or Turks? A millennium ago neither Iran nor Turkey existed as a state. Peoples who spoke the diverse languages and dialects belonging to the Iranian and Turkic families of languages were spread over a vast territory that extended far to the east of present-day Iran and, until the eleventh century, did not include any part of what is now Turkey. Modern Turks would have trouble understanding the Turkic languages of tenth-century Central Asia, just as a citizen of Tehran would surely have been unable to comprehend Sogdian, Bactrian, or Khwarazmian, even though they were all Iranian languages. These diverse Iranian and Turkic peoples met and mingled above all on the territory of Greater Central Asia where, from the earliest days, they acquired a pluralistic but very real and distinctive identity of their own.

To distinguish Central Asians of Iranian stock from the inhabitants of what is now Iran, scholars have applied the terms "Persianate" or "Iranic" to the former. The geographical location of Central Asia played a significant role in forging this special identity. Proximity put its inhabitants in direct trade contact with India and China, as well as the Middle East. By contrast, even speakers of Iranian or Turkic languages further west looked mainly to the Middle East, the Caucasus, and Europe.[31] Thus, to speak an Iranian or Turkic language in AD 1000 meant something quite different from what either means today.

From earliest times it was understood that people of Persianate stock in Central Asia were different from Persian speakers in most of what is now Iran. Herodotus noted that the Persian empire of Darius and Xerxes did not tax people it accepted as "Persian." But Central Asians whose languages belonged to the Iranian language group were considered sufficiently different that the Persian state taxed them as foreigners.[32] Today the difference between citizens of Iran and speakers of Dari and Tajik in Central Asia is reinforced by the fact that the former are all Shiites, while their Central Asian and Afghan cousins are mainly Sunni. Similarly, Turkic peoples who came as nomads to settle in Central Asia adopted new relationships and patterns of life that distinguished them increasingly from the larger body of Turkic peoples, not to mention those in the remote Altai homeland region of what is now Siberia and East Kazakhstan.

A second common characteristic that was nearly universal among both Persian and Turkic writers and intellectuals is that they were formed mainly by urban environments and spent their careers in cities. Unfortunately, the Central Asian cities in which they lived can scarcely be imagined, let alone seen, today. This is due to the fact that the chief building material across Central Asia was impermanent sun-dried brick, which, like adobe, is cheap and strong but subject to erosion by rain and wind. Nearly all the monumental and more humble buildings that medieval writers amply describe have long since dissolved, leaving only a mound of dirt. Had they been built of stone they would still be standing, with the result that tourists would be flocking to Central Asia and Afghanistan the way they do to Italy or India, where most buildings were of stone. An equally formidable enemy of built structures in Central Asia was earthquakes, which hit with alarming frequency across the length of this seismically active region. Earthquakes devastated the great city of Nishapur twice in one generation (in 1115 and 1145),[33] and even the clever antiseismic techniques devised by Central Asia's medieval architects and engineers could not prevail when a big one struck.

Thanks to extensive archaeological work in the region, we can now begin to form a picture of the medieval Central Asian city. Like great business centers everywhere, they were hives of industrial and commercial activity, teeming with traders and offering no corners of tranquility and repose. Typically, the great religious scholar Burhan al-Din al-Marghinani did most of his writing not in a rural monastery but at his urban residence only a few paces from the main east-west caravan route through his native city of Marghilan, now in Uzbekistan. Marghilan and other Central

Asian cities exhibited a few of the characteristics of the generic "Islamic city" invented by Western Orientalist scholars, but Central Asian cities were in fact quite distinctive within the Muslim world, in both form and structure. This is not surprising since they had existed for up to three millenniums before the Arab armies arrived and had had plenty of time to develop their distinctive spatial planning and architectural styles.

Who paid the intellectuals of medieval Central Asia to sit and think deep thoughts? Biruni believed that kings should do this, "for they alone [can] free the minds of scholars from the daily anxieties for the necessities of life and stimulate their energies to earn more fame and favor, the yearning for which is the pith and marrow of human nature."[34] A few of the great minds of the era found royal backers, but relations between patron and thinker were rarely tranquil. More often our thinkers failed to find a royal patron or deep-pocketed Mycaenas who could enable them to work in peace. Lacking steady support, such learned men became "wandering scholars," to cite the title of Helen Waddell's 1927 book on medieval European lyricists who moved from court to court in search of patronage. Some had worldly skills that they could put to use as administrators: Ibn Sina, whom we met as a precocious scientist and philosopher, enjoyed the patronage of the Samanid dynasty and later served for a few years as what we might call a prime minister to the Buyid ruler Shams al-Dawla. Nizam al-Mulk, author of a famous volume of advice for his prince, occupied the same position in the Seljuk empire and was its most powerful political figure. Still others, among them the Turkic writer Yusuf Balasaguni, wrote fat volumes in the hope that a ruler would "discover" them and reward them with a pension. In Yusuf's case the support actually materialized. But the great poet Ferdowsi, author of a national epic many times the length of Homer's *Iliad*, waited a lifetime for his patron to pay him the money he had been promised, and even then it arrived only after his death.

Whether or not things worked out to the thinkers' satisfaction, the intellectual history of Central Asia is in part a story of patronage, of the rich and powerful who were prepared to spend part of their wealth on the support of science and the arts. Fortunately, over the course of several centuries nearly all rulers in the region, among them several certifiably brutal tyrants, acknowledged that the patronage of wise men was one of the obligations that came with kingship. At its worst, such royal patronage descended to a type of exhibitionism, with an ambitious ruler convening writers and thinkers in elegant soirees to show off his own wit. Yet there were also royal or aristocratic patrons who truly understood the life of the mind and had a rare ability to identify true talent for their entourages. Their generous financial support, combined with a broad outlook and patience, enabled a few brilliant scientists and thinkers to toil in peace for years, without concern for their daily sustenance—an amazing stroke of good fortune in any society.

Many of these patrons were purely local, the heads of ruling houses or dynasties that held sway in a single city, valley, or district. For support on a larger scale, the intellectuals and artists looked to the rulers of the various empires that claimed control over the territory of Greater Central Asia. Some of these, like the early Kushans, Bactrians, and Khwarazmians, or the later dynasties of Ismail Ibn Ahmad Samani, Mahmud of Ghazni, and Tamerlane (Timur), were locally based. Others, including the Baghdad Caliphate, originated outside the region and by force of arms asserted claims of suzerainty over the local courts and territorial dynasties. Many were ruthless and blood-drenched rulers, but among them were sultans or monarchs who quickly grasped that support for thinkers and artists would glorify their rule and be a source of strength, not weakness.

All the intellectuals listed above, and many more not yet mentioned, are generally grouped under the rubric "Muslim or "Islamic." Most, but not all, were indeed adherents of Islam, and some were deeply devout. But is this a defining feature of their identity, or merely a convenient label? Going deeper, were they orthodox Sunni, Shiites, heterodox, or, like many during the eighteenth-century Enlightenment, mere deists who acknowledged God as a First Cause but not necessarily as a presence in the material world? We know that Islamization in Central Asia proceeded slowly over some three hundred years, during which many other religious and intellectual currents continued to flourish. Is it therefore accurate to characterize all art from this time and place as "Islamic," or is the notion of "Islamic art," as a reviewer of a major London show argued in the New York Times, a "groundless myth" perpetrated by Western orientalists?[35] What, if any, was the influence of other faiths, and what about the skeptics, freethinkers, agnostics, and atheists among the scientists and philosophers?[36]

THREE QUESTIONS: EASY TO POSE, DIFFICULT TO ANSWER

These and many other questions inevitably arise when one sets out to identify those creative thinkers who, over several centuries, made Central Asia the center of the intellectual world and whose work profoundly affected science and civilization in both the East and the West. Rather than allow such queries to proliferate indefinitely, and thereby lose the connecting thread in a welter of details, it is useful to reduce them to three. This turns out to be quite simple. First, what did Central Asian scientists, philosophers, and other thinkers achieve during these centuries? Second, why did this happen? And third, what became of this fecund and tumultuous movement of ideas?

Each of these questions poses a serious challenge. The first leads into a dizzying array of fields and disciplines, from astronomy to epistemology and to Muslim theology. Not all these fields and disciplines flowered equally or at the same time. By what standard should advances in one be weighed against stagnation, or worse, in another? And should advances be evaluated in terms of their long-term impact or on the basis of their influence on the thought of contemporaries? The latter approach, which is entirely legitimate, would cause us to devote the same amount of attention to Abu Mashar al-Balkhi, who was the most renowned astrologer in the Muslim world and venerated equally in the West, as to an astronomer like Khujandi or Farghani, whose accomplishments are still recognized today.

The second question—why did it happen?—is yet more demanding, for it plunges us into the fundamental questions of causation in human history. Tolstoy, in the second epilogue to his novel *War and Peace*, ventured onto this dangerous territory in his effort to account for Napoleon's actions at the Battle of Borodino in 1812. Yet it is a far simpler matter to account for a single European battle in comparatively recent times than to elucidate the causes of an intellectual and cultural effervescence at a far-off time and place.

Why, we might equally ask, did Periclean Athens achieve such incandescent intellectual vigor, or Renaissance Florence, Restoration London, Classical Weimar, Nara in its golden age, or, for that matter, Concord, Massachusetts, in the era of Emerson, Thoreau, and the Alcotts? Underlying each of these specific instances of cultural greatness—and Central

Asia's Age of Enlightenment as well—are timeless imponderables about the sources of human creativity and the motives for human action. One might as well ask what it is that brings out the inquiring and thoughtful side of each of us today!

The third question—what happened to it?—is particularly compelling, for it bears directly on current events in the region and in the world. Thoughtful men and women within the region ask this of each other, and analysts further afield raise it whenever discussion turns to the arc of lands extending from Central Asia westward to the Middle East. The same question has been asked concerning periods of brilliance at other times and in other places. This is the simple question that impelled Edward Gibbon to pen six volumes on *The History of the Decline and Fall of the Roman Empire*. Not shy in his judgments, Gibbon advanced so many brilliant hypotheses, ranging from the erosion of public morals, the decay of specific military units, and the influence of other-worldly Christianity, that one comes away from his tomes as from a feast with six main courses. The late Joseph Needham and his colleagues produced an exhaustive twenty-six-volume masterpiece on *Science and Civilization in China* and found themselves compelled to add a final volume of *General Conclusions and Reflections* that is more reflection than conclusion. Having so deeply considered the developments that caused China's rich tradition in science and technology to wane, this inquiry is justly called "the Needham Question." It remains open.

Roads Not Taken

These, then, are the great questions to which this work is dedicated, and which this or any other attempt to delve into the Age of Enlightenment that occurred in Greater Central Asia must eventually address. Firm answers may prove elusive. But the proliferation of questions within questions poses a challenge of its own. The danger is that the inquiry will come to resemble a prickly bush that has not been pruned, with many thorny branches and twigs but no overall shape. Lest the inquiry have so many foci that, like a kaleidoscope, it leaves us with nothing more than a memory of infinite colors and shapes, it is necessary to indicate what will *not* be included in what follows.

Figure 1.1. Graceful musicians from a Kushan-era limestone frieze, Airtam, Uzbekistan (first to third centuries). Central Asian musicians pioneered plucked string instruments and invented the bow. ▪ Airtam Frieze-Musician, detail of the Airtam Frieze, 1st c., limestone. From Bactria, the Kushan period, inv. no. CA 3199. The State Hermitage Museum, St. Petersburg. Photograph © The State Hermitage Museum. Photo by Alexander Lavrentiev.

First, there is the matter of music, which, along with poetry, was considered the king of the arts. In few areas did Central Asians sweep more dramatically beyond their Hellenistic Greek mentors and blaze the trail for later Europeans than in music making and especially in theorizing about music.[37] Long before the Islamic era, Central Asians invented the bow as a means of eliciting sound from a string; thanks to this invention, which quickly spread to China, India, and the West, Central Asia can be considered the genetic homeland of the violin.[38] Rudaki, a poet of timeless appeal, was a brilliant musician. The philosopher Farabi, who was himself a talented lutenist, was the author of *The Great Book on Music*, generally considered the premier theoretical work on music from the medieval period, a work that, in Latin translation, deeply influenced European thinking about music.[39] Other Central Asians built on Farabi's foundation. Yet the absence, until the seventeenth century, of a systematic system of notation prevents us from hearing the music of Farabi's era. Worse, the psychological gulf between Central Asian music, with its modes and semitones, and the Western twelve-tone scale thwarts comprehension and real appreciation, even if it is heard. For this reason, music does not play the role it should in what follows.

Popular culture, too, finds little place in the following pages. The same literature that produced treatises of philosophy and the sciences also gave rise to the writings of storytellers and exorcists, jugglers and magicians, not to mention whole collections of anecdotes, incantations, tricks and talismans, and books composed on everything from freckles to twitching. There were even compendiums of sexually titillating tales drawn from Persian, Indian, Greek, and Arabic books, not to mention both large and small books on *The Capable Woman*, concubines, and homosexuals.[40] Yet such manifestations of popular culture seem rarely to have affected the high culture that is the subject of this study,[41] although further investigation could well change this judgment. The one clear instance of popular values being the driver for an intellectual shift is the case of Sufism, the mystical and ecstatic form of Islam that seeks to strip away all worldly concerns to put the believer into direct communion with God. In this case a movement "from below" eventually forced itself onto the attention of the intellectuals, who responded in ways that changed the religion of Islam forever and affected Christianity as well.

Some readers may wish that the following exposition dealt more fully with the culture of the many nomadic peoples, whether Iranian, Mongol, or Turkic, who swept through the territory of Central Asia from the first millennium BC through the fifteenth century AD. The Turkic group of people who gained suzerainty over Central Asia in the pre-Islamic sixth century were so serious about protecting their realm that they entered into formal diplomatic contact with both Byzantium and China. Other nomadic empires embraced similarly huge territories and diverse peoples, the management of which required constant attention and immense expenditures of energy. Also, it is no slight to say that the nomads' intellectual faculties found expression in the elaboration of their complex cosmological systems and beliefs, and their expression in song and poetry, rather than in the fine points of Aristotelian epistemology. Increasingly, though, these communities produced intellectuals who participated in the ecumenical and multicultural life of the mind that prevailed in the region's cities and who made their own contribution to that discourse. But the many intriguing questions regarding the religion, worldview, social outlook, and literary monuments of the nomads range beyond the bounds of our inquiry, which will be defined in terms of formal texts and deliberate works of art produced in the settled cities.

Space does not permit a more detailed analysis of all the complex cultural and intellectual interactions between Central Asia and the major cultures that lay just beyond its southern, southwestern, and southeastern borders. Beyond doubt, some of the most persistent and productive stimuli to fresh thinking were the new ideas that flowed in from India, China, and the Middle East from the fifth century BC through the era of Tamerlane (Timur) in the fifteenth century AD. Some, like the concept of zero from India, had to do with mathematics or science; others, such as the wavelike ornamentation that artists in Herat, Afghanistan, borrowed from Chinese colleagues in the fifteenth century, were in the aesthetic realm. The subject is the more intriguing because such influences operated in both directions. Edward H. Schafer devoted a whole book to enumerating the exotic products from Central Asia that lent color and excitement to the court of Tang China.[42]

The subject of intellectual and cultural interchange with these great civilizations, and the balance between them, is important to our inquiry, for it goes far toward defining the special character of Central Asian life and thought. But its dimensions are so huge that it can only be dealt with telegraphically, rather than exhaustively. What can, and will, be considered is how Central Asians processed those ideas from abroad and whether, and how, they may have reworked some of them in the process.

We will also treat only superficially the many ways in which specific works and ideas of Central Asian thinkers found audiences in both the East and West, mainly through translations into Hindi, Chinese, or Latin. Much study has been devoted to this important question, but the work has barely begun. But it is worth noting that the distinguished art historian Oleg Grabar and his colleagues assert that during the period of flowering, culture flowed from East to West, that is, from Central Asia into the rest of the Islamic and Mediterranean worlds, not vice versa.[43] Leaving aside the immense impact of the Central Asians on Islamic thinkers from elsewhere—figures like Averroes, Ibn Khaldun, and scores of others—the subject of their profound influence on the Christian West, including everyone from Abelard to Thomas Aquinas and Dante, would alone fill many volumes. Joseph Needham went further than anyone in documenting their influence in China, but the analogous research on India remains at a preliminary stage, even though many thinkers from Central Asia actually lived and worked there.

Other readers, reviewing the parade of male thinkers presented below, may well ask "Where are the women?" Where is the Central Asian woman comparable to Hildegard of Bingen, the learned and capable abbess of a large German convent in the twelfth century and a composer of genius, or Rrroswitha of Gandersheim, a Benedictine canoness who in the tenth century wrote six plays in the classical style, thereby anticipating the revival of theater by two centuries?[44] To be sure, there is Rabia of Balkh, the cosmopolitan metropolis in north-central Afghanistan, whose ardent and subtle poetry earned her the admiration of all.[45] But one looks in vain for females from the region who left significant legacies in the realm of systematic thought. The closest any came to this were the later mystic religious poets who, in their quatrains, explored the Sufi experience.[46]

Some have traced this situation to the status of women in Islamic societies and, presumably, to their status in pre-Muslim society as well. Manuela Marín, writing on "Women, Gender, and Sexuality" in early Muslim societies, concluded that "It was considered dangerous for women to write because they could use this skill for unlawful communication with men."[47] This led to a situation in which the scholarly vocation came to be exclusively male, as were the ranks of those who interpreted Islamic law.

Of course, women could own and inherit property and in fact often served within the family as bankers. The fact that Zoroastrian inheritance law, like Jewish law, was far more favorable to women than the Islamic law that replaced it may also have strengthened the role of Central Asian women in areas other than learning and scholarship.[48] Women were certainly prominent behind the scenes in Central Asian politics. Thus when Arab armies arrived at the gates of Samarkand bearing their new religion and seeking plunder, they encountered a steel-willed local woman who was ruling on behalf of her young son.[49] During the tenth century, when the Samanid dynasty in Bukhara reached its intellectual zenith, another woman ruled quite successfully as a wife of the former ruler. In a Central Asian city further west a widowed queen, known ironically, as "The Lady," faced the ruthless Mahmud of Ghazni; rather than back down, she issued a frontal challenge, throwing in his face that if she won she would have defeated the greatest commander of the era, but if he won he would only have defeated a woman.[50]

plots. Then there were suburban houses, markets, and quarters for visiting traders in the so-called *rabat* quarter. Soon the visitor confronted the looming walls of the city proper. At Balkh, as at scores of other Central Asian cities, these were not the simple vertical walls with parapet that were standard in the Mediterranean world and most of Iran, but a massive sloping construction of sun-dried bricks faced with fired bricks, atop which was a further high wall of fired brick crowned with long galleries interrupted with loopholes for shooting, and frequent towers that protected the galleries and offered panoramic views of the surrounding countryside (see plate 1). Such ramparts defined the *shahristan*, a densely built inner city of one- and two-story homes, bazaars, and temples of various faiths. Within the *shahristan* was the main citadel or *ark*, with its own even higher walls. Here were situated the ruler's palace and the principal offices of government.

To appreciate Balkh's size, the citadel alone, called Bala Hisar, was twice the size of the entire lower city at Priene, a typical Hellenistic city on the Turkish coastline, and ten times the total area of ancient Troy.[3] And the citadel comprises less than a tenth of Balkh's total area! Everything about Balkh exuded the immense wealth amassed from a booming agricultural sector based on wheat, rice, and citrus fruits; the manufacture of metal tools and ceramic housewares, turquoise gemstones, and fine leather goods; and from international trade that reached as far as India, the Middle East, and China. Indeed, Balkh was perfectly positioned along the main route across Afghanistan to India and westward to the Mediterranean.[4]

Even today one finds on the surface at Balkh shards of pottery that are identical to both Roman and Indian ware of the period AD 100–400. No wonder that Roman writers already described Balkh as fabulously rich,[5] and that later Arab visitors, who knew well the bazaars and palaces of Damascus, Antioch, and Cairo, would refer to it as the "mother of cities."[6] These later writers, it should be noted, were describing a city that had suffered an economic decline in the generation immediately before the Arab invasion and had then been ruthlessly sacked by the Arab armies.[7] Yet even after this, visitors from the Middle East continued to pay tribute to Balkh as the mother of cities, and "beautiful Balkh."

A number of other great cities of Central Asia rivaled Balkh in size. One was Afrasiab, the predecessor to Samarkand in what is now

Uzbekistan, which had grown rich from its mass-produced cloth and other goods and covered over five hundred densely built acres.[8] Another was the river port of Tirmidh (Termez), which covered a thousand acres on the Uzbek side of the Amu Darya, across from Afghanistan.[9] Still another was Merv, in what is now southern Turkmenistan, an enormous urban complex that was already ancient in AD 500.[10] Several of these cities rivaled Xian (Chang'an) in China, said to be the largest city on earth at the time, with its walls extending for sixteen miles. Unlike Chinese cities, Central Asian cities had several rings of walls, the outermost to keep out invading nomads and the encroaching sand. At the Merv oasis the outermost rampart ran for more than 155 miles, three times the length of Hadrian's Wall separating England from Scotland. At least ten days would have been required to cover this distance on camelback.[11] This wall protected a region of intensive agriculture, many small towns with diversified manufacturing, and the core city, which surpassed Balkh in size and population.[12] Satellite towns and villages like those that surrounded Merv were to be found at all the other metropolitan centers. Otrar, in what is now southern Kazakhstan, boasted nearly a hundred surrounding towns and villages, all of which were loosely linked together in a single local economy.

Equally important were the scores of only slightly smaller cities that were sprinkled eastward from the shores of the Caspian Sea deep into what is now Xinjiang in China, and southeastward across Afghanistan to the Indus Valley.[13] Some of these still exist today, centers like Chach (now Tashkent);[14] Sarakhs in Iran; Kashgar, Khotan, and Turfan in Xinjiang; and Kabul, Herat, and Ghazni in Afghanistan. Others, like Akhsikent in the Ferghana Valley, Tus and Nishapur in the northeastern Iranian province of Khurasan, Gurganj (Kunia Urgench) in Turkmenistan, Otrar or Suyab in Kazakhstan,[15] or Gissar in Tajikistan, hang on only as depopulated villages or have died out completely.

Large numbers of cities of this middling rank had been well-established centers of commercial and civil life for several millenniums before the Arabs showed up. Excavations at a dozen of them have established that their residents did not need to travel to the megacities to gain access to the latest amenities and fashions. Typical was Isfijab, now Sayram, in southern Kazakhstan, where a highly diverse population had access to fashionable products brought from the Mediterranean, India, China, and points between. The earliest estimate of Isfijab's population—forty thousand inhabitants—came several centuries later, but the surviving walls indicate that this market town was already ancient by the time of the Arab conquest. Isfijab's population, typical of many smaller market centers in Central Asia, was similar to that of early medieval Paris.[16]

Besides the large and middling centers, the western and northern reaches of Central Asia were dotted with solidly built castles and fortified manors belonging to the large landowners, or *dihkans*. At least three different types of such noble residences existed. Atop rocky outcroppings in the northern desert area were scores of high-walled fortresses that enclosed small towns, miniature versions of the citadel-and-town model that existed at Balkh and other cities. On the flat deserts of what is now southern and western Turkmenistan were dozens of large, block-like structures with crenellated exterior walls. Called *koshks*, these massively constructed brick residences were owned by grandees connected with the nearby cities.[17] And on hilltops in Tajikistan and Afghanistan were citadels where a local ruler or nobleman and his court held forth. Mighty-walled Khulbuk in Tajikistan is a particularly impressive example of this kind of secondary fortress. Archaeological research on all three types confirms that the inhabitants of these outposts enjoyed the same high standard of living as those in large urban centers.[18] The 250 finely made clay goblets excavated in one kitchen area at the trading center of Paykand on the Bukhara oasis attest to the fact that the good life was not confined to the largest metropolises.[19]

Central Asian cities were densely populated—one expert estimates that 230–270 persons per acre was typical[20]—and the footprint of four-fifths of the houses was as small as 380 square feet, even though they typically housed up to six people on two or three floors. These figures may reflect the prevalence of slavery, which was to expand greatly under Muslim rule and the increasing militarization of states, but had deep roots in local life tracing back centuries. As early as the second century AD we find a wealthy Central Asian household of four served by no fewer than seventeen slaves![21] A few such families lived in much larger residences, some with up to fifty rooms.

Residents, including slaves, had good access to running water[22] and could sleep on built-in brick beds, which in winter were made comfortable by heat generated by charcoal fires and piped through neatly

Figure 2.1. Kyz Kala, a grandee's residence at Merv, sixth to seventh centuries. The aristocratic *dihkans* who inhabited such fortresses across Khurasan were active patrons of culture. ▪ Photo by Brian J. McMorrow.

built channels inside the masonry. In hotter parts of Afghanistan and Khurasan, cooling towers were built to draw the summer's heat out of the residences, while elsewhere specially built basements provided refuge from the sun and heat.

Of great help in assuring the circulation of air and light were the brick and often ribbed domes with which Central Asians roofed everything from palaces and commercial buildings to the more lavish private residences. Many centuries before domes became a hallmark of Islamic architecture, they had been applied to all categories of architecture across the Persian-speaking world, and especially in the highly urbanized world of Central Asia. At the great Buddhist center at Mes Aynak and at several other sites in Afghanistan and the Amu Darya valley, one encounters not only the customary round arches but pointed arches as well.

In Europe the pointed Gothic arch is usually traced to the eleventh-century abbey church of Saint-Etienne at Caen in Normandy.[23] Slightly earlier examples at San Ambrosio in Milan and other churches in northern Italy hint at the possibility that this arch might have been an import from the East. But from where? Slightly pointed arches appeared in a handful of ancient buildings of the Middle East, in several early buildings of the early Islamic era, and in a seventh-century church in Armenia.[24] Such pointed arches were more numerous in Iran but appear far more frequently in Buddhist Afghanistan and Central Asia than anywhere to the west. One of many such sites with conspicuous pointed arches is Guldara near Kabul. These regions, of course, were in constant trade contact with the Iranian lands and Middle East. Thus one line of the genealogy of Gothic arches may trace to Buddhist Central Asia. But this would not be the end of the trail, since an Indian origin for this feature of Buddhist architecture is all but certain.

Returning to urban architecture across the region, the high plastered walls of major chambers were ornamented with bright designs and painted figures, a practice that spread from the rich to middling burghers and was to continue into the Islamic era (see plate 2). And the walls and floors of even modest dwellings were rendered softer and less severe by ubiquitous woven carpets and hangings dyed with rich colors. Many Central Asian urbanites were comfortable sitting on the floor, but chairs were common.

Aside from domestic comforts, civic life had achieved a high level of development during the millennium before the Arab conquest. Streets were paved, public baths were commodious, and extensive retail areas existed, usually close by temples and shrines and often connected with hostels for traders visiting from abroad.

These and other amenities reflect the existence of a deeply rooted and sophisticated urban way of life in Central Asia. In fact, the region's tradition of urbanism stretches back nearly five millenniums, when stock breeders began grouping together in large communities. By four thousand years ago, walled cities of the Bronze Age like Gonurdepe and Margush, both on the Merv oasis in Turkmenistan, were thriving.[25] Recent excavations have uncovered these large, rectangular walled towns, as well as their palaces, temples, public buildings, bazaars, and residential areas. These finds show that architecture had long since moved beyond the merely practical. Only a few centuries later, fanciful Bronze Age townspeople at Mundigak near Kandahar in Afghanistan were building a massive temple that resembled nothing so much as a Mesopotamian ziggurat.[26]

All this took place only slightly later than the emergence of the great civilizations of the Harappa in the Indus Valley and the Sumerians in Mesopotamia. Indeed, the archaeologist Victor Sarianidi from

Figure 2.2. This reconstruction of Gonurdepe on the Merv oasis in Turkmenistan suggests the high level of urbanism and civic life attained in Central Asia four thousand years ago. ▪ From Viktor Sarianidi, *Gonurdepe, Turkmenistan: City of Kings and Gods* (Ashgabat, 2006), 182–83. Gonurdepe reconstruction by M. Mamedov.

Turkmenistan, who excavated Margush and Gonurdepe, argues that they prove that the Amu Darya (Oxus River) valley in Central Asia constitutes a fourth point of origin of urban civilization, along with the Nile, Indus, and Tigris-Euphrates valleys. Excavations proved that the very early Central Asians already maintained extensive trade and cultural contacts with all three of these centers of world civilization. And in at least one sphere the Central Asians led the pack. Thanks to the research of two ingenious American archaeologists, Raphael Pumpelly around 1900 and Fredrik Hiebert a century later, we know that Bronze Age Central Asians were the first humans anywhere to cultivate grain for baking bread.[27]

Of course, there were large cities elsewhere in ancient and early medieval times, whether in the Middle East, China, India, or the Americas. Archaeologists report the existence of no fewer than twenty-five complex urban centers in Mesoamerica by 3000 BC.[28] The distinctive achievement of Central Asian cities was to have combined the organizational sophistication required by large-scale irrigation systems with export-oriented agriculture and manufactures, and to have nurtured large cadres of traders who traveled the world and businessmen who managed their trade.

Was There a Climate Boost?

To approach Balkh today is a sad experience. Where ancient visitors reported on vineyards, citrus groves, and fields of sugar cane, there is only sagebrush and dust, relieved by an occasional hollyhock in the lower-lying areas. Similarly, far to the north in Central Asia, the vast reaches of Khwarazm in Uzbekistan and Dehistan in Turkmenistan were once alive with castles surrounded by farmland but are today bleak deserts, utterly devoid of plant life. Was the cultural and intellectual flowering of Central Asia fostered by a temporary era of moderate weather and generous rains, with neither the summer nor winter extremes of temperature that are common in that region today? Perhaps the burst of creative life coincided with a moist and comfortable phase in this stark climate, during which an effervescence of cultural life briefly became possible?

However attractive such a theory, there is as yet little evidence to support it. On the contrary, most experts assert that Central Asia's climate, including annual rainfall, during the era 100 BC to AD 1200 was not only consistent throughout the era but very much like what exists there today. Some are convinced that a xerothermic crisis occurred from Greece to India in the middle of the third millennium BC and that this imposed a harsh drought that lasted for centuries.[29] But following that event they believe that something like the climate we know today came into being and persisted through ancient times and down to the present.[30] A century ago a Dutch Orientalist, Michael Jan de Goeje, published the works of tenth-century Arab geographers that discredited the notion that the desiccation of modern times had been preceded by a relatively more verdant era that coincided with the Age of Enlightenment.[31] The only dissenting voice is that of Richard W. Bulliet from Columbia University, who has recently suggested that growing desertification in Khurasan may help explain the rise of the Seljuks in the eleventh century.[32] For the time being, though, this hypothesis must be considered unproven. Overall it seems that Central Asia's climate throughout the era that concerns us was as dry and demanding as it is today. Modern farmers in the region can well sympathize with their ancestors at Margush, who four thousand years ago built a temple to Water![33]

What, then, caused the obvious changes that have occurred? The disappearance of wood for roof beams forced builders at some of the earliest cities, such as 3,000-year-old Gonurdepe in Turkmenistan, to roof their buildings with brick domes.[34] Even at Afrasiab, Merv, and Gurganj, wood was once fairly commonly available but then became scarce. What happened to the forests that once flourished on the lower mountainsides? And what happened to the Balkh River, the stream that once flowed from the city of Balkh to the Amu Darya with enough water to carry boats, but which has now dried up completely?

Both changes trace not to climate shifts but to the actions of human beings. An American scholar, Naomi Miller, has traced the deforestation to the Bronze Age, 2,400 years ago, when the opening of many forges created a huge demand for firewood.[35] This, along with the cutting of forests for construction, heating, and cooking, goes far toward explaining the disappearance of woodlands across Greater Central Asia.

Grazing sheep and goats, both of which have teeth that are able to clip grass so short as to kill it, further hastened the water crisis. Over the centuries these grazing animals stripped the lower mountainsides of grasses and other forms of plant life that held the soil. This led to massive erosion that denuded the terrain and exposed the rock beneath. Where showers once caused grass to grow and fed steady-flowing streams, the waters from spring rains now rush in torrents down the hillsides, to be followed by drought.

In short, the environment of Central Asia *has* changed dramatically over the centuries since the cultural golden age. But the agent of these changes has not been nature itself but humankind, and especially the relentless prowling for wood to be used for fuel and the shepherds' ceaseless quest for green pastures to fatten their sheep and goats. Superficially one might conclude that these changes justify the application of Jared Diamond's theory in *Collapse: How Societies Choose to Fail or Succeed* that civilizations die when people destroy the environment on which they depend.[36] But in this case the changes do not suffice to explain either the onset or the end of the Age of Enlightenment.

An "Intensive" Civilization

Besides these negative factors, an important positive force made possible the development and maintenance of civilization and a high culture across Central Asia. Again, the agent was not nature but humankind, specifically, people's gradual mastery of the arts and technologies of irrigation. It was irrigation, and only irrigation, that made possible the rise of civilization on some of the otherwise barren land of Central Asia. In this sense it is fair to call Central Asia a "hydraulic civilization," one in which the main focus of social energies was on the construction and maintenance of complex systems for the conservation, distribution, and overall management of a scarce resource: water.[37] This term was first introduced by the German American scholar Karl Wittfogel in a highly controversial volume entitled *Oriental Despotism* (1957). Although he defined societies from China and India to Mexico and Mesopotamia as "hydraulic" in character, his notion would also appear to fit certain aspects of medieval Central Asia. Over time the stress on irrigation created highly disciplined social orders and strictly hierarchical political cultures—which Wittfogel called despotisms. The governments assumed full responsibility for the large and complex irrigation systems, including the critically important task of mobilizing and managing the labor force that maintained them.

However, it must be noted that not all irrigation leads to governmentalization, centralization, and top-down management. Ancient Greece was also deficient in water, but its hilly terrain prevented the construction of the kind of large-scale, government-organized irrigation systems that prevailed in Central Asia. Instead individual farmers worked with their neighbors at the local level to solve their water problems through small-scale projects. By fostering a sense of communal responsibility and citizenship, this had important implications for Greece's civic and political life.

Overall the irrigation systems of Central Asia had much in common with Wittfogel's template, but with one important difference. In China, which was his main focus, and in Mesoamerica as well, the "hydraulic" civilization embraced the state as a whole, and not just individual oases, communities, or city-states. But in Central Asia the high degree of organization and governmentalization rarely extended upward beyond a single oasis. The great distances between oases, combined with the organization skills needed to manage each of the separate hydraulic systems, created an intense public life on each oasis but a much thinner and more narrowly military governmental presence at the regional or international level. This opened the way for a series of empires—many of them originating in Central Asia—to claim hegemony over the region. In spite of this, the major Central Asian hydraulic systems appear to have been maintained with few serious interruptions for over two millenniums, extending down to the Mongol invasion in the thirteenth century.[38]

As early as the Iron Age in Central Asia, men and women were beginning to construct the irrigation systems that were essential for urban life.[39] Long before the advent of Persians, Greeks, and other invading outsiders, these irrigation systems had come to define the life of these oasis civilizations. The technicians who oversaw these operations, who must certainly be considered highly qualified hydrologists and engineers, employed two basis techniques.

The first was to dam mountain rivers at the point where they emerged on the plain to create ponds or lakes. These dams were often large masonry structures lined with clay, but smaller, covered ones were also constructed. On the Balkh River at Balkh, the Zarirud at Bukhara, the Murghab at Merv, the Zarafshan at Afrasiab, and the Amu Darya at Gurganj, the dams were fitted with huge gates or valves that could be opened and closed to assure a steady flow of water through the cities in all seasons. Obviously, an enemy army could flood the city simply by destroying the dam, as occurred at both Balkh and Gurganj.[40] These dams in turn fed the half dozen or so open trunk channels that were dug to and through the city and surrounding agricultural lands. At Balkh there were twenty such channels. It was not uncommon for these principal canals to run for sixty miles and to involve carefully engineered aqueducts.

In an effort to minimize evaporation, Central Asians came to dig these channels deeper and deeper and thus reduce the area exposed to the sun. They also lined them to prevent loss through seepage—both of these being techniques that modern Soviet engineers ignored, with disastrous consequences. The resulting riverlets were often channeled underground through baked clay pipes that fit neatly into each other. At

Afrasiab's inner city, the main feeder pipes were made of lead and were described by one early visitor as nothing less than "the eighth wonder of the world."[41]

The second method was to collect water on high ground near the city and to channel it to or through the settled areas and agricultural lands by means of carefully dug underground channels or *kerezes*. This system, developed to water fields, involved the digging of lengthy underground passages, as well as the construction of periodic vertical shafts to provide ventilation and access. Considering that these channels often ran for many miles, reached three hundred or more feet in depth, and passed directly under whole cities, they, too, must be considered an engineering marvel.

Both types of hydraulic systems required the maintenance of precisely calibrated grades to assure steady flows, as well as the smooth functioning of the various types of carefully engineered lifting mechanisms that were placed at regular intervals. Within the cities the maze of underground pipes of baked clay that served public baths and private homes became yet more complex, for they included valves, catch basins, and access points for cleaning, as well as exceedingly complex changes of gradients.[42] All the skills necessary to design, construct, and maintain these systems existed in abundance. Suffice it to say that during the twelfth century one city, Merv, had a staff of twelve thousand to maintain the hydraulic system, among them three hundred divers![43] Granted that at this time Merv was the largest city in the world, outpacing even Hangzhou in China,[44] but in pre-Islamic times it was also a very large urban center with an already ancient and highly developed water system.

Some writers on Central Asian cities invoke the generic concept of "the Islamic city" to explain their subject.[45] There is some justification for this. By the twelfth and thirteenth centuries, Central Asian cities came in many respects to resemble their counterparts elsewhere in the Islamic world. But down to that time, the great cities of Central Asia had a character of their own, formed centuries before the Arab conquest and maintained for centuries thereafter. This distinctiveness arose above all from the irrigation systems that made life possible and the rigorously hierarchical and strictly regulated social systems that enabled those systems to function.

This oasis civilization produced agricultural abundance and high-quality manufactures, and hence wealth. But it was founded on scarcity: of water and of irrigated land. Available water had to be carefully captured, channeled, and deployed, which posed huge challenges to Central Asians in many fields of endeavor. They responded with focus and imagination. Suffice it to take note of the several hostels for caravans (caravanserais) that employed clever and highly efficient technologies for gathering dew, or the intricate underground pipe systems that provided urban dwellings with potable water. Such resourcefulness epitomized a civilization that exploited resources *intensively* rather than *extensively*. Intensive civilizations, of which traditional Japan is a prime example, increase productivity by getting more "bang" from existing limited resources, rather than by seeking to get their hands on more resources. Tsarist Russia and the USSR, by contrast, were prime examples of extensive resource users, increasing agricultural productivity by the endless addition of land or labor, rather than by more productively farming the existing fields and more efficiently deploying the existing farm laborers. Needless to say, the intensive quality of Central Asians' oasis agriculture affected every aspect of their lives and culture.

High-Value Traders

The second source of Central Asian wealth—long-distance trade and commerce—also depended on a combination of geographical realities and human initiative. A glance at the map reminds us of the utter uniqueness of the region's geographical location: all the great civilizations on the Eurasian landmass are accessible *from* Central Asia, and those same civilizations are accessible to one another by land only *through* Central Asia. From the standpoint of transport and trade, Central Asia is indeed central and was so from the dawn of history. To take advantage of this windfall of Creation, residents of the region needed only to discover a means of overcoming distance.

They accomplished this by the eighth century BC. Prior to that era, the wheel was in wide use, mainly on war chariots but also on ox-drawn wagons. Camels were used as draft animals. The rise of mounted cavalry at the dawn of the first millennium BC rendered the chariots useless.

Later the decline of Roman roads in the Middle East further reduced the value of wheeled vehicles. In a highly unusual reversal of the normal course of things, the camel, described by an admirer as "900 pounds of muscle, hauteur, and, for those who can come to appreciate it, grace," replaced the wheel.[46] Camels, it turned out, offered the most efficient means of transporting goods and people in the more austere geography of the region. Soon large, single-humped camels, called dromedaries, were being bred for transport in the Middle East. But the virtues of the two-humped camel made that beast far more useful than the dromedary to Central Asians. For one thing it was more impervious to cold, with the longer-haired, hybridized version being especially cold resistant. For another, it was more sure-footed in the mountain defiles that are common in parts of Central Asia. This native "Bactrian camel," not the Middle Eastern dromedary, became the backbone of regional and continental transport.

But what were these animals to transport? The answer turned on issues of both weight and mass. A Bactrian camel can carry up to 500 pounds, and a caravan of a thousand could therefore carry about 500,000 pounds. Assuming that a modern standard container carries up to 50,000 pounds, this means that even a moderate-sized caravan would have been able to carry as much as a ten- or twelve-car freight train. Needless to say, excessive weight or volume could kill profits. The ideal cargo was therefore high in value and low in weight and mass. In recent times this kind of calculation lured many Afghans and Central Asians into the drug trade. But in 3500 BC the single most profitable trade product on the Eurasian land mass was the brilliant blue lapis lazuli mined in Afghanistan.

Five millenniums ago Afghan lapis lazuli was well-known and treasured in both pharaonic Egypt and the Harappa civilization in India. Other precious stones and minerals were similarly prized, which made Afghanistan a premier source of luxury goods in both East and West. Jade from Khotan in present-day Xinjiang, emeralds from the Badakhshan region of what is now Afghanistan and Tajikistan, gold from rich mines in what is now Uzbekistan, and copper from Afghanistan joined lapis as high-profit export goods. From this origin in the export of precious stones and metals, the commerce expanded to embrace all commodities and manufactures that could be profitably traded.

Figure 2.3. Central Asians, not Chinese, were the principal traders along the Silk Road, as they were on the southern routes to India. This Chinese earthenware sculpture from the Tang dynasty, circa AD 618–907, depicts a Bactrian camel, a Central Asian rider, and the rider's dog. ▪ Courtesy of Iliad, New York City.

Soon caravans of hundreds, then thousands, of Bactrian camels were wending their way toward India,[47] China, and the Middle East, delivering high-value commodities and products from the workshops and markets of Central Asia and bringing back whatever goods would find a market at home or at more distant markets on the opposite points of the compass. Thousands of different items soon filled the saddlebags. The long caravans moved at up to about twenty miles a day and, in hot weather, by night.[48] Because camels do not need paved roads, the caravan leaders could shift routes constantly in response to changing weather, markets, and politics. The immense flexibility of the caravan trade gravely undermines the many recent efforts to pin down specific rights-of-way of the great east-west and north-south routes.[49] Meanwhile, to a greater extent than we can imagine today, Central Asia's traders also moved their goods by large, solidly built boats on the region's three main rivers.[50] The finely constructed nineteen-foot-long cargo boat preserved at the museum in Otrar, Kazakhstan, gives a hint of the kind of wooden vessels that once plied the Amu Darya and Syr Darya.

This tangled web of routes and modes of conveyance is what a nineteenth-century German geographer inappropriately named the "Silk Road" (Seidenstrasse). Baron Ferdinand von Richthoven (1833–1905), who coined this term, was right to note that silk from China traveled westward over these routes, from about 100 BC to AD 1500. But he erred in implying that silk was the sole or primary trade good: he could just as easily have spoken of a "Lapis Lazuli Road" from Afghanistan to Egypt and India, a "Jade Road" from Khotan to China, an "Emerald Road" stretching east and west from the Pamir Mountains of Tajikistan and Afghanistan, or a "Gold Road" or "Copper Road" to the capitals of the Middle East. He also erred in assuming that the great corridors of transport ran mainly to China and not, equally, to India as well. He further erred in supposing that silk came only from China; in fact it did not take Central Asian entrepreneurs long to figure out that they would do better producing silk on their own than transshipping the silk of others. By the tenth century the city of Merv was the single major producer and exporter of silk to the West and even had a kind of "institute for sericulture" for the study of silk production.[51] And finally, he erred by implying that there were no equally valuable goods being transported to both China and India from Central Asia and points west.[52]

In considering how this continent-wide trading complex shaped the culture of Central Asia, it is helpful to focus on three distinct functions in which people of the region were intimately involved: first, the emergence of regional cities as commercial entrepôts; second, the creation of a skilled class of professional traders with networks extending to distant lands; and, third, the development of export-driven economies based on high-quality local industries and manufactures.

Continental trade, by its nature, involves freight forwarders and salespeople from different lands. Indian merchants, for example, were a constant presence in the major cities of Central Asia. Even in northern Khwarazm, with its main routes heading east and west rather than southeast, they were so common that Khwarazmians were familiar with the Indian decimal system long before it was known in the Middle East. Later a scholar from Khwarazm was to play a key role in convincing the Arabs in Baghdad to adopt that system.[53] No less frequent were traders and other visitors from Syria. These were almost exclusively Nestorian Christians, many of whom settled throughout Central Asia after about AD 400.

Surprisingly, in light of the enormous volume of trade to and from China, Chinese merchants themselves played little role in the caravan trade. Needham, in a striking passage, speaks of "the marked disinclination of Chinese to travel far outside what they felt to be their natural geographical boundaries."[54] This Chinese sense of cultural borders created an extremely important opening for Sogdians, and also Khwarazmians, Uyghur traders from East Turkestan, Nestorian Christians based in Central Asia, and all other Central Asians.

The presence from early times of all these merchants—most of them locally based—assured that Central Asian cities would become the major center of banking and finance for trade between China, India, and the Middle East. One ancient city, Taraz in what is now Kazakhstan, was so closely identified with trade that its very name means "the scales." Numerous service industries arose as well, including hostels or caravanserais, bazaars, and storage facilities. These enabled the cities to become the main international entrepôts on the entire Eurasian continent, the gathering place of all.

It was inevitable that Central Asians themselves would become skilled merchants and traders. After all, with the Chinese out of the picture, they had a huge advantage over all their rivals. By the third century BC they were frequent visitors to India and to the great centers of Iraq, Syria, and the Mediterranean coast. Geography created a degree of specialization. Merchants from Balkh dominated the Indian trade, traders from Merv look westward, while those of the more northern regions around Samarkand, called Sogdiana, and around Kath, called Khwarazm, controlled much of the trade with East Turkestan and China. Everyone got in on the act, so that businessmen in a center like Akhsikent in the Ferghana Valley were scarcely less at home in Damascus or Lahore than those from Merv or Balkh. It is revealing that a medieval Chinese writer, Li Yanshou (618–676), thought that the ruler of Bukhara sat on a throne in the form of a camel.[55]

During the four centuries before the Arab invasion, it was Sogdian merchants from Samarkand, Panjikent, and neighboring towns who stood at the head of Eurasian commerce.[56] Sogdian traders seemed to pop up everywhere. After gaining early footholds in East Turkestan and then in Inner Mongolia,[57] they then established sizable diasporas—it would be accurate to call them colonies—along the routes to China. This positioned them to dominate the China trade for centuries.[58] The present eastern territories of Kazakhstan and Kyrgyzstan are dotted with the ruins of ancient towns that began life as Sogdian merchant colonies.[59] The Sogdians followed the same practice along all the main routes to India.[60] The boundless ambition of their merchant houses extended also to the sea lanes, which had the advantage of requiring fewer middlemen. This led Sogdians to open up routes across the Black Sea to Constantinople, and from Basra in Iraq across the Indian Ocean to Sri Lanka and Canton, in both of which they maintained offices.[61] The only drawback to this trade was that its timetables were dictated not by the market but by the monsoons, which meant that a round-trip required nearly a year.[62]

Sogdians certainly knew the ways of the world, including the arts of hucksterism. One showman-salesman who had traveled to China was said to have dressed himself in Daoist robes to perform alchemistic rituals that culminated in the production of a potion that assured immortality to all who drank it. He then proceeded to peddle this elixir of life to gullible members of the Chinese public.[63]

The result of this centuries-long surge of commercial initiative was to create in all the cities of Central Asia a large class of active and rich merchant traders. These men knew the world better than their governments, were used to making their own decisions, and paid sufficient taxes to make the governments dependent on them.

A further consequence of the centuries-long trade boom that preceded the Arab invasion was to stimulate local manufactures. In city after city, specialized industries arose to serve the export trade. Thus the cities of the Ferghana Valley, which had easy access to rich deposits of coal and iron, produced steel blades that could be exported profitably to the Middle East and India.[64] At Akhsikent, Pap, Merv, and other Central Asian centers of metalworking, their technique required crucibles that could sustain heat up to 1,600 degrees Celsius; in this part of Central Asia, people had a sophisticated knowledge of local clays and could therefore produce such vessels. Pioneered in the heart of Central Asia, crucible steel spread from there to Damascus and eventually to the West.[65] Lost-wax casting, which the French archaeologists Benoît Mille and David Bourgarit traced to an area of northern Pakistan adjoining Central Asia, was also a specialty of Central Asian metal workers.[66] Jewish craftsmen, who brought glassblowing from Egypt to Merv and other centers, were exporting their goods from Balkh to China by the end of the fourth century AD.[67]

Other Central Asians were to introduce to China such technical achievements as screws, force pumps for liquids, and crankshafts.[68] Less obvious items that were exported in quantity from Central Asia to China included lutes, harps, transverse flutes, both plucked and bowed stringed instruments, and even Central Asian dances,[69] all of which took root there and became staples of Chinese culture. The fact that these originated among peoples who spoke Sogdian, Bactrian, and other Iranian languages has led some writers to conclude that they were exported to China from "Iran." To be sure, the territory of modern Iran was not irrelevant to this process of technological and cultural export. But the key point of dissemination was Central Asia, not Iran. Even the chair appears to have been an import to China—from Bactria![70] Many crops, like pomegranates, sesame, jasmine, peas, and broad beans, entered Chinese cuisine from Central Asia, as did the seed drill and harrow with which they were planted and plowed.[71]

Entrepreneurship and opportunism were at a premium in the world of Central Asian commerce. Local firms that produced everything from perfumes to drugs, finished jewels, fine metals, and diverse utilitarian objects could all find ready buyers in China, India, and the Middle East; exporters of horses, wild animals, and exotic birds thrived equally.[72] The first silk reached East Turkestan from China between the second and fourth centuries AD and the rest of Central Asia shortly thereafter.[73] But it did not arrive in a vacuum. Locally woven textiles of all sorts had been hugely profitable export items for all Central Asian cities beginning at least from the first millennium BC. Kabul, Bukhara, Merv, and other regional centers were all exporting their distinctive fabrics to the East and West centuries before the arrival of silk in their region. There were therefore many knowledgeable professionals in the textile business who could easily figure out how to produce silk on their own and to reap profits from its export. By the fourth century AD they had accomplished this, and within another few generations Central Asians had developed their own production capacities and were moving vigorously to elbow Chinese producers to the sidelines.[74]

This story was repeated with another Chinese invention, paper. It was long believed that papermaking in Samarkand and elsewhere in Greater Central Asia did not begin until Arab armies defeated a Chinese force at Taraz (Talas) on what is now the Kazakhstan-Kyrgyzstan border in AD 751. Among the captured Chinese were said to have been papermakers, who introduced their art to Samarkand. But archaeology has proven that such East Turkestan cities as Turfan, Khotan, and Dunhuang were all producing paper by the third century AD. These cities were in close commercial contact with Central Asians to the west of the Tian Shan thanks to the activities of the Sogdian trading houses. These merchants would have eagerly seized on the new invention and moved quickly to ferret out details of its production so as to be able to replicate the process back home.[75]

No sooner did the Central Asians begin producing their own paper than they dramatically improved the product. Early Chinese paper was made from either mulberry fibers, bamboo fibers, or a combination of the two. These yielded a product that was both stiff and flimsy.[76] The Central Asians saw at once that their own long-fibered cotton fibers could produce a paper that was more durable and more flexible than what the Chinese were selling. And since their supply of cotton was virtually unlimited, they could also produce their improved paper at less cost than the Chinese product. For many centuries thereafter, paper from Samarkand, not China, set the world standard for quality. Indeed, paper itself was seen as a Central Asian product. True, craftspeople in Baghdad, Damascus, Cairo, Fez, and Cordoba were soon producing paper of their own, but the Central Asian product long continued to dominate the high end of the market. Europeans, by contrast, did not begin manufacturing paper until the thirteenth century.[77]

A millennium before it was invaded by Arab tribes, Central Asia already boasted a successful export-driven economy. As in Japan after World War II or China in the late twentieth century, its manufacturers studied foreign products passing through their markets and identified those they could produce better or cheaper. This meant knowing the materials and relevant technologies and mastering the production processes. While this did not in itself create the inquiring spirit that prevailed during Central Asia's Age of Enlightenment, it certainly placed a premium on open-mindedness and innovation.

IMPERIAL RENT-SEEKERS

The Arab conquest of the late seventh century was a cataclysmic event in the history of Central Asia but by no means an unprecedented one. In fact, external powers had repeatedly conquered the region's city-states and subjected them to their rule. Among these invaders were some of the most powerful empires of the classical age and late antiquity. Yet none of them succeeded in fully controlling, let alone governing, the territory they had gained through force of arms. Their experience— like that of more recent aspiring hegemons in the region—confirms the wisdom of Gibbon's remark that conquered territories are invariably a source not of strength but of weakness. The reason for this is clear: in the course of their long and difficult history, Central Asians had mastered the art of managing their conquerors. These talents were to be brought into play after the Arab conquest as well.

The first of these invasions to be recorded in history took place in 523 BC, when Darius the Great, the Achaemenid king of Persia, marched his armies into Balkh and the surrounding region. The Greek historian Herodotus reported on the upshot of this conquest, namely, that Darius required Balkh and neighboring territories to send tens of thousands of young men to fight with the Persian armies in the West. Thus, when the fate of Greek civilization was being determined at Thermopolae and other epochal battles against the Persians, the armies of Athens and the other Greek city-states were up against Central Asians, among others. The Persians dragooned both city dwellers and Scythians (Sacas) from the surrounding countryside, and the two groups from Central Asia fought side by side in the West.[78] Later, however, the cities and their rulers discovered that it made more sense for them to hire steppe nomads to do the fighting for them, which became common practice throughout the Age of Enlightenment and down to the sixteenth century.

Darius's Persian empire is often described as "centralized."[79] But in the ancient East, to rule did not mean exercising full control over a given territory and its government, let alone establishing there a system of law, as did the Romans. Rather, it implied the ability to extract tribute. Tribute established a clear hierarchy of power extending from small towns or territories up to an imperial capital. Even China found itself paying tribute as a means of keeping the fierce Huns at bay.[80] But the extraction of tribute did not imply day-to-day control. In fact, tribute was nothing more than a systemized form of what economists call "rent seeking," imposed through force.

In the case of the Persians, subject territories were called satrapies and became largely self-governing, provided they met the demands for tribute, which were high. Central Asia, for example, was compelled to pay the equivalent of 25,000 kilograms of silver, only slightly less than required of Mesopotamia or Asia Minor.[81] This figure would doubtless have been higher had Central Asia not been so much more distant than these other territories. Thanks to distance, governors or satraps appointed to rule the region were easily tempted to pocket payments meant for Persepolis and even to threaten secession.[82] This tendency of imperial governors to "go native" in Central Asia and champion local autonomy was to be manifest repeatedly, first by the Greeks, then by the Arab conquerors, and later by the Mongols. Typically the Achaemenid Persian state claimed a monopoly on the minting of gold coins, yet largely failed to get them into circulation in Central Asia. However, businessmen in Balkh and other cities in the region liked this innovation— metal coinage—and prevailed on their Achaemenid governors to issue and use their own coins, which long remained the currency of choice among the locals.[83]

After destroying the Persians' capital at Persepolis, Alexander of Macedon swept into Central Asia and Afghanistan in 329 BC and waged a bloody three-year war against Bactrians, Sogdians, and Margianans (e.g., the people of the Balkh region, Afrasiab, and Merv), in short, against the indigenous inhabitants of the region's main cities.[84] Later, after Alexander's death in 323, his generals divided his realm among themselves, with one of them, Seleucis, gaining control of the vast territories stretching from Mesopotamia to India, including most of Central Asia. The Greek rulers of the kingdom of Bactria, with its capital at Balkh, were initially subordinated to this Seleucid empire, but they quickly realized that Seleucis and his heirs in Babylon had no power to enforce their demands for tribute. Perceiving this, they claimed de facto independence and began minting their own handsome gold coins. The armies of independent Greek Bactria successfully invaded India in 180 BC, which transformed their realm into a continental state, a Greco-Indian empire based in Central Asia. Soon Bactrian merchants began showing up in markets as far afield as Alexandria in Egypt and southern India. Such wide-ranging contacts continued even after the Greek state in Central Asia had faded from history. Thus the Roman lyric poet Propertius considered it quite normal to write of a young girl whose swain had traveled several times to Bactria.[85]

This proud Hellenistic Greek kingdom in Central Asia was dead by 129 BC, having come to depend on the Chinese for protection against invading nomads.[86] It had lasted just short of two centuries. The final ruler of Greek Bactria, Eucratides, sensed danger and, before fleeing, hid his stunning collection of Indian gold and utilitarian objects in the palace treasury. The invading Huns captured Eucratides and murdered him. His was a sad fate, but one that garnered sympathy from future writers, including Giovanni Boccaccio in his *On the Fates of Famous Men* and Geoffrey Chaucer in his *Knight's Tale*.[87] Two millenniums after Eucratides's flight, French archaeologists discovered part of his treasure.

Between 100 BC and AD 100, both China and Rome began to eye Central Asia with interest. Later, Tibetans were also to figure in this competition.[88] China's first concern was to prevent the region from becoming a staging ground for the nomadic tribes that threatened China proper.[89] Beyond this, travelers had informed the Chinese court of the commercial potential of the area, which led directly to the opening of the silk trade in about 114 BC. Even though the Chinese briefly dispatched several armies to the region, China never became a true hegemon in early Central Asia, and local self-governance there remained intact. But China's cultural impact was significant, extending to such practical items as the Chinese-type cast coins with square holes that

among competing heirs to this throne gradually undermined the Parthians' political power. This, along with continued conflicts with Rome and nomad incursions, brought down this Central Asian powerhouse by the start of the third century AD.

Contemporary to the Parthians in Central Asia were the Kushans, another group of tribal nomads who had been pushed westward and south by other nomads. The Kushans soon forged their own extensive empire based in Afghanistan. Between 100 BC and AD 200 they ruled a vast territory stretching from Khurasan to Punjab in India, and briefly including parts of Xinjiang as well. They extended their commercial reach to Europe, providing rich fabrics for Roman senators and their wives, thus creating a true "Silk Road."

Like many conquerors before and after them, the Kushans "went native," settling into the chief cities and establishing their own capital at Begram[95] in Afghanistan. They also adopted many local beliefs and customs, including Zoroastrianism and the two rising religions in the region, the Greek cults and Buddhism. The Central Asian territories they ruled would eventually become the major route by which Buddhism found its way to China.[96]

Though they had no prior experience of urban life, the Kushans took to it with a passion. A huge urban expansion took place under their rule, with new cities that built on the Hellenistic heritage in planning and architecture.[97] Closely connected with this was a great expansion of irrigation, to which Central Asians in the Kushan era applied a number of innovative technologies.[98] The locals also convinced their Kushan overlords of the importance of a stable and widely available currency[99] and were probably responsible for the fact that the Kushans pegged the value of their currency to the Roman *aureus*.[100] The resulting economic boom was reflected in the spectacular golden horde of jewelry, sculpture, and ornaments dug from one of the lesser Kushan capitals in northern Afghanistan.[101]

While Rome was entering its decline and fall, the Kushan empire stood at a peak of development and sophistication. Its leaders, transformed by their Central Asian subjects into thoroughgoing cosmopolitans, felt it their right to flaunt lofty Greek and Indian titles like "Basileus" and "Devaputra" (Son of Heaven) and to send their diplomats as far afield as Rome, China, and India.[102]

The final traditional-type empire to seek to control Central Asia was that of the Sassanian dynasty in Iran, which had its noble capital at Ctesiphon, just south of the future site of Baghdad. After centuries spent fighting their way to the top, the Sassanians defeated Parthia's army in AD 224–226. Over the following decades they established their suzerainty in Khurasan and Sogdia but managed to gain only partial control over the rest of the former Kushan territories in Central Asia and the Indus Valley. In spite of their warlike past, the Sassanians favored commerce, which led them to remove most of the trade barriers between Central Asia and the Middle East. Down to its demise in 651 at the hands of the invading Arab armies, the government of this new Persian empire extracted taxes from local rulers across Central Asia—partly in the form of fabrics—but otherwise left the region alone. And while it minted money at all the major cities of the empire, including Balkh, Samarkand, Merv, and Bukhara, local rulers in most parts of Central Asia did not hesitate to issue their own coinage as well.

HORSES, STIRRUPS, AND NOMADS

The traditional empires that made their presence felt in Central Asia were impeded by distance and shortages of staff from taking an active role in local life. At their best, they protected the currency and removed local barriers to trade. Otherwise, they were content with the role of imperial rent-seekers. But they were by no means the only players on that field. Throughout the entire period from the second millennium BC down to AD 1500, Central Asia was repeatedly overwhelmed by groups of mounted nomads, who filtered, and then stormed, into the region from the area north of the Black Sea, the Altai Mountains in what is now Kazakhstan and Russia, from East Turkestan, and from northern China and Mongolia. No less than the populations of the great cities, these nomads shaped the character of Central Asian history.

The extent and intensity of nomadism in the urban world of Central Asia requires some explanation.[103] Geography once more created the essential condition. Beyond the perimeters of the irrigated oases,

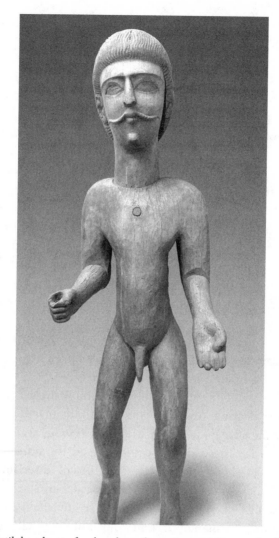

Figure 2.6. Until the advent of archaeology, the ancient steppe nomads of Central Asia were known mainly through the writings of urban dwellers. This early wood sculpture by a nomad artist shows how they viewed themselves. ▪ Courtesy of the National Museum under the Institute of History, Archaeology, and Ethnography of the Academy of Sciences, Tajikistan.

most of the region consisted of either deserts, grassy steppes, or some combination of the two. Even in the great mountain zones of the Hindu Kush, Pamirs, and the Tian Shan, those valleys that were not irrigated continued as grasslands or desert. Even in the populous Ferghana Valley, nomads continued to roam throughout the unirrigated expanses between the many cities.

Had it not been for the horse and stirrup, these areas would have remained a domain of tranquil shepherds and their flocks. But this changed dramatically when people began riding horses sometime after 1000 BC. Everything connected with the early history of the horse seems wrapped in controversy. One recent researcher pinpoints the earliest domesticated horses to a site near Astana, the capital of modern Kazakhstan,[104] fixing the date at 3500 BC. Other researchers propose a date more than a thousand years earlier.[105] But most agree that this epochal event occurred on the steppes of Central Asia.

For several thousand years Central Asians used horses as draft animals and to pull war chariots. Not until some time after 1000 BC did people start to ride them.[106] When eventually riders mounted their horses, they brought about a revolution in speed comparable to the invention of railroad steam engines in the nineteenth century. Many heretofore sedentary people of the steppes now turned to nomadism. But the really decisive change occurred with the invention of the hard saddle and the stirrup, probably as late as 300 BC, again, somewhere in Central Asia. These simple devices freed the horseman's hands, enabled him to shoot a bow and arrow in any direction, including backwards, and made it possible for the rider to stay in the saddle all day. Instantly nomads gained a decisive advantage of speed, mobile firepower, and maneuverability, not only over city dwellers but over imperial armies, with their old-fashioned war chariots and foot soldiers.[107]

These were the most epochal military innovations until the fifteenth century, when Europeans discovered how to use gunpowder effectively. Suddenly Turkic or Mongol-Turkic tribes from the steppes became the most potent military forces on earth, with the ability to dominate the empty territories between cities, to devastate irrigated farmland, and to lay siege to the cities themselves. This continued down to the sixteenth century. Central Asian nomads also became the world's premier horse breeders, which created a huge new industry exporting horses to China, India, and the Middle East.

It is not necessary here to recount all the tribal groupings and federations whose mounted warriors threatened Central Asia's urban civilization in the centuries prior to the Arab invasion.[108] One of them, the Huns, overran the Greek city of Ai Khanoum in northern Afghanistan, one of many cities in the region to be destroyed by nomadic invaders. Kushan rulers had built a huge turreted wall across the Hissar region of what is now Tajikistan to defend against nomads.[109] Even in the relatively peaceful fifth century AD, Merv's ruler felt it necessary to construct a defensive wall extending toward the Caspian Sea to keep the Huns at bay.[110] Rulers in Balkh, Samarkand, and other cities did the same, throwing up an outer ring wall to protect the entire oases; Samarkand's ran for nearly forty miles, but oasis walls of sixty miles or more were common.[111] The high and crenellated ramparts that surrounded every city and town were directed less against threatening empires than against mounted nomadic warriors.

The effectiveness of the nomadic warriors stemmed also from their characteristic form of organization, which centered around the tribal lord and his loyal band of supporters. These hearty fighters derived their power not from their genealogy but from their absolute loyalty to the ruler. This arrangement, called a *comitatus*, would eventually be transferred from the nomad conquerors to the Islamic states they ruled, but this was much later. The weakness of this system was that the post of supreme leader continued to be influenced by bloodlines, so that each time a leader died, his male relatives fought over the succession, which they usually resolved by dividing the patrimony. Time and time again this process of tribal fragmentation was to prove fatal to the empires formed by nomadic conquerors.[112]

Many formerly nomadic groups settled down and became part of the urban or agricultural life of the region, Kushans and Parthians being only the most conspicuous participants in this process. By the fifth century AD a few formerly nomadic Turks in the Ferghana Valley began to take up farming and a settled life. During the sixth century two groups of a single Turkic people briefly established their control over the entire territory from the Black Sea to Korea. Thereafter, the ruler or *kagan* of the group that was in Central Asia became an active participant in political and urban life. Indeed, in the murals on the palace walls at Samarkand, the Turkic *kagan* is depicted as a friendly and respected overlord. Perhaps too friendly, for by the seventh century the new Tang dynasty in China asserted its suzerainty once more over Central Asia, against the claims of both Sassanians and Turks.

In spite of these exceptions, most nomads continued to dwell in their yurts and tents on the open steppes, as did their heirs down to modern times. To the end of their days, both Chinggis Khan and Tamerlane (Timur) preferred tents to palaces. But they needed urban products like pots and cotton fabrics, just as townspeople wanted access to the fresh meat, carpets, and saddles that only the nomads could provide. And so there arose a very practical modus vivendi, a mutual dependence between steppe and city, herder and farmer, Turk and Persian. Over time this transcultural exchange transformed both groups, drawing them together in a thoroughgoing symbiosis.[113]

But this process did not gain real momentum until several hundred years after the Arab invasion. Until then, the nomads for the most part kept to themselves on the steppe and confined their interaction with cities to the collection of tribute and visits to urban markets. In return for tribute, the nomads could promise that tranquility would reign in the vast interurban spaces and that caravans would not be molested.[114] Technically a form of rent seeking, this arrangement could also be considered an outright bribe in the name of stability and peace.

And it worked. Unlike ancient Rome, which eventually collapsed under repeated blows from the Gauls, Franks, Huns, and Goths, Central Asian cities usually managed to work out a modus vivendi with the invading nomads that enabled each to survive and all to coexist. Perhaps because of their deep involvement with trade by land and with the constant negotiations and deals this involved, the burgeoning cities of the East had greater absorbent powers than those of the West. The comfortable interaction between urban dwellers and nomads that would have been evident at any bazaar in the region reflects the kind of mutual accommodation that enabled Central Asian cities to thrive down to the arrival of the Arabs and beyond.

The Urban Culture

In ancient Rome, as in Central Asia, it fell to the poets and historians to give voice to the lives and aspirations of the public at large. Such elegant Latin poets as Catullus, Virgil, Horace, and Ovid were all men of the city, steeped in the bustling life of the imperial capital. But all of them dreamed bucolic dreams and fled whenever possible to their country houses, whence they surveyed the world in rural tranquility.[115] In early medieval times rural monasteries in the West offered the same escape from chaos. However, such bucolic idylls were rare or nonexistent among the poets and intellectuals of Central Asia. The stark contrast between the hospitable world of the irrigated oases and the inhospitable environment of the surrounding desert or steppe caused them to focus all attention on the quotidian life of their cities. To a greater extent than even Rome, and far more than early medieval Europe, Central Asia's was an urban culture.

Even though they told and retold the traditional epics common to all those peoples who spoke languages of the Iranian group, writers of the region rooted their identities squarely in the contemporary city. This is quite natural since Central Asian cities were more numerous and much larger than most of those to the west. As the British historian Peter Brown wrote, "The towns of the Mediterranean were small, fragile excrescences in a spreading countryside."[116] By contrast, even before the Arab conquest, writers in Central Asia had begun to pen the histories of their cities and to sing their praises in verse.[117] After the conquest, they did so as a means of asserting their own identities as opposed to the one the Arabs sought to impose on them. The poets directed no corresponding efforts to their khanates, princedoms, or empires—unless someone paid them to do so.

Theirs was a *patriotisme de clocher*, or "bell-tower patriotism," that treasured the local and the specific. This is not in itself exceptional, but in the case of Central Asia, this urban nationalism involved a blend of intense localism with thoroughgoing cosmopolitanism. In this respect, the great cities of the region anticipate medieval Venice far more than they do such later inner Asian centers as Isfahan. They did not use the term "self-government," but thanks to the strong rule of local monarchs, the power of *dihkans* or noblemen, and the wealth of the burgeoning commercial classes, they would have been justified to do so. One of the most dreaded punishments in pre-Islamic Samarkand was to be excluded from the city.[118]

Vasilii Bartold, the great Russian historian of the region, went so far as to assert that the powers of local lords created a kind of liberty within the top-down and despotic system of the khans and local rulers.[119] This certainly reinforced the intensity of urban life and consciousness, but at the price of weakening regional feeling. Only in the northern territory of Khwarazm, perhaps, did the flame of regionalism burn, but even there it was focused on Khwarazm itself and not on Central Asia as a whole. For the most part, Central Asians were too divided among themselves to resist successfully the Arabs' assaults. But they were to prove very effective at maximizing their specific interests once the fighting was over.

Beyond this, the Central Asians who greeted the Arabs' invasion knew that theirs was an ancient land. Ancient ruins were everywhere, a reminder that empires had risen and fallen there. The process was very much under way in the seventh century, when Balkh, Tirmidh, and Bukhara all seemed to be in eclipse, but when other urban centers like Tus and Merv in Khurasan were thriving. This feeling that their own past was important and contained lessons for the present doubtless lay behind the decision of the eleventh-century poet Ferdowsi to write his regional epic, the *Shahnameh*, and, slightly later, to Biruni's decision to devote a long book, *Chronology of Ancient Nations*, to the region's deep past and to examine that past for insights on how to solve pressing issues of the present.

Central Asian urban dwellers could not have welcomed the fact that external powers coveted their region, or that they were locked in a permanent embrace with the nomads who inhabited the steppe and desert land between the cities. Yet they had mastered the art of deriving benefit from both relationships. Together, foreign suzerains and regional nomads reduced the need for Central Asian cities to mount large armies of their own. Since the suzerains were glad to accept payment in lieu of service, and since the nomads were pleased to hire themselves out, the urban folk were left to do what they did best: manufacture, trade, make money, and, it turns out, think creatively.

Being at the center of Eurasia and surrounded by the civilizations of India, China, and the Middle East, Central Asians were constantly confronted with new ways of doing things, and new ideas. Over the centuries they became adept at finding what was useful in whatever showed up on their doorstep and what was not. With strong skills and an equally strong sense of themselves, they learned how to *adapt* rather than *adopt* what they learned from abroad. This was especially important in the realm of ideas, to which we now turn.

CHAPTER 3

❀

A Cauldron of Skills, Ideas, and Faiths

LITERACY AND NUMERACY

A Chinese visitor to Samarkand in the century before the Arab invasion wrote in his notes the following observation on young people there: "All the inhabitants [of Samarkand] are brought up to be traders. When a young boy reaches the age of five they begin to teach him to read, and when he is able to read they make him study business."[1] Another Chinese visitor, equally astonished, observed that young Central Asian men were not allowed to participate in trading trips abroad until they were twenty, prior to which time they were expected to be absorbed in study and training.[2]

These observant contemporaries enable us to understand something very important about the lost world of Central Asia before the Arab conquest: the high level of literacy that prevailed there. The mass destruction of books and documents carried out by the Arabs leaves us particularly dependent on the reports of outsiders like these. As it happens, archaeology confirms what the two Chinese wrote. The Hungarian British explorer Aurel Stein, prowling in the ruins of a Chinese watchtower in a very dry region of Gansu province a century ago, came across a remarkably well-preserved horde of ancient documents. Among them was a personal letter written around 313 BC by a very angry young Central Asian woman in Dunhuang to her husband in Samarkand. She accused him of stranding her and her mother for three years in the middle of nowhere and then prescribing her every movement: "[Even] in my paternal abode I was not subjected to the kind of restrictions you impose. I obeyed your command and went to Dunhuang, ignoring my mother and my brothers. Surely, the gods were angry with me on that day . . . ! I would rather be married to a dog or a pig than to you!"[3] This stormy blast came not from a government official or businessman in the course of his duties but from an ordinary young woman who, reduced to being a house servant, used her literacy most effectively to lambast her negligent husband. Such documents, and there are many of them, led a Tajik Soviet scholar to conclude that literacy was more widespread in pre-Islamic Central Asia than under later rule.[4]

The people of Sogdiana, of which Samarkand was the capital, were also highly numerate, and for a very practical reason: they had to be. Another letter from 313 BC, this one written by a Samarkand merchant in Xinjiang to his partner back home, requested crisply that "You should remind Varzakk that he will withdraw this deposit; count it, and if he is to hold it, then you should add interest to the capital and put in a transfer document."[5] Similar letters were doubtless being written daily by merchants from Bactria, Khwarazm, Margiana, and all the other regions of Central Asia. How could one function as a trader without the ability to read a bill of lading sent by the shipper, a letter of credit, and the accompanying documents? How could one use the weight of one pack to calculate the weight of a few hundred of them without the ability to multiply? And if products were to be divided by a certain ratio, one had to be able to compute the shares.

Above all, one had to understand how to make and enforce contracts, attract investors, shift resources from one currency to another, and execute complex financial transactions over thousands of miles. These skills were to become common in Europe, but only after more than half a millennium had passed. In China, which left the conduct of distance trading mainly to foreigners, they also developed slowly, while the natives of Central America, for all their skills, failed to develop these capacities in a way that stimulated innovation and change.[6] That they should have been honed to such a fine degree in Central Asia is not surprising, for this was a competitive world in which even a slight advantage or error meant profits or loss.

By the nature of their business, merchants and traders, whether consciously or not, became transmitters of knowledge as well as of goods, bringing to bear their pragmatic eye on every question before them. This practical turn of mind created among the merchants, and even in the culture as a whole, what the Harvard historian Richard N. Frye called a "mercantile secularism."[7] Centuries later, when Muslims were having to choose among no fewer than four competing systems of Islamic law, Central Asians overwhelmingly opted for the Hanafi school, which took the most practical approach to issues of daily life and was the most accepting of existing norms for trade. Indeed, no part of the Muslim world more ardently favored the Hanafi approach nor does so today.

Demand for literacy and numeracy went far beyond the world of trade. Written laws regulated social and economic life in most regions of Central Asia.[8] Chinese sources report that the laws of pre-Muslim Sogdiana were written down and kept in one of the temples.[9] Matters relating to marriage, divorce, property, and taxation were all regulated by written norms. As complex societies, the states of Central Asia required sophisticated laws. An economically crucial function like irrigation would have been defined in terms of property rights, eminent domain, compensation to landowners, and so forth. The fact that polygamy was widely practiced long before Islam no doubt complicated inheritance law, which was also written down.[10] A small army of officials kept careful records of every legal transaction. When necessary, they corresponded with each other much the way officials do today, and also with members of the public at large.

How do we know such details about a world that has vanished? Because in the year 1933 a shepherd on a mountaintop in southern Tajikistan spotted in the dust what appeared to be the lid of a pot. It covered the mouth of a large pottery vessel that had been buried one and a half millenniums earlier by the ruler of Panjikent, named Dewashtich, who was fleeing before the approaching Arab cavalry. The pot on Mount Mug contained not gold and silver but scores of official records written on parchment. Carefully sealed with waxes and resins, it had preserved its contents down to that day in 1933.[11]

Reading these ancient disputes over laws and regulations, one quickly understands that Central Asian civilization before the Arab invasion placed a high value on technical competence and knowledge as such. This is quite understandable since the very survival of the oasis cities depended on it. Trade, manufacturing, construction, and urban governance all possessed their own body of technical knowledge. The irrigation systems, for example, required some rational means of calculating the width and depth of irrigation canals, the diameter of underground channels (*kerezes*), and the size of exit sluices so that they could handle the volume of water that needed to pass through them.

Respect for knowledge and technical proficiency was natural in a society that required hundreds of tons of water to be raised daily to water fields, serve domestic needs, or supply public baths. For this task alone Central Asians employed nine different types of machinery, including windmills, which they either invented on their own or borrowed from others.[12] A Soviet engineer in the 1920s calculated that the ancient Central Asian waterwheel (*chigir* or *charkh*), of which there were once thousands, achieved the same level of irrigation as a gravity-based system but required 30 to 50 percent less water.[13] And by concentrating irrigation on raised beds rather than distributing the water evenly and inefficiently, these ancient systems prevented the salinization that destroyed much Central Asia cropland during the Soviet era. The technologists responsible for such devices were inevitably respected members of society.

It might be objected that other peoples, whether in the Middle East, Asia, Mexico, or South America, were also constructing elaborate irrigation systems during these years. It cannot be doubted that the Central Americans, for example, also maintained large forces of specialized laborers to manage their water systems. But in neither Central America, Asia, nor the Middle East did this activity give rise to anything like a systematic field of hydraulic engineering, let alone one that could be sustained over many centuries.

The maintenance of the region's status as a source of high-value exports required an array of further skills. Central Asia's ironmongers, coppersmiths, and bronze factors all gained distinction across Eurasia for the quality of the tools, utensils, and armaments they produced. Each generation had to master the skills involved and pass them down to successor generations by means of effective pedagogy. This was easier when a specific subgroup or ethnicity monopolized a specific craft. Glassmaking, for example, which was established as a major regional export industry by the fourth century AD, with furnaces in Afrasiab (Samarkand), Chach (Tashkent), and the Ferghana Valley, was almost exclusively a Jewish industry.[14] But even such ethnic monopolies, which were exceptional, contributed to the general interest in and respect for specialized technical skills. Each of these competencies rested on a solid knowledge base in metallurgy, mining, geology, chemistry, or other fields. This is not to say that the region was teeming with metallurgists, geometricians, or astronomers. But the realities of the economy required high levels of competence and knowledge in all these areas.

Shifting into more philosophical language, one might say that Central Asia was a land of Aristotelians without Aristotle, a place where many people were concerned with what the Greek thinkers called tekhne (τεχνη), or "the way things are made or the manner in which a goal is attained." Today it has become fashionable to be dismissive of this quality, on the grounds that it has become the *only* concern of many modern men and women. But as we shall see shortly, this was scarcely the case in Central Asia fifteen hundred years ago.

SCIENCE

No sharp line separates the kinds of technical expertise that were so widespread in Central Asia from what in the modern world is defined as "science." With practitioners, the missing element is usually the urge to generalize and, in some cases, to speculate on the causes of the observed phenomenon. The shift from practice to theory can be seen in surveying.

Irrigation made arable land a valuable and scarce commodity, to be parceled out with precision. In the course of centuries of work on the land, surveyor-practitioners in the various oases had worked out methods of measuring regular and irregular fields, calculating total areas, and other functions. It was a natural step for Central Asians to compile the known techniques of geometry, work out the first system of practical algebra, and create the field of trigonometry. The major step in this direction occurred in the ninth century, on the basis of the accumulated experience of Khwarazm, the most intensely irrigated region of all. The ease with which later mathematicians from this region and elsewhere in Central Asia were able to set down principles for solving second- and third-degree algebraic equations owed much to the quantitative facility they had amassed to meet the needs of practical field work.

In much the same way, those caravan traders who navigated by night accumulated a large body of practical knowledge on the seasonal circuits of the sun, moon, and stars. This knowledge brought them to the threshold of astronomy, which seeks to measure and understand those movements. It also inspired them to explore the arcana of astrology, with its focus on the links between observed phenomena in the heavens and human affairs. Easily dismissed as quackery, astrology demanded the preparation of precise tables of planetary movements, called ephemerides, and the ability to predict precisely both solar and lunar eclipses.

Unfortunately, the thorough destruction of documents from the pre-Islamic era in Central Asia leaves us with mere fragments of information on astronomy and other sciences before the ninth century. But these fragments are intriguing. Was a local scientist responsible for the sophisticated and precisely calibrated hemispherical sundial, hewn out of limestone, that was installed more than two thousand years ago at a prominent place at the Greek city of Ai Khanoum in Afghanistan?[15] What about the delegation of AD 719 that is known to have gone from the Balkh area to China and was headed by a man named Modjo, who prided himself on his mastery of many fields, including astronomy?[16] Were there writings on astronomy or other sciences among the forty books written by Sogdians and translated into Chinese?[17] What can be learned from the seventh-century astrological documents that were carefully folded into the horde from Mount Mug? Or from the astronomical tables prepared under the Sassanians, which made their way to the empire's eastern capital at Merv, where a school of astronomy was forming?[18] Or, finally, what do the highly sophisticated calendar systems that were in use in Khwarazm, Bactria, Parthia, Tokharistan, and Sogdiana tell us about the state of Central Asian science in the pre-Islamic centuries? It is surely worth noting that Biruni's research on calendar systems, which he undertook in the early years of the eleventh century, took as its point of departure the Khwarazian calendar.[19] Biruni concluded that this Central Asian system, based on thirty-day months, was equal or superior to the much better known Babylonian, Hebraic, and Indian calendars.[20]

No less important than the calendars themselves was the drive to systematize, generalize, and achieve ever greater accuracy in all areas of astronomy. This was to become a grand quest in Central Asia beginning in the tenth century, when local scientists first confronted the principal works of Greek science. The passion with which generations of later Central Asian astronomers pursued these goals owed much to the solid foundation that had been laid down in pre-Islamic times. Even the speed with which they took up such new challenges as spherical geometry traces to the mastery of practical geometry that had long been part of the regional culture.

Similar processes were under way in the field of medicine. In all ancient societies medicine was close to power, since doctors could ward off dangers to the ruler's health and that of his family. One can imagine the gratitude felt by the Indian ruler Ashoka when a physician from Bactria cured his son of blindness by means of cranial surgery.[21] Besides such occasional miracles, the day-to-day practice of medicine in pre-Islamic Central Asia meant calling on a large body of received wisdom and being willing to improvise when necessary. A fifth-century apothecary's shop that was excavated at the ancient commercial center of Paykand, near Bukhara, was fully equipped for the preparation of remedies or for the development of new medications.[22] It is clear from its location in the heart of the bustling trading center that the clientele of this pharmacist's shop included the public at large. Equipment similar to that of the Paykand pharmacists has been found at other sites from Khurasan to Khwarazm in the North, suggesting that the medical profession was large and widespread.

Practitioners could pore over both translated and locally produced tomes on medicine, fragments of which have been found in Xinjiang.[23] Pre-Islamic doctors in Central Asia were well aware of Western practice and also of the Vedic medicine of India. Twenty-seven medical texts in Sanskrit turned up in the dry sands that cover the ruins of ancient Kucha in Xinjiang.[24] Balkh, with its intimate trade links with India, was especially well placed to gain a knowledge of Indian medicine. Over many years the richest family in Islamic Baghdad was the Barmak clan, converts from Buddhism with deep historical roots in Balkh. When they donated a new hospital to the capital of the caliphate, they brought in an Indian physician, probably from Balkh, to head it.[25]

WRITTEN WORDS, LOTS OF THEM

Intending high praise, many outsiders rhapsodize about Central Asia as having once been a "crossroads of civilizations." In one sense this is not only true but more valid for Central Asia than for any other region on earth. It is the point of juncture between Middle Eastern, European, Chinese, and Indian cultures and hence, in the phrase of the American Owen Lattimore, the "pivot of Asia."[26] But strictly speaking, a crossroads is simply the abstract point between four real places, with no identity of its own. This is emphatically not the case for early Central Asia. While it was assuredly a "crossroads *of* civilizations," it was, even more, a *crossroads civilization*, with its own distinctive features as such. From the earliest days this was evident in many areas, but in none more than in language and religion.

The region's core linguistic stock was Iranian, which included both the ancient language of the Zoroastrian scriptures and separate languages like Bactrian, Sogdian, and Khwarazmian. Until the arrival of Greeks at or before the time of Alexander the Great (356–323 BC), none of these language groups had developed its own alphabet or script. Thereafter the Seleucid kingdom of Bactria, founded by Alexander, adapted Greek script to the local Bactrian tongue spoken at Balkh.[27] The Sogdians at Afrasiab (Samarkand) and Khwarazmians in the many towns that bordered the northern steppes followed a similar process, but instead of Greek letters they adapted the script used for Aramaic, the chief language of the Jewish Talmud and the language spoken by Christ. The use of the Aramaic alphabet in Central Asia is not as surprising as it may at first seem; Syriac, a form of Aramaic, was widely known across the region because it was the language of the thousands of Syrian merchants, settlers, and, later, Christian missionaries who came there.

Early Turkic peoples, too, had contrived a quite sophisticated runic script, which turns up on bowls and other household articles, but which was also used to present beautiful poetry in the Turk and Uyghur languages. The nomadic peoples participated fully in this process of assimilating foreign alphabets. In recent years fragmentary writings in Scythian and Turkic have turned up, carefully transliterated into Aramaic. Still other scripts remain to be deciphered.[28] It is impossible to determine levels of literacy among the early Turkic nomads, but the evidence that is steadily coming to light suggests that it was high and must certainly be duly acknowledged in any evaluation of Turkic culture during the centuries before the advent of Islam.

That indigenous languages could adapt foreign scripts and then flourish rather than be overwhelmed by the languages of vigorous outside powers attests to a fundamental characteristic of Central Asian civilization in both its Persianate and its Turkic forms. It offers further evidence that Central Asia was less a crossroads of cultures than a cross-roads culture, influenced by all its international contacts yet in the end defined still more by the indigenous strengths it had built up through the centuries.

It is impossible to estimate the number and distribution of books in Central Asia prior to the Arab invasion. Suffice it to note that the world did not even know of the existence of written Sogdian, Bactrian, and Khwarazmian until as recently as the 1950s, when Soviet archaeologists and Soviet and Western linguists began to plumb the problem. Now, however, the number of fragments of writings multiplies every decade. One recent find was of a line of Sogdian verse. Inscribed on the lip of a clay vessel, it struck its Russian translator, Vladimir Lifshits, as a strik-ing anticipation of verses by the eleventh-century poet Omar Khayyam: "The one who fails to discern damage will also never see wealth. Then drink, oh Man!"[29]

There is every reason to think that books were numerous and wide-spread throughout pre-Islamic Central Asia. A century ago Aurel Stein came across fifteen thousand volumes in the dry caves of East Turkestan, where they had been left by settled Turkic Uyghurs and other groups in the area. Written in Sanskrit, Hebrew, Persian, Syriac (Aramaic), and Sog-dian, they included both translations and original works. Most of these were from the ninth and tenth centuries, but others dated to the arrival in the region of Buddhism, Manichaeism, and Christianity. Valerie Han-sen, in her recent study, *The Silk Road*, details the story of how Buddhist migrants from Gandhara in Afghanistan and Pakistan brought writing to the long-forgotten realm of Kroraina in what is now Xinjiang by the third century AD.[30] Long after the invading Arabs began imposing their lan-guage and values on the area west of the Tian Shan, East Turkestan con-tinued to maintain a lively and highly pluralistic intellectual life and to be a center of reading and writing. However, we know the great commercial centers of the Central Asian heartland were larger and more prosperous than these East Turkestan cities. So it is reasonable to hypothesize that in pre-Islamic times, the major Central Asian cities boasted even more books than did the cities of East Turkestan. Beyond doubt, theirs was a society that valued the production and exchange of knowledge.

not merely a crossroads for the cultures of others. But this is not to say that the region was impervious to religious ideas from abroad. Quite the contrary, three of the great religious and philosophical traditions all took root in Central Asia and flourished there: Indian Buddhism; worship of the Greek pantheon; and Christianity from the lands of the eastern Mediterranean.

Only one of these—the Greek pantheon—entered Central Asia as a result of military conquest. All three, including worship of the Greek gods, were borne on the shoulders of traders and caravan merchants, who often doubled as missionaries.[47] Equally important—and generally neglected in accounts of the religious life of the region—is the fact that all three of these systems were promoted by colonies of émigrés, whether from India, Greece, or Syria, settlers who had moved to Central Asia from elsewhere and brought their religious beliefs with them.

Far from feeling threatened by what were at first alien ideas, large numbers of Central Asians were attracted to these imported faiths. In the remarkable era that preceded the Arab invasion, all three of these faiths, along with Manicheanism, Zoroastrianism, and Tengrianism, found active adherents in Central Asia. Did any other region on earth experience greater religious diversity? But this pluralism did not result in a series of religious ghettos. Even within a single ethnic or territorial group, one could find practitioners of diverse religions. The Sodgians, for example, included in their number Manicheans, Christians, Buddhists, and worshippers of traditional local divinities.[48] In most case these diverse religious systems and cults seem to have coexisted without conflict.

More positively, the various religions borrowed freely from one another in what can fairly be described as an ecumenical spirit. In the process, they created many instances of religious symbiosis. This probably occurred because many sensed that all the faiths were engaged in the same quest and had concluded that each belief system offered serious answers to questions about what shapes the fate of humans: the nature of good and evil; the definition of the good life; and the origin and destiny of the universe. The process of comparing and evaluating the answers from the various faiths imparted to Central Asian intellectual life in the pre-Islamic era a depth, seriousness, and ecumenical openness that was later to find expression in the Age of Enlightenment.

Greeks, Bearing Pantheons

Students everywhere learn about Alexander the Great's invasion of Persia and his army's advance through Central Asia clear to the banks of the Indus. They may remember, too, that he left behind nine colonial cities in Afghanistan and Central Asia.[49] Specialists see this as the starting point of Greek influence in Central Asia. But that point must now be dated 150 years earlier, for Herodotus, in his *Histories*, records that the Persian king Darius (550–486 BC) forcefully resettled in Bactria the entire population of a coastal town in Libya. The people of Barca, as the town was known, had murdered their ruler because he had supported the Persian general, Cambysus. For this crime Darius destroyed the city and moved the entire population to what is now Afghanistan or an adjacent area of Turkmenistan.[50] Here they doubtless established the kind of colony that Barca itself had been, an outpost of Hellenism in a new land.

Another such colony was founded by priests from the temple of Apollo at Miletus, the great Greek center on what is now the Mediterranean coast of Turkey. Several ancient writers report that these pioneers built a new city, which is variously thought to have been at Kalif in Turkmenistan or in the Kashka Darya area of southern Uzbekistan. Many Greeks migrated there, and when Alexander's forces came through a century and a half later, the inhabitants were still speaking a curious blend of Greek and Bactrian.[51]

By the time Alexander finally departed Central Asia in 326 BC, he and his army of 60,000 to 100,000 men had been fighting there for three years, crisscrossing large parts of Afghanistan, Uzbekistan, Turkmenistan, and Tajikistan. Unlike the Persians, Alexander was interested in much more than extracting tribute from the populace. Before leaving, he planted detachments of Greek soldiers in at least seven cities, created an eighth ex nihilo at Ai Khanoum, and established a permanent capital of Greek Bactria at Balkh. Alexander modestly renamed all these cities after himself.

How "Greek" was Greek Central Asia? Since he began his Asian campaign with barely fifteen thousand Macedonian and Greek cavalry, Alexander could not have left contingents of more than fifteen hundred Macedonians at each of his nine towns.[52] This assumes that all fifteen thousand survived the preceding campaigns against Persia, but we know that many perished before arriving in Afghanistan. This means that over the century and a half that the Greek colonies existed in Central Asia, new immigrants must have substantially enhanced the Greek contingent. For now, we know next to nothing about this important human flow from West to East. Archaeological remains indicate that among their number were priests from the various cults, as well as artists, sculptors, architects, linguists, craftspeople, and thinkers.

Shrines dedicated to the Greek gods soon began appearing from western Central Asia to India. Ionic columns were added to the portico of a temple in Tajikistan,[53] and a notable Ionic temple was constructed at Takshashila, now Taxila in Pakistan, near the Pakistan-Afghanistan border.[54] Greek sculpture also took root in the region, with artists at Taxila and other sites across the region achieving a stunning synthesis of Greek and Indian styles that ranks as one of the world's great achievements in the plastic arts. At one site in central Afghanistan archaeologists found sculptures of Heracles, Aphrodite, Tyche, and other deities.[55] At another, in southern Uzbekistan, they discovered statues dedicated to Greek river gods, which had become linked with the nearby Amu Darya.[56] Heracles appears to have been especially popular across Greek Central Asia. Also, a sculptural representation of the Roman wolf suckling Romulus and Remus turned up at Panjikent in Tajikistan, along with Buddhist, Greek, and Zoroastrian deities.[57] Even Greek erotic art found a receptive audience in the Eurasian heartland.[58]

And the Hellenic influence persisted: Parthians at their first capital at Nisa, in what is now Turkmenistan, continued into the first century AD to issue coins bearing Greek inscriptions. Ample evidence supports the conclusion of Uzbekistan's Edvard Rtveladze that "for at least 200 years the Greek language in Central Asia played the same role as a lingua franca as Latin did in medieval Europe."[59] And in philosophy the Hellenistic Greek movement of Hermeticism, which explored the magical and occult properties of objects and would later greatly influence both Arabs and Europeans, found its most ardent champions in Merv, where it thrived as a branch of natural history down to the Arab invasion.[60]

But there were limits to this widespread Hellenism, as was clearly evident at the one city in the region that the Greeks built ex nihilo: Ai Khanoum in Afghanistan (see fig. 3.2).[61] Here, twenty-five hundred miles

Figure 3.1. In the cities of Greek Bactria, artists and thinkers from the Mediterranean world rubbed shoulders with their local counterparts and immigrant talents from India. Was this horseman the work of a Greek immigrant, a local Central Asian sculptor, or an artisan from India? ■ From Fredrik Hiebert and Pierre Cambon, eds., *Afghanistan: Hidden Treasures from the National Museum, Kabul* (Washington DC: National Geographic Society, 2008), 205. Courtesy of the National Museum of Afghanistan.

from Athens, was a true Greek polis, complete with a market place or agora, a theater with seating for six thousand spectators, a temple to Zeus, and a gymnasium that was one of the most commodious in the ancient world.[62] The palace's vast forecourt, big enough to hold two football fields, was flanked by 108 columns with pure Corinthian capitals, creating a colonnade that led the visitor to the main receiving rooms and to the royal residence, which featured ninety more columns, this time in the Doric order.[63]

But if the temples show powerful Greek influence on their design, their overall plan was more Persian than Greek, while the gods to which they were dedicated, with the exception of Zeus, were Indian or Eastern. This was the case everywhere in Central Asia, where Greek forms combined with Greco-Indian-Central Asian content in a syncretism among seeming opposites.

Figure 3.2. Virtual reconstruction of the classical Greek theater and palace area at Ai Khanoum, founded by Alexander the Great in 327 BC. With seating for up to six thousand spectators, the theater was a center of Greek culture in the East. In the background is the Amu Darya, which now forms the border between Afghanistan and Tajikistan. ▪ Restored view of the city (CG) with in the foreground the theater and in the background the palace courtyard. Between both *propylaea* giving access to the palace area. © NKH-TAISEI, image O. Ishizawa - G. Lecuyot.

Figure 3.3. Capital of a Corinthian Greek column from the Bactrian palace at Ai Khanoum, northern Afghanistan. ▪ From Fredrik Hiebert and Pierre Cambon, eds., *Afghanistan: Hidden Treasures from the National Museum, Kabul* (Washington DC: National Geographic Society, 2008), 120. Courtesy of the National Museum of Afghanistan.

This is not to say that the values of Hellenism did not resonate in Central Asia. It is tempting, and probably partly justified, to imagine the audience in the theater at Ai Khanoum ruminating over Euripides's (now lost) tragedy *Heracles*, a god popular locally, or slapping their thighs and guffawing at Aristophanes's comedy *The Babylonians* (also lost), which depicted Greeks as slaves under Persian rule. But we can be sure they thrilled equally, or even more, to the antics of gymnasts, conjurers, wrestlers, and other purveyors of low-brow entertainment.

Directly affirmative of the presence of Greek thought in ancient Central Asia is the fragment of a philosophical dialogue that was found embedded in a brick wall at Ai Khanoum. Experts debate whether this remarkable document, which analyzes the connection between the realm of ideas and material reality, is from a lost work of Aristotle or was written by a local philosopher at Ai Khanoum itself.[64] And then there is the inscription discovered in the ruins of the main square by members of the Délégation archéologique Française en Afghanistan, which excavated at Ai Khanoum from 1964 to 1978. On the stone base for a large but long-vanished statue dedicated to the city's founder were engraved the following maxims:

> As a child be well-behaved (Παις ων κοσμιος ισθι)
> As a youth be self-disciplined (ηβων εγκρατης)
> In middle age, be just (μεσος δικαιος)
> In old age, be sensible (πρεσβυτης ευλογος)
> On reaching the end, be without sorrow (τελευτων αλυπος)[65]

These words, exuding the serene spirit of Hellenism, were accompanied by a further inscription indicating that they had been placed there by a Greek immigrant named Clearchos. In fact, Clearchos had borrowed them directly from the wall of one of the shrines at Delphi. Did Clearchos, while still in Greece, know that he would soon be relocating to Afghanistan and therefore visit Greece's most holy shrine before departing? As he made the arduous journey across Eurasia, the text was in his baggage, to be offered as a philosophical gift from West to East.

BUDDHIST MONASTICISM

Except when we are reminded of the enormous statues of Buddha in Bamiyan, Afghanistan, that were destroyed by the Taliban, it is hard to imagine that Central Asia was for nearly a millennium as deeply Buddhist as it is Muslim today. But from the first or second century BC to the Arab conquest, this was the case. As early as the mid-third century BC, the Indian emperor Ashoka erected large stone pillars or signboards in Kandahar, Afghanistan, as well as in other cities, on which he displayed edicts on how good behavior or the Buddhists' dharma could be spread. He counseled his subjects that "piety and self-control [exist] in all philosophical schools. But the most self-possessed are [those people] who are the masters of their tongues. They neither praise themselves nor belittle their fellows in any respect, which is a vain thing to do. . . . The correct thing is to respect one another and to accept the lessons of each other. Those who do this enlarge their knowledge by sharing what others know."[66] These uplifting words came from a ruler who candidly acknowledged on another engraved signboard that he had killed

100,000 men in war before converting to pacifism, vegetarianism, and Buddhism. Now he looked with great good will toward Afghanistan and Bactria. To reach the largest local public, he had his message inscribed in both Greek and Aramaic—an astonishing testimony to the importance of these languages in the southern reaches of Central Asia at the time.[67] It may have helped that a physician from Bactria, mentioned earlier, had cured Ashoka's son of blindness.[68] Whatever the cause, Ashoka's homily to Central Asians marked the start of a powerful wave of the new Buddhist faith that swept across the region beginning at least by the early second century BC.[69]

Buddhism transformed the religious topography of Central Asia.[70] Where once there were only boxy shrines to local deities, Zoroastrian fire temples with open courtyards, and temples to diverse deities in a Greco-Persian style, now the high-pointed peaks of Buddhist stupas punctuated the skyline. One, Takht-e Rustam at Haibek in central Afghanistan, was carved deep into the ground out of living rock. Again, syncretism and what a noted Russian archaeologist called "creative assimilation" were in the air.[71] Greek elements appear in many of these temples, and Ashoka's words were translated for the benefit of the local Greek-speaking population by someone who was obviously deeply familiar with Greek philosophical terminology.[72]

The idea of representing the Buddha visually came from India, but there he was depicted as a barely clad ascetic. The Central Asians transformed this image by clothing him in flowing robes with deep folds, in other words, the garb that Greek sculptors had been masterfully depicting for centuries. This cultural blending of East and West was a product of collaboration between Greek sculptors or Central Asian sculptors trained in the Greek school, probably themselves Buddhists, and Buddhist monks, both Indian and local.[73] It occurred all over the region but reached its highest flowering in the kingdom of Gandhara, which extended along the trade routes near what is now the border of Pakistan and Afghanistan north of Islamabad. It was this image of the Buddha, early examples of which existed at dozens of Central Asian stupas and on Central Asian coinage, that subsequently spread across China, Korea, and Japan. It still shapes our visual image of Buddha today.[74] Not only was it "made in Central Asia," but it could not have appeared anywhere else.

Figure 3.4. Terracotta Buddha from Hadda, Afghanistan. Long hidden from Taliban rule, it was rediscovered in the presidential palace in Kabul in 2004 by American archaeologist Fredrik Hiebert. ▪ Courtesy of Fredrik Hiebert.

No less altered was the religious geography of Central Asia. Far from being confined to north- and east-central Afghanistan, where Bamiyan, Mes Aynak, and other impressive Buddhist sites are located, or the great centers at Balkh or at Tirmidh (Termez) in Uzbekistan, the ruins of large Buddhist stupas can be found at Merv in Turkmenistan,[75] at Dalver-zin Tepe in Uzbekistan northeast of Tirmidh, and at many other sites

Figure 3.5. A statue of Buddha being excavated at Sahri Bahlol, Pakistan, in 1904. The combination of a Greek robe (*himation*) and Indian facial features henceforth defined Buddha's image across Asia.

throughout the region.[76] In northern Kyrgyzstan alone, archaeologists have found four monastic complexes and three temples,[77] while in Kazakhstan monasteries and stupas have been discovered along the entire 375-mile route from the Chinese border in the East to Kizilkent near Karaganda in the middle of the country.

Many of the Buddhist monastic centers and pilgrimage sites around Ghazni, Kabul, and in northern Afghanistan trace their existence to shifts of the major trade routes from Central Asia to India that took place in the century and a half before the Arab conquest. Earlier, traders had avoided crossing the Hindu Kush, but from the mid-sixth century they found passes through the mountains that took them to the heart of Central Asia.[78]

The numerous new types of structures featured statuary and ornamental plasterwork. At Airtam, near Tirmidh, on a low bluff now being eaten away by the fast-flowing Amu Darya, a stunningly beautiful frieze, currently at the Hermitage Museum in St. Petersburg, depicted smiling musicians with Persianate faces playing Indian musical instruments, alongside other musicians with Buddha-like faces playing Greek and Persian-type instruments (see fig. 1.1).[79] The forty-three-foot-long

recumbent statue of Buddha from Ajina-Tepe ("Witches Hill") in Tajikistan and now in the National Museum, Dushanbe; the Bamiyan Buddhas; and scores of other statues were all brightly painted, for Buddhism
was, and is, a religion of brilliant colors.

Buddhism received an enormous boost from the powerful second-
century AD Kushan ruler Kanishka I. Born a nomad, Kanishka adopted
both Buddhism and the worship of certain Greek deities, all the while
forging a core link with Zoroastrianism.[80] Kanishka's grand synthesis
can be seen everywhere at his summer capital at Begram in Afghanistan
and at the remarkable dynastic sanctuary complex he built—mainly in
honor of himself—on the hillside temple at Surkh Kotal, on the north
face of the Hindu Kush, also in Afghanistan.[81]

Buddhist monastic communities sprang up across the region. The
city of Bukhara even took its name from one such Buddhist monastery or *vihara*. One city, Kabul, was surrounded by a ring of some forty
Buddhist monasteries. A later Muslim poet rapturously described one
of these, the opulent Subahar monastery, effusing that it was "the spring
of beauty. The pavement was of onyx; the walls of marble; the gates of
sculpted gold; the floor of native silver; the stars were everywhere depicted; crenellations were ranged around the great arch [*iwan*]; the sun,
in the middle of the Sign of Leo, was picked out in brilliant rubies and
freshwater pearls; and in the vestibule, seated on a throne of gems, a
golden idol as beautiful as the moon."[82]

At first immigrants from India doubtless played prominent roles
in the intellectual life of the great centers of Buddhist monasticism at
Balkh and Tirmidh. Trilingual inscriptions from the sprawling Buddhist center at Tirmidh in Bactrian, Brahmi, and Kharoshti (two Indian
languages) suggest that among the monks there, all of whom were adept
translators, were some from India. But the many Bactrian Persian names
discovered in graffiti on the walls of one of these same monasteries leave
no doubt that natives of the region were no less active in Buddhist life
throughout the region.[83]

Central Asian Buddhists also engaged in architecture and the arts,
using their native heritage to develop new types of temples that differed
from those in India.[84] They attended the teaching institutions associated
with the monasteries, joined the monastic communities, and engaged
in disputes among the various factions of Buddhism. One monk who

participated in such activities, named Ghosaka, became an authoritative and deeply respected Buddhist theologian and traveled from his native Balkh to play an important role in the deliberations at the Fourth Buddhist Council in Kashmir.

Which schools of Buddhism did Central Asians favor? Many different schools were active, but the Vaibhasika school was predominant at Balkh and at several other leading centers throughout the region. Ghosaka was an exponent of this school, as was another distinguished Central Asian, Dharmatrata.[85] Of the four main schools, the Vaibhasika was the most engaged with the world, asserting that people have direct knowledge of external reality, including both past and future, as opposed to rival schools, which argued that only the present truly exists.

No activity of Central Asian Buddhists was more consequential and significant for the era following the Arab conquest than the collection, editing, and, in some cases, translation of Buddhist texts from the Sanskrit originals into Sogdian, other local languages, and especially Chinese. The Buddha had himself ordained that all should recite the sacred texts in their own language. The Soviet scholar Litvinsky calculated that among known translators of Buddhist writings into Chinese, six were Indians, six or seven were Chinese, and fully sixteen were Central Asians.[86] Many achieved such renown for their translations and overall piety that contemporaries penned biographies of them.[87] Central Asians particularly favored the writing of *vibhasa*, the Sanskrit term for "commentaries,"[88] for which they gained renown. One translator and commentator, a Parthian known to the Chinese as Chi Ch'ien and famed for his translations into their language, is said to have read "books of the West" and have spoken six languages.[89] Another, K'ang Serng-Hui, a Sogdian, combined his work as a translator and editor with expertise in astronomy and astrology.[90] Others produced original works. Tirmidh-born Dharmamitra wrote a commentary on the *Vinayasutra*, the text that regulated life in Buddhist monastic communities. The revered Ghosaka, meanwhile, wrote a huge commentary on the *Abidharmapitaka*, the basic principles governing mental and physical processes, which is now lost.[91]

Prominent among Central Asian missionaries was a prince from Parthia who, after the death of his father, renounced the throne, became a Buddhist monk, and traveled to China to spread the faith. Beginning in

AD 148, this An Shi Kao preached Buddhism in the Chinese capital and wrote a book on Buddhist doctrine. Another Parthian prince who converted, became a missionary monk, and wrote several books, now lost, under the Chinese name of An Fagin.[92]

It is not known whether this Central Asian passion for codification, which was to become so crucial in the Islamic era, began with Buddhism or traces still further back in time. What we do know is that Buddhist monks from Greater Central Asia figured prominently among the leading editors and codifiers of Buddhist texts and the most active translators. The editorial process sometimes led them to amplify the text, or to add passages that explained apparent difficulties. It is important to stress that this activity began centuries before the start of the translation of works of classic Greek thought into Arabic, or the codifying of Muslim texts and traditions, in which movements Central Asians were also to be centrally involved.

The later fate of Buddhism in Central Asia is unclear. Until recently it was thought that decline set in when the Sassanian Persian state, in an attempt to impose its new "official" Zoroastrianism late in the third century AD, cracked down hard on all religions other than Zoroastrianism. The Sassanians' special venom, like that of later Muslims, was directed against the Buddhists, several of whose monasteries were closed at this time. But this campaign against Buddhists, and also Manicheans, Jews, and Christians, turned out to be as ineffective as most other Sassanian efforts to control their willful and independent eastern satrap.[93] Once it subsided, Buddhism revived and thrived again. Others have argued that Buddhism was fast dying out prior to the rise of Islam, while still others blame its decline on the arrival of yet another tribe of nomadic invaders, the Hephthalites, who are said to have suppressed it.[94] Yet the discovery of a large Buddhist stupa in Munk, Tajikistan, that was built in the seventh or eighth century suggests that Buddhism by no mean ceased to exist in Central Asia.[95] Indeed, we know from carbon 14 dating that the great 185-foot-high Bamiyan Buddha was created around AD 600, during the same years that Muhammad was delivering his revelations.[96]

The Arabs, once they managed to subdue the region, attacked Buddhism and Buddhists with special ferocity because of the "idols" they found at shrines of the faith. Nonetheless, while they destroyed and defaced scores of Buddhist shrines, it was not they who hacked away the faces of the Bamiyan Buddhas; this was the result of an independent effort to mount portable metal masks on them, a long-standing practice in the region.[97] Indeed, Deborah Klimburg-Salter argues convincingly, on the basis of her close study of the archaeological evidence, that Buddhism continued to figure prominently in the life of regions south of the Amu Darya down to the end of the tenth century.[98] Other communities of Buddhists hung on in the Seven Rivers (Zhetisu) area of northeastern Kyrgyzstan, and probably elsewhere as well, down to the Mongol invasion. As Arabs and Muslims intensified their attacks on Buddhism, its geographical center of gravity shifted to the east of Central Asia, beyond the reach of the invaders. Finally, the great caravan cities of East Turkestan emerged as the new heart of Central Asian Buddhism, and of Manichaeism as well. In Kashgar, Khotan, Turfan, and other monastic centers, Uyghurs, Sogdians, and others from the region continued to codify, translate, and disseminate the tenets of Buddhism to the whole of North and East Asia.[99]

CHRISTIANS AS CHAMPIONS OF HELLENISM

Just as the intellectual and spiritual history of Central Asia required a look deep into India, it now calls for close attention to developments on the Levantine shore of the Mediterranean, thousands of miles to the West. Christianity had first appeared within the borders of the Roman Empire: recall Paul's declaration that "I am a Roman citizen." But by the second century, large Christian communities had also sprung up further east in the cities of the emerging Sassanian Persian empire. By the early fourth century, communities of Christians had appeared in all the main cities of Khurasan, including Herat, Tus, Merv, Sarakhs, and Balkh. However, the earliest known Christian text from the region is much older, dating to AD 250.[100] It is revealing that this work, written in Bactrian and entitled *A Book of Laws of the Countries*, was the same kind of compilation of laws and rules of conduct that later Muslim scholars from the region were to produce in great number.

Many early Christians on both sides of the Rome-Persia border took a very independent view on theological questions. Their differences with the orthodox mainstream reached a climax when one of their number, Nestorius, was elevated to the episcopal throne in Constantinople, where he proceeded to champion views to which many other Christians in the eastern Mediterranean objected on the grounds that they reduced the holiness of Mary. Worse, while Nestorius certainly accepted Christ's dual nature, he emphasized Christ's humanity to such an extent that some believed he diminished his divinity. Finally, Nestorius's enemies drove him from office in 431 and persuaded the Fourth Ecumenical Council of Chalcedon in 451 to condemn his views. Sensing danger, many of his Syriac-speaking followers moved across the Roman-Byzantine border into Persian territory.[101] There they proceeded to establish their own seats of learning, first the great Christian center at Nisibis in Syria and later at Gundeshapur in Mesopotamia.

Finding themselves now in a Persian milieu, the Nestorian Christians, as they came to be called, looked to spread their vision of the faith to the eastern lands that opened before them. During the fifth and sixth centuries they founded communities in the major cities of Khurasan; bishoprics at Samarkand, Balkh, Herat, Nishapur, and many other cities; and a metropolitan see or archbishopric at Merv.[102] While many Syrian Nestorians migrated to these "new" territories, the ranks of parishioners were soon filled by local converts from Zoroastrianism and Buddhism. Many active parishes grew up in smaller centers, and especially in the many trading cities on the northern and southern routes across East Turkestan (Xinjiang) to China.[103] In short order they had translated the Creed and other main statements of the faith into Sogdian and other regional languages.[104] Many Turkic peoples converted to the new faith. One to do so was a prince or *kagan* with his entire army, whose conversion occurred in 644, barely a generation before the arrival of Muslim armies in the region.[105] The Nestorians' influence continued to spread; their highly distinctive cross has even been found carved on an escarpment in distant Ladakh, on the road from India to China.[106] And within two centuries, the Uyghurs in Turkestan adopted the Nestorians' Syriac (Aramaic) script for writing their language.

The spread of Christianity in Central Asia was not accompanied by an influx of new approaches to architecture or the arts and crafts, as had occurred with Buddhism. But sizable Christian communities sprang up in every section of the region, sometimes within the old cities, as at Merv, and in other cases, as at Chach (Tashkent) and Samarkand, in the fertile nearby countryside. All built churches or meeting places for their congregations.[107] Monasteries thrived in Urgut south of Samarkand, Merv, and many other centers, and both parish churches and diocesan cathedrals were constructed. The archaeological footprint of Christianity in Central Asia has barely been explored, but large numbers of Christian objects have turned up from Turkmenistan (Gokdepe) to Kyrgyzstan and beyond, deep into Xinjiang.

If little is known of the physical remains of Christianity in Central Asia, the same cannot be said of the Nestorians' intellectual footprint. Born of doctrinal controversy, the Nestorian Christians naturally took education and hence writing very seriously. In spite of the Arabs' destruction of pre-Islamic libraries throughout the region, fragments of books of their writings in Syriac have turned up in many places, while whole volumes were preserved in Xinjiang, which remained beyond the Arabs' reach.

In the centuries immediately preceding and following the Arab conquest, thousands of students, including many from the Christian communities in Central Asia, enrolled at the Nestorians' university at Nisibis. Meanwhile, in the course of an early battle between the Sassanian Persians and Byzantine Greeks, the Persians took captive several thousand Nestorian Christian Syrians. Subsequently settled in Gundeshapur, three hundred miles southeast of Baghdad, this community, enriched by many later settlers, became another, more eastern center dedicated to the advancement of Nestorian Christian scholarship and thought. Persian emperors liked the idea of helping any Christians who opposed Byzantium and generously supported Gundeshapur, which thrived from the mid-sixth century.

All three of these centers offered instruction in theology and philosophy, and also in medicine and science. At Gundeshapur the Persian emperor encouraged the scholars to translate Greek and Syriac texts into the prevailing Pahlavi Persian. Soon these eager translators had turned their attention also to Indian works on mathematics, astrology, and astronomy and to Chinese medical texts. Their specialty, however, became the translation of classical Greek texts for Persian readers. Later, under Muslim rule, they were to shift to translating into Arabic. As we shall see, this was to bring about a revolution in thought.

For now, it is important to take note of the impact on Central Asia of the type of translation and scholarly activities that were going forward at Gundeshapur and elsewhere in the world of eastern Christianity. It is known that one of the most highly acclaimed archbishops at Merv, Theodore (appointed AD 540), was a thoroughly trained expert on Aristotle in general and on his *Logic* in particular. That Theodore gained this high post in spite of holding theological views that some considered suspect, attests to his high standing as a student of Hellenism.[108] Here, in the most senior Christian ecclesiastic post in Central Asia, was a linguist, scholar, and philosopher who focused on the great questions of epistemology that four centuries later were to preoccupy an even greater Merv-trained student of Aristotle's *Logic*, Farabi. A scholar like Theodore would have gathered a library with all basic texts and also the latest translations by his coreligionists; he would also have assembled students and assistants to help him in his work. In his scholarly endeavors, Theodore followed in the footsteps of Buddhist translators from Merv, who had rendered classic Sanskrit texts and fables into the local language.[109]

Since none of these early Central Asian translations from Greek into Syriac have survived, we can only speculate on whether they were true translations or merely paraphrases, as was sometimes the case, and whether they contained commentaries added by the translator. We can be confident, however, that the great work of study and translation going forward at Gundeshapur, and which Theodore was fostering at Merv, was also taking place at the other principal bishoprics of Central Asia. As at Gundeshapur, this would have included the translation of works in both philosophy and medicine.

Epitomizing these Syrian Christians in Central Asia and attesting to their importance was Ali ibn Sahl Tabari (838–870), a Syrian (Nestorian) Christian from Merv. Tabari lived a century after the Arab conquest, yet his work built directly on the foundations that had been laid during earlier centuries. It was Tabari who first translated Euclid's *Elements* from Greek into Arabic and who wrote the first comprehensive medical encyclopedia in Arabic. In the field of medicine, the line of descent runs from Tabari through his great student, Razi, to Ibn Sina and his *Canon*. Tabari also became an early enthusiast for the study of human psychology and its impact on health.[110]

Many of these same traits also characterized Jewish settlers, whose entry point into Central Asia was also Merv. The book of Esther, composed in the second century BC, confirms the presence of Jews "in all the provinces of Persia,"[111] which would have included Bactria, Parthia, Sogdiana, Margiana, and Khwarazm. In the early fourth century we find a reference in the Babylonian Talmud to a well-known explicator of Jewish texts, Samuel bar Bisena. Visiting Merv from the religious academy at Pumbedita (now in Iraq), Bisena complained about the lax observation of Kosher law in the East.[112]

Centuries before the Arab conquest, Jews had settled in some numbers in most of the major cities of Central Asia, from Khwarazm to Afghanistan,[113] with some venturing as far as East Turkestan. A trove of manuscripts recently discovered in an Afghan cave confirms the existence of a large and highly literate Jewish community there more than a millennium ago.[114] The tradition of learning dating from Bisena's time flourished to such an extent that a later medieval traveler, Benjamin of Tudela, confirmed the presence of many eminent scholars among the seven thousand Jews at remote Khiva, or among the fifty thousand at Samarkand.[115]

IDENTITY AND INHERITANCE

Shortly after the Arab conquest of Central Asia, the new overlords ordered the first mosques constructed. Drawing on their understanding of what a proper house of worship looked like, the local architects constructed cubic structures surmounted with domes, exactly what they had built for Zoroastrians over the centuries.[116] Three centuries later the ultra-orthodox Muslim Mahmud of Ghazni decided to build himself a palace at his capital in Afghanistan. Here, too, the architect called on the ancient local prototypes that were known to him, ornamenting his palace with objects looted from Indian temples.[117] Both projects remind us that people everywhere respond to new challenges by mobilizing whatever they have at hand and know best.

Architecture was but one of many areas in which Central Asia's past bequeathed a useful legacy to the future. In law, the Arab conquerors had no choice but to adopt many Central Asian approaches, especially in such critical areas as irrigation, for which neither the Quran nor the Arabs' nomadic experience offered guidance. Even in areas like the law on marriage and divorce, on which the Quran had much to say, actual practice in the postconquest era continued to be shaped in part by pre-Muslim—in this case, Zoroastrian—principles.[118] It is worth noting that not only did the great Ibn Sina have Zoroastrians as students, but he himself translated a Zoroastrian text into Arabic.[119]

The Arab conquerors brought with them their traditional lore concerning the heavenly bodies, human health, and other subjects. Such folk wisdom could occasionally inform the study of astronomy, medicine, and other fields, but it could not provide the basis for fundamental inquiry in these areas. However, long before the Arabs arrived, the Central Asians, as well as Persians, had begun the work of systematizing these and other branches of knowledge. Thus the study of astronomy had already achieved an advanced level in the region of Khwarazm, while medicine thrived in Khurasan and Bactria, partly thanks to their ready contact with the great medical centers in India and at Gundeshapur. We can be sure that Merv, as the region's unofficial capital, would have had the same quality of hospitals and astronomical observatories as Babylon. Mathematics seems to have thrived wherever traders demanded it, and especially in those regions that were in steady contact with India, with its highly developed mathematical heritage. Indeed, if there is one causal factor that may account for the sophistication of early Central Asians in many fields of formal enquiry it is that they were in contact *both* with the Middle East and with India, and at a time when peoples further west had only sporadic intellectual contact with the Indian subcontinent. This cross-fertilization was not unknown in Persia but it was a constant reality in Central Asia from long before the Arab invasion, due to the close trade ties that existed.

In this context, it is all the more strange that Central Asians had practically no awareness of Chinese science and thought generally, even at a time when Chinese goods were readily available in their markets. Because of this, the savants of Central Asia who had played the key role in transmitting Buddhism, Christianity, and Manichaeism to the East did not become the transmitters of Eastern learning to the West. For this fateful lacuna the West itself is partly to blame. Not only did it fail to tap into Chinese and Indian learning, it even failed to avail itself of the scientific and medical knowledge of the Nestorians at Gundeshapur.[120] Whatever the causes, the failure of Chinese learning to penetrate Central Asia and hence the West had fateful consequences for both.

Anyone drawing a single straight line connecting the heritage of pre-Islamic times in Central Asia with postconquest intellectual achievements would be going beyond the evidence. But in some areas the relationship is striking. The later development in Balkh of a pioneering school of geography surely owes something to that city's ancient status as a continental entrepôt and the constant presence there of people from all points of the compass. When the Bukharan vizier and geographer Jayhani wrote his important compendium on roads and kingdoms, he had simply to tap into the fund of geographical knowledge built up by commercial travelers from his hometown over the centuries.[121] And the great Muslim astronomers of Khwarazm and Khurasan were generous in their praise for predecessors who had prepared astronomical tables, established sophisticated calendars, and predicted lunar and solar eclipses with great precision. The preoccupation of many later Central Asian astronomers with the design and construction of astrolabes and astronomical quadrants traces to the dual traditions of orienting in vast desert spaces and of mechanical tinkering that had long thrived in the region.[122] And the heritage of the pre-Islamic era emerged again in the writings of the ninth-century thinker Abu Mashar from Balkh, who tapped into old local knowledge of the Greco-Indian-Iranian occult sciences to become the leading astrologer in the Muslim world and to exercise a profound influence on European astrology as well.[123]

Among the most sharply etched thinkers of pre-Islamic Central Asia, and one of many who helped shape the world of those who followed, was Bozorghmer (531–578). A native of Merv, Bozorghmer propounded ideas on ethics that influenced thinkers deep into the Muslim era. Trained in the rich Central Asian traditions of astrology, he interpreted a dream for the Persian shah so successfully that he was invited to the court at Cstesiphon on the Tigris and quickly rose to the rank of chief counselor, or vizier.

Like many later Central Asian thinkers, Bozorghmer of Merv tried his hand at many fields. When a visiting Indian ruler introduced the Persian court to the Indian pastime of chess, Bozorghmer proved his quick mastery by beating the guest at his own game. He then proposed certain ways the Indian game could be improved and, for good measure, invented the game of backgammon (*nardy*) as a kind of reciprocal gift to the Indian.[124] This encounter marked the launch of both modern chess and backgammon into the Persianate world, whence they spread to the Arabs and eventually to the West. Four centuries later, another native of what is now the border area between Turkmenistan and Iran, Assuli, penned the first classic analysis of chess.[125] Thus, before and after the Arab conquest, Central Asians had established themselves as the main champions and arbiters of the Game of Kings.

Even though only fragments of Bozorghmer's writings survive, he ranks as a thinker of note. His ethics were saturated with the dualism that underlay Zoroastrianism.[126] He worshipped the good but was equally moved by the folly of human aspirations and the transitory nature of humankind's lauded achievements, whether great empires or cities. Worldly activities, he wrote, utterly fail to prepare men and women for life's only certainty, the Last Judgment. Salvation, Bozorghmer believed, was achievable solely through the exercise of reason and understanding, and through a life of good thoughts, good words, and good deeds. He was eventually condemned to death, probably for having moved too close to Christianity. It is not hard to discern the stamp of Bozorghmer in the rationalism of later Central Asian thinkers or in their frequent denunciations of worldliness.

More important than any specific individuals or achievements of pre-Muslim times was the fact that Central Asia was firmly established as an inquiring society, highly literate and also numerate, worldly, and self-confident. Prosperity earned through hard work in trade and agriculture created leisure for a fortunate few and for those whom they patronized. Many devoted at least part of that freedom to the arts and music, or to intellectual endeavor in astronomy, mathematics, medicine, and other fields. Still others were drawn to ponder the great questions of the purpose and objectives of human existence, and of humankind's relationship to time and the cosmos.

The geological layering of religions in Central Asia made all such existential questions particularly compelling there. The fundament that was Zoroastrianism, with its dramatic vision of struggle between good and evil and of the possibility that men and women could exercise their free will to prevail in that struggle, made the entire region fertile ground for religions of salvation. Buddhism, Manichaeism, and Christianity, for all their differences, all provided guidance for daily life that was accepted as being compatible with the divine order of things. And all three believed that ordinary people could participate in the rites of the faith and be open to salvation.

During the millennium before the Arab invasion, Central Asians were presented with four or five new systems of faith, not even counting those of the nomads. They became skilled at comparison shopping, identifying the distinctive elements of each and also the common denominators among them. Such evaluations led them to consider the sources of human knowledge, the inquiry that formal philosophy calls epistemology. It is not surprising that during the later Age of Enlightenment, Central Asians like Farabi led the world in this field and in classical logic as well. And if some of them hailed these tools as the instruments par excellence for establishing truth, others became the world's most effective critics of logic and champions of faith.

It is not hard to detect common elements in the way Central Asians received the various religions that appeared on their doorstep between 300 BC and AD 670. Each time they showed themselves to be remarkably open and genuinely inquiring. With the exception of the "official" Zoroastrianism of Sassanian Persia, no religion prior to Islam sought to inhibit the practice of the others. Central Asians, open to the world and facing the new without visible fear, were as committed to free trade in the religious sphere as in commerce. They accepted dynamism and change as normal and considered it quite natural to pick and choose, combining desirable elements of various religions into unexpected and original syncretisms. We may be surprised at the combination of a Zoroastrian eternal flame and the worship of Greek gods at a single temple at Panjikent in Tajikistan,[127] but such innovative blending was hardly unusual.

In their effort to make sense of the religions and ideas that swept in on them, Central Asians became adept at codifying, analyzing, editing, and commenting on the holy books of each new faith. Zoroastrianism had grown up in their midst, so they knew its tenets intimately. They wrestled over Greek texts on religion and philosophy and may have produced some of their own. Buddhism elicited from Central Asians a many-sided effort to edit and translate the major texts. In the process they expressed their strong preferences and added their own commentaries and amplifications.[128] The same process took place as they absorbed, analyzed, and passed along the main texts of Christianity and Manicheanism.

All this suggests the possibility that the religious/philosophical history of Central Asia is not divided into neat thematic or temporal compartments of Zoroastrianism, Buddhism, Hellenism, Christianity, Manichaeism, or Islam but is one, and that its evolution is a single process. Seen in this light, the codification of the Hadiths of the Prophet Muhammad, in which Central Asians were more prominent than all other Muslims combined, is a continuation of the great editorial project of Buddhism and, to a lesser extent, of Manichaeism and Nestorian Christianity.

Is it coincidental that Central Asians, after favoring the more practical Vaibhasika school of Buddhism, with its acceptance of sensory perception, the embrace of past and future, and stress on authoritative commentaries, would later gravitate toward the more practical Hanafi school of Muslim law, and toward the writing of commentaries on the translated works of ancient Greek thinkers?[129] Or that the physical design of the classical madrasa, a Muslim institution that was regularized and popularized by a Central Asian official, built directly on the Central Asian tradition of Buddhist monasteries, with their rows of cells and assembly rooms?[130]

The eleventh-century Shiite philosopher Miskawayh, whose father had converted from Zoroastrianism, did not hesitate to cite many sources from the old faith in defense of his own views on ethics.[131] The doctrine of free will, equally important to Zoroastrianism, Buddhism, and Christianity, became so fixed in the Central Asian mentality that long after the arrival of Islam, Central Asians were notable among those who championed the doctrines of the Mutazilites, who defended free will and the unfettered exercise of reason to such a degree that more orthodox Muslims severely censured them. Similarly, the tombs of Muslim saints that dot the Central Asian landscape are later manifestations of the same urge that had led to the identification of Buddhist and Christian saints across the region and the celebration of their lives.

The earliest known Sufi Muslim mystic was Abu Yazd Bistami (c. 804–874) from the town of Bistam in Khurasan. Bistam, today a town of seven thousand that is devoid of Buddhist monuments, must once have supported a thriving Buddhist life. Bistami was clearly from a family of seekers. His father was a Zoroastrian convert to Islam, but Bistami himself sought to enrich his new faith with mystical teachings and the practice of yoga, to which he had been introduced by a local Buddhist master.[132] How different, in the end, is the mental and spiritual discipline of Central Asian Sufism, which seeks the obliteration of the ego, emancipation from worldliness and the temporal, and an embrace of the eternal, from the Buddhist tradition that had led to the construction of the huge sculpture of *Buddha Entering Mahaparinirvana*, the highest state of Nirvana, in Tajikistan? Similarly, one might suggest that the religious poetry of Central Asian Sufis is a continuation of the early spiritual poetry of the Zoroastrian Avesta, while the rich and complex music of Samanid Bukhara in the tenth century was a lineal descendant of the Buddhist musicians depicted in the sculpted frieze at Airtam in southern Uzbekistan a millennium earlier.[133]

If Buddhist mysticism found its way into Islam via Central Asia, so did the editing of scriptural texts, which was an important activity at the many Buddhist monastic centers across the region. Among the most prominent early collectors of the Hadiths of Muhammad was Abu Isa Muhammad Tirmidhi (824–892), who was born and died in the great Buddhist center at Tirmidh, now Termez, in Uzbekistan,[134] where Buddhist monks had long labored at the same task of codification for their religion. Even today Tirmidhi's collection is accepted throughout the Muslim world as one of the canons of the faith.

Even dissidence and heresy may exist in a continuum that bridges the pre-Islamic and Islamic eras. After the Arab conquest, the old Buddhist center of Balkh produced a well-known Quranic commentator whom orthodox clerics harshly denounced for anthropomorphism.[135] This was the standard attack on anyone deemed sympathetic to Buddhism, of whom there were many in Balkh. A later Balkh native, Hiwi al-Balkhi, was one of the most thoroughgoing skeptics in the Islamic world, who attacked the authority of the Quran and of revealed religions generally.[136] Writing in the ninth century, he recited the claims of the region's other revealed religions, including Zoroastrianism, Christianity, and his own Judaism, to debunk the claims of all religions, including Islam. Over the next few centuries the region from Balkh to Nishapur was to nourish a rich tradition of skepticism, nonconformity, and heresy. No facile explanation will suffice to explain this, but it is worth noting that this was a region where every one of the many contending religions was busy making its case and attracting ardent believers. Since all such claims could not be equally valid, such an environment tended naturally to spawn its own antibodies in the form of thoroughgoing skepticism and free thought.

By far the most conspicuous feature of the spiritual life of Central Asia's crossroads civilization was its pluralism and diversity. This did not end with the Arab conquest but, as we shall see, continued to thrive for nearly four centuries after the arrival of Islam. Conversion proceeded very slowly. Indeed, Muslim theologians themselves acknowledged that many people of other faiths nominally adopted Islam but did not abandon their prior beliefs.[137] The British classicist Peter Brown speaks of Islam resting "lightly, like a mist" over the highly diverse religious landscape.[138] Only in the combative eleventh century did pluralism come to be seen as an evil and as a threat to the prevailing orthodoxy. By the time such a view took hold, the Age of Enlightenment was already approaching its end.[139]

Where do fresh thinking and innovative ideas come from? The question has been the subject of after-dinner philosophizing and scholarly analysis for centuries. A recent author on the subject, Steven Johnson, suggests that many good ideas arise when people take only one or two mental steps beyond what is already known, or through the intensive workings of "liquid networks" of reflective individuals interacting with one another.[140] Central Asian intellectual life was ripe for both types of innovation in the centuries preceding the arrival of Islam. This was a complex, yeasty world of practical thoughts and abstract ideas, of reason and faith, where many literate and numerate people interacted with one another concerning the affairs of the world but also concerning ideas as such. The destruction of libraries leaves us unable to speak with certainty about the status of new thinking in philosophy, science, or religion prior to AD 670. And then the confusion that followed the Arab invasion prevented its emerging for a century thereafter. But when life once more returned to normal, more fresh and innovative thinking blossomed in Central Asia than practically anywhere else on earth.

CHAPTER 4

❁

How Arabs Conquered Central Asia and Central Asia Then Set the Stage to Conquer Baghdad

By the seventh century the inhabitants of Greater Central Asia were all too familiar with conquest by foreign powers. Over the past millennium the armies of the Persians, Greeks, Kushans, Hepthalites, Parthians, Chinese, and Turks had all swept in and forced local rulers to submit to them. But then the conquerors faced the hard part. Each time, manpower shortages and other constraints forced them to turn governmental functions back to those whom they had just defeated, leaving behind just enough troops to keep the locals in line. If the victors' goal was to extract tribute or collect taxes, there was no alternative to this. And so, beginning anew after each conquest, Central Asians gradually reclaimed their self-government, their religions, and their values.

The Arab invasion that began in 660 brought the usual destruction, suffering, and temporary collapse of trade. But it differed from earlier invasions in three important respects. First, the local populations this time offered fierce and sustained resistance. Never in all their wars of conquest had the Arabs faced anything like the tenacious opposition that the Central Asians mounted against them.[1] Second, the invaders themselves—the Arabs—were soon at each other's throats, with warring factions and groups seeking alliances with various local forces against their fellow Arabs. Third, because of these two factors, the conquest proceeded very slowly and uncertainly, requiring nearly a century to achieve its immediate aims, by which time local self-government and cultures had already begun to reassert themselves. Thoroughgoing Islamization required several centuries thereafter.

Notwithstanding all the problems, the Arabs' onslaught was ultimately successful in two important respects: it left behind a new language for official communication and intellectual interchange—Arabic—and a new religion, Islam. The Arab conquest of Central Asia entailed massive cultural destruction in many spheres, but this was offset, in part at least, by important and enduring elements of cultural enrichment. The flowering of Central Asian thought and culture that took place over the following three and a half centuries would never have happened without these revolutionary changes introduced from the Middle East. This chapter therefore focuses on the dynamics of conquest over the century after the first attacks in the 650s, and the locals' response.

Central Asia on the Eve of the Conquest: A Summary

Central Asia on the eve of the Arab conquest was a self-governing region. Still largely of Iranian stock but with a growing Turkic population, and speaking a variety of Iranian and Turkic languages, its populations were grouped into a network of local kingdoms that owed formal obeisance to the Persian (Sassanian) empire and to the Turks who had swept into the area a century before the conquest began. But their subordination to both Persians and Turks, and Tang China's protectorate over the eastern parts of the region, was by now only nominal. Both the Turks and the Sassanians were in steep decline by the seventh century, and the Tang empire increasingly focused its energies further east, in present-day Xinjiang.

The Turks had firmed up their hegemony in the 550s by defeating the so-called White Huns, or Hephthalites, yet another nomadic people who had ridden into the region from beyond China's Great Wall. After their initial successes, many Turkic leaders in the heart of Central Asia served comfortably as overlords of their subject kingdoms, enjoying all the elaborate ceremonies of state but not interfering in commerce or culture.[2] Meanwhile, the Turks themselves had done much to weaken the Sassanian empire's claims on Central Asia. More than once the Persian Sassanians and Turks fought it out in Central Asia, leaving the Central Asians as bemused spectators.[3] But the Turks were constantly looking over their shoulder in fear of new threats from the East. Back at their sumptuous capital of Cstesiphon on the Tigris, the Sassanians realized their grip on Central Asia was also evaporating as a result of the massive financial problems that arose from their military campaigns against the Greeks in Constantinople.

Forced by these circumstances to cut back their claims on the Central Asians, both Sassanian Persians and Turks were content to extract what tribute they could but otherwise leave the region to itself. The Tang emperors, too, preoccupied with a campaign against Tibet, were satisfied to be acknowledged as suzerains, without placing further demands on their nominal subjects. Indeed, by the seventh century there was not a single viceroy of a foreign ruler in all Central Asia.[4] From the third to eighth centuries, the Central Asian kingdoms were functionally independent. This created a vital period of local rule under the watchful eyes of local kings and landed aristocrats (*dikhans*). Continental trade flourished and cities burgeoned. Confident rulers cut back expenditures for maintaining fortifications. The same relaxed mood existed in the sphere of religion. Images on surviving examples of coinage from many cities reveal that large numbers of Central Asians continued to worship the ancient local gods, even as they explored new faiths.[5]

All these trends were evident at Merv, already the largest urban center east of Mesopotamia. This great entrepôt spread far beyond the walls that had been constructed several centuries earlier.[6] Whole industrial quarters existed, where skilled workers manufactured, among other products, the hardened steel that the Greek historian Plutarch had lauded in the first century.[7] Here, too, were produced bolts of Merv's colorfully printed cotton fabric (*tiraz*) that found markets from China to the Mediterranean. Side by side with all this mercantile activity, Zoroastrian temples flourished both within the central city and in the adjacent countryside,[8] while Christian and Manichean missionaries based at Merv ranged as far afield as China.[9]

Astronomers, mathematicians, physicians, and other scientists of Merv benefited from the city's close scientific ties with the late classical Mediterranean world and also with India. Research in astronomy was going forward at the city's long-established observatory. Supplementing the several major libraries that had long existed at Merv, the last Persian ruler, fleeing the Arabs, was to bring to the city a large collection of books from the imperial library at Cstesiphon.[10]

Nothing brings us more dramatically face to face with this lost world than the painted murals that adorned the walls of a chamber in the royal palace at Samarkand, the receiving rooms of both royal and private residences in Panjikent (now in Tajikistan), and Balalyk-tepe in southern Uzbekistan.[11] Like photographs, these miraculously preserved masterpieces document the elegant court life, sumptuous domestic environment, and rich literary and religious milieu of those members of the regional elite who were soon to feel the brunt of the invasion. They make little claim to edify, joyfully celebrating instead the good life that was about to end.

Some of the Panjikent murals depict religious ceremonies, while others portray scenes from folklore and the Persian epics, especially the tale of Rostam from the *Shahnameh*. The latter, many of them in the reception rooms of aristocratic dwellings, include sequentially arranged scenes that are much like modern comic books.[12] At the somewhat earlier royal palace in Panjikent, the viewer is confronted by confident-looking Sogdian men and women in close-fitting costumes with large lapels bordered with contrasting fabrics at the hemline.

In the room from the Samarkand palace, the enthroned local ruler, Varkhuman, surrounded by his Turkish guards, is shown receiving what most analysts have taken to be foreign emissaries during the New Year (Novruz) festival (see plate 4). Except for the Korean in his feathered headdress, all are bearing gifts. The black-capped Chinese brings silks, the Persians in lavish caftans present necklaces and embroideries, while rugged mountaineers from the Pamirs offer yak tails. On another wall an aquatic scene depicts an Indian tale of a fight between cranes and dwarfs, further evidence of the extent to which preconquest Central Asia looked east as much as west. Accompanying these scenes are musicians playing lutes and *santurs*. Nor is religion neglected amid these images of secular life. Not only are Zoroastrian priests depicted with cloths over their faces to protect the sacred fires from their breath, but an inscription in the Sogdian language reminds viewers that it is their duty to "protect the gods of Samarkand."[13]

As a group, these murals have a "last days of Pompeii" quality, poignant images of a final flowering before the collapse. But they can also be seen quite differently, as reflecting the vitality of Central Asian culture and the people who lived it (see plate 5). This was the vitality of people who, when they came under direct assault from Arab armies a few decades later, chose to fight rather than submit and then sought to protect and preserve what was theirs, even as they adopted valuable innovations from their new overlords.

Faith and Plunder: The Arabs' Eastward Tide

About the time artists were painting these murals, the future prophet Muhammad was born in Mecca, a small town on the high desert plateau of western Arabia. The new monotheist faith to which his prophesies gave rise swept up tens of thousands of Bedouins in a whirlwind of conquests. Driven by religious zeal and visions of earthly riches, these Arab armies swept westward along the coast of North Africa, northward toward the Caucasus, and eastward into Persia. The exhaustion of the Sassanian empire and its army assured that the Arabs, after two initial assaults and a major battle in 636, encountered little resistance as they advanced into Persian territories. The last Sassanian ruler, Yazdegerd III, gathered together his entire court and as much treasure as a long caravan could carry and fled to Merv, where he hoped to rally local loyalists and mount a counterattack. But an impoverished miller, threatened with death, betrayed him, and Yazdegerd was murdered in 651.[14]

The desperate Persian shah had tried to enlist the support of Tang China against the Arabs, but the court at Chang'an (now Xian) rebuffed him.[15] And for good reason, since the Chinese were experiencing troubles of their own, as were the Turkic tribes in Central Asia. This perfect storm of decaying empires opened a huge power vacuum across the region, which the Arab troops rushed in to fill.[16] Advancing along the main road to Central Asia, they overwhelmed all of western and northern Persia, sending a shock wave of fright and terror eastward into Central Asia. With little time to mobilize, Khurasan was unable to mount a strong resistance. By the end of the year 651 Arab armies were camped at all the major cities of Khurasan, including Nishapur, the old capital of Tus, Sarakhs, and the gem of them all, Merv.

The inevitable next step was to cross the Amu Darya (Oxus River) and carry the banner of Islam into Central Asia's heartland. But events back in Arabia forced the Arab forces to pause after taking Khurasan and, more important, changed the leadership of those forces in ways that greatly prolonged the conquest. This was all precipitated in 661, when rival Muslims murdered the Arabs' fourth leader or caliph, Ali, the cousin and son-in-law of Muhammad. The ensuing civil war was a struggle between those favoring Ali's heirs and those who backed another but more distant line of the Prophet's Hashimite clan, the Umayyads. The fact that the Umayyads' grandfather had initially opposed Muhammad's revelations and mounted battles against the Prophet's forces gave the internecine struggle a sharply religious character as well. In the end, the Umayyads won out. But from the first day of the new caliph's rule, there was a taint of illegitimacy about the Umayyad dynasty. The legacy of the bitter schism of these years is still with us today, in the form of the Sunni-Shiite split, with the Shiites still holding aloft the banner of martyrdom.

Their bloody and fractious rise did not prevent the Umayyads from advancing their cause. Celebrating their fresh conquests in the West and East, they moved the capital to Damascus and launched a campaign that would, within a generation, bring almost all of the Iberian Peninsula under their rule. Regarding the eastern front, there is disagreement over whether Arab armies began their assault across the Oxus in the mid-650s or slightly later, after 661.[17] What is clear is that the first phase of the conquest consisted of nothing more than brief raids for plunder.[18] Even though conquest was the key to the Muslim state,[19] these raids convinced Central Asians that the invaders were impelled mainly by avarice and led them to mount a staunch resistance. No less ominous for the Arabs was the emergence of armed opposition within the ranks of their own army in Khurasan, namely, from religious dissident warriors who considered the new caliphs, and hence their generals, illegitimate.

Both sources of opposition should have been entirely predictable, yet these early rebellions caught the Arabs by surprise. The mutiny of Arab contingents at both Merv and Sistan in Khurasan forced the Arabs' main force briefly to retreat and regroup. As it did so, the focus of local opposition broadened from anger at pillaging to the defense of the local religions. Significantly, the Sistan revolt erupted when the Arab governor broke the arms off an image of a local divinity and carved out the rubies from its eyes. The people of Sistan fought back wildly and with such success that three centuries later the restored idol was still being worshipped there.[20]

This same arrogance and cultural obtuseness on the part of Arab governors ignited further violent outbursts from within their own garrison. In the ensuing strife neither side demonstrated much of the fortitude and nobility for which the later chroniclers praised the Bedouin warriors. At one point the Arab governor of Khurasan, who had declared himself virtually independent of Damascus, flogged to death two of his soldiers simply because they were from an Arab tribe opposed to his own. To avenge this sin, forces from the offended tribe seized and bound the governor's son at Herat in Afghanistan. Hugh Kennedy reports that as the governor's son "lay bound in their camp that night, [the rival Arab tribesmen] sat about drinking, and whenever one of them wanted to urinate, he did it over the prisoner. They killed him before dawn."[21]

The caliph in Damascus responded to all these alarming developments by pouring more troops into Central Asia. The death, during an early raid on Samarkand, of a first cousin of Muhammad added an element of piety to the Umayyads' resolve. Soon, a massive force of fifty thousand Arab horsemen drawn mainly from the Persian Gulf port of Basra was heading for Merv. This time the goal was conquest.

But the Arabs' expanded mission and newfound resolve gave rise to a series of utter fiascos. A new governor at Merv, said by his friends to be easygoing and generous and by his enemies to be effeminate, tried to mount a campaign against Bukhara. More than one of the Arab officers went heavily into debt to pay for what they hoped would be a lucrative season of plunder. But scarcely had the army left Merv and crossed the Oxus than the governor's deputy rebelled, recrossed the river, and then burned the boats before setting out to seize Merv.

The worst debacles occurred in the lands of present-day Afghanistan. This had begun as soon as Arab armies appeared at Sistan and began marching eastward down the Helmand River valley. Even before they reached Kandahar, the whole of Afghanistan had taken up arms. What started as a promising move against Balkh in the North ended in a fiasco. A large detachment of the expanded Arab army managed to overwhelm Balkh, but the populace of the Queen of Cities immediately erupted in rebellion and the Arabs were forced to withdraw. In response, they organized an "Army of Destruction" but this, too, encountered a solid wall of resistance.[22] Twenty-five thousand of the Arabs' force of thirty thousand men were killed, and once more those who survived had no choice but to sound a retreat. The Afghan campaign ended in the worst setbacks Arab arms had ever suffered.

The Arabs' failure to establish a solid foothold in Central Asia gave the region a respite. Understaffed, the invaders had no alternative but to make whatever deals they could with local leaders and otherwise leave them in peace. This arrangement, adopted out of sheer necessity, was to become the Arabs' permanent default strategy across Central Asia, including in Khurasan. Over the following century and a half, the Arabs were to mount many frontal attacks on local kings and the aristocrats and rich traders who supported them. But without exception, the initial onslaught was followed by a fall back and withdrawal. And in every case, this created a big opening for the forces of political and cultural continuity throughout the region to assert themselves.

QUTAYBA'S WAR OF CIVILIZATIONS

At first the Arabs had deceived themselves into thinking that the conquest of Central Asia would be easy. Their smooth entry into Khurasan and the success of their early raids across the Oxus for booty had led them to undertake more ambitious campaigns. But these efforts ended in disaster. Worse, they prompted both dissident members of the Arab administration and indigenous leaders from Central Asia once more to take up arms against the caliph's forces.

A son of the Arab governor of Khurasan who had caused the flogging incident had now gathered around himself a large band of brigands at the old Buddhist monastic center and soap-exporting city of Timirdh (Termez), on the border between Uzbekistan and Afghanistan. From there, Musa, as he is remembered in folklore, launched an unsuccessful raid on Bukhara and moved on to Samarkand, where the gullible local king unwisely opened the city gates to him. During a brief stay at Samarkand, one of Musa's men managed to affront local sensitivities by imposing himself as guest of honor at an annual feast honoring the "Knight of Sogdiana." Sent packing, Musa returned to Tirmidh, raised a larger army, and declared himself the leader of all anti-Arab forces in the region. Ironically, this claim brought him face to face not only with the infuriated king of Samarkand and the ambitious ruler of Tirmidh but also with those Arab forces who had not joined the mutiny, and even with the Turkic leaders, who saw themselves, not Musa, as the leaders of the anti-Arab opposition.[23]

Through wile and brute force, Musa prevailed against all these opponents. For a decade and a half, this dissident Arab and his ever-expanding band of supporters ruled the heart of Central Asia, driving the caliph's governors and their backers out of the land. Musa's revolt made him a legendary hero comparable to El Cid, the eleventh-century Spaniard who waged war against the Arabs. The longer-term consequences of Musa's revolt were significant. First, it confirmed that at least some Arabs could be accepted as champions of the local cause and leaders of the regional opposition to the caliphs in Damascus. Second, it finally convinced the caliph that the entire Arab effort to now had been halfhearted and ineffective. Without a more decisive escalation, the cause of Islam in the Eurasian heartland was doomed.

Such fears led to the appointment in 705 of a new viceroy for the region, a hardened Arab general named Qutayba.[24] From his very first appearance before his assembled troops at Merv, he made clear that his mission was to wage jihad, a holy war directed against the nonbelievers of Khurasan and the Central Asian lands beyond the Oxus. Conquest and plunder were necessary means to that end, but the goal, as he proclaimed with ample citations from the Quran, was for the united Arab forces to convert or wipe out all infidels. Those who perished in this holy cause would spend eternity with their Lord.[25]

Over the next decade Qutayba, aided by his zealous brothers, waged a relentless war against the region's great centers of disbelief, beginning with Balkh and moving on to Bukhara, Samarkand, and distant Khwarazm, Chach (Tashkent), and Ferghana. He was an able field commander, as shown by the way he mined the walls of the great commercial center of Paykand, thirty-seven miles west of Bukhara. But his tactical acumen paled before two other skills that he demonstrated with ruthless brilliance.

First, no Arab leader before him was more adept at exploiting local rivalries, playing one group of Central Asians off against another and meting out rewards and punishment on the basis of their readiness to collaborate with him. Once he brought a Central Asian leader to his side, he demanded soldiers for his army. Even at the start of his campaign, one out of eight of Qutayba's soldiers was local.[26] By the end it must have been much higher.

Second, Qutayba was a master at the tactical use of terror. In his early campaign against Paykand, for example, he slaughtered the entire force of defenders and took all the women and children into captivity. When he finally broke into Samarkand after laying siege to it over four years, he took thirty-nine thousand people as slaves. At another locale he offered his troops one hundred silver pieces for every enemy head and then piled the skulls in a tall pyramid to intimidate the remaining locals.[27] And when the ruler of a city near present-day Dushanbe in Tajikistan offered resistance, Qutayba wore him down to the point that he accepted safe passage in accordance with a law of war that was universally accepted throughout the region. No sooner did the ruler arrive in the Arab camp than Qutayba ostentatiously murdered him. What Qutayba did not do himself, his generals did for him. One Arab general crucified an entire defeated army, and another stripped all the losing forces and left them to die.[28] Such actions sent a clear message to all those contemplating resistance. Summarizing this approach, an Arab governor in Khurasan declared that the region could be ruled "only by the sword and the whip."[29]

All these tactics were means to the end of wiping out local religions and spreading the faith of Muhammad. To this end, Qutayba tore down the main Zoroastrian temple at Samarkand and melted down its treasure, repeating the process in other cities. At Balkh and Samarkand he forced the residents themselves to erect mosques and then used various means to compel them to attend services. When outright compulsion failed at Samarkand, he adopted a "pay for pray" scheme, rewarding attendees at the Friday services with cash.[30] Qutayba did not hesitate to kill four thousand captives turned over to him by his brother,[31] and at the Silk Road city of Isfijab (Sayram) he ordered his troops simply to slaughter the ten thousand Christians who offered resistance.[32]

Of more lasting consequence was Qutayba's systematic destruction of books and religious literature. In Bukhara he destroyed an important library, but in Kath, the capital of Khwarazm (near the Aral Sea in present-day Uzbekistan), he succeeded in wiping out an entire literature in the Khwarazmian language, including works on astronomy, history, mathematics, genealogy, and literature. Writing in the eleventh century, the great scientist Biruni rued this destruction as a crime against an ancient culture.[33] Qutayba's special animus was directed against Zoroastrianism. Besides killing various writers from this faith, he obliterated much of the corpus of Zoroastrian theology and letters, a tragic loss to civilization.[34]

In his assault on the cultures of Central Asia, Qutayba went so far beyond the normal practice of the Arab conquest elsewhere that one is prompted to inquire into his motives. Except for his speech to his troops at Merv, there is no surviving evidence that Qutayba was especially pious. Besides, there were pious Muslims among the less brutal Arabs who both preceded and followed him in Central Asia. Qutayba's main concern, at least at this moment, was more practical, relating to his own career. Unlike most Arab generals or governors, he came from a very minor Arab tribe with no strong lobby at the court of the caliph in Damascus. Moreover, he was beholden for his command not to the caliph personally but to the caliph's governor of Iraq and de facto viceroy of the East, Hajjaj ibn Yusuf. Hajjaj had fought in Khurasan and knew well what Qutayba was up against. But he expected results. It is fair to hypothesize that Qutayba was an insecure underling who had convinced himself that the only path to success was to outdo his boss in zeal and severity. Whatever his motives, his war in Central Asia was nothing less than an assault on culture as such, a civilizational crusade.

Qutayba's end provides further support for this view. No sooner did he receive news of Hajjaj's death, in 714, than he assembled his troops at Merv and demanded that they swear loyalty to him personally. He had never felt secure among his fellow Arabs in Central Asia and had even moved his family to Samarkand to avoid them. Now, in demanding their fealty to his person rather than to the "Commander of the Faithful," he placed himself in the same position as Musa and all the other Arabs who had rebelled against the caliph. When the soldiers responded with stony silence, Qutayba flew into a blind rage, abusing each of the Arab tribes arrayed before him with obscene invective. Within days Qutayba lay dead, murdered by his own men.[35]

CENTRAL ASIA COUNTERATTACKS

It was plain to all that Qutayba had failed to impose Arab and Muslim rule on Central Asia. Central Asians knew it, as did dissident Arabs, whether those aroused by tribal rivalries or those who believed that the Umayyads had kidnapped their faith. In the end, Qutayba left Central Asia as he found it: a region of unresolved conflicts both within and against the world of Islam and the Arabs. These tensions eventually burst forth in a whirlwind of civil war and change. Beginning at several points in the Middle East, it enormously gained momentum in Central Asia, whence it eventually engulfed the entire Arab world and put in place a new dynasty of leaders of the faith, the Abbasid caliphs. Paradoxically, the Abbasids found themselves significantly beholden to the very region their predecessors had so conspicuously failed to control or govern.

This conflict, one of several "civil wars of Islam" in the early history of the faith, exploded in the years 744–751, but its crucial Central Asian components were all in place as Qutayba's body lay on the field at Merv in 715. Among these, none was more important than the divisions among the Arabs themselves. Their numbers at Merv had by now grown to at least fifty thousand,[36] which was several times larger than the next largest garrison in the region. Modern armies seeking to hold and control occupied territories have at hand devastating weapons and instant communications, yet they still rely on sheer numbers. In eighth-century Central Asia, the Arabs' occupying force was far too small and too weak to succeed at its mission of conquest.

Among those Arabs who moved east were doubtless many anonymous warriors who were sincerely inspired by the vision of jihad. Members of this group could look up to at least four true "Companions of the Prophet," that is, members of that select group who had been in the entourage of Muhammad himself[37] and had then joined the crusade against Central Asia. The bulk, however, were impoverished Bedouins in search of plunder and a better life, with or without jihad, By 700 all the major and minor Arab tribes were well represented among both groups. It was probably inevitable that they transferred to their new home all the internecine feuds that had existed back home in Arabia.

Many Umayyad loyalists could be found among the Arab forces, but opponents of the caliph were increasingly ascendant. These accused the Umayyads of being lax and godless usurpers from a clan that had fought against Muhammad and then stolen the caliphate from its legitimate heirs, the direct descendants of the Prophet's cousin and son-in law. Many who were oblivious to these genealogical points simply hated the Umayyads as a band of avaricious, corrupt, and tipsy playboys. It did not help when word spread that a relative of Caliph Uthman, who had extended Muslim conquests and gathered a committee to produce a standard edition of the Quran, had tried to conduct religious services while drunk.

Even though a few of Central Asia's landed aristocrats had made their peace with the Arabs, most remained in grim opposition. The great noble houses of the western province of Khurasan remained loyal to their Zoroastrian faith, as did most in Bukhara and Samarkand.[38] Buddhists continued as before, and Turks clung to their shamanism. Worse, those locals who converted to Islam quickly divided along the same line as the Arab believers, further swelling the ranks of the Umayyads' opponents. Large numbers in both camps turned against the regime when it failed to honor pledges of money it had made to those who converted.

Taxes and misrule, not religion per se, were the main foci of anti-Umayyad sentiment. Every kingdom in Central Asia was forced to pay tribute in whatever way it could. The ancient city of Isfara (now in Tajikistan), for example, had to deliver ingots of iron to the tribute collector, while Kabul, Samarkand, and several other centers each had to turn over several thousand slaves.[39] With no salary from the caliph and short terms of office, Arab governors grabbed whatever they could.[40] One Arab governor from Khurasan paid his annual taxes to Baghdad by sending two thousand slaves from Kabul to the capital. It is not surprising that many urban sophisticates in Central Asia looked on their Arab conquerors as savages.[41]

Taxation and religion under Arab rule were closely intertwined. At one point the governors announced that converts would henceforth be exempted from taxation. The populations of entire cities promptly announced that they had become pious Muslims, thereby freeing themselves from the hated levies.[42] It did not take long for the Umayyad

accountants to realize that they had unintentionally destroyed the caliph's tax base. Compounding their mistake, they then imposed the ter-
rifying demand that all new Muslims provide proof of their conversion
in the form of circumcision.[43]

Beyond such instances of ham-handed governance was the view pre-
vailing in the cities that the Umayyads' new Islamic order was mainly for
Arabs, with the Central Asians left to pay the bills. It did not help that
Arab settlers tended to live apart in their own neighborhoods and stick
to themselves.[44] The Arabs' demands for tribute and taxes had a different but equally negative impact on nomadic Turks in the countryside.
Besides the geopolitical reality that the Arabs proposed to replace the
Turks as regional hegemons, their demands for tribute effectively cut
off the Turks' main income stream. No wonder that the Turks fought
back, and that the Arabs responded by hitting them all the harder. At
Bukhara alone the Turkic and Arab armies fought three major battles
in the years immediately following Qutayba's fall.[45] Many Turks, eager
for vengeance, signed on to fight with the local Persian aristocrats, while
others joined the Chinese army.

These potent resentments could not have boiled over at a worse time
for the Arabs. At the same time that Qutayba first attacked Bukhara,
Umayyad armies had blazed through much of Spain and were penetrating into southern France. But by 732, when the Central Asians were
taking up arms once more to defend themselves, Frankish and Burgun-
dian forces under Charles Martel had destroyed the Muslim army at the
Battle of Tours. Even before this historic turning point, however, Central
Asians smelled blood.

In earlier days the Arabs at Merv had been able to take out loans
from local merchants to finance raids into Central Asia.[46] Putting aside
old animosities, the autonomous rulers of the cities of Sogdiana now
began to cooperate with one another by holding annual conclaves in far-
off Khwarazm, safely beyond the Arabs' reach.[47] Typical statesmen, the
assembled rulers gravely announced that they had agreed to work together and to continue meeting. On a more practical level, they also sent
ambassadors to China to enlist the support of the Tang emperor against
the Arabs. Between 713 and 726 alone the Central Asians sent no fewer
than ten missions to the Chinese court seeking aid.[48] But Qutayba had

beaten them to the mark by sending his own mission to the East. The emperor responded to the Central Asians' pleas by doing nothing.

Acknowledging that their Chinese gambit had failed, the assembled kings of Central Asia also reached out to the eastern Turkic tribes, who, to now, had kept their distance from the region on the grounds that their western Turkic cousins were already established there. The eastern Turks now responded by preemptively occupying all but the western-most part of the region and then linking up with a new coalition formed by the Sogdian rulers and the kings of Chach (Tashkent) and Ferghana.[49] These diplomatic maneuvers emboldened the coalition members, and their new confidence soon spread to the populace at large.

In 721 much of Central Asia erupted in rebellion. Arab officials were immediately expelled from the eastern regions.[50] Meanwhile, at the heartland city of Panjikent, the local ruler, Dewashtich, had decided to tear up the peace accord he had signed with the Arabs and, with the help of the eastern Turks at Chach, throw them out of his realm.[51] De-washtich's ambassador wrote a pitiful letter back to his boss explaining that the Turks at Chach had by now cut their own deal with the Arabs, and that "Oh, my master, I am greatly afraid for you."[52] Arab troops be-sieged rebellious Panjikent. Dewashtich and his court took refuge at a nearby fortress atop Mount Mug, in present-day Tajikistan. Convinced that he would be returning shortly to his restored realm, Dewashtich took with him a load of official documents, the same stash of papers that was discovered in 1933 (see chapter 3).[53] Fighting was fierce but in the end the Arabs prevailed, sacking and burning Panjikent and cruci-fying Dewashtich, whose severed head they shipped off to the caliph in Damascus.

The new anti-Arab resistance again centered in Samarkand, which the eastern Turks had retaken by 730.[54] The Arabs again fought to retake the city they had struggled to control for half a century, and once more a combined force of Turks and local Sogdians drove them out. When the region finally settled down in the 730s, Arab forces, besides Merv and parts of Khurasan, exercised secure control over only that handful of centers they had managed to regain since Qutayba's death. Beyond this, their presence in Central Asia, Afghanistan, and even Khurasan remained precarious at best. A fair observer might have concluded that

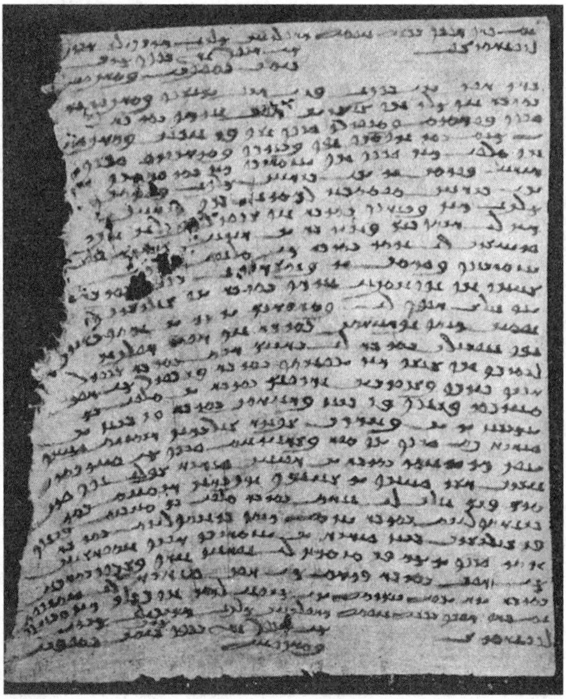

Figure 4.1. Writing on the eve of the Arab destruction of Panjikent, a loyal civil servant warned his king of the looming danger. Written in Bactrian, his epistle, shown here, was among Sogdian manuscripts discovered by a Tajik shepherd in 1933. ▪ From *Drevnosti Tadzhikistana: katalog vystavki*, Academy of Sciences of the Tajik SSR, State Hermitage Musuem, Leningrad (Dushanbe, 1985), 246.

the Arabs' struggle over three quarters of a century had been in vain. Worse, during the civil war the Arabs had inadvertently mobilized anti-Arab sentiments among most of the population of Central Asia, which was to have grave consequences a generation later.

DEAL MAKING DURING THE CENTRAL ASIAN THERMIDOR

Accepting reality, Arab administrators concluded that Greater Central Asia could never be secured through military means, and that diplomatic, religious, and economic concessions were called for. In 739 they therefore struck a series of deals with the rulers of Chach and Ferghana, the Sogdian kings, and their Turkic allies.[55] Here and elsewhere these concessions confirmed the authority of the traditional local rulers.

With respect to Islam, the Arabs abandoned their ill-advised circumcision test for new converts and lifted the taxes they had attempted to impose. Going still further, a new governor named Nasr, ignoring the Quran's demand that apostasy be punished by death, in 741 struck a deal with the Sogdians that abolished all punishments for those who had reverted to their former religions. Further, he nullified prior debts to the state and even abolished all private debts. The caliph strongly objected to these concessions, but his politically more astute governor knew that anything less would have called forth renewed rebellion. Instead, he emphasized the more normal tax on property (*kharaj*), which did not differ greatly from levies that had long existed in the region. Such measures brought about a renewal of commerce and assured that the coinage that the Arabs began minting in several cities remained stable.[56]

At lordly Balkh, this combination of political and economic concessions produced one deal that was to prove enormously important a generation later. The initial Arab assault had left large parts of the city in ruins. As the civil war wound down, the Arabs agreed to construct a new city a few miles east of the ancient center. Spearheading this ambitious development project was the wealthy Barmak family, long the administrators and protectors of Balkh's renowned Buddhist monastery and scholastic center, Naubahar, or New Vihara. Not only did this arrangement seal the alliance between this influential family and the Arabs, but it led directly to the Barmaks becoming the center of political might and the world's most powerful cultural patrons during the greatest days of the caliphate.

Such adroit concessions inspired other local kings and aristocrats to strike their own deals with the Arab governors. So successful were these that a Persian grandee at Herat in Afghanistan declared that he had experienced "nothing deserving of blame" in his dealings with the Arabs and that they were, in fact, "excellent overlords." Proof of this newfound comity is the fact that local rulers did not hesitate to call on the Arabs to support them against their enemies.[57] A more lasting fruit of this respite from strife was the Arabs' decree requiring all official documents to be written in Arabic. Issued in 741, this seems to have met with general acceptance.[58]

The Arab pullback, coupled with timely concessions, created a welcome mood of normalcy. In the end it was to prove illusory, a kind of Muslim Thermidor, but for the time being it was real. Nowhere was this belief that life had returned to normal more poignantly evident than at the palace of the kings of Bukhara at Varakhsha, situated on the edge of the Bukharan oasis twenty-eight miles west of Bukhara.[59] At this heavily walled fortress along the Zaravshan River, the Sogdian kings of Bukhara had long maintained their court. There had probably been a palace at this site since ancient times, but the grand edifice excavated by Soviet archaeologists between 1936 and 1991 was erected *after* the Arab invasion and was extensively expanded during the post-Qutayba thaw.

Murals in several of the audience halls at Varakhsha present the king in the same vivid colors that had been employed at Samarkand and Panjikent during the last years before the Arabs attacked. The technique the artists employed—painting on dry plaster—had been imported from India.[60] Here were realistically depicted jaguars and mythical beasts, hunting scenes, ceremonies, rulers and heroes from the national epics—in short, precisely the sort of murals that were being painted before the Arabs showed up. The main difference—and an important one—is that the Varakhsha murals insistently repeat the image of the king. To one of the Soviet archaeologists, it was obvious that the ancient artists were using what he called an "Aesopian" language to proclaim the continuing vitality of local rule.[61] Set in borders painted to resemble architectural ornaments, these large wall paintings present a bewildering array of elements drawn from diverse religions. Bukharan kings were shown worshipping at a Zoroastrian fire altar. The paintings portrayed naked Indian divinities, probably Buddhist, mounted on elephants and doing battle with wild beasts. And the murals included also a triumphant Sogdian god of war. Whoever commissioned these paintings clearly felt no need to soft-pedal or encode his beliefs.

We will never know if this was a show of defiance or proof that the Bukharan kings really believed that the Arabs had now left them in peace. The absence of any new defensive walls from the time the murals were painted supports the latter view. In the end, they were to be engulfed by the fresh whirlwind that swept all Central Asia and then the entire Muslim world after the arrival in Khurasan of the missionary, general, and

Figure 4.2. Tashkent architect V. Nilsen's reconstruction of the palace at Varakhsha, home of the rulers of Bukhara before and after the Arab invasion. ▪ Courtesy of Aleksandr Naymark.

charismatic leader Abu Muslim. But even then, the old faith defiantly endured at the Bukhara court for another century, in other words, two centuries after the Arabs launched their invasion.[62]

ABU MUSLIM AND THE DESCENDANTS OF ABBAS

The unexpected death of the Umayyad caliph Abu Hashim in 743 brought to the fore all the unresolved problems of Central Asia and of the caliphate itself. Who would succeed him? Those Muslims who had always viewed the Umayyads as usurpers now made their move. An obvious contender would have been a descendant of Muhammad's grandson, Husayn, but the Shiites who championed this course were almost everywhere out of favor. A second band of dissenters championed the line of descent from Muhammad's youngest uncle, Abbas. Never mind that Abbas seems never to have become a Muslim, since his son had proven his piety by passing down many Hadiths of the Prophet. More important, the Abbas faction, concentrated in Mecca and the Iraqi city of Kufa, was ready to fight.

Among their first moves was to dispatch a missionary to advance their claim in the critical power centers of Khurasan and the rest of Central Asia. The man they chose was a proven propagandist and warrior but otherwise a complete enigma. His nom du guerre, Abu Muslim, conveniently disguised his origins. Most assumed him to be the son of an Arab convert, but modern scholarship has produced convincing evidence that he was a native Central Asian of Iranian stock and from a village near Merv.[63]

Soon after reaching Khurasan in 747, Abu Muslim won over the local Arabs to his cause, and some of the local aristocrats as well. He accomplished this by making a show of repressing a dissident Zoroastrian sect that both local Arabs and the Zoroastrian elite considered a threat. He then moved on Merv, where he easily dispatched the Umayyad governor, Nasr, the same man who had engineered so many concessions to local rulers and nonbelievers a few years earlier. But now it was the ambitious Abu Muslim's turn to compromise on doctrine. He extended a hand to the large Christian community, welcomed the support of Zoroastrians, and courted heterodox Muslims as well. From local folk beliefs or Hinduism he picked up the idea of the transmigration of souls and postmortem reincarnation. As Abu Muslim seized on these elements from other faiths, some people began to question his devotion to Islam. Both supporters and detractors had every reason to suspect that he harbored ambitions that went far beyond his mandate from the Abbasids in Iraq.

These suspicions gained credence when Arabs at Bukhara rose in revolt against him.[64] They, too, had benefited from Nasr's deal with the local king and saw that Abu Muslim now threatened to undercut their position on that key oasis. Abu Muslim promptly sided with the local forces that arose against the rebellious Arabs. What had begun as an Arab-led revolt against the Umayyads now showed signs of evolving into a regionwide rebellion against any form of Arab rule from Damascus.[65]

ABU MUSLIM'S WONDER YEARS, 750–751

In 750 and 751 Abu Muslim devastated all the forces arrayed against him. To accomplish this stunning feat, which had eluded the Arab invaders for a century, he had first to secure Central Asia itself, then to finish off the Umayyads in Damascus, and finally to destroy the rising threat posed by China.

When Abu Muslim arrived in Central Asia, his most obdurate foes were the western Turkic tribes, most of which had long fought Arab rule, and the eastern Turks to whom local rulers had successfully appealed for support against Damascus. His forces swelled with Arab and local recruits, Abu Muslim was able to deliver several decisive blows to the western Turks and then chase the eastern Turks out of the Central Asian heartland. Fighting in the east was to continue for several more decades, but Abu Muslim had broken the back of Turkic power.[66] Acknowledging this, and perceiving that the Arab who defeated them was actually championing the locals' cause, thousands of Turkic warriors now enlisted in Abu Muslim's army, further diluting the Arab character of what was fast becoming a regionwide crusade.

Taking note of Abu Muslim's prowess in battle, the followers of the Abbasid family in Iraq summoned him back to the Middle East so that he could help finish off the now dispirited Umayyads. After a forced march across Central Asia and Iran, Abu Muslim confronted the entire Umayyad army on the banks of the Great Zab, a tributary of the Tigris in Iraq, in January 750. No sooner had he devastated this force than he moved on Damascus itself, where he handily overthrew the last of the Umayyad caliphs.

But all was not well back at Abu Muslim's power base in Central Asia. Even before his defeat of the Turkic tribal armies, the Chinese had moved swiftly to take advantage of the emerging power vacuum in the eastern regions of Central Asia. By 748 the Tang forces, supplemented by Turkic troops, had taken all of Xinjiang, as well as the Seven Rivers (Zhetisu) region of present-day Kazakhstan and Kyrgyzstan, Chach (Tashkent), and most of the nearby cites.[67] Written off as irrelevant a generation earlier, Tang China was now but a step away from establishing itself as the new regional hegemon in much of Central Asia. Abu Muslim and his army rushed eastward to confront this danger.

The turning point was reached on the rolling steppe land in the Talas River valley southeast of the old trading city of Taraz, now Talas, on Kazakhstan's border with Kyrgyzstan. Here, in July 751, the Chinese army suffered a crushing defeat at the hands of forces led by Abu Muslim's field commander, Ziyad ibn Salih. But even though this forced the

Tang emperor to refuse help to a coalition of Central Asian cities that had again appealed to him, the struggle was not yet over. In fact, the Chinese threat did not die until 755, when a massive rebellion and attempted coup d'état arose on China's northeastern border. The Chinese army general who led this rebellion, An Lushan, was himself the son a Sogdian immigrant to China (his Chinese name means "the Bukharan") and a Turkic mother. The human cost of the uprising was appalling, with estimates ranging up to thirty-six million; one recent writer claims it was proportionately the single greatest atrocity in history.[68] Whatever its true scale, the An Lushan revolt forced the emperor to hurriedly withdraw his entire army from Central Asia, thus handing final victory to Abu Muslim's forces.

Who Won?

China's defeat in the Talas Valley is usually seen as an Arab and Muslim victory. While true in some general sense, it was at least as much a specifically Central Asian triumph. After all, the nominally Arab armies by now comprised mainly Central Asians of both Iranian and Turkic ethnicities. True, the field commander, Ziyah ibn Salih, was an Arab, but their ultimate leader, Abu Muslim, was from the region and had increasingly accommodated himself to its interests, as opposed to the interests of his initial Arab sponsors. And few of the troops besides the core Arab cadre were actually Muslim, for the process of conversion in Greater Central Asia proceeded very slowly; even two centuries later, Muslims constituted barely a tenth of the region's population.[69] Given this, it is safe to assume that most of the supposedly Arab and Muslim forces on the Talas field were impelled more by the desire to clear the region of foreign armies than by any spirit of religious jihad. Had their goal been pure conquest, whether in the name of Islam or of Arab rule, Abu Muslim's commanding general would surely have taken advantage of the victory and pushed eastward into Turkic Xinjiang. But he did not and in fact turned back without even subduing the Ferghana Valley or the large Seven Rivers region—all of which lay open before him.

The evolution of Abu Muslim's ideology in the aftermath of Talas provides strong evidence that he had shifted his first loyalty from Damascus to Central Asia and in the process embraced many views that were anathema to most orthodox Muslims. Of course, Abu Muslim's army marched under the black banner of the Abbasids, whose religious goal was to entrust the caliphate to a dynasty that was fully legitimate. Yet increasingly Abu Muslim had taken over regional leadership of the emerging Shiite cause as well. Moreover, he had shown himself sympathetic to many starkly heterodox local beliefs drawn from Buddhism and other faiths. This is not surprising, since from the moment of his arrival in Khurasan, Abu Muslim had been like a mighty tornado, sucking all culture forces into his vortex, absorbing them and gaining strength from the process.

Abu Muslim's embrace of heterodoxy did not stop with doctrines of the soul. To many contemporaries it seemed that this charismatic and seemingly infallible leader was moving toward embracing the Eastern notion that he himself was some kind of incarnation of the divinity.[70] There could be no doubt that Abu Muslim was conducting himself as an independent potentate, the lord of Central Asia and much of Iran. The palace he built for himself at Merv placed the ruler's throne at the center of four domed rooms, as if to say that he ruled all points of the compass.[71] Needless to say, this aroused grave concern among his now triumphant Abbasid sponsors back in Iraq.

In 754 the Abbasids appointed as caliph one of their number, Mansur, who was single-mindedly dedicated to crushing all rival powers and all manifestations of heresy. In an effort to separate the ruler of Central Asia from his power base, he appointed Abu Muslim governor in Syria. Abu Muslim knew the net was tightening around him and penned a groveling letter retracting his many heretical views.[72] This pitiful step did not satisfy Mansur. Still fearing a coup, he treacherously murdered his rival from Central Asia and had his body thrown into the Tigris. He then hunted down and killed as many surviving descendants of Husayn, the son of Ali and grandson of Muhammad, as he could. In a gruesome show of the man's character, Mansur installed the corpses of these heirs of the Prophet, all dried and neatly labeled with tags affixed to their ears, in a sealed vaulted chamber in his palace.[73] By these moves the new Abbasid caliph believed he had decapitated the rising power of Central Asia and stamped out Shiism and all other forms of heterodoxy to boot. He was wrong on both counts.

Word that the new caliph had murdered Abu Muslim quickly reached Khurasan and the rest of Central Asia. To a degree that no one could have predicted, Abu Muslim's charisma survived his death. For thirty years it floated like a cloud over the region, to be seized and appropriated by each new populist leader or dissenting preacher in the land. Indeed, the early Abbasid years—the era when culture flowered at Baghdad—was also a period when much of Central Asia fell under the spell of a series of would-be liberators.

An early rebellion was mounted in Khurasan by one Sinbad (or Sunpad), a Zoroastrian who raised an army of 100,000 on the proposition that Abu Muslim had not died but had turned into a white dove.[74] Another Central Asian rebel was called Ishak, "The Turk," an illiterate follower of Abu Muslim who declared himself the successor of Zoroaster.[75] Then an ethnic Persian cloth worker from Merv named Hashim, son of Hakim, who had risen to general rank under Abu Muslim, declared himself the successor to every prophet from Moses and Christ through Muhammad and, significantly, Abu Muslim himself. Wearing a green cloth over his presumably disfigured face, al-Muqanna, or "The Veiled One," rallied the populace in the region from Bukhara through present-day Tajikistan to avenge the Abbasids in the name of his one true faith.[76] So colorful was the memory of this apocalyptic rebel that it inspired a poem by the nineteenth-century Irish poet Thomas Moore, a story by the Argentine author Jorge Luis Borges, and a carnival organization in the American city of St. Louis. Meanwhile, for nearly a century after Abu Muslim's death, Central Asians continued to support anyone who raised the banner of populist rebellion against Arab domination.

No region played a more crucial role than Central Asia in bringing down the Umayyad Caliphate in Damascus and installing the Arab dynasty of Abbas in its place. The Umayyads had garnered enough strength to capture Central Asia but not to hold it, in spite of the relentless and brutal efforts of their best general, Qutayba. Their only recourse was to make whatever concessions to local rulers and cultures as were called for. Abu Muslim transformed this tactical necessity into a strategic priority, thanks to which his brief but astonishing rule gained a strongly Central Asian cast. It was a Central Asian army that defeated the Chinese at the Talas River, and it was Abu Muslim's Central Asian forces that destroyed the last Umayyad army in 750. It is hard to imagine the Abbasids' rise, or the dramatic changes in the Islamic world that followed, without the power and authority wielded by Central Asia.

The Arab and Islamic conquest of the Eurasian heartland is traditionally viewed as a political and cultural wave rolling from west to east and finally crashing over the expanses of Central Asia. But this misrepresents what actually occurred, for an "equal and opposite" Newtonian counterforce surged from Central Asia toward the Arab West after the initial Arab incursions. Qutayba and his troops destroyed priceless treasures of Central Asian civilization, but at the same time they released mighty cultural and political energies from that same source, and these eventually prevailed. These energies found an outlet in the caliphs' new capital, Baghdad, as well as within Central Asia itself.

Thanks to this cultural explosion, and to the strange but effective manner in which Abu Muslim engaged and channeled its energy, the Abbasid caliphs who served after his death were singularly beholden not just to their Arab base in the Middle East but to the Iranian and Turkic East, the heartland of Greater Central Asia. Of course, this also included the people of Iran proper, who brought to bear on the new caliphate many practices and attitudes inherited from the Sassanian Persian empire. Even though this state had collapsed under Arab pressure, its memory was omnipresent, not least because its ancient capital, Cstesiphon, was a mere twenty miles southeast of Baghdad.

The Abbasids' dependence on Persianate and Turkic Central Asia was even stronger. Unlike the Iranians, the Central Asians under Abu Muslim had forged a powerful regional army and used it to destroy two of the main threats facing the Abbasids, namely, the armies of the Umayyads and of China. The Abbasid regime would continue to depend significantly on Central Asian forces and personnel. The caliphs' dependence on Central Asia and Central Asians enabled that region, alone among all the territories the Arabs conquered, to maintain its social and economic system virtually intact. This assured that Central Asian culture, too, would continue to thrive, that it would maintain much of its old strengths and character, and, eventually, that it would exercise a powerful influence on the caliphate itself. Much the way that Greek culture eventually conquered a triumphant Rome, Central Asian culture was now positioned to conquer the Abbasids and their new capital, Baghdad.

CHAPTER 5

❀

East Wind over Baghdad

Who does not know of Baghdad's golden age, the storied era of the *Thousand and One Nights*? The very name of the reigning caliph, Harun al-Rashid, conjures up images of Sheherazade telling tales and of court poets whiling away limpid evenings in opulent gardens. Our more earnest friends will remind us that ninth-century Baghdad was also a center of intellectual life, the place where purportedly "Arab sciences" reached their splendid apogee. All this was sustained by the caliphate's unparalleled economic and political might. So rich were the early Abbasid caliphs that one of them, as a trifling gift to his far-off Christian contemporary, Charlemagne, sent a gorgeously equipped war elephant. The poor beast, which had actually been regifted from an Indian rajah, ended its days fighting heathen Scandinavians at the North Sea in Denmark.[1] For his part, Charlemagne knew that Baghdad's armies could do what he could only dream of, namely, attack bejeweled Constantinople.

The story of Baghdad's flowering under the Abbasid rulers has been told so many times in so many languages that it has taken on a canonic form. Scholars have recently reevaluated, or are in the process of reevaluating, many of the themes that run through this story. But the main narrative, reflected in numerous general studies, including some very recent ones, characterizes the era as a golden age of "Arab" learning. While it nods to the fact that this was a time when Arabic and Persian traditions of thought came together in a mutually enriching process, the former is still the predominant element, the driver. Many studies have sought to give the Persian component its due, but these have yet to be adequately reflected in the general literature. More to the point, these rarely, if ever, disaggregate their subject in such a way as to distinguish the specifically Central Asian contribution. And while there is increasing recognition of the religious diversity of the age, the movement as a whole is seen mainly as Islamic in character, with other religious traditions, as well as currents of thoroughgoing secularism and skepticism, assigned to minor roles. Overall the great achievement of Arabic thinkers of the Abbasid era is seen as having translated and hence preserved classic Greek texts, which they then passed on to a grateful western Europe which, emerging from its long centuries as a backwater, used them to help spark a general Renaissance.

In perspective drawing, what is nearest to the observer appears large and what is distant appears small. In the eyes of Western writers—both Europeans and Mediterranean Arabs — Central Asia is somewhere near the vanishing point, and its role in the narrative on the caliphate has been diminished accordingly. However, if the perspective is shifted from west to east, that is, from the Mediterranean to Baghdad itself or, better still, to Central Asia, a quite different picture emerges.

We have seen how this problem of perspective presented itself during the fall of the Umayyad caliphs and the rise to power of the Abbasids. It is true that the descendants of Abbas first raised their call for change in southern Iraq, but their campaign gained traction only when the charismatic Abu Muslim successfully engaged large numbers of his fellow Central Asians in the movement. The new caliphs were indeed Arabs by ethnicity, but their power base was now in the East—in Iran and, still more, in Central Asia. Caliph Mansur was so acutely aware of the power that Abu Muslim and his Central Asian army could wield against him that he murdered the charismatic leader from Merv and dumped his body into the Tigris. When even this failed to weaken the power of Khurasan and Transoxonia (the land beyond the Amu Darya), Mansur and his successors resorted to a policy of concession and cooptation. In the process, the entire caliphate gained a markedly Central Asian cast. Elton Daniel affirms that it is "fairly clear that the issue of central authority and regional autonomy was at the heart of these events."[2]

The eastward shift of the economy, and hence of the tax base, of the Islamic world reinforced this trend. And the largely Turkic Central Asian army that brought the Abbasids to power continued as the backbone of the caliphate's military might. The caliph fully understood that these fighters from the East had more power over him than he did over them.

We will see that culture, too, felt this powerful wind from the East, and that a significant number of those brilliant "Arab" scientists were not Arabs at all, but Central Asians who chose to write in Arabic. Earlier we suggested that the cultural impact of Central Asia on the new caliphate bears comparison to the triumph of classical Greece civilization over Rome. But there was one important difference: the Greeks shaped the cultural life of Rome but controlled neither the military nor the economy. Central Asians under the caliphate dominated the intellectual class but also the army and much of the economy.

BAGHDAD

On July 30, 762, a crowd assembled on a desolate spot along the Tigris River fifty-five miles north of the ruins of ancient Babylon and twenty miles from the old Persian capital at Cstesiphon to dedicate a new, circular-shaped capital of the Muslim world, Baghdad, called by its founder Madinat al-Salam, or "City of Peace." The caliph, Mansur, had ruled for eight years and had gained a reputation for pursuing his interests with single-minded dedication to the faith, often with gratuitous brutality. It is curious that he delegated the selection of the site, near several Syrian Christian monasteries, to a team of astrologers headed by a highly respected man of learning from Merv. This was Abu Sahl al-Fadl Nawbakht, who was among the many scholars involved in translating books at Merv long before Baghdad existed.[3] Nawbakht's son was later to be placed in charge of the caliph's new library. With the entire Muslim world to choose from, it is revealing that Mansur turned to an astrologer from Central Asia when he needed an intellectual heavyweight.

Nawbakht was not the only Central Asian on hand that day. Many of the 100,000 construction workers had come from Syria and Iraq, to be deployed by the army in platoons.[4] But the skilled work was done mainly by a phalanx of artisans and craftspeople who made the long trek from Merv to the building site. The latter were so numerous that they were assigned a whole quarter in the new Round City, a distinction accorded to no other regional group. So important was Merv to the Abbasids' victory and the foundation of the new capital that a principal road in the Round City was named for the Central Asian capital, the only city in the caliphate to be thus honored. Other streets were named for the caliph's guard, the police, water carriers, and prayer-callers.[5] Among the group from Merv was a certain Hanbal, who died soon after the move but whose son was later to play a central role in a theological controversy that nearly overwhelmed the caliphate.

The influence of Merv on Baghdad's foundation was not limited to astrologers and artisans. The plan itself was said to have been the work of yet another man who had come from Merv, Khalid ibn Barmak.[6] This recent convert from Buddhism was born in Balkh in Afghanistan, where his family had long been custodians of the vast and wealthy Buddhist center of Nawbahar, just over a mile west of the city. The Barmaks were said to have descended from a Buddhist immigrant from Kashmir, but since Nawbahar was already several centuries old, they had had ample time to rise to the top of the establishment at Balkh. At any rate, the Barmaks' rich shrine was among the earliest targets of Arab destruction,[7] leaving the family no choice but to consider collaborating with the new powers that be.[8] Significantly, when the Arabs finally got around to rebuilding the old city, they turned to Khalid ibn Barmak to manage the job.[9] In the process of rebuilding Balkh, Khalid discovered that he could work with the Arabs. He therefore betook himself to Merv, where he rose to prominence just as Abu Muslim was raising his revolt against Umayyad rule from his base in that city.[10]

Rich and accustomed to viewing power as a birthright, Khalid promptly announced his conversion to Islam and adroitly aligned himself with the rising regional force that Abu Muslim was directing against the Umayyads and their strong Arabian and Syrian orientation. In an attempt to come to terms with Abu Muslim, the future caliph Mansur had actually traveled to Merv. His mission failed, but while in Merv he met the intellectual and astrologer Nawbakht, and also the competent and ambitious Khalid ibn Barmak, whom he marked as someone to be drawn to his side.

Nothing is known of Khalid's role in the planning of Baghdad besides the fact that he was directly involved. By the time of the dedication, the alignment of walls, palaces, and mosques had all been marked on the ground with ashes. The plan itself impressed with its utter simplicity: Baghdad (in this, its earliest form) was to be a geometrically perfect circle defined by a tall circumferential wall and a lower outer wall, with four gates leading to the central edifice. At the precise center of the city was not the mosque but the caliph's green-domed palace, atop which was mounted the anthropomorphic figure of a rider with a spear in his hand.

Whence came the idea for this circular and utterly rational plan for the new capital? Of course, it could have arisen de novo, the invention of one of Mansur's advisers or of the caliph himself. But had this been the case, contemporary chroniclers would surely have mentioned it, which they do not. Instead they reported only that Khalid ibn Barmak had been involved in the process. In the absence of a known planner or designer, attention has shifted instead to possible prototypes. Three alternative sources have been proposed.

First, the plan has been ascribed to the several cities with circular walls that already existed in the Tigris Valley. The most prominent of these, the circular city of Ardeshir-Khwarrah (now Firuzabad, Iran), had been the Persian capital half a millennium earlier but lay in ruins even before Muslim armies swept across the province of Fars.[11] It is difficult to imagine that the mightiest empire of the era would model its capital on the ruined ancient capital of an empire it had vanquished.

Both the second and third hypotheses take us back to Central Asia. During the same visit to Merv in which Mansur met Khalid ibn Barmak, he would have become familiar with the massive Erk Kala, the thousand-year-old citadel of the ancient capital. Its mighty walls, which even today rise over eighty feet, define a perfect circle that protected the palace and principal civic buildings.[12] The future caliph could not have failed to be impressed by this citadel, the more so since it was from here, at Merv, that the upheaval that brought his Arab family to power had erupted. If there is a question regarding this otherwise promising theory, it is that Erk Kala, at 94 acres, is much smaller than the 740 acres sometimes claimed for Mansur's Baghdad.[13] But this estimate for Baghdad has now been shown to be greatly exaggerated; indeed, it is fully three times larger than Damascus at its peak.[14] Mansur's Round City was in fact not a city at all but a palace complex and administrative center[15]—precisely the function that Erk Kala served at Merv. Hence the size of Erk Kala matches that of Mansur's City of Peace.

A third theory traces the circular plan back to the Barmak family's ancient Buddhist center at Balkh.[16] Built in the form of a circle, the Nawbahar monastery was even smaller than the citadel at Merv. At dead center of the Buddhist shrine stood the spire of the stupa surmounted with banners. The main differences between Nawbahar and Mansur's city, besides their size, is that Nawbahar was built to a strict hub-and-spokes

Figure 5.1. Erk Kala, the 2,500-year-old citadel at Merv and apparent model for Caliph Mansur's Round City of Baghdad, founded in 762. ▪ Courtesy of OrexCA, Tashkent, Uzbekistan, www.OrexCA.com. Photo by DN Tours.

plan while most of the interior space of the Round City was taken up by a rectilinear palace complex disposed, as at the Erk Kala, in a curiously asymmetrical manner within the circle.

Both the Erk Kala and Nawbahar hypotheses are consistent with the known role of Khalid ibn Barmak in the planning process. But it is hard to imagine that as severe a Muslim as Mansur would model his capital after a Buddhist monastery, let alone one that he had never actually seen. Hence, Erk Kala is the more likely prototype. Merv, after all, was the base from which the Abbasids came to power. Its role as a model fits with the several other acts of deference Mansur made to that city, such as naming a street and a quarter after it, alone of all cities in the caliphate. Besides, Mansur had actually visited the Erk Kala.

Thus the genealogy of the world's greatest city over the next two centuries traces directly to Central Asia. Even though Mansur's circular city was soon engulfed by the world's fastest growing metropolis, it remained a living reminder of the power of the new cultural breeze blowing over Baghdad from the East. Far to the west, Venice, the future queen of European cities, did not gain its independence and begin serious development until sixty years after Mansur founded Baghdad. The whirlwind of cultural activity that engulfed Baghdad during the first century and a half of its existence had no equal anywhere.

Harun al-Rashid: Playboy and Jihadist

Mansur died in 776 and a peaceful succession left his twenty-year-old son, Harun al-Rashid, as the "Leader of the Faithful." Harun, Arabic for

Aaron, was to rule for nearly a quarter century, leaving behind the commendable sobriquet "Harun the Just." He could just as well have been remembered as "Harun the Wealth-Maker," for during his reign Baghdad emerged as the richest city on earth and his government as a paragon of munificence. Harun put his riches to good use. As a result, historians who are otherwise at sword's point with one another hail him with one voice as one of the main patrons of Baghdad's golden age and a prime force in bringing about the greatest intellectual boom between antiquity and the Renaissance.[17]

What they are less clear about is whether Harun actually caused the cultural effervescence with which his name is forever associated or simply allowed it to take place under the guidance of others. At one level this question may seem to arise from mere petulance, the inevitable posture of historical revisionists who are eager to cut heroic figures down to size. As such, it need not concern us. At another level, however, it is of real moment, for it opens a window onto the crucially important activities in Baghdad of that great Central Asian family whose patriarch we have already encountered, the Barmaks.

One thing is for sure: none of the preceding caliphs, whether the fourteen Umayyads or the four prior Abbasid rulers, had benefited from so thorough an education as Harun had. His father had had the good sense to entrust him to the care of Khalid ibn Barmak's erudite and perspicacious son, Yahya.[18] The tutorial program of this scion of the Barmak clan included wide reading in ancient and modern texts and, equally important, long discussions of the application of theory to practice. Yahya ibn Barmak brought formidable experience to both tasks. Besides his classical education, Yahya could draw on his service as vizier, or prime minister, to a previous caliph.

Before ascribing to Harun the role of culture-maker, however, it is well to note the very unserious side of his personality.[19] Even if we dismiss as apocryphal half of the stories about him recorded in the *Thousand and One Nights*, the other half remain. These and other reports have him catting around Baghdad by night in search of willing females. On these nocturnal missions, always carried out incognito, the young Harun was often accompanied by one of his tutor's sons, Jafar ibn Barmak, a contemporary and close friend. There is no reason to believe that Jafar reported to his father on their adventures, which included trips down the Tigris to meet beautiful singers, drink, and carouse. The specificity of some of the surviving tales, as, for instance, their finding the dismembered body of a woman done in by her husband, leaves little doubt as to their authenticity.

Even if we reject all these reports as slander, there remains the fact that Harun was intimately associated with the rise of an entire new genre of poetry about drinking. And as to his womanizing, more reliable sources record the names of a whole bevy of ladies from Harun's large harem, among them "Beauty Spot," "Charm," and "Splendor."[20]

To the extent that Harun was serious about anything, it was about his commitment to the principle of jihad, the spreading of the faith through force of arms.[21] He invested huge sums in his army and personally led every campaign. His idée fixe since before he became caliph was the conquest of the Byzantine Christian capital of Constantinople. Even though his siege failed, the struggle against Byzantium continued to preoccupy him. To this end, he moved his own headquarters to the city of Raqqah in northern Syria and abandoned Baghdad for more than a decade. He paid a heavy price for all these heroics. Not only did Harun neglect the management of the empire, but his various campaigns sapped the caliphate's resources to the point that discontent and centrifugal stirrings arose throughout the realm, from North Africa to Central Asia. Thus, the seeds of Abbasid decline appeared at the very time of the caliphate's golden age.[22]

Not only did the Barmaks help bring Harun to power in the first place through what amounted to a coup, but thereafter Harun's prolonged absences from Baghdad left the city in the hands of his deputies, specifically, the Barmak clan. Under three caliphs they had filled the post of vizier and chief adviser to the throne, and they continued to do so under Harun. Thus Central Asians were the rulers of the caliphate in all but name.

Their connections with power made the Barmaks fabulously rich, which would eventually prove their undoing. But for now, they spent with abandon. Yahya is said to have paneled a room in his mansion with gold tiles, while his son Jafar, remembered for his prowls with Harun, spent twenty million gold pieces on his residence. The lavishness of their entertainments is recalled in the expression "a Barmacide feast," which appears in the *Thousand and One Nights*. Nor did the family abandon its connections in Central Asia. When that region again began to exhibit restiveness, Yahya himself returned there as governor of all Central Asia, which included the family's ancestral seat at Balkh. He proceeded to demonstrate his local loyalty by keeping all tax revenues within the region—an act of stunning defiance of Baghdad that led to his replacement by a man who returned the province to the former cycle of exploitation and resistance.[23] More important, Central Asian intellectuals were frequent guests of honor at Barmak salons and at the many discussions evenings organized by members of this illustrious family.

In the area of culture, an informal division of labor arose between Harun al-Rashid and the Barmaks. While the caliph patronized poets and musicians, nearly all of whom were ethnic Arabs, the Barmaks concentrated on the natural sciences and humanities, in which people of Iranian or Turkic stock—mainly Central Asians—figured more prominently. Yahya Barmak, his son Jafar, and his second son, Fadl, the most serious of the lot, became Baghdad's most active patrons of new thinking in philosophy, mathematics, astronomy, and medicine.[24] Many, and in some cases most, of the leading lights in these fields hailed not from what is now Iran's heartland but from Central Asia.

The Barmaks' most significant contribution to world civilization was as the patrons of translations from ancient Greek. Caliph Mansur had sponsored a few translations, but his interests scarcely reached beyond astrology.[25] Harun al-Rashid might have continued along these lines had it not it been for the Barmaks and the much broader perspective they brought to Baghdad. For centuries the Buddhist Barmaks had served as cultural intermediaries between India and Central Asia. Along with other centers in Sogdiana and Khurasan, the monastic community at Nawbahar had been a center for translating and editing Buddhist texts from India and passing them eastward to China. Now the Barmaks continued this tradition as Muslims, forging a link between Eastern learning and Baghdad. It was Yahya ibn Barmak who oversaw the translation of Indian works on medicine and who subsequently served as a patron of the medical sciences at Baghdad.

The Barmaks' experience over the centuries had made them deep-dyed cosmopolitans. As Central Asian Buddhists, they had had no doubt that the achievements of certain other cultures, especially India, were in advance of their own and that the locals would benefit from a broader approach to the world. They were certainly aware also of translations from Sanskrit, Babylonian, and Greek documents that had been done in Merv, and they knew how exposure to them could enrich life.[26] Once in Baghdad, they learned that the great predecessor empire in the region, Sassanian Persia, had for centuries taken exactly the same approach to the achievements of other peoples. They learned that as early as the third century the Sassanians had established at Gundeshapur a center for the translation and study of ancient texts, mainly from Greek and Syriac. Finally, they learned that the Sassanians had had no problem patronizing this activity, even though the translators were Christians, albeit ones who had been driven out of the Byzantine lands and had gratefully found refuge among the Persians.[27]

Because the task of these translators was to render classic Greek and Syriac texts into Pahlavi, the language of the Persian court, they provided a nearby model of the same kind of cosmopolitanism the Barmaks had always championed. It did not hurt that the main focus at Gundeshapur had been medicine, the field in which the Barmaks took a special interest. In short order the first Nestorian Christian translators began arriving in Baghdad to work under the protective wings of the Barmak family.

The Barmaks also sponsored translations from the Sanskrit. Kevin van Bladel, who has closely examined the Barmaks' activities, concluded that "almost all" the translations from Sanskrit during the Age of Enlightenment were carried out at the initiative of this family of former Buddhists from Balkh, especially Yahya.[28] While the list of works translated included several astronomical handbooks, medical texts that could be used at the Barmaks' hospital were especially prominent. Thanks to this Central Asian channel, major works of Indian thought entered the Islamic world at the same time as the Greek classics.

Did the translations sponsored by the Barmaks have a religious agenda as well? Their family's translating work at Nawbahar certainly had had such a purpose, namely, to advance Buddhism. Zoroastrians, too, had translated certain texts of their faith into Arabic as a means of disseminating their beliefs.[29] But the Barmaks and the Baghdad court selected for translation mainly books in medicine, science, and philosophy. While the claim that the campaign to translate Greek thought into Arabic had no overt religious agenda may be true, however, it does not mean that the translations had no implications for religion. After all, as philosopher John M. Cooper points out, "the ancients made philosophy the . . . only authoritative foundation and guide for the whole of human life. . . . Philosophy would be the steersman of one's whole life."[30] Now, this was precisely what Christians had claimed for their revealed faith and what Muslims were now claiming for their guiding revelation. Classical Greek philosophy and Islam, in other words, were on a collision course. Over time, both Christian and Muslim thinkers came to focus more on the later followers of Plato, with their emphasis on the immaterial world and the soul, as opposed to the more materialistic thinkers of Greece's classical age. Both Christians and Muslims thought they were co-opting the works of these Neoplatonists, but the reverse was at least as true. From the very outset, the pious of both faiths worried that philosophy was less a supplement to religion than an alternative to it. To be sure, such thinking was slow to take hold among many Muslims, notably those of Central Asia. When eventually it did, it marked the waning, not only of the classical tradition of philosophy in the Muslim world, but of Greek thought there generally.

As sponsors of translations, the Barmaks soon found themselves in the book business. Translations they had commissioned became the subject of presentations at their salons, and attendees naturally wanted their own copies. To meet this demand, the Barmaks set up the first paper mill in Baghdad.[31] A type of fine paper was even named for one of the Barmaks.[32] Rising demand for books also caused the copyist's trade to flourish. Bookshops and even book auctions followed in quick order, so that Baghdad soon emerged as a center for both the production and marketing of the written word.[33]

BLIND SPOTS OF THE BARMAKS' TRANSLATION MOVEMENT

The translation movement launched by the Barmak family in Baghdad was a cultural achievement of the first order. For the first time the rapidly expanding circle of Arabic readers had access to classical authors of the ancient Mediterranean world. Among those whose works were to be translated over the following century were Aristotle, Ptolemy, Euclid, Aristarchus, Archimedes, Nicomedes, Hipparchus, and Heron. To achieve this, the caliphate borrowed the model of a center for translation and study that their predecessors, the Zoroastrian Persian empire of the Sassanians, had long sponsored at Gundeshapur. This prototype focused on works that were deemed useful, as opposed to history and belles lettres.

No less important than the Persian example was the crucial role played by the Barmaks. Thanks to their cosmopolitanism and their long experience at both Balkh and Merv with translation and scientific cross-fertilization, they appreciated the value of creating a Muslim Gundeshapur or Nawbahar at Baghdad and used their ample resources to bring it into being. Speaking of these various centers of translation in Baghdad and Persia, historian Peter Brown wisely observed that these represented a last flowering of classical Mediterranean civilization.[34] He might have added that Central Asia and Central Asians played a central role in this flowering.

In contrast to the Arab Christians of Syria, very few Arab Muslims or Central Asians ever bothered to learn Greek, let alone Latin. Now, for certain fields at least, they did not need to do so. Thanks to the Barmaks and to a number of other patrons, scholars had at hand almost the complete extant corpus of classical learning in mathematics, geometry, astronomy, physics, geography, and medicine. Especially prominent among those ancient authors translated into Arabic were the Greek-speaking scientists from Alexandria in Egypt at the time of Ptolemy I (323–283 BC). Among these, Euclid bears special notice because his *Elements* were the essential starting point for all subsequent work in mathematics and geometry. It was the Barmaks' translation of *Elements* that enabled readers of Arabic for the first time to become acquainted with the "Father of Geometry." Euclid, it turns out, was translated by yet another Central Asian scholar from Merv, Ali ibn Sahl Rabban al Tabari (ca. 810–870).[35]

While crediting the Barmaks, and especially Harun's friend and vizier Jafar ibn Barmak, with the sponsorship of translations that literally changed the world, it is important to take note also of what they and their successors did *not* choose to translate. It has been suggested that Baghdad translators followed their predecessors at Gundeshapur in focusing on those books that were of practical and utilitarian value and ignoring the rest.[36] This is clearly an overstatement, but the list of those excluded is nonetheless impressive. Among Greek writers, they favored the mathematicians, scientists, and philosophers but systematically ignored the great tragedians: Aeschylus, Sophocles, and Euripides. How might the Islamic world have been different had they developed a sense of tragedy, and of the fatal flaws that appear in human lives as the working of Destiny? Would they have bent over laughing at the impolitic, irreverent, and at times pornographic humor of Aristophanes's comedies, and even produced such works of their own? We will never know because the very idea of tragedy contradicts all three of the monotheistic faiths whose adherents led the translation movement: Zoroastrianism, Christianity, and Islam.

More perplexing is the failure of the Barmaks and their contemporaries to deal with Greek historians, especially Herodotus and Thucydides, and also the most important work of Greek political thought, Aristotle's *Politics*. Maybe the neglect of the historians traces to some lingering Persian prejudice against a people who threw off Persian rule and then, under Alexander the Great, destroyed Persia's first empire and its capital, Persepolis. Baghdad may have acquired this prejudice from the many Persians there, or developed the hostility on its own. Or maybe the problem arose simply from the lack of access to Greek texts of these writers' works. Whatever the cause, the effects were profound. Reading Herodotus, the Arabs might have better appreciated the actual fragility of such seemingly all-powerful leaders as Croesus or Midas, while from Thucydides they could have studied the terrible consequences for states that are organized around their armies and war, precisely the condition that befell the caliphate under Harun al-Rashid.

The failure to translate Aristotle's *Politics* was equally consequential. This did not arise from an opposition to political thought as such, since in a sense all Muslim thought is political. By the mid-ninth century there were translations or at least paraphrases of Plato's *Republic*, with its apotheosis of the state as moral guide under the rule of a "philosopher king." But for some reason Aristotle's *Politics*, with its solid grounding in the author's detailed study of practical political life and of citizenship in many settings, did not reach readers of Arabic until modern times. By contrast, Thomas Aquinas was mulling over the *Politics* by the twelfth century, thanks to Latin translations from the Greek.[37]

Finally, it is worth asking why the Persian-Arab translation movement that arose at Baghdad under the Barmaks ignored almost the entire corpus of writings in Latin. Latin texts had slipped from the intellectual luggage of the Orthodox Christian eastern Mediterranean by the time the Syrian Christians, on whom both the Persians and then the Arabs drew so heavily for their texts, began their translations. Nonetheless, it is interesting to speculate on the impact in Baghdad or Bukhara if key Latin texts had been translated into Arabic—works like Cicero's deeply civic orations before the Roman Senate; or Seneca's essays, with their humanistic concern for good conduct; or Marcus Aurelius's stoic meditations on the need for individual fortitude in an otherwise meaningless world.

However engaging, such questions are surely unfair, considering the vast body of writings that the Baghdad experts did succeed in translating.[38] These translations were to exercise a profound influence on thought from Xinjiang to the Atlantic coast. And nowhere was their impact felt more immediately than in Baghdad itself and in Central Asia.

FALL OF THE BARMAK CLAN

Power brings distraction, and Harun al-Rashid had more than his share of both. Early on he was distracted by pleasures of the flesh and then, increasingly, by the austere demands of his jihad against the Byzantines and other enemies to the west, which caused him to abandon Baghdad for a decade. During these years the power of the Barmaks soared, with Harun's old friend Jafar now commanding an army of his own, which he employed to throw his weight around in the capital. Word reached Harun that the populace increasingly viewed him, the caliph, as a puppet of the Barmak clan, whose members dared surpass the caliph in worldly opulence and show. Worse, the economic stress brought about by Harun's huge army and ceaseless campaigns caused many provinces of the empire to withhold taxes. A spirit of revolt was especially common among the landed aristocrats of Central Asia, which Harun inevitably blamed on the Barmaks.

Impelled by these and other motives that are less clear, Harun in the year 803 destroyed the Barmak family. He summarily jailed his old tutor and long-serving chief minister, Yahya, and also Fadl. Worse, Harun al-Rashid had his former crony, Jafar, murdered as a prelude to exhibiting his sliced up body parts on the Tigris bridges.[39] Power now shifted decisively back to the throne. But who would be the caliph? Forever distracted and hence impetuous, Harun set off on his final mission, rushing off to Merv and Samarkand to resolve an embarrassing scandal that had arisen there, thanks to the follies of a local administrator, and also to pacify the region after yet another bout of treasonous unrest, which the flailing caliph blamed on the Barmaks.[40] Harun al-Rashid did not survive the trip and died at Tus in Khurasan.

Mamun: Dogmatic Defender of Reason

One of Harun's final acts was arguably his most foolish. Instead of designating as his successor one of his two sons, either Amin or Mamun, he divided the caliphate and commanded both brothers to swear an oath at the Kaaba in Mecca that they would rule together. This sure formula for civil war had a cultural dimension, since Amin's mother was an Arab and Mamun's a Persian.[41] Caliph Mamun had already taken up residence at Merv, where the power of the Abbasid caliphate had been born and whence, he hoped, it could now be revived.

Mamun's arrival there had been greeted with a versified salutation by a poet from Merv named Abbas. Though Abbas was a fluent writer of Arabic, he chose his native tongue, the emerging New Persian language, for this important assignment:

> No poet before me has ever sung an ode in this fashion.
> Persian speech is lacking in how even to begin this manner of verse;
> Yet that is the reason I chose to sing your praises in this language;
> That by thus lauding and praising Your Highness my verse will gain
>   in grace and true charm.[42]

Central Asians joined Abbas in welcoming Mamun as a true champion of their region against the powers that be in Baghdad, a new Abu Muslim but with a royal crown. When his brother and rival, Amin, dared to march an army toward Central Asia, Mamun's largely Turkic forces handily defeat him. With Amin out of the way, Mamun pushed his army toward Baghdad. But local Arabs and townspeople combined to resist this invasion from Central Asia and the confrontation settled into a siege, which lasted from August 812 to September 813. It was a ghastly year, with one confrontation pitting besiegers against besieged and an equally bloody combat being waged among conflicting classes within the city itself. When finally the city fell to Mamun's forces, its population had been decimated and its homes and palaces were in ruins. Mamun meanwhile continued to bide his time in Merv, as he was to continue to do down to 819. Thus for a key decade, Merv in Central Asia was the seat of the caliphate and the capital of much of the Islamic world.

Mamun, like Caliph Mansur and Abu Muslim before him, found Merv to his liking. He had already erected for himself a new palace just outside the thousand-year-old city walls.[43] Surviving fragments confirm that its scale was immense. Clearly Mamun at first had no intention of moving the capital back to Baghdad. Nor did his chief adviser, vizier, and army chief, an Iraqi convert to Islam named Fadl ibn Sahl, who immediately set about drumming up support for Mamun among the landed aristocrats or *dihkans* of Khurasan. Little is known of this powerful, "Svengali-like" figure,[44] although it is claimed that he, too, was a protégé of the Barmaks. When Mamun in 819 finally decided to move the capital back to Baghdad, he first gave orders for Ibn Sahl to be killed. Clearly, the thirty-three-year-old caliph had no intention of playing second fiddle to any rich and willful vizier.

During his years in Merv, Mamun revealed a sincere interest in learning that went far beyond that of his father, Harun al-Rashid. The ancient and distinguished local observatory continued to thrive under his patronage, and he supported research in a number of other fields, including engineering and pneumatics: one instrument produced with his support was not surpassed until the introduction of pneumatic instrumentation in the twentieth century.[45] At Merv he also advanced the work of translation that had been going on there since before the founding of Baghdad. Thanks to this, many seminal translations of scientific works that are credited to Baghdad should in fact be ascribed to the learned circles of Merv. Indeed, as the late DeLacy O'Leary affirmed on the basis of all known evidence, "Khurasan was the channel through which astronomical and mathematical material came to Baghdad."[46] Mamun eventually realized that if he did not move to Baghdad he risked losing the entire empire. When he finally took to the road, he included in his entourage that traveled to the city a large retinue of scientists and scholars from Central Asia.

The biggest difference between Mamun and his father as patrons of culture was that now the caliph himself, rather than the Barmaks or anyone else, was leading the enterprise. This decision reflects an aspect of Mamun's character that was to prove important in the years to come, namely, his absolute confidence in his own abilities. He was an attractive man, intelligent, and dedicated to his office. Yet he was also egotistical and lacked the patience to think through projects before acting. His exalted view of himself was on full display when he ordered his name to be emblazoned on the walls of the Dome of the Rock in Jerusalem.[47] Both in Merv and in Baghdad, he was to raise the intellectual life of the caliphate immensely, but his impetuousness and intolerant self-assurance were to bring about a major crisis in the Islamic world, a crisis that was to reverberate powerfully across all Central Asia.[48]

During Islam's early centuries, caliphs thought it self-evident and natural that they should exercise both temporal and spiritual authority. This was assumed whenever a caliph was referred to as "Commander of the Faithful." True, personal inclination and the demands of war and day-to-day leadership caused some caliphs to limit their religious activity to fulfilling their formal duties. This was definitely not Mamun's approach to the job. From the outset he felt it his responsibility to be the leader and chief arbiter of all things religious. In his view, this meant exercising spiritual and intellectual authority over the entire class of religious scholars, the *ulama*. This set him at odds with the very men who believed they knew most about Islam and were therefore responsible to guide the faithful. Mamun was to push his case relentlessly, but in the end he called forth a backlash, which empowered the religious scholars to stand up to what they considered the temporal powers. Islam was never to be the same.

Since the days of Abu Muslim, the Abbasids' banner had been black. This was the color of the banners that Harun al-Rashid had carried to the very walls of Constantinople. However, while still in Merv, Mamun abruptly changed the hue to green, which he knew full well was the color under which marched the forces that would in time become the Shiites. It may be that Mamun did not intend to favor either side in the emerging schism between Sunni and Shiite Muslims and that he simply wanted to balance the two contending factions. But few doubted that this sudden shift was a crass attempt to mobilize those who would become Shiites in

Central Asia and Persia against the Arab and Sunni-leaning establishment in Baghdad that had backed his brother and rival, Amin.

Going further, Mamun, while still in Merv, had summoned from Medina the prominent religious scholar Ali al-Rida (known in Persian as Ali Reza) and assembled a multitude of thousands for a ceremony naming Ali Rida as the successor to the throne. This unexpected move solidified regional support for Mamun, but at the price of utterly polarizing the caliphate as a whole and alienating Baghdad. And so Mamun abruptly reversed course once more. Just as he began his trip to reclaim the Abbasids' capital, he paused in the city of Tus and a short distance from his own father's recent grave. There he apparently arranged for Ali Rida to be poisoned by underlings, whom Mamun then killed in order to destroy the evidence of his crime. This brutal volte-face shocked everyone and created a potent symbol of martyrdom that later became a core element of Shiite lore. Today the shrine of Ali Rida at Mashhad is one of Shiism's holiest pilgrimage points and Iran's greatest tourist attraction.

MAMUN'S HOUSE OF WISDOM

When he is remembered at all, Caliph Mamun is best known for the "House of Wisdom," which he is said to have founded and patronized. The very name conjures up images of bearded and robed Middle Eastern sages peering at their astrolabes or convened in a solemn conclave or *majlis* to discuss fine points of mathematics or metaphysics.[49] Said to have existed for two centuries, this institution has been portrayed as a kind of Academy of Sciences where scientists and scholars could pursue fundamental research without constraints, thanks to generous support from the throne.[50]

The Arabic original of the name of this institution traces to a Saying of the Prophet Muhammad that was collected by the ninth-century scholar from the old Buddhist center at Termez, Tirmidhi. Criticized for being inauthentic, the Hadith was purged from later compilations. Seemingly everything else about Mamun's House of Wisdom has been subjected to similar scrutiny, and with similar results. Was the House of Wisdom really a formal institution, like a research university or academy? Almost surely it was not. The histories record the names of no heads of the House of Wisdom, nor do they provide lists of members, mention a building, or refer to regular convocations.

What emerges instead is an informal and constantly shifting coterie of scholars supported by the caliph or his viziers and grouped around a major library. This was almost certainly the Barmaks' collection of books, transformed after their fall into the caliph's library. An early librarian was Fadl ibn Nawbakht, son of the astrologer from Merv who had participated in the dedication of Baghdad. In addition to following his father's profession of astrology, Fadl also translated from the Greek. Under Mamun, if not earlier, the caliph's budget paid the salaries of Ibn Nawbakht and certain other scholars and also covered the cost of translations. These were made by a cosmopolitan group of specialists, among whom Syrian Christians were prominent for the simple reason that they knew both Greek and Arabic; several of these translators had been lured there from the old center at Gundeshapur. From time to time the court also funded trips to acquire new manuscripts, including at least one expedition to Constantinople, which took place in the early years of Harun al-Rashid's reign. Besides the library, the caliph also funded the astronomical observatory, which Mamun established at Baghdad in imitation of the old observatory that had thrived at Merv during his years there. Since astronomy and the other learned fields were still fully integrated with each other, it is safe to assume that relations among scholars at both centers were close.

This, then, was the famed House of Wisdom. Institutionally more modest than later imaginations would have it, this "library and translation circle" nonetheless provided the one thing that was most essential to the scholarly life: time to concentrate and freedom from material want and outside interference.

However generous, such patronage can succeed only when both the identification and recruitment of talent is at the very highest level, and it was at these tasks that the leaders of the House of Wisdom proved themselves masters. It is no exaggeration to say that for a generation, at least, they assembled in Baghdad the greatest aggregation of scientific luminaries between the third to first centuries BC, when the Library of Alexandria in Egypt maintained a large team of editors and scientists, and 1660, when the Royal Society for Improving Natural Knowledge in

London began gathering the likes of Locke, Boyle, Wren, Goddard, and Hooke under the watchful eyes of its president, Isaac Newton.

How important was Mamun's role in creating the splendid environment for learning? Direct evidence is thin and inconclusive. But the lives of three remarkable brothers from Merv, and the central role these Central Asians came to play at the House of Wisdom, provides a glimpse into Mamun's activism on the intellectual front. The brothers, Muhammad (Abu Jafar), Ahmad, and Hasan ibn Musa, worked so closely with each other in the fields of geometry, astronomy, and mechanics that they are known collectively simply as the "Sons of Musa" (Banu Musa). Their father, Musa ibn Shakir, had specialized in robbing caravans in Central Asia until a change of heart led him to take up scientific studies. Mamun had encountered the family during his years in Merv and was so struck by the sons' exceptional abilities that he volunteered to become their guardian after the father's death. When Mamun left for Baghdad in 819, the Sons of Musa were among his entourage and remained linked with him until the caliph's death. They played central roles both in the development of Mamun's library and in his translation projects. Few, if any, of the scientific initiatives mounted during Mamun's reign did not bear the strong stamp of these remarkable brothers. Yet their organization work pales in comparison to their direct contributions to learning.

In geometry, the starting point of the Banu Musa was the great Sicilian mathematician and inventor Archimedes. But unlike their ancient teacher, who conceived the products of two lines as a plane, the Banu Musa assigned them numbers and used these to calculate areas. Stated differently, where Greek mathematicians expressed the areas and volumes of more complicated figures by setting them in a ratio to simpler figures, the Banu Musa expressed them as the products of certain lines. They were the founders of a specifically "Arabic" school of mathematics. Two centuries later a copy of their treatise on *The Measurement of Plane and Spherical Surfaces* turned up in a library in Spain and was translated into Latin by Gerard of Cremona.[51]

It would be wrong to assume that the life of Baghdad scholars was one of uninterrupted research. Authorities would call on them from time to time to solve practical problems, as when the Banu Musa were deployed to design canals for a new city. There were also more lighthearted projects. For example, Ahmad, one the Banu Musa, took time off to produce a *Book of Ingenious Devices*, in which he described some one hundred mechanical toys and automatons. This neglected but astonishing document bears comparison with Leonardo da Vinci's sketchbooks for the bold and intricate devices it describes. Here were pioneered one- and two-way pneumatic valves, automata that responded to feedback, and a host of ingenious devices that demonstrate a wildly imaginative yet disciplined and practical engineering mind.[52] Ahmad's mechanical flute, driven by steam, has been hailed as the first programmable machine, a title that is contested by the brothers' own hydro-powered organ, which was programmed through interchangeable cylinders.[53] In both devices Ahmad showed the same inventiveness that led a millennium later to the Jacquard loom, player piano, and eventually the punch card–programmed computer. The brothers also invented the clamshell excavator, a bellows to clear air from mines and wells, a gas mask for use when the bellows failed, and hurricane lamps, self-feeding lamps, and self-trimming lamps.

Ahmad's great book of inventions found readers throughout the Muslim world but never seems to have reached the West. The closest it came was thanks to al-Muradi, an eleventh-century Andalusian tinkerer who produced his own collection of astonishing mechanical devices. Two centuries later al Jazari (1136–1206), a Kurd from Anatolia, followed in the same tradition with a further collection of mechanical devices that still challenge engineers today. Unlike Ahmad ibn Musa, whose book was illustrated with primitive sketches, Jazari was a talented artist who could depict his inventions with great precision.[54] Both of these better known engineer/inventors from the Arab world took direct inspiration from Ahmad and did not hesitate to acknowledge that they were following in the brothers' footsteps. And on important points their later devices still lagged behind those of the eldest Banu Musa brother in sophistication and subtlety.

Amid all this activity there was also time for the internecine conflicts that infect academia like lice on a show dog. In the case of the Banu Musa, their bête noire was their great Arab colleague, the philosopher and polymath Kindi. All evidence suggests that this native of Basra in Iraq was a quiet, apolitical, and even stoic personality. A follower of the Greek sage Epictetus, Kindi cultivated inner serenity and at the same

Figure 5.2. In the mid-ninth century Ahmad, eldest of the Banu Musa brothers, invented this clamshell device for recovering valuables from the seabed. It was one of a hundred inventions described in his *Book of Ingenious Devices*. ■ Inv. ms. or. qu. 739 fol. 74a. Staatsbibliothek zu Berlin, Stiftung Preussischer Kulturbesitz, Berlin, Germany. Photo: bpk, Berlin/Staatsbibliothek zu Berlin/Christine Koesser/Art Resource, NY.

time was a tireless champion of reason as the ultimate standard for truth. He was also prolific, turning out more than two hundred treatises in fields as varied as geometry, medicine, music, and cryptology. Hailed as "*the* Arab philosopher," Kindi is seen by many as the only Arab to gain distinction in that field.

Just why the Banu Musa came into conflict with this mild-mannered man may never be known. Ethnic rivalry cannot be ruled out, nor can ideological and political differences. A few years after Mamun's death, the three brothers from Merv made their move, launching a campaign that ended in Kindi suffering a beating and the confiscation of his library. The shadow of the father of the Sons of Musa, the former caravan robber, hangs over the entire affair.

MUTAZILISM

Far more serious and lasting in its effects was the ideological controversy that Mamun himself stirred up. Starting from a centuries-old debate on some abstruse points of monotheist theology, it soon exploded into a full-blown confrontation between contending defenders of reason and of tradition as the basis for religious truth. By the time the conflict finally burned out, a protective barrier had been placed around the realm of faith, a caution flag had been waved before the champions of unconstrained reason, and the power of the caliphate in the realm of ideas had been destroyed forever.

Muslims, like Jews and Christians, had long debated the extent and limits of human reason, the degree to which human beings enjoy free will, the origins of evil in the world, and whether holy scripture should be taken literally or allegorically. As early as the first Muslim century, a school of thinkers in the Iraqi city of Basra adopted what was considered a radical stance on all these issues.[55] However, the first to systematize their position and their most effective proponent was Jahm ibn Safwan al-Tirmidhi, or simply "Jahm," whose ties were with the city of Tirmidh, now Termez, on the Afghan-Uzbek border in Central Asia.[56] This Central Asian was to be severely criticized for what might be called his minimalist version of the faith. Yet Tirmidhi's vision spread rapidly throughout Khurasan, Khwarazm, and much of the rest of Central Asia. The fact that many local princes supported it, as did most of those who backed the practical-minded Hanafi school of jurisprudence, added to its viability regionally.

Called "Mutazilites," from the Arabic verb "to separate oneself from something," those who followed Tirmidhi's doctrine claimed that God's creation was accessible to human reason, that God endowed men and women with freedom to will both good and evil, that evil arises from bad choices rather than from the nature of things, and that allegorical readings and rationalistic interpretations of scripture were both permissible and desirable. Accordingly, the Mutazilites rejected the doctrine of predestination, stating that every individual is responsible for his or her transgressions. Obviously this posed a frontal challenge to all those who, following the religious scholars (the ulama), embraced revelation and faith as the sole paths to higher knowledge and guide to daily life. At the same time it opened new vistas for intellectuals who were eager to apply to all questions the rigorously rational method they had come to admire in the writings of ancient Greeks. Like their Greek mentors, the Mutazilites were "atomists" in philosophy. But even though their belief in an irreducible substance could have led them to a form of materialism, they retained a role for creation and hence for a divine role in worldly affairs, albeit a more limited one.[57]

The Mutazilites were certainly believers, while the majority of Muslims who opposed them was far from denying some kind of role for reason in the realm of religion.[58] On a doctrinal level, at least, they may well have found some common ground, had it not been for a final argument advanced by the Mutazilites. Mainstream Muslims held that God's word, as embodied in the Quran, was as eternal as the Creator and therefore "uncreated." Mutazilites, following a line of argument first propounded by Tirmidhi, objected that God's unity excluded the possibility of "co-eternals."[59] This meant that the Quran had been "created" and was not co-eternal with God. This being the case, it could be interpreted both rationally and allegorically.

It is a mystery why Mamun chose to place himself at the very epicenter of this abstruse and highly controversial doctrinal feud.[60] Most likely three different factors figured in his decision to plunge into the fray. First, the many translations sponsored by Mamun personally and by the Barmaks before him had inundated the intellectual world with fresh models of reason in action. Many intellectuals lent support to the Mutazilites, among whom those from Central Asia were particularly conspicuous.[61] Mamun, whose theology was generally eclectic, probably saw in Mutazilite doctrine the basis for a Muslim defense of the new knowledge emerging from ancient Greece as a result of the translations he was supporting, and in which he took a personal and active interest. The prominent Kindi, who figured centrally in the Baghdad library and research center, stood aloof from the Mutazilite controversy, but his ringing defenses of Greek thought put him squarely in the camp of reason.[62]

Second, Mamun's aborted adventure with Ali Rida back in Merv may have been far in his past, yet he remained sympathetic to certain aspects of what was to become the Shiite school of theology, and especially its assertion of the need for a wise imam to interpret holy writ for all.

Third, Mamun's long-running battles with Baghdad's religious scholars had left a deep residue of resentment and even paranoia in the caliph's mind. He had tried buying off these traditionalists and literalists with grants of patronage, but they had taken the money and stuck to their guns.[63] Now it seemed that they were gaining the upper hand, with their claim that the sole source of authority was God's word as recorded in the Quran and Hadiths of the Prophet. Mamun responded by repeating ever more insistently the old demand that both secular and religious authority be concentrated in the person of the caliph. He, Mamun, would be both caliph and imam,[64] an absolute ruler with authority to declare religious doctrines and to define the contents of the law, or Sharia. Mamun saw this as a return to the established norm. Traditionalist religious leaders viewed it instead as a usurpation of their own role and denounced the caliph's defense as the ravings of a heretic.

Brooding on all this as he campaigned with his army at Tarsus in what is now Turkey, Mamun decided it was time to act. He would subject each of the religious scholars of Baghdad to a simple test that would reveal clearly whether he was with or against the caliph. The key question of this Inquisition or *mihna* would be utterly direct: did the scholar believe the Quran to have been created or uncreated?[65] If the former, the respondent would have proven himself a man of reason and hence likely to back the caliph's campaign to become the spiritual leader of Muslims everywhere; if the latter, the respondent would have shown himself to be a slave to tradition and unreason and a foe of Mamun's quest for spiritual authority. In a series of letters, he charged his staff back in Baghdad to launch this "rationalist inquisition"[66] in which he himself would be the grand inquisitor.

Like the medieval Catholic inquisitions, which began three centuries after Mamun's, this was no mere academic exercise. Mamun saw his inquisition not just as a test of personal loyalty but as a concrete exercise of the spiritual authority that was rightly his. Since neither side could defend its position beyond endless appeals to the assumed authority of reason or of tradition, this was ultimately a war of opinions. Opinions in turn became dogmas, the validity of which was affirmed not by reason or tradition but by the depth of conviction with which they were advanced and the power wielded by those advancing them. What began as a theological confrontation quickly ripened into a test of raw power.

Over the course of two decades, first under Mamun and then under his nonintellectual but compliant younger brother, Caliph Mutasim, the inquisition ground on. Some defendants were led in chains before the panel of two or three inquisitors. As a recording secretary took notes, most quickly averred that they of course believed the Quran to be created and hence that it was subject to interpretation by qualified authorities. Only two defendants dared to oppose this proposition. One of them, Ahmad ibn Hanbal, was first jailed for several years and then flogged.

Ibn Hanbal was an Arab whose family had been settled in Merv since the early days of the Muslim conquest of Central Asia;[67] his grandfather had risen to the rank of governor of Sarakhs on the present border between Turkmenistan and Iran. After Mansur founded Baghdad, the family moved from Central Asia to the new capital and the boy's father promptly died. Following a brief stint in a government office, the young Ibn Hanbal discovered his true vocation as a collector and student of the Hadiths of the Prophet Muhammad. He became an ascetic who traveled the world in search of the authoritative words of guidance that his father's early death had denied him. No one surpassed him in loyalty to the words of the Prophet and his Companions and in rejecting the conclusions of reason and all other competing sources of authority in matters of religion and law.[68] It is revealing that, in clear contradiction to Muslim doctrine, Ibn Hanbal's view of God was unapologetically anthropomorphic.[69]

Even before the inquisition, Ahmad ibn Hanbal had acquired great spiritual authority, not just among other religious scholars but among the same pious public that Mamun, in one of his letters, had dismissed as rude and ignorant. Ibn Hanbal's sufferings at the hands of authority now transformed him into a martyr. Long before his death a few years after the Inquisition ended, it was evident that the circle of opponents of unfettered reason that had formed around him would become one of the four great juridical schools of Islam. The school's great caution about the state flowed directly from Ibn Hanbal's grim experience with the Inquisition.[70]

By their arrogant and ruthless handling of the Mutazilite controversy, Caliph Mamun and his successor, Mutasim, called forth a powerful and understandable defensiveness on the part of orthodox traditionalist. Mamun, like Rousseau a millennium later, genuinely championed reason and free will as the essential tools for overcoming the fact that "man is born free but is everywhere in chains." But like Rousseau, too, he came to see his task as to "force them to be free." With this, "God's caliph"[71] hit a wall. What had begun as an effort to enrich and expand Islam with the insights of learned men of the past ended as a narrowing of the faith and the polarization of Muslims with respect to freedom and the use of reason. The final resolution of this controversy was to be postponed for several centuries, during which the life of the mind flourished as never before in the Islamic world, and especially in Central Asia. But the seeds of a powerful reaction against reason had been sown as the staunch and resolute martyr Ahmad ibn Hanbal of Merv quietly stood up to Mamun's tyranny.

POSTLUDE

Caliph Mamun died in 833, only months after his facedown with Ibn Hanbal. His successor, Mutasim, besides continuing the persecution of Ibn Hanbal, devoted himself to pursuing the old jihad against Christian Byzantium and building twenty palaces. Like Mamun, Mutasim depended overwhelmingly on Central Asians, with his chief adviser and general being one Afshin, the former ruler of Ustrushana, near Samarkand. He achieved some notable successes, but at a high price. He watched helplessly as the sources of wealth declined. The resulting social unrest on the streets of Baghdad drove Caliph Mutasim to move his residence to a new site at Samarra, seventy-eight miles north of Baghdad. Whether or not the unbridled behavior in Baghdad of his Turkish slave soldiers figured among Mutasim's motives for moving to Samarra, a large contingent of these same enslaved fighters accompanied him there.[72] The vast and precisely planned rectilinear city he established at Samarra suggests that Mutasim had some kind of plan to surmount the various challenges before him. But he did not. Having fled Baghdad, he allowed the irrigation system of the old capital to decay, which drove up food prices and led to a fresh round of rioting and soon outright rebellion, with Central Asians once more prominent among the leaders. But Mutasim thrashed about helplessly, as the power of his office slipped away.

Neither Mutasim during his short reign nor his brutish successor, Mutawakkil, could ignore the centrifugal pressures that were pulling the Islamic caliphate apart. Central Asians, who had figured prominently on both sides of the Mutazilite controversy, were particularly clever at whittling down the influence of the caliphate over their lives, while at the same time nodding from time to time to Baghdad. Within a generation of Mamun's death, all the Central Asian provinces that Arab troops had subdued with such difficulty over two hundred years were functioning once more virtually as independent states.

The rising influence of Central Asia was felt in other areas as well. After the death of Mamun, Baghdad gave up even the pretense of maintaining a normal army of freemen. To maintain themselves in power, the caliphs now depended entirely on Turkic slave soldiers from Central Asia.[73] A few of them rose to general rank while others advised the throne on policy. The mass of Turkic troops, however, formed what was in effect a mob, which from time to time took full control of Baghdad. Less than a century after the rise of the Abbasids, the caliphate was in full decline.

Barely were Mamun and Mutasim in their graves than the rollback of Mamun's ill-starred support for Mutazilism began. Caliph Mutawakkil went so far as to jail translators whose work was judged to be supportive of rationalism. Clearly Baghdad's great age of intellect had passed its zenith. As doctrine hardened into dogma, intolerance set in. Over the century since its founding Baghdad had been open to people of other faiths. Zoroastrian literature continued unhampered, synagogues abounded, and important Christian institutions, including the Monastery of the Virgin and several cathedrals, flourished within Baghdad's walls.[74] But in this new, less secure environment, authorities closed churches and synagogues and repressed their adherents.

Anxious over the decline of social cohesion in Baghdad and elsewhere, and eager to regulate more closely the social and economic life of their flagging empire, Abbasid authorities worked hard to align all civil activity with Islam. This signified a great expansion of the purview of Islamic law (Sharia). Soon Muslim jurists, with Ahmad ibn Hanbal's traditionalists in the lead, were laboring to define correctly Muslim positions on every detail of personal, family, and civil life. Like authoritarian rulers everywhere, the Abbasids hoped to regulate society from above by merging their laws with the strictures of faith. This effort polarized society, proved a gag on innovation, and was among the most important forces driving the epicenter of knowledge eastward toward the increasingly independent societies of Central Asia.

Amid this mood of defensive retrenchment, two thinkers emerged who took as their task the development of doctrinal rationales for the new orthodoxy. The first of these was Abu al-Hasan al-Ashari from Basra in Iraq, who brought to his mission a convert's zeal. Down to his fortieth year, this Baghdad resident had been among the most zealous proponents of Mutazilism. Then he saw the light—or saw which way the political winds were shifting—and penned some hundred tracts attacking his former colleagues. The Asharite school he founded lay semidormant for generations but was to emerge triumphant two centuries later when the great Central Asian theologian and philosopher, Ghazali, used his brilliant polemical gifts to hack away at the pretensions of rationalism and "philosophy."[75]

The other great opponent of rationalism to emerge in the wake of Mamun's disastrous inquisition was Muhammad al-Maturidi (853–944), known as the "Imam of Guidance," an inquisitive and widely read Central Asian thinker from near Samarkand. Like Ashari, he was accommodating on the issue of free will, but in other respects he was a hard-liner, defending orthodoxy from a host of perceived threats, including Manichean dualism and every variety of polytheism, as well as Mutazilism. He thrived on confrontation, loved to write "Refutations" of one evil or another, and favored such combative titles as *A Book Exposing the Errors of Mutazilism*. As to those who failed to use their God-given powers of mind to grasp the truths of Islam, Maturidi was sure would end up in Hell. Whether because of his truculence or in spite of it, the writings of this Central Asian were popular for half a millennium among Muslim literalists and traditionalists from Turkey to India.

In spite of this hailstorm of criticism, many continued to embrace reason as a key to unlock the most important knowledge, and to affirm free will as God's gift to mankind. Mutazilism as such may have waned, but more than a few thinkers continued to develop its main concepts. The chief heir to the Mutazilite tradition in Baghdad was again a Central Asian, Abu al-Rabban al-Balkhi from Balkh, who railed more against those rationalists who were led into doubt and irreligion than against his more orthodox critics. Epitomizing this radical wing of the camp of reason and free will was still another Central Asian, Abu Hasan Ahmad Ibn al-Rawandi, whose outright atheism we will take up when our focus turns to his native region, Khurasan, and its rising capital, Nishapur.

Beyond the avowed heirs of Mutazilism were many keen thinkers who drew prudent conclusions from Mamun's inquisition but continued to hold high the banner of rational enquiry. Most were good Muslims in that they accepted the truths of revelation. Like John Locke in a later age, they acknowledged the realm of faith and considered it above reason. But like Locke, too, they looked on the rest of creation as a realm accessible to reason and set about exploring it with zeal, intellectual rigor, and delight. In this great adventure of the mind, no one exceeded the Central Asians, to whom we now turn.

CHAPTER 6

❀

Wandering Scholars

Few places were more intimately linked with physical mobility than Greater Central Asia. Nomads were seasonally on the move, and the economies of the great urban entrepôts rested on a foundation of continental trade. The long-distance routes to the Middle East, India, China, and Europe had been well established for nearly a millennium by the time Mamun began to staff his library. Anyone restless to travel had simply to attach himself to one of the frequent caravans that set off like slow-moving trains across the wasteland between the oases. As early as the fifth century bc, Central Asians from places like Sogdiana and Khwarazm were turning up in the distant Persian capital at Persepolis, at Memphis in Egypt, and even at the remote Elaphantine Island of the Upper Nile on the border with Nubia.[1]

Poets, astrologers, savants, musicians, and dancers were no strangers to travel. Astrologers gladly moved to whatever courts were in need of someone who could foresee the future, and scores of lutenists and dancers from Central Asia are known to have found their way to the Chinese court. The idea of the wandering scholar was to take deep root in medieval Europe,[2] but the norm was to stay put in one's home valley, town, or monastery. Not so in Central Asia. The arrival there of Greek thinkers and then Indian savants after 300 bc, and the travels of Jewish astrologers, Manichaean sages, and Nestorian monks thereafter, all gave rise to reciprocal visits by Central Asians. For centuries before Caliph Mamun developed his library and center for research, physical mobility, and the mixture of idealism and opportunism that underlay it, defined the life of the mind throughout Central Asia.

What impelled these peripatetic thinkers to travel? Political insecurity and upheavals drove more than a few from their homes. Khwarazm, the native region of mathematician and astronomer Khwarazmi, had yet to recover from the Arabs' invasion when that young student left home. The Samanid state was in its death throes when the youthful doctor Ibn Sina departed Bukhara, and when the Sufi poet Rumi later fled Khurasan that region was writhing under the Mongol conquest.

Prospects for research and sheer curiosity also drove Central Asian thinkers onto the high road. Abu Zayd al-Balkhi was to use his travels to gather information for his seminally important maps, while Bukhari, compiler of the Hadiths of the Prophet Muhammad, spent years on the road conducting oral interviews for his renowned compendium. And even though he had no choice in the matter, the polymath Biruni clearly welcomed the opportunity to spend years traveling and conducting research in India.

A fortunate few of the great thinkers enjoyed private fortunes, while several more secured lucrative administrative posts. Most, though, were dependent on the royal courts for support, the more so since other Muslim patrons were not in the business of supporting what many considered learning that was irrelevant or even hostile to the faith. Granted, the region's generally warm climate reduced the physical needs of scholars and scientists. But they still needed support and security, and also the companionship of like-minded intellectuals. During the era of the Barmaks, the reign of Caliph Mamun, and for a couple of generations thereafter, Baghdad grandly met these needs, in the process becoming a magnet for Central Asians.

Cultural factors specific to Central Asia reinforced this process. The establishment of relative peace under Abbasid rule after a century of chaos gave rise to a great flowering of intellectual activity across the region. The fact that Central Asian thinkers had quickly mastered Arabic facilitated this boom, but they were well positioned even without this, thanks to their command of Persian. No less important was that a Central Asian intellectual heading for Baghdad did not feel himself to be a country bumpkin venturing into alien territory. After all, Central Asians had played a key role in placing the Abbasids on the caliph's throne, and their heirs reasonably believed that Baghdad had no choice but to attend to their views. Moreover, the Barmak family and then the Banu Musa brothers had opened wide their doors and purses to scholars and scientists from their land of origin, while their style of patronage assured anyone contemplating a move that in Baghdad all careers were open to talent.

No wonder that for a time the capital on the Tigris River was the center of Central Asian intellectual life. Not only did scholars and scientists from the region rank among the capital's leading lights but, as a tenth-century Arab writer observed, "There is not a scientist or poet in Baghdad who does not have a student from Khwarazm."[3] This occurred in spite of the simultaneous influx of ambitious Arab scholars to Baghdad: at least one Arab savant traveled all the way from al-Andalus (Spain) to sit with the learned men of the capital.[4] Yet in spite of the Arabs' numbers and distinction, in most fields of science and philosophy the Central Asians outshone them.

BAGHDAD'S CENTRAL ASIANS

Thanks to pioneering research by a German scholar from Leipzig, it is possible to appreciate the scale of this Central Asian inundation of the salons of Baghdad. Heinrich Suter, writing at the dawn of the twentieth century, laboriously compiled the places of origin of some 515 mathematicians and astronomers during the Islamic Middle Ages, the largest contingent being from the early Abbasid era.[5] Because they chose to write in Arabic, it has been all too easy for all but specialists to conclude that they were in fact Arabs, not Central Asians.

Suter discovered that of his list of high achievers in just these two fields, the overwhelming majority were Central Asians, nearly all of them of Iranian ethnicity. His method was simple: to trace the *nisbas* or hometown names that are affixed to personal names in Arabic. Even if we generously discount his findings, they more than suffice to call for a wholesale redrawing of the intellectual map of the world in the Middle Ages.

Suter's effort has recently been amplified by a scholar from Tashkent, Bakhrom Abdukhalimov. His study of the "House of Wisdom" (he uses the term) focuses on fifteen of the best-known Central Asian scientists and scholars who worked in Baghdad. He shows that they were at the very core of the city's intellectual life.[6] This confirms that the large quantity of Central Asian scholars in Baghdad was more than matched by their quality.

Was the Central Asian genius spread equally across the various fields and disciplines? Few Central Asians in Baghdad achieved distinction for their poetry, since the official language of the court was Arabic and poetry was above all declaimed, not read. Most of the well-known musicians were also Arabs, although it was a Central Asian, Farabi, who pioneered the theoretical study of music and established its main questions and propositions in a manner that defined both Eastern and Western music theory for half a millennium. Many Central Asians emerged as innovators in architecture, too, and also as miniature painters for books. But many of the best of such artists found patrons at home and had no need to travel to Baghdad for commissions.

This leaves two broad areas of activity—mathematics, astronomy, and the sciences, and what might be lumped under the heading "philosophy"—as the fields in which Central Asians especially stood out. Among the sciences might also be included certain areas of the social sciences, including cultural anthropology, and also scientific instrumentation and technology generally. It is no accident that these fields together happen to comprise the main categories of scientific and humanistic learning as they existed between ancient times and the rise of the modern world.

In some cases Central Asians made seminal contributions to technologies generally associated with Arab inventors. For example, it has been assumed for a millennium that the astrolabe, invented by ancient Greeks (hence the Greek origin of its name, *astrolabon*), had been perfected by Arab scientists and technologists. Held aloft by a qualified technician, precisely constructed and elegantly ornamented brass instruments could fix and even predict the location of the sun, moon, planets, and stars; measure time at specific latitudes; and calculate the height of mountains or the depth of wells. Fazari, a Persian scientist from Fars, is credited with constructing the first "Arabic" astrolabe, while an Arab scientist from near Baghdad, Battani (Latinized to Albatenius), first worked out the mathematics of the instrument. But it was Saghani, a native of the Merv region in present-day Turkmenistan, who opened new vistas for the astrolabe with his discovery of how to project a sphere onto a plane perpendicular to its axis, thereby earning for himself the sobriquet "The Astrolabist."[7] And it was other Central Asians who did most to refine further this important instrument.[8] They played equally seminal roles in developing the astronomical quadrant in the same period.[9]

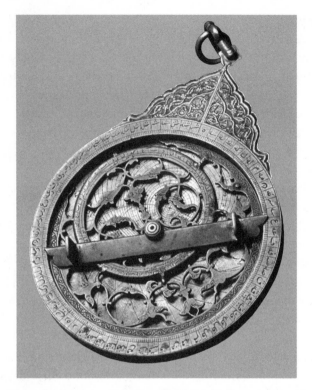

Figure 6.1. Eleventh-century astrolabe from Central Asia. Central Asian scientists led in the refinement of this essential instrument of medieval astronomy and in its use in pathbreaking research. ■ Photo by Anvar Iliasov, Aiphoto, Uzbekistan.

In a telling passage, Saghani of Merv boldly summed up the difference between ancient Greek scientists and those of his own era:

> The ancients distinguished themselves through their chance discovery of basic principles and the invention of ideas. The modern scholars, on the other hand, distinguish themselves through the invention of a multitude of scientific details, the simplification of difficult [problems], the combination of scattered [information], and the explanation in coherent form of [material which already exists]. . . . How much did the former leave for the latter to do![10]

In the areas of philosophy, science, and the humanities, the work of Central Asians in Baghdad was facilitated by the weekly seminars on humanistic issues convened by one of their number, Abu Sulayman al-Sijistani (932–1000). Sijistani was one of those human magnets without whom intellectual progress is impossible. Every Friday evening he assembled at his home a diverse group from many lands, including many fellow Central Asians. So interesting and engaging were the comments he offered at these soirees that one attendee noted them down and issued them in a five-hundred-page book, the title of which nicely encapsulates the mood that prevailed there: *A Book of Pleasures and Conviviality.*[11]

Sijistani hailed from the town of Sijistan near what is now the border between Afghanistan and Iran. It was in this rural backwater that he first encountered a manuscript in Greek. As he matured he took an interest in all the component fields of philosophy, including mathematics, medicine, logic, music, and cosmology. Sijistani reveled in reason and the life of learning and was consumed by the open questions that constantly arose during his evening discussions. At the same time he showed fulsome respect for religion and the revelations on which it is based, going so far as to declare that prophets rank higher than philosophers and scientists.[12]

One suspects that this utterance included a big element of diplomacy. For having made this declaration, Sijistani went on to argue that religion and philosophy belong to totally separate realms and are in fact mutually incompatible. Religion, which he equated with religious law (Sharia), has neither need nor use for reason and inconvenient "Why?" questions. Its purpose, after all, is to affirm humanity's subordination to God and to promote fear before Him, not to engage in argumentation. Conversely, the world of learning ("philosophy") has no need or place for religion. In his *Storehouse of Wisdom,* Sijistani offered the striking conclusion that philosophers and scientists should completely separate themselves from religion, just as adepts of religion should distance themselves from all rational thought.

It is clear that Sijistani had a tougher side. At one of his sessions a young scholar from Bukhara had the temerity to complain about the financial aid he was receiving, pontificating that someone focused on learning should not have to bother about his sustenance. Sijistani archly replied that the Bukharan should consume less, and concentrate instead on achieving wisdom and virtue. That this young Bukharan dropped from history at this point suggests that his grant was not renewed.[13]

Arab Colleagues

To acknowledge the role of Central Asians in the "Arab Renaissance" is not to diminish the achievement of Arab scientists at Baghdad. They were there from the outset and played a noteworthy role. Beyond doubt, the doyen of Arab scholars in both the sciences and philosophy at Baghdad was Kindi, whom we have already encountered in connection with his troubles at the hands of the Banu Musa brothers during the Mutazilite conflict. An Arab from Kufa, Kindi was an early example of the kind of intellectual omnivore that soon multiplied in the Arab lands and, still more, in Central Asia. Although known as "*the* Philosopher of the Arabs," Kindi left a worthy record in the sciences as well. His work on geography inspired his renowned student from Balkh in Afghanistan, Abu Zayd al-Balkhi (850–934), who pioneered an innovative school of terrestrial mapping, while Kindi's two studies in the field of optics still found favor seven hundred years later when the English polymath Francis Bacon (1561–1626) came across them.

The intellectual biographies of several prominent Arabs in Baghdad reveal strong links with Central Asia. Thus the founder of the philosophical movement in Baghdad was the complex and enigmatic figure of Jabir (Abu Musa Jabir ibn Hayyan, 721–815), known in the West as "Geber." While Jabir was of Arab descent, he had been raised and educated in Tus, the ancient capital of Khurasan. Further complicating his identity is the fact that many of his writings date from his Central Asian period and reflect the concerns of scientists in Khurasan. Jabir believed that philosophy included all life, and he proceeded to carry out groundbreaking work in chemistry, astronomy, physics, metallurgy, pharmacology, and medicine. As a son of Khurasan, he was called to Baghdad by the Barmaks, to whom he dedicated many of his writings.[14]

Jabir's core discipline was alchemy, which down to the eighteenth century was considered a legitimate part of science rather than a magic-tinted pseudo-science. He was not above mystifying his readers, however, and often used a coded language that was so impenetrable as to give rise to the English term "gibberish." But there was method to Jabir's madness, and he succeeded in advancing a new and more

Figure 6.2. Distilling apparatus designed by Tus native Abu Musa Jabir ibn Hayyan (721–815), an alchemist and the "father of chemistry," known in the West as "Geber." His prolix writings gave rise to the word "gibberish." ▪ Courtesy of the History of Science Collections, University of Oklahoma Libraries.

mystical Muslim cosmology that was rich with the Pythagoreanism of the ancient world.[15]

At heart Jabir was an experimentalist, a practical inventor who came up with such innovations as waterproof paper and rust-free steel.[16] While mixing his potions, Jabir also discovered that crystallization is an effective purification process. It is no wonder that he is honored as the "father of chemistry" and also the father of laboratory-based experimental science.[17] So great was his influence in Europe that many of the 2,500 works attributed to him were Western forgeries, known as the writings of "Pseudo-Geber."

The Central Asian Edge: Medicine, Mathematics, and Astronomy

Within the sciences, certain areas particularly attracted the attention of Central Asians at Baghdad. High up on the list was medicine, which, from the era of the Barmak family onward, attracted first-rate minds on account of its combination of theoretical and practical knowledge. Indeed, it is possible to speak of a distinctly Central Asian school of medical experts, which culminated with Ibn Sina and his *Canon*. The pioneers of this movement came once more from Merv, specifically from the large and intellectually sophisticated community of Nestorian Christians there. Ali ibn Sahl al-Tabari's father had been one of Merv's many distinguished medical men and took a keen interest in the other sciences as well. Working along the same lines as the Syriac translators at Gundeshapur, he made one of the earliest translations of Claudius Ptolemy's epochal *Almagest* into Arabic. After studying with his father, he went to Baghdad, where the reigning caliph converted him to Islam. His education at Merv had given Tabari a solid knowledge of both Syriac and Greek, thanks to which he had direct access to the rich medical lore of both languages, without the need for translations. Once in Baghdad he set about compiling his *Paradise of Wisdom*, one of the earliest medical encyclopedias in Arabic.[18] To assure a larger audience, he issued the *Paradise* in Syriac as well.

Reading randomly through Tabari's huge tome,[19] one is struck by his oft-repeated conviction that physical and mental health are intimately interconnected. His diagnostic vocabulary is rich with psychological insights, and his prescriptions often call for a kind of proto-psychotherapy that included the counseling of patients. The same concern for the link between mind and body caused Tabari to stress child development and pediatrics. Given that these were all to become the preoccupation of a number of leading medical experts in Central Asia (see chapter 7), it is fair to characterize this as a specifically regional interest tracing to the work of Nestorian Christian doctors in Khurasan.

Aside from medicine, the fields in which Central Asians were to become absolutely preeminent were mathematics and its related disciplines, and astronomy. Suter's careful research confirms this in a quantitative sense but does not venture into the intriguing question of why this should have been the case. Was it due to behind-the-scenes maneuvering by the Banu Musa brothers, who drew freely upon the caliph's treasury to cultivate and subsidize fellow mathematicians and astronomers from Central Asia? Or was there something in the culture of Central Asia that gave special urgency to questions of number, spatial relations, and the heavens?

Rather than plunge into mere speculation, it is useful to look at several of the chief Central Asian exponents of these fields at Baghdad. There is no better starting point than Habash al-Marwazi from Merv, known simply as Habash, who is acknowledged as a pioneer in both "Arabic" mathematics and astronomy. Born and educated at the great scientific center in what is now Turkmenistan, Habash was another Central Asian who traveled to Baghdad in the entourage of Caliph Mamun in 819. At fifty-five he had already outlived most of his contemporaries, but he was to continue working to age one hundred. At Baghdad he immediately joined other astronomers working at the observatory that Mamun had set up on the model of the one at Merv.

A decade after his arrival, Habash for the first time used an eclipse to devise a means of telling precise time from the altitude of the sun.[20] Over a decade he was a key figure in the team of scientists who developed three sets of astronomical tables to facilitate the calculation of planetary positions, eclipses, phases of the moon, and precise calendrical information. To aid his research, he helped develop a totally new instrument, the so-called melon astrolabe, which used azimuth-equidistant mapping through a system that simulated the daily motion of the celestial sphere against the stationary horizon.[21]

Beginning with his participation in the team of researchers seeking to measure one degree along Earth's circumference, Habash calculated the circumference at 20,159.77 miles (versus the actual 24,901.55 at the equator), its radius at 3,201.06 miles (versus the actual 3,847.57), and its diameter at 6,417.025 miles (versus the actual 7,926.4098). Biruni and other Central Asian astronomers soon refined these figures to such an extent that their conclusions were very close to modern calculations. He went on to develop like figures for the moon and sun, their distance from Earth, and the size of their orbits. The great Khwarazmi would soon significantly improve all these estimates, but it was Habash and his team who launched this quest for precision in Baghdad.

Meanwhile, as a mathematician Habash posited the concept of an umbra or shadow that functioned the same way as a tangent in modern trigonometry and then proceeded to introduce the notion of a cotangent and work out the first tables for it. Habash was also the first to calculate tables of auxiliary trigonometric functions, which are essential for science, engineering, and navigation.[22]

Two generations younger than Habash was another Central Asian astronomer, Ahmad al-Farghani (797–860), who hailed from the Ferghana Valley, southeast of Tashkent in present-day Uzbekistan. Like Habash, he was closely allied with Caliph Mamun and also with his patrons and coworkers the Banu Musa brothers, and probably also arrived in Baghdad from Merv with the new caliph. Farghani was involved with Habash and several other Central Asians on Mamun's astronomical projects. His most well-known book, *The Elements*, capably summarized Ptolemy's second-century Greek treatise, *Almagest*, on the complex motions of the stars and planetary paths, updating Ptolemy's data on the basis of the latest findings from Baghdad. Farghani's was among the earliest works on astronomy to be written in Arabic.[23] His great virtue was clarity, with the result that his book attracted immense interest in the West, where Farghani, known as Alfraganus, became the most widely read "Arab" astronomer, including by Christopher Columbus. *The Elements* was translated repeatedly from the twelfth to the sixteenth century and became the source of Dante's many references to astronomy in his *Divine Comedy* and his *Convivio*.[24]

Following Ptolemy, Farghani proposed to explain the astronomic basis for Earth's seven inhabited climes. Unlike both Western and Arab writers, he proceeded from east to west, identifying three of the zones with Greater Central Asia. Later it was Farghani whom the Banu Musa brothers sent to Cairo to construct a new canal to the city. The project nearly ended in disaster when the slope of Farghani's canal was too gradual to prevent backup. In this instance Central Asian science proved less than practical. But Farghani made up for this misstep by constructing his famed Nilometer, thanks to which Egyptians could precisely gauge the depth and flow of the river on which their survival depended.

KHWARAZMI, MASTER SCIENTIST

The careers of Habash and Farghani attest to the seminal role of Central Asians in what is frequently identified as Arab mathematics and astronomy in the Abbasid era. But both of them, and the many other scientists from the region who had relocated to Baghdad, as well as their Arab colleagues, were mere planets circulating around the sun of the great Abu Abdallah Muhammad al-Khwarazmi, or simply Khwarazmi (780–850). As his name implies, Khwarazmi came from Khwarazm, the great expanse of desert and rolling land on the northern border of Central Asia south and east of the Aral Sea.

Toiling as an émigré for nearly a half century in Baghdad, this grandly gifted scientist systematized and named algebra, in the process offering an accessible method for solving linear and quadratic equations that virtually defined algebra for half a millennium. He opened the field of spherical trigonometry and was also the key advocate for persuading Arabs and then Europeans to adopt the Hindu (now wrongly called Arabic) decimal system of numbers and its innovative concept of zero (which the Olmecs in Mexico had also hit upon) and negative quantities. The method he developed for doing arithmetic using Hindu ("Arabic") numerals caused his name—distorted from an already distorted Latin form—to be given to the concept of an algorithm, best known today as the precise instructions that enable a computer to sift through vast amounts of data to reach useful outputs. And among his many other achievements, Khwarazmi compiled data on the precise latitudes and longitudes for 2,402 locations on Earth, far more than anyone else before him.

The Belgian scholar George Sarton, who pioneered the study of the history of science, called Khwarazmi "the greatest mathematician of the time, and . . . one of the greatest of all time."[25] Scribes in Khwarazmi's time tediously copied out scores of copies of the Arabic originals of his four main works on mathematics, astronomy, and geography, and these found avid readers from India to Spain. Yet over time all copies of Khwarazmi's original texts on arithmetic perished, while only one copy each of the Arabic originals of his books on algebra and geography survive to our day. The sad fate of all the others attests to the turmoil, cultural discontinuity, and decline that characterized much of the Arab world and Central Asia between the end of the Age of Enlightenment and the modern era. Indeed, had it not been for three inquisitive and intrepid medieval scholars, two English and one Italian, who lived three century after Mamun's heyday in Baghdad, we might not have any of these masterpieces today.

Adelard of Bath (ca. 1080–1152) began his career as a typical medieval scholar, poring over old texts. But when he caught wind of unexpected new insights to be gleaned from scientists writing in Arabic, he, too, became a wandering scholar. Garbed in his signature long green cloak, he devoted years to travel. His wanderings took him first to France and Italy and then to Antioch and other cities of the eastern Mediterranean. He returned with a copy of Khwarazmi's book on arithmetic and also a shortened version of Khwarazmi's astronomical tables that had been drafted by an Arab in Spain. In short order he translated both into Latin. They became instant classics in England and on the Continent, establishing Khwarazmi's book on mathematics as the chief European textbook on that field down to the sixteenth century. In the course of this work, Adelard became an ardent champion of his Arabic-language mentors, proclaiming to his many readers that he would "defend the cause of the Arabs as my own."[26]

A generation after Adelard, another English enthusiast for the "new" learning (which was already three centuries old), Robert of Chester, traveled to Spain in search of manuscripts. There he came across an Arabic original of Khwarazmi's *Algebra*. Robert's command of Arabic was less than perfect: he mistransliterated the Arabic version of the Hindi word for the function of an angle, leaving us with the trigonometric term *sine*. Nonetheless, Robert's translation unleashed a tide of European interest in geometry and trigonometry as both practical and theoretical disciplines.[27] About the same time, an Italian from Cremona had taken the high road to Spain to learn Arabic and absorb the same "new" learning from the East. At the center of Arabic studies in Toledo, once more under Christian rule, Gerard of Cremona did his own translations of the *Algebra* and of Khwarazmi's astronomical tables, as well as eighty-five other works in Arabic.[28]

Thanks to these three great translators, Khwarazmi's main works were preserved in Latin through all the centuries when nearly all the Arabic originals perished. Their editions, along with the two Arabic originals, enable us to appreciate Khwarazmi's achievement in greater depth.

Turning first to the *Compendious Book on Calculation by Completion and Balancing*, known as his *Algebra*, one is struck at the outset by the way Khwarazmi defined his target audience. He intended his book not for scholars but for practitioners and therefore strove to include in it "what is easiest and most useful in arithmetic, such as men constantly require in cases of inheritance, legacies, partitions, lawsuits, and trade, and in all their dealings with one another, or where the measuring of lands, the digging of canals, geometrical computations, and other objects of various sorts are concerned."[29]

It was useless, he knew, to overwhelm such readers with pages of technical manipulations of symbols and numbers if they did not truly understand the concepts underlying them. Six centuries earlier the Greek mathematician Diophantus had achieved a breakthrough with his invention of mathematical symbols,[30] but either Khwarazmi was unaware of this or, more likely, he did not find it useful for his more practical purposes. Instead he tended to speak of *dirhams*, the Abbasids' unit of money. More important, his practical and pedagogical purpose led Khwarazmi to present his entire *Algebra* in clear prose, without a single number or equation. His writing is so simple and direct that it immediately engages readers from any field.

How, for example, do you divide an inheritance when the deceased left behind a widow and three sons and wanted the sons together to receive two-thirds of the sum for the widow and the eldest son to receive twice the amount allotted to each of his brothers? Or, to take another example, involving slaves (he included more than a dozen problems of this type): suppose an ailing man emancipates a slave but has already been paid a sum by that slave toward her emancipation, and then the slave dies before the master, leaving two children, one of whom is to receive a third of the sum going to the other. Who owes what to whom?[31] In language anyone could understand, Khwarazmi explained the process by which these simple equations could be set up and solved and then proceeded to postulate another problem, which turned out to require a quadratic equation.

Khwarazmi's *Algebra* turns on two basic processes, which he called "reducing" and "balancing." Reduction is the procedure, known today to every school student, by which negative terms are subtracted from both sides of an equation. Balancing is the process of reducing like positive terms that occur on both sides of an equation. For example, $x^2 = 40x - 4x^2$ is reduced and balanced to $5x^2 = 40x$. Proceeding from the simple to the complex, Khwarazmi then set forth—again in clear prose—six different types of linear and quadratic equations (squares equal to roots, squares equal to numbers, etc.) and suggested simple but effective processes for solving each.

In some respects Khwarazmi was no innovator. Ancient Chinese had discovered a method for solving linear equations, and ancient Babylonians had understood quadratic equations. Khwarazmi himself acknowledged his debt not only to Greek thinkers but to the Hindu mathematician Brahmagupta (598–668), who solved several categories of linear equations almost two centuries before he did. But Brahmagupta's method did not catch on, nor did those of the other ancient pioneer mathematicians. The reason for this is that all of them obscured algebra behind some other branch of mathematics, whether geometry in the case of the Greeks or number theory in the case of Brahmagupta.[32] Khwarazmi's great achievement was to have brought algebra into its own, explained it with stunning clarity, and offered original solutions to a number of important problems. Overall he transformed algebra into what two Scottish authorities on medieval mathematics, John J. O'Connor and Edmund F. Robertson, describe as "a unifying theory which allowed rational numbers, irrational numbers, geometrical magnitudes, etc., all to be treated as 'algebraic objects.' It gave mathematics a whole new development path ... much broader in concept to that which had existed before."[33] Even though Khwarazmi's starting equations may strike us as rudimentary, they were "the first attempt at algebraic calculation as such" and led directly to the theory of quadratic equations, algebraic calculation, indeterminate analysis, and the application of algebra to many practical problems.[34] This striking innovation opened vast horizons for the future.[35] As such it became the solid foundation upon which later science was erected.

In a separate and untitled work on mathematics, Khwarazmi made the case for adopting the ten-digit system of Hindu numerals and presented the rules for using it in mathematics. The Latin translation of this work by Adelard of Bath opens with the phrase "Dixit Algoritmi," or "Thus spoke al-Khwarazmi." Because of this, medieval savants who went over to this "new math"—as opposed to using the old Roman numerals, which were hopeless for calculations—called themselves "Algorithmists."[36] Khwarazmi's demonstration of the use of Hindu numbers came to be seen as an apologia for mathematics as such, and it was in that role that his name became connected with algorithms.[37]

Khwarazmi was a central figure, along with Habash, Farghani, and others, in the team Mamun assembled to measure a degree of Earth's longitude. This linked his name with calculations of the sizes of the celestial bodies and their distance from Earth that were far advanced compared to any prior estimates, including those by Habash.

There is no doubt that Khwarazmi was a hands-on astronomer. Mamun's great project required him and his team to make precise field measurements of land distances and angles in the heavens. His fascination with the challenge of producing precisely calibrated data from real life found expression in his two books (both lost) on the design and function of astrolabes and in his efforts to perfect the quadrant, another measuring device that used astronomic data to fix time and location. Once all the data were gathered, it became a matter of geometry and trigonometry. It was almost certainly Mamun's project that prompted Khwarazmi to develop the field of spherical trigonometry, which addresses questions of triangles involving curved surfaces like Earth.[38]

Acknowledging this, Khwarazmi's contribution to astronomy arose less from field observation than from his study and analysis of recent works by the seventh-century Indian astronomer Brahmagupta from Rajastan. In such related areas as algebra, trigonometry, and negative numbers, he could also have benefited from the work of Chinese scientists, but these were as yet unknown beyond China.[39] Brahmagupta and other Indians had long been familiar with ancient Greek astronomy and in many ways surpassed it in their own computation of astronomical tables. We think of the era of Khwarazmi as a time when scientists and scholars writing in Arabic rediscovered ancient Greek learning. No less, it was a time when translators and scientists in Baghdad and other Eastern centers came across the vast but neglected corpus of Indian algebra, trigonometry, geometry, and astronomy; translated and analyzed it; and disseminated it to colleagues. Khwarazmi and other Central Asian scientists were the key drivers of this important intercultural project.

The title of Khwarazmi's astronomical work is *Astronomical Tables from Sind and India*. It is the earliest work on astronomy in the Arabic language to have come down to us in its entirety.[40] Its scale is impressive, with more than a hundred tables covering everything from the movement of the heavenly bodies and the times of the moon's rising to sine values, tangents, and astrology. With *Astronomical Tables* in hand, one could compute eclipses or solar declinations, pinpoint the position of the sun, moon, and the five known planets, or solve a problem in spherical trigonometry. Curiously, Khwarazmi saw his task in this opus as chiefly reportorial, which caused him in certain cases not to correct Indian values with more accurate figures he himself had developed.[41] Yet this did not prevent him from proposing highly innovative uses for the tables, and means of connecting them with the fruits of direct observation.

It has sometimes been argued that much Muslim astronomy can be traced to the need in Islam to define the precise hour for prayers in terms of the rising and setting of the sun, or to specify precisely the location of Mecca for the more accurate orientation of the direction of prayers (*qibla*).[42] It is true that several later astronomers focused on these questions and no doubt garnered financial support and readers by doing so. They treated these issues as useful applications that were worth pursuing because they were of interest to their patrons. As such, the questions were neither more nor less important than the other practical questions Khwarazmi used as case studies in his *Algebra*. In spite of claims that several manuscript fragments on these topics trace to Khwarazmi, their authenticity appears doubtful.

A grand compilation of the latitudes and longitudes of 2,402 localities throughout Eurasia constituted Khwarazmi's contribution to yet another field of study, geography. Once more he stepped forth as an editor, in this case of Ptolemy's third-century *Geography*. This time, however, he did not hesitate to improve on the original. Drawing on a large body of recent research by himself and others, he corrected Ptolemy's measurement of the Mediterranean, more accurately positioned the Canary Islands, presented the Indian and Atlantic Oceans for the first time as bodies of open water rather than inland seas, entered the Pacific Ocean for the first time on a world map, and added hundreds of locales in Central Asia and the Middle East, thus greatly expanding the known world.[43] He also discussed the seven climatic zones and anticipated Farghani in

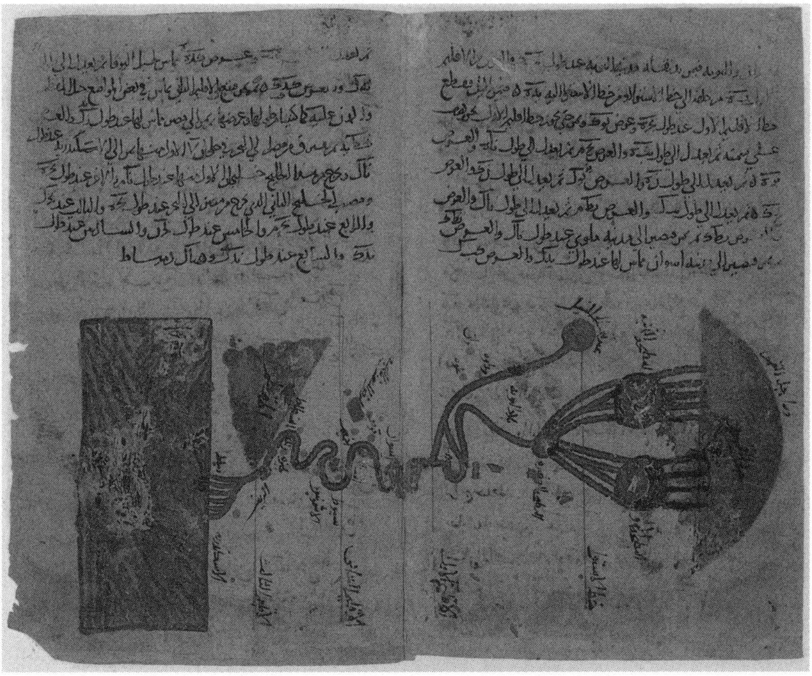

Figure 6.3. Khwarazmi plunged into the work of correcting and improving the works of ancient geographers, as he did in this map of the Nile, from his *Book of the Map of the World*. ▪ Courtesy of the Bibliotheque Nationale et Universitaire, Strasbourg. Ms. 4247, fols. 30b–31a.

providing useful new information on Central Asia.[44] Owing to its clear and comprehensive summary of past and present geographical knowledge, Khwarazmi's *Book of the Map of the World* became the cornerstone of all subsequent geographical studies in Arabic or Western languages.

WAS KHWARAZMI REALLY A SON OF CENTRAL ASIA?

After this telegraphic summary of Khwarazmi's opus, one is prompted to inquire into the route by which he rose to become the most innovative thinker of his age. In terms of our larger concerns, was he connected with Central Asia and its culture in any meaningful way, or was he merely born in its northern frontier kingdom of Khwarazm but otherwise fully a product of Baghdad life? His early formation and education are particularly important in addressing this question. But astonishingly, given Khwarazmi's eminence, less is known of his life than of almost any other leading medieval thinker, East or West. Besides his birth and death dates, his association with Mamun's circle of scientists and scholars in Baghdad, and a mission he is reported to have undertaken to the Jewish empire of the Khazars in 841, his life is an utter blank.

But this is not to say that there are not solid hints, scraps of evidence from which can be pieced together a plausible outline of his education and early formation. We know Khwarazmi was born in 780, which made him an exact contemporary of the Muslim jurist Ibn Hanbal, the Merv native whom Caliph Mamun so ruthlessly humiliated and thereby challenged to action. His place of birth was Khwarazm, the sprawling province along the lower reaches of the Syr Darya on what is now the border between Uzbekistan and Turkmenistan. People in the cities of Khwarazm still spoke the local Khwarazmian language, a branch of Old Persian with links to the ancient language in which the Zoroastrians had written their Avesta. This Persianate language was Khwarazmi's mother tongue, although it is certain that from an early age he learned Arabic and probably Middle Persian as well.

The Arab general Qutayba had laid waste the Khwarazmian capital in 712, and what local power remained shifted then to the cities of Dehistan along the great north-south caravan route that stretched from the northern steppes to the southern shores of the Caspian. With the rise of the Abbasids in the 750s, Khwarazm was included into the large province of Khurasan, with its capital at Merv and then at Nishapur.[45] Khwarazm rebuilt, however, and its trade with the Middle East, India, China, and the Jewish Khazar kingdom north of the Caspian Sea revived as well.

When did Khwarazmi leave Khurasan for Baghdad? The question is critical to Khwarazmi's identity as a thinker. There is no evidence that he arrived in Iraq before 819, the year Mamun finally decided to move his court from Merv to the caliphate's capital. The fact that Khwarazmi dedicated both of his most famous works to Mamun indicates a close link between that caliph and the great mathematician-astronomer. Khwarazmi was twenty-nine years old in the year Mamun's caravan of administrators, scientists, and intellectuals departed Merv for Baghdad.

For an ambitious young scientist in the years 802 to 819, Merv was arguably the best place on earth. Harun al-Rashid had named his intellectually gifted son, Mamun, as governor of Khurasan, and the young governor immediately set about establishing his well-financed and splendid court there. Baghdad, meanwhile, was shaken by civil war, which culminated in the bloody siege of 812–813 and the beheading of Mamun's brother and rival, Amin. For some years thereafter, Baghdad lay in ruins, scarcely the kind of place to which a rising intellectual in search of patronage would have been drawn. More to the point, unofficially from Caliph Harun al-Rashid's death on March 24, 809, and officially after the death of Amin in 813, Merv, not Baghdad, was the capital of the caliphate and hence of nearly all the Muslim world, a status it enjoyed down to the moment Mamun decamped for the Tigris in 819.

From these facts we can confidently hypothesize that the young Khwarazmi arrived in Merv at some point between 802 and 810, made contact with Mamun and leading intellectuals there, and departed Merv only when the caliph and his court did, in 819.[46]

We have noted the practical and utilitarian orientation that Khwarazmi demonstrated when he dedicated his *Algebra* not to scholars but to practitioners in disciplines as diverse as law, commerce, and even hydraulic engineering. This suggests a man who, at some point in his life, had garnered direct knowledge of these practical fields. None of this was likely to have happened in Baghdad during those years of turmoil, but all would have been natural in the rising capital of Khwarazm, at Dehistan, or in the great commercial center of Merv. Note particularly his inclusion of engineers engaged in "the digging of canals." Historic Khwarazm, where Khwarazmi was born, had for centuries maintained what one expert has called "the most highly developed of all the ancient irrigation systems."[47] As we have seen, Merv employed thousands of people to maintain its complex irrigations systems. This called for a high level of engineering skill, not only in calculating water volumes and the depth and fall-rate of channels, but in constructing dams, diversion channels, and so forth. Practical experience gained in Khwarazm and Merv enabled Khwarazmi to target his masterpiece.

Beyond such practical experience, did Khwarazmi's formation as a mathematician and astronomer also take place in Merv? This conclusion is all but inescapable. The old observatory there thrived under Mamun's patronage, and the astronomers Habash, Musa ibn Shakir, the highwayman-turned scientist and close friend of the young caliph, and probably Farghani as well were all working there at the time. No city on earth rivaled Merv in astronomy and related areas of mathematics during those years. It is even likely that Khwarazmi's deep interest in Indian science and mathematics arose in Merv. Buddhism appears not to have spread to Khwarazm,[48] but other aspects of Indian culture and art exerted a powerful influence there, even to the point of possibly influencing key details of the Khwarazmian calendar.[49] Khwarazmi's great successor from Khwarazm, Biruni, was to be so preoccupied with Indian learning that he seized on the first opportunity to pursue his studies in India. And while it is certain that the great Indian work on mathematics that so inspired Khwarazmi, the *Sindhind*, was first translated into Arabic at Baghdad, this may have occurred via an earlier Persian translation from Khurasan.[50] Recalling that Yahya ibn Barmak rose to prominence in Merv in precisely these years, and that the Barmaks were consistent champions of Indian learning, it is not excluded that Khwarazmi also launched his study of Indian math and science while still in Merv.

One further solid fact on Khwarazmi's life bears mentioning. In 841 Caliph Wathiq sent him on an expedition to the Khazars, the Turkic people who had converted to Judaism and established a steppe kingdom in what is now southern Ukraine. The Arabs had fought two wars against the Khazars and now feared a third. Almost nothing of the expedition's purpose or results survives.[51] But the fact that the caliph considered the Khwarazm native to be ipso facto an expert on a people with whom the Khwarazmians had maintained close commercial relations[52] provides yet one another piece of evidence that the great mathematician and scientist had remained in his homeland long enough not only to be educated there but to gain broad experience as a mature adult.

Decades before Khwarazmi's death in 850, he was recognized as the doyen of scientific Baghdad. Wherever aspiring astronomers and mathematicians could read Arabic, Khwarazmi was hailed as a model polymath. He inspired more than a few Arab followers but also many astronomers from his native Central Asia. Of the latter, none was more prominent than Abu Mahmud Khujandi, a native of Khujand, the ancient and independent city of vineyards in the north of present-day Tajikistan where Alexander the Great had found his future bride, Roxanna. Khujandi's absolute distinction was to be the first Turkic practitioner of classical astronomy. Nothing is known of the path by which this scion of the Turkic leadership class from the heart of Central Asia launched a career as an ambitious designer first of astrolabes and armillary spheres and then of the world's largest astronomical instrument. It is more than likely that he studied with the astronomers at Bukhara and then moved on to the court of the Shiite Buyid kings at Rayy in the suburbs of present-day Tehran. Khujandi, too, was a wandering scholar.

Near Rayy he constructed his new type of sextant to determine the angle between the plane of Earth's orbit and the plane of Earth's equator (called the "obliquity of the elliptic"). By measuring the meridian height of the sun at the two solstices, Khujandi reached conclusions on the axial tilt that were more precise than anyone before him but were flawed by what another Central Asian savant, Biruni, concluded was the settling of a corner of the building in which the sextant was housed.[53] Besides his astronomy, Khujandi is also hailed as the probable discoverer of the law of sines for spherical triangles, the equation that relates the lengths of the sides of a triangle to the sines of its angles.

PHILOSOPHICAL QUESTIONS

The rediscovery and translation into Arabic of ancient Greek texts posed as many questions as it solved, for the ancient writers raised anew many philosophical questions that Muhammad's revelation was supposed to have laid to rest. How was the world created? Is the soul immortal? What is the ideal human society and how is it ruled? And above all, what, if any, are the limits of human reason? Curiously, these questions did not arise when Baghdad scholars read the translated works of Hindu writers, perhaps because the cultural distance between the Indian and Arab worlds was simply too great. But they arose with urgency and force in the minds of all who read the new translations from Greek. Nor was it possible for readers of the translations to sidestep these questions by cherry picking the works of Aristotle and the other Greeks, embracing their scientific and practical views but leaving aside the vexing philosophical questions they posed. The two sides of Greek thought were too inseparably interlinked for that.

And so arose the centuries-long debate over what in Arabic came to be knows as "Falsafa," or "Philosophy." The very name reflects the overpowering impact of ancient Greece on Muslim Baghdad. It assumes a thoroughly humanistic and cosmopolitan approach to life and a staunch belief in the idea of progress.[54] With respect to method, champions of Falsafa ardently favored reason. Reason, rather than custom, tradition, or faith, was humankind's best tool for gaining an understanding of the cosmos and its place in it. Most of those who embraced Falsafa also accepted the principles of religion and even assumed that reason would prove them. But from the outset it was unclear whether the point at issue was to revise the classical heritage so as to make it fit the revelations of Islam or to modify Muslim thinking to make it more precisely accord with the insights of the Greeks. Either way, the reception of Greek thought by the world of Islam was anything but passive. Every step of the way was accompanied by sharp debates, which called forth yet more polemics over the meaning of Falsafa.

That these were contentious matters had already become clear during Caliph Mamun's misguided effort to purge all who disagreed with the strict rationalism of the Mutazilites. The battle was to be waged in fields as diverse as philosophy, epistemology, metaphysics, moral philosophy, physics, and theology. Its high point was reached in the writings of the greatest philosopher of the Muslim world, Farabi, famed in the West as Alfarabius, from present-day Kazakhstan. It was to culminate three centuries later when yet another brilliant and combative Central Asian, Ghazali, strove to lay waste the arguments of the rationalists.

KINDI'S CAN OF WORMS

In light of the bitter debates that were to follow, it is ironic that the launching point for this great battle of minds was the writings of the benign al-Kindi, the pride of the Arabs in science and philosophy.[55] Kindi noted with approval Aristotle's argument for the unity and eternity of God but understandably tripped over his Greek mentor's contention that matter, too, was eternal. This, after all, contradicted the dogma, common to Judaism, Christianity, and Islam, that God had created the universe from nothing. Even this presented problems, however, for it implied that God, prior to this act, was somehow incomplete. The

Mutazilites, Kindi, and his heirs all pondered this conundrum and came up with an adroit solution that turned on a fine distinction between *essence* and *existence*, with God alone embodying essence and relegating the physical world to a somewhat lower order of existence. Beyond this, Kindi welcomed Aristotle's affirmation of the human soul but, as a Muslim, could not accept the Greek's contention that the soul could not exist independently of the body. Finally, he had to address the fundamental question of truths obtained through direct revelation, essential to Islam but absent from the Greek tradition of thought.

To these and other core points of apparent conflict between Greek thought and Islam, Kindi brought the twin tools of reason and faith. On reason he conveniently glided over the fundamental differences between Aristotle and Plato and between Plato and the later Neoplatonists from Egypt. Instead he simply affirmed, again and again in his treatise *On First Philosophy*[56] and in other works, that reason is one of the two legitimate means of establishing truth, the other being the revelations of prophets. Not mentioned in his work was intuition, which several of Kindi's successors proposed as yet a third "way of knowing." Meanwhile, in what became his most important affirmation, Kindi asserted that reason and faith are fully compatible with each other.

Reason, he argued, could unveil the truths of nature as surely as could revelation and theology. The exercise of reason is the very essence of what it means to be human and is the element of the soul that is immortal. Comparing the world of thought pre-Kindi and post-Kindi, it is obvious that he succeeded in greatly expanding the realm in which reason could operate.[57] His critics, and all those who later attacked Falsafa, did not fail to note that this occurred at the expense of revelation. It is ironic that this Arab philosopher, who valued his serenity above all else, would give rise to fierce debates that dragged on for centuries.

It was some while after Kindi's death before these intellectual fireworks commenced. All three of his leading disciples were Central Asians, and all worked safely within the parameters Kindi himself had observed. Since we will encounter this trio in the next chapter, suffice it here to summarize their relation to Kindi's legacy. Abu Mashar al-Balkhi, a native of Balkh, pursued a career in astrology, but with a twist. Rather than offer up a trove of arcana, Abu Mashar, reverting to the solid rationalism of Aristotle, included long passages and paraphrases of the master's works in his own voluminous writings (see plate 6). Thanks to these, "Albumasar," as he was known to later Latin readers, became the prime sources of information on the Greek teacher throughout the medieval West in the century before Aristotle's original texts were translated from the Arabic around AD 1100.[58]

A second protégé of Kindi's was from Sarakhs, the old trading city south of Merv on the present border of Turkmenistan and Iran. A manuscript discovered in Florence describes this man, Ahmad al-Sarakhsi (833/37–899), engaging a Christian bishop in Baghdad in finely honed debate on theology, making his points on the basis not of dogma but of close reasoning and logic.[59] So effective a pedagogue was Sarakhsi that he was named tutor to the caliph's son. Later a boon companion to the same Caliph Mutadid, Sarakhsi was accused of Mutazilite and Shiite sympathies and charged also with accumulating a small fortune on the side. For these misdeeds, real or imagined, he was beaten and executed.[60] A third Kindi student was Abu Zayd al-Balkhi, another wandering scholar who traveled by foot from his birthplace on the Afghanistan-Iranian border to Iraq. Inspired by Kindi's example as a polymath, Abu Zayd achieved distinction both as a geographer and, as we shall see, as a pioneer in the field of mental health.[61]

The flash point over Kindi's legacy arose not from his earnest and mainstream disciples but from others who were inspired by the expansion of the realm of reason that Kindi helped bring about. If, as Kindi maintained, reason and prophetic revelation were equally valid avenues to knowledge, why should reason not be applied to the truths of religion? It was inevitable that someone would advance this perfectly logical proposition, and in fact many thoughtful people did so. Most, like Kindi, were desk-bound philosophers who advanced their views through earnest treatises written for other scholars. But others pressed their point very publicly and with burning zeal. As we shall see in the next chapter, the Khurasan region of what is now northwestern Afghanistan, northeastern Iran, and southwestern Turkmenistan was to become a hotbed of such radical freethinking.

The person who showed the greatest boldness in applying the tools of logic and reason to religion was Muhammad ibn Zakariya al-Razi

(865–925). Known in the West as Rhazes, Razi is best known, and justly honored, as the first true experimentalist in medicine and the most learned practitioner of medicine before Ibn Sina. Since he spent most of his life in his hometown of Rayy, near modern Tehran, Razi might seem to lie outside the geographical scope of our study. Yet his entire education took place at Merv; his teachers were all Central Asians; it was the governor sent by the Samanid state in Bukhara who commissioned the writing of his great compendium of medicine; and, finally, it was another Central Asian—Ibn Sina—who triumphantly developed and transformed Razi's legacy in medicine. In short, Razi lived his entire life within the intellectual orbit of Central Asia.

Razi's eight-volume encyclopedia of medicine was not the first such compendium. That distinction goes to his second teacher, another Central Asian from Khurasan, Ali ibn Sahl Rabban al-Tabari (838–870), who, as we have noted, also pioneered in the field of pediatrics. In one respect Tabari, a Muslim scion of a well-known Jewish family from Merv, went beyond his student: he focused squarely on mental health, the psychological origins of certain maladies, and the possibility of treating certain conditions through a therapeutic process of discussion between doctor and patient—in short, psychotherapy.[62] Razi was the father of immunology, the first to distinguish smallpox from measles, the first to write on allergies and, through his *Diseases of Children*, on pediatrics as well. By all accounts, Razi was dedicated to the welfare of his patients and was deeply admired for his commitment to the community he served. Even today there is a university named for him in Iran, and Iran's pharmacists annually celebrate "Razi Day."

In sharp contrast to his medical research, Razi's ventures into religion earned him nothing but abuse.[63] The very names of Razi's three treatises on religion say it all: *The Prophets' Fraudulent Tricks*; *The Stratagems of Those Who Claim to Be Prophets*; and *On the Refutation of Revealed Religions*. Razi did not mince words:

> If the people of [a given] religion are asked about the proof for the soundness of their religion, they flare up, get angry and spill the blood of whoever confronts them with this question. They forbid rational

Figure 6.4. Muhammad al-Razi (865–925), the first true medical experimentalist, examines a urine sample. Razi's sharp-tongued religious skepticism led Ibn Sina to wish that he had stuck with the "examination of boils, urine, and excrement." ▪ Gerard of Cremona, *Recueil des traités de médecine*, 1250–1260. Reproduction in Samuel Sadaune, *Inventions et découvertes au Moyen-Âge dans le monde*.

speculation, and strive to kill their adversaries. This is why truth became thoroughly silenced and concealed.[64]

And about the Quran itself he raged:

You claim that the evidentiary miracle is present and available, namely, the Quran. You say: "Whoever denies it, let him produce a similar one." Indeed, we shall produce a thousand similar, from the works of rhetoricians, eloquent speakers and valiant poets, which are more appropriately phrased and state the issues more succinctly. . . . By God what you say astonishes us! You are talking about a work which recounts ancient myths, and which at the same time is full of contradictions and does not contain any useful information or explanation. Then you say: "Produce something like it?"[65]

Such passages from Razi's pen, and there were many like them, enraged pious Muslims. Never mind that Razi later had second thoughts on some subjects and overall became more temperate; he even turned out a weighty commentary on the Quran. But the damage was done. He was permanently cast as a heretic.

Ibn Sina would later seethe at the two lengthy tracts in which Razi had attacked philosophy and tossed aside free will as an illusion. Razi, he declared, should have limited himself to the examination of boils, urine, and excrement "and refrained from delving into matters beyond the range of his capacity."[66] Two of Razi's earliest critics were from Balkh.[67] Their attacks caused him to pen a candid but argumentative *Philosophical Biography* to defend himself. But to fulminate was not to refute. Razi, after all, had simply applied to religion the same rational and analytic approach that he had brought to medicine. His credo on the role of reason was absolutely uncompromising. Reason, he wrote, "is the ultimate authority, which should govern and not be governed; should control and not be controlled, should lead and not be led."[68] Left unanswered, Razi had created a situation in which rationalism and faith were heading for a frontal confrontation.

Farabi Squares the Circle

The man who clipped Razi's wings and confounded all those who shared his absolute and uncompromising devotion to reason was Abu Nasr Muhammad al-Farabi (ca. 870–ca. 950), known in the West as Alfarabius. Farabi's sphere of thought embraced all knowledge, which enabled him to become "the first person in Islam to classify completely the sciences, to delineate the limit of each, and to establish firmly the foundation of each branch of learning."[69] His great achievement was to have defined a specifically spiritual realm while preserving and even expanding the sphere in which rational inquiry could go forward. For this he is revered in the Muslim world as "the Second Teacher" (the first being Aristotle himself), and by modern philosophers as the first logician and the father of Islamic Neoplatonism. Advancing a notion of God as the First Mover, Farabi exerted a significant influence, directly and indirectly, on St. Thomas Aquinas, Dante, and even Kant, as well as on the medieval Jewish thinker Maimonides.[70] A writer of global import, Farabi was a prince of medieval thought, East and West.[71]

Farabi was also a wandering scholar par excellence. Born in Central Asia, he traveled as a young man to Askalon (now Ashkalon, in Israel), Harran in upper Mesopotamia (now Turkey), and Egypt. He spent four decades in Baghdad and then moved on to Aleppo and finally to Damascus, where he died.[72] By modern standards he lacked worldly ambition. No evidence links him with the caliph's court in Baghdad or the circle of scientists and thinkers assembled there. Farabi had no known patron in Baghdad, and in Damascus he is said to have briefly accepted work as a gardener, even though by then his writings were famous. A rare human touch that has come down to us is that he loved music and played the lute. This we know from his *Great Book of Music*, acknowledged as the premier work in Arabic on the mathematics of tonal systems and the theory of music generally; Farabi's study of music was to became the foundation stone of Western musicology.[73]

This detail sheds light on one of the most hotly debated yet trivial points of Farabi's life: was he of Persianate or Turkic ethnicity? Such a question about a deeply cosmopolitan humanist would never have arisen had not an enthusiastic Turkic biographer, Ibn Khallikan (1211–1282), writing more than two centuries after Farabi's death, claimed that Farabi was a Turk.[74] Such tribal claims were common by the twelfth century, when Turkic dynasties ruled most of Central Asia and Persia.

In the absence of any solid evidence supportive of this Turkic thesis, it might be noted that the musical instrument that Farabi played, with the tuning system he analyzed so meticulously, is clearly the Iranian lute, native to Central Asia and Iran, which became the model for fretted plucked instruments from Gibraltar to China. His musical analysis employs many terms common to Sogdian and other Iranian languages, but none of Turkic origin. A distinguished scholar of Greek and Arabic philosophy, Dimitri Gutas, has noted that in Farabi's other works he employs a number of Sogdian words but none that traces to Turkic languages.[75]

As with Khwarazmi, it is worth asking whether Farabi's links with Central Asia extend beyond the mere fact that he was born there. And as with Khwarazmi, too, this turns on his education and the age at which he moved to Baghdad. The paucity of hard evidence has led to claims that he was raised in the capital on the Tigris and received his entire education there. But certain facts would appear to undermine this. Farabi himself wrote that he studied with the Nestorian Christian logician and scholar Yuhanna ibn Haylan.[76] Now, Ibn Haylan was from Merv and taught there down to the year 908, at which time Farabi was over thirty, scarcely an age to undertake studies with a new teacher. The obvious conclusion, shared by most scholars, is that Farabi began his studies with Ibn Haylan at Merv and then ended up with him at Baghdad, perhaps after an interval of several years devoted to travel.[77] By this reckoning, Farabi spent a full quarter century in Central Asia and would have received nearly all his education there and launched his professional career there as well. This would explain his knowledge of the Sogdian language, which, as we have seen, was long a lingua franca in Central Asia but was largely unknown in Baghdad.

This also reflects Farabi's birth in a town near the great center of continental trade, Faryab (now Otrar), a city of seventy-five thousand on the ancient Jaxartes River (now Syr Darya) in southern Kazakhstan.[78] Faryab was located amid salt marshes at a crucial ford on the Jaxartes, dominating the central section of the great East-West corridor over which passed much of the traffic between China and Europe.[79] Heading westward from Faryab, one reached Khwarazm, and to the east the traveler would pass through the Turkic capital at Balasagun and then on to Kashgar in what is now China. Ethnically, Otrar in Farabi's day was basically Sogdian, that is, Persianate, but over the succeeding centuries its population became overwhelmingly Turkic, which explains the later assumption that Farabi, too, must have been a Turk.

One final detail worth exploring concerns Farabi's education. Several writers assert that he was trained initially in Bukhara, which could be easily reached by the main route leading south from Otrar. This would have been a logical decision since Bukhara at the turn of the ninth century was the booming capital of the Persian-speaking Samanid empire and its leading intellectual center. It was also known as a place where Shiite Islam, while by no means not the official faith, was tolerated and strongly influenced cultural life. It may not be coincidental that Farabi's later wanderings took him to Egypt just as it was beginning to flower under the Ismaili Shiite Fatamids, that he was often accused of pro-Shiite sympathies, and that he spent his later years at Damascus at a time when it was also under Fatamid, that is, Ismaili Shiite, rule.[80]

A CAUTIOUS INNOVATOR

As a personality, Farabi was the very oppositive of Razi, earnest, steady, and retiring as opposed to dynamic, impulsive, and commanding. His writing style, honed by a knowledge of six languages, exactly suited someone who prefered to announce even his boldest ideas without stirring a ripple and who saw himself merely as the one who was putting them in logical order. A card-carrying *faylasuf*, or adherent of Falsafa, he produced lengthy tracts first on Aristotle, then on Plato, and finally on the Neoplatonists, the mystically oriented ancient heirs of Plato.[81] Unimpeded by modern notions of footnoting, Farabi lifted wholesale or paraphrased whole sections from his sources.[82] But for all his low-keyed tone and avowed reverence for the ancient masters, Farabi revised the antique classics as much as he summarized them, causing one scholar to argue that he was "breaking with Athens."[83] Whether this was the case, Farabi undeniably broke with important elements of prevailing Muslim thought.

That Farabi was a good Muslim was not in question, but a mainstream Sunni could well have asked, "Of what kind?" His definition of the divinity as the First Cause came straight from Aristotle, and from this God "emanated" the whole universe. "Emanation" was the Neoplatonic term that the more rationalistic Islamic thinkers appropriated to finesse the whole debate over creation. Where Farabi burst the bounds of Muslim dogma was in his defense of the primacy of reason. He asserted that humankind's highest virtue is reason, which he ranked above both sensation and "imagination," or intuition. God is omniscient, yet humans enjoy freedom to act or not act, think or not think. The mission of the "soul" is to rise above perception and imagination and reach its perfection in "reason and disciplined thinking."[84]

Both in his definition of the scope of reason and of humans' freedom and responsibility for their actions, Farabi found himself at odds with what mainstream religious scholars at the time considered core Muslim doctrines concerning God's omnipotence. On both these important points and others, the great thinker was dangerously close to the much-criticized Mutazilites.

But what of the truths enunciated by the prophets? In a brilliant passage, German-born Richard Walzer, the greatest modern expert on Farabi, captured the measured boldness of Farabi's answer. For Farabi, he wrote, prophesy was

> A man who in waking life has reached the utmost perfection of his imaginative powers can be called a man gifted with prophesy (*nubuwwa*). Since he is aware of particulars, present and future, and visualizes things divine in symbols of outstanding beauty and perfection. "This [says Farabi] is the highest perfection which 'imagination' can reach, and the highest level accessible to man on the strength of this faculty." Thus, prophesy is understood in rational terms and, moreover, as "auxiliary to the rational faculty." Philosophy is in a higher place than the different religions and has everywhere the same truth, whereas the religious symbols produced by the imaginative power of sectional prophets vary from land to land.[85]

The First Cause rules the universe, but it is reason that defines and governs the life of human beings. Religion at best provides a symbolic rendering of the truths of reason.

This is not the place to delve into the possibility that this astonishing stance traces to a strong Shiite strain in Farabi's thought.[86] What counts is that he defined the life of the mind, not faith, as the pinnacle of human experience. But he did not stop there. In a long text on "virtuous religions," Farabi noted the mutual contradictions among the prophetic religions.[87] Sounding for all the world like Razi, he pointed out that they could not all be right. But then he backed off, proposing the same kind of hierarchy among religious revelations that he had employed in analyzing the human faculties. Islam emerged at the top of the list, for its truths fully mesh with those of philosophy. The day was saved.

It is impossible to summarize the many areas into which Farabi delved with his imposing intellect. Even an incomplete list of headings would have to include metaphysics, epistemology, eschatology, ethics, physics, astrology, and psychology, as well as music. Each has been the subject of scores of studies in as many languages written between the tenth century and today. Yet more difficult it would be to encapsulate the formidable method he brought to bear in each area. Suffice it to say that his prime tool, the skeleton key he used to unlock the entire world of knowledge, was logic. Logic, both for Farabi and his great successor, Ibn Sina (Avicenna), provided a rigorous method of reasoning that could be applied equally to all the sciences and to theology in order to arrive at the underlying truth of things. Building on his studies with Yuhanna ibn Haylan, Farabi not only employed logic more effectively than anyone before him but left us what long remained, in both the Middle East and Europe, the definitive treatment of the subject.[88] Farabi's powers of logic, combined with the low-keyed and moderate voice in which he presented his thoughts, won him deep esteem wherever Falsafa was valued.

FARABI'S DECEPTIVELY NONPOLITICAL POLITICS

Shortly before his death in 950, Farabi completed work on a thick tome entitled *Principles of the Views of the Citizens of the Best State*.[89] He had touched repeatedly on politics before this, notably in a collection of aphorisms.[90] But in this volume, usually called simply *On the Perfect State*, he treated the subject systematically, beginning with relevant notions from metaphysics and the natural sciences and ending with four chapters on the civic realm as such. Even his most admiring modern student admits that *On the Perfect State* offers "a thoroughly bookish philosophy."[91] This charge arises from Farabi's habitual thoroughness and also from the fact that he wrote under the powerful influence of Plato's *Laws* and *The Republic*, that abstruse and perplexing diaolgue on the purpose of society. Making it yet more "bookish" and superficially apolitical is the fact that Farabi reached his understanding of *The Republic* by plowing through the dense tomes of sixth-century Neoplatonist writers, mainly Christians from the eastern Mediterranean. *On the Perfect State* represents Farabi's careful attempt to "naturalize" *The Republic* and Plato's other political writings within the context of Islam.[92]

Farabi's announcement that the purpose of society is the "happiness" of its citizens may cause a modern reader to expect a medieval outburst of Jeffersonianism. But Farabi's "happiness" is not pleasure at all but the full exercise of reason. Such happiness, he took pains to point out, cannot be attained by more than an elite few. He ranked citizens in much the way he ranked mankind's abilities and the various religious revelations. Like Plato, Farabi was no democrat, for democracy is bound to drag society down to its lowest common denominator. But he was an intensely civic person, praising voluntary cooperation among citizens and holding that all human beings reach their highest potential only through participation in urban life. The aim of society is to make that possible.

At least the few can reach for the stars, provided they are under the constant and austere guidance of a wise and omniscient leader. Like Plato, Farabi posited a top-down world. His ideal leader receives advice from a council, but in the end it is he alone who defines, orients, and empowers the lives of all whom he rules.

Some of the most compelling passages of *On the Perfect State* are devoted to the figure of this leader.[93] Plato had argued that society's goal of "justice" can be realized only through the leadership of a moral and all-wise "philosopher-king." Farabi saw the philosopher-king as the human embodiment of the divine order of things. He is free from all sin, and his judgments are unerring. He teaches through "strict demonstration" and not through "symbols"[94] (i.e., religious dogma), for these give rise to endless objections and debate and eventually to the false belief that there are no ultimate truths. Absent such a wise leader, the "ignorant city" wallows in the pursuit of honor and wealth, and its citizens descend to the level of dumb animals. Worse, it engages in endless wars with its neighbors and deludes itself into believing that whoever succeeds in overpowering his enemies is most happy.[95] Justice is thereby reduced to "defeat[ing] by force every possible group of men which happens to be in one's way."[96]

In their ignorance, citizens of "cities that miss the Right Path" embrace common bonds of ancestry, ethnicity, or residence as the basis of human solidarity. Some even corral religion in their effort to create the bonds among people that could lead them to succeed on the battlefield of life. Such "tricks and contrivances" are the last resort of those "who are too weak to gain these goods by force, in tough and open fight."[97]

This, then, is the fate of the city that is unguided by the wisdom of a true leader. Who can be such a leader? Farabi's version of the philosopher-king is none other than the leader of the faithful, the political and spiritual head of the community of Islam, the imam.[98] By this he did not mean the awaited imam of the Shiites, but simply the perfect ruler. Understood thus, *On the Perfect State* can be read as a searing indictment of the degenerating Muslim caliphate of the mid-ninth century. It was also a frontal attack on the principle of jihad as Baghdad practiced it at the time.[99] Is it really possible, Farabi asked, for the caliph, as "Commander of the Faithful," to expound an uplifting religious and political order when his own house is so obviously in disarray? But *On the Perfect State* also presented a daunting recipe for what was needed to redeem and regenerate the state of Islam and its caliphate. We will never know whether the always-cautious Farabi considered reform possible, but he strongly implied that it was not. What we do know is that he held off writing *On the Perfect State* until after he had departed Baghdad and was safely in the Fatamid-controlled city of Damascus.

Farabi's Lost Opportunity

Let us now briefly turn back to the work of those many scholars who were doing translations from the Greek. Perusing the list of their translations, one sees that one particular book is conspicuously absent: Aristotle's *Politics*. If Farabi had any contact with this magnificent treatise, there is no evidence of it in *On the Perfect State*, which is not surprising since it was not translated into Arabic until modern times.[100] Indeed, our arch-Aristotelian did not even mention Aristotle's name in his own book on politics. True, there are many similarities between Aristotle's view of politics and those of Plato, Farabi's constant mentor on the subject. But at bottom they were profoundly different.

Where Plato was abstract, Aristotle was practical.[101] Where Plato was theoretical, Aristotle was empirical, drawing on both his firsthand knowledge of Athenian democracy and his field research on a dozen or more city-states in the Greek world. Where Plato approached his subject as a philosopher, Aristotle proceeded like a medical clinician, identifying various pathologies and prescribing workable remedies. Aristotle did not separate religion (philosophy) and politics, but he did not merge them, as do Plato and, far more explicitly, Farabi. Aristotle railed against "false utopias" such as Plato's. Above all, Aristotle, unlike Plato (and Farabi), concentrated on the unwritten understandings—in Greek, the *politeia* or "constitution"—that guide the interactions of citizens and that law-givers are responsible to protect. For Plato, as for Farabi, the leader was a quasi-divine thinker, but for Aristotle he was more of a craftsman, exercising practical skills of lawmaking and leadership. He does not do this alone, however, for there should also be deliberative bodies in which all qualified persons participate as free citizens and from which the leaders themselves are selected.

There is much in Aristotle's *Politics* that is utterly impractical. And it would be wrong to see in Aristotle's aristocratic state an apologia for representative government, let alone for democracy, which he severely criticizes. Yet the utter absence in Farabi's great treatise of the kind of empiricism and hands-on approach that suffuses every page of Aristotle's *Politics* is, to say the least, regrettable. Missing from Farabi's text is any appreciation of the practical skills needed to govern, the incremental steps by which improvements are made, and the constitutional understandings that define the polity—all matters that states guided by Islam, or any other states, for that matter, needed to master. Blame it on the fact that the translators in Gundeshapur, Baghdad, Merv, and other centers, amid all their achievements, fell short in their failure to render Aristotle's *Politics* into Arabic. By any reasonable measure, Farabi's *On the Perfect State* was a major achievement. Yet in the end it was a lost opportunity.[102]

DID CENTRAL ASIA SUFFER FROM BRAIN DRAIN?

Khwarazmi and Farabi may stand in the top rank of wandering scholars from Central Asia, but we know that the honor role of restlessness neither began nor ended with their names. Both of the other two leading minds of the Middle Ages, Biruni and Ibn Sina, spent their lives on the move, the former to the east as far as India and the latter westward to the cities of what is now Iran. Scores of others competed with them for the number of days whiled away as hangers-on to slow-moving caravans, and the number of nights spent at noisy caravanserais or on the ground under the stars.

The departure of so many scholars from their native cities and even from the region as a whole raises the question of a possible brain drain. Did the flowering of Baghdad in the end impoverish the intellectual life of Central Asia? The travels of Central Asian savants attest to the gravitational pull exerted by certain centers of learning, Baghdad above all. But if Baghdad's attraction was immense, so was its capacity to repel. Ibn Sina had nothing good to say about Baghdad life and left as soon as he could. Farabi quit Baghdad without having any clear further plan in mind. As soon as the young Bukhari finished conducting his field interviews on Muhammad's Hadiths, he made only one brief stop in Baghdad, to present a draft of his opus to Ahmad ibn Hanbal. Mahmud Kashgari also tired of the capital, where he observed Turks being everywhere demeaned, and returned to his homeland in present-day China to write his magnum opus on the Turkic peoples. Before his involuntary removal to Afghanistan, the great Biruni worked abroad for a few years but never in Baghdad, and as soon as he could he returned to his home in Khwarazm. All the major urban centers of Central Asia were full of locals who, like Bukhari, had returned from abroad.

With only one or two exceptions, not one of these returning intellectuals, or any others from the region, seems to have been motivated by a romantic longing for hearth and home. They were, as a group, thoroughgoing cosmopolitans and citizens of the world. Their departure from Central Asia, whether temporary or permanent, was a product of ambition, mobility, and sometimes necessity. So, too, were their decisions to return, stay there, or relocate within the region, as became increasingly common. The rise of Baghdad produced a short burst of intellectual immigration that gave the salons of the new capital a decidedly Central Asian cast. For a brief period this removed important talents from the Central Asian scene. But in the end it enriched the region and became yet one more factor in the intellectual flowering of that ancient land. It makes little sense to characterize such a process as a "brain drain."

Within a generation after the death of Caliph Mamun, Central Asian cities emerged once more as the major centers of learning, generating together and even individually an intellectual gravity that surpassed Baghdad's. Over the next two centuries, Nishapur, Bukhara, Balasagun, Gurganj, Merv, and Ghazni in Afghanistan all had their day of glory, as later did Samarkand and Herat. In each case new sources of patronage and new venues for research and writing emerged, reinforced by great libraries, scintillating salons, thriving circles of copyists, and well-stocked bookshops. Arabs and people of other ethnicities joined the ranks of those who headed east in order to be in on the action, further enriching these Central Asian centers as Central Asians had formerly enriched Baghdad.

Let us turn, then, to the first great Central Asian center of the intellect to emerge after Baghdad reached its apogee: Nishapur in Khurasan, the capital of what had become de facto the independent state of Central Asia.

❀

CHAPTER 7

Khurasan: Central Asia's Rising Star

Misty-eyed readers of the *Thousand and One Nights* and Islamic fundamentalists have next to nothing in common, but both idealize the era of the Abbasid caliphs and its capital, Baghdad. Yet that city's most brilliant phase proved very brief, and its decay prolonged. Within a few generations the mantle of leadership in philosophy, science, and the arts had shifted decisively eastward to the Central Asian cities. This process was hastened by endless internecine feuds, bloody coups, and civic strife within Baghdad itself. But it was caused equally by a new political and cultural dynamism within Central Asia.

Perhaps this move eastward was inevitable, since so many leaders of the Abbasids' "Arab renaissance" were in fact not Arabs but various people of Iranian stock and other Easterners drawn mainly from Central Asia. Had it not been Central Asians of all ethnicities who had rallied to the side of Abu Muslim to overthrow the feckless Umayyad caliphs? And had they not also established someone intimately connected with their region, Mamun, on the throne of the refounded caliphate? And finally, did not the caliphs themselves owe whatever security they enjoyed to the thousands of enslaved Turkic soldiers from Central Asia who constituted their army?

This shift of power from Baghdad to Central Asia began the moment Caliph Mamun arrived in Baghdad from Merv. He well understood that the region he left behind—now on the borderlands of Iran, Turkmenistan, and Afghanistan—was both the most vital and the most volatile of all the Muslim lands. Persian-speakers knew it as Khurasan, the "Land of the Rising Sun." It was now the linchpin of the entire Muslim empire.

Mindful of the raw power that Khurasan could exert against the caliphate, Mamun understood that he had to strike a deal that would give the region a high degree of self-government in exchange for its continuing to pay tribute to Baghdad. He also knew from personal experience that if he tried to buy off Khurasan with fine words and something less than real autonomy, all Central Asia might once more rise in revolt. Seeing all this, the caliph acted decisively and as successfully as the difficult situation allowed.

First, Mamun expanded Khurasan to include virtually all Central Asia except for the northern steppe country and Xinjiang. He reasoned that it would not do to leave key cities like Bukhara and Samarkand under the control of others, for whatever happened in Merv and Nishapur would affect the whole region anyway, and vice versa. He therefore established a single large entity that integrated under one ruler virtually the entirety of Central Asia up to the northern steppes and east to the Ferghana Valley. Second, he designated Merv, then Nishapur, as the capital of this new province—or, better, state—of Central Asia. Third, he entrusted control of this "state within a state" to a powerful and effective leader whom he knew local elites would accept: Tahir ibn Husayn, the general who had commanded the troops that had brought Mamun to power in the first place. Finally, since Tahir, as governor of Khurasan, would be the second most powerful figure in the entire caliphate, Mamun had to give him real authority, which he did.

Tahir, born in a town outside Herat in present-day Afghanistan, was a genuine son of Central Asia who enjoyed great political legitimacy throughout the region. He was highly Arabized and understood well the subtle arts by which a client can control his nominal master.[1] The minute he took up his governorship, initially at the old capital at Merv and then at Nishapur, Tahir set about establishing order, winning the support of the same population that in earlier times had rallied to Abu Muslim and then to the mercurial Sinbad the Magi and the other-worldly al-Muqanna. Tahir showed himself to be just and effective, which further strengthened his hand both at home and in Baghdad.

Fully conscious of his power, Tahir asserted his independence in 822 by striking the caliph's name from the shiny new coinage issued by his mints. At the same time he took the brazenly aggressive step of deleting all mention of the caliph, the "Commander of the Faithful," from its normal place in the Friday sermon throughout Khurasan. Thus, a mere three generations after the Arab conquest, Central Asia had once again asserted its autonomy, even while continuing to pay tribute to Baghdad.

Through the person of Tahir, the caliph's own appointee, Central Asia thumbed its nose at the caliphate.

Tahir died suddenly on the evening he issued the order to strike the caliph's name from the Friday sermon. Mamun well understood that Tahir had frontally defied him, but he had no choice now but to appoint Tahir's son to succeed him. This assured that the ruler of Central Asia would henceforth be hereditary, effectively sealing Central Asia's victory. To be sure, Tahir's heirs not only recognized the caliphate but in certain respects bolstered it. They paid tribute to Baghdad, showered gifts on top officials there, and displayed outward respect toward the person of the caliph. Yet they did so in the full knowledge that the caliphate had fragmented and that they, the heirs of Tahir, ruled its most powerful province virtually as an independent state. Tahir's dynasty ruled with a sturdy prudence and effectiveness that was very rare in that era of political drama and caprice.[2]

Nishapur, the new capital, was situated 160 miles to the southwest of Merv in what is now the northeastern corner of Iran. It sat astride the great route from India and China to the West, thanks to which it burgeoned in its new role. From Tahir's death in 822 down to the fall of his dynasty in 873, Nishapur was second only to Baghdad in political might in the entire Muslim world and at least its equal in intellectual and cultural eminence. It was universally acknowledged to be the center of philosophy in the Muslim world,[3] and within a century it had become the literary capital as well, thanks to one writer, Abolqasem Ferdowsi. Down to its destruction by the Mongols in 1219, Nishapur remained also a preeminent center for religious study and speculation, political thought, mathematics, and the sciences generally. No wonder that a local publicist churned out promotional books with titles like *The Subjects of Pride of Khurasan* or *The Good Deeds of the Tahir Dynasty*.[4] As often happens with such boosters, the author was himself an enthusiastic transplant, in this case from Balkh.

What was the key to Nishapur's cultural eminence? One thing is certain: its cultural riches did not arise from political power alone. The city served as the political capital of Central Asia for a mere fifty years, during which time it experienced constant political upheavals. Thereafter it fell to the status of a provincial capital, a "second city" subordinated to a political center elsewhere: first Zaranj, along the modern border between Iran and Afghanistan; then Bukhara; Ghazni and then Ghor in Afghanistan; and, once again, Merv. Foreign invaders conquered the city and the entire province not once but three times in the century after the fall of Tahir's dynasty. Yet for several hundred years this sprawling metropolis in the desert was among the world's major centers of culture and thought.

Nishapur was rich, at least initially. It boasted a large tax base thanks to continental trade, manufacturing, and agriculture and also controlled the lucrative slave trade, sending thousands of Turks westward each year, some as tribute to the caliphs, others as a kind of natural resource export to benefit the local regime.[5] Trade of all sorts defined the city, just as it was later to define Antwerp, Brussels, and Amsterdam. With trade came the same kind of respect for practical skills and preference of reason over dogma that was to characterize these European centers a half millennium later.

But following an initial boom, Nishapur's economy remained torpid during most of its years of greatness. In fact, the dominant theme in the social life of this great center was not growth but decline, a prolonged but culturally fecund twilight. With this came attitudes and ideas that contrasted fundamentally with the concerns of the solid local burghers who continued to thrive on a more limited scale. Defying the stereotype of such economically expansive cultural capitals as Rome, Alexandria, Venice, and Baghdad, Nishapur and its neighboring city, Tus, confirm Hegel's quip that the owl of Minerva flies at night.

Learned Nishapur: The Physical and Material Setting

The population of Nishapur was mainly of Iranian stock. True, some Turks had settled in both the city and surrounding countryside, but they did not set the tone locally. The Arab invaders had also left a garrison there following the conquest. Many of the Zoroastrian elite resented the Arabs and their religion, and we have some nasty Persian verse to prove it.[6] But the Arabs were not as numerous as in Merv or Bukhara, and over time they intermarried and blended into the local culture. In the end the Arabs' main contribution was to have set in motion changes that transformed a country town into a major urban complex and gave the locals access to the rising tide of writings, both religious and secular, in Arabic. Many of the local Zoroastrians eventually converted to Islam, and those who did not worked out a modus vivendi with the Muslims.

Just how big was Nishapur? Calling on the rich but largely ignored medieval literature on the city, the American scholar Richard W. Bulliet carried out a masterful reconstruction of Nishapur's population and topography, identifying its many quarters and districts and locating its palaces, garden suburbs, and slums. He concluded that Nishapur at its zenith in the tenth and eleventh centuries sprawled over nearly eight square miles and that its core city claimed from 200,000 to 500,000 residents. This estimate does not include inhabitants of the many satellite towns or the tens of thousands of traders and transients who must have rubbed shoulders in Nishapur's bazaars at any given time.[7] To put this in perspective, even the lower figure is four times that of medieval Rome and also far larger than Paris in the same period. It dwarfs London, which at the time of the Norman Conquest numbered barely 10,000. The lower figure is about the same as medieval Nara in Japan, while Bulliet's upper figure exceeds that of Constantinople and is only a third less than the population of Chang'an (now Xian) in China or Baghdad at the time.

In terms of creature comforts, Nishapur could compete with any city on earth. A sophisticated irrigation system brought water by underground channels from the nearby hills through the city's center, the channels sometimes reaching a depth of 175 feet. Excavations conducted by New York's Metropolitan Museum in the 1930s revealed a densely built urban complex with grand palaces, noble mosques, stately urban residences for the rich, and endless quarters of two- and three-room dwellings for the general populace.

Since the site of Nishapur is just short of a mile above sea level, its seasonal climate changes are severe. To accommodate this, members of the aristocracy heated their urban residences through ducts under the floors. Braziers for cooking and heating were also built into the floors of even the most modest homes, and these were fed with air channeled through underground vents. The better dwellings had ample running water and spacious, richly painted underground rooms that provided respite from the summer heat.[8] They also featured intricately sculpted stucco panels and window frames painted in bold red, yellow, and blue.[9]

The many human figures adorning the walls of both private residences and large official buildings indicate that Muslim prohibitions against depicting people were taken more as gentle recommendations than as firm commands (see plate 7). Domestic piety was evident in the form of small shrines in the corners of rooms, as well as the many mosques, Christian churches, and Zoroastrian temples.

In terms of material culture, the good folk of Nishapur lived well by any measure. Its potters produced glazed ceramic ware that is so prized by the world's museums and collectors that local treasure-hunters have destroyed much of what remains of the city to find it. The finest such pieces featured slip painting under a transparent glaze, a technique developed by Nishapur artisans, and elegant writing in Kufic script.[10] Khurasan was also the heart of the inlaid bronze and brass industries, with the leading artists proudly signing their works.[11]

When members of the upper crust wrote letters, they turned to their locally made glass inkwells and signed with intricately carved seals that included portraits of the owner and such exhortations as "Be happy!"[12] Local glassmakers also turned out elegant lamps, lanterns, and alembrics for distilling rose water and date wine.[13] Most of the better homes had glass window panes. Nishapur's doctors boasted carefully crafted medical instruments of glass or brass. So large and highly regarded was the local medical community that a contemporary wrote a lengthy *History of Medicine in Nishapur*, now unfortunately lost.[14]

EDUCATION, THE LIFE OF THE MIND, AND CIVIL STRIFE

A degree of prosperity may have been a condition of Nishapur's cultural eminence, even a necessary condition, but it was not its cause. Undeniably, enlightened leadership played a crucial role. The first ruler to organize the new Central Asia from his base at Nishapur was Tahir's son, Abdallah, a thoughtful and wise leader who would stand out in any era. Abdallah wrote poetry when he was not attending to affairs of state. He also took the position, highly unusual then or at any other time, that a state's prosperity rests ultimately on the peasantry. "God," he said, "feeds us by their hands, welcomes us through their mouths, and forbids their ill-treatment." This led him to an extraordinary conclusion: that

Figure 7.1. Large glazed ceramic bowl from Nishapur. The stylized calligraphic inscription reminded diners that "Planning before work protects you from regret; good luck and good health." ▪ H: 7 in. (17.8 cm). Diam: 18 in. (45.7 cm). Rogers Fund, 1965 (65.106.2). Image © The Metropolitan Museum of Art. Image source: Art Resource, NY.

the state must promote general education. "Knowledge," he affirmed, "must be accessible to the worthy [i.e., the elite] and the unworthy [e.g., the peasantry]." Under Abdallah's rule this applied not just to the city of Nishapur but to the entirety of Central Asia. The eminent Russian scholar Bartold looked into Abdallah's policy in practice and found evidence that during his reign even the children of poor peasants were being sent to the towns to study.[15]

Abdallah's impressive expansion of education, which persisted through subsequent centuries, coexisted with what can only be described as tumultuous social conditions. The key actors on the social stage of Nishapur were the landed aristocrats or *dihkans*, on the one hand, and the artisans and proletariat on the other. Each warrants closer attention.

Thanks to the indefatigable Richard Bulliet, who examined the many surviving genealogies and local and family histories, we can peer into the lives of the *dihkans* who dominated the politics and cultural life of Nishapur.[16] These rich extended families had amassed their fortunes early and then asserted their local preeminence by claiming the high ground of culture. Some found their calling as patrons. Others chose to establish themselves as thinkers or writers, especially in the field of religious law. In this important area, which touched on every aspect of civic life, the patricians split into two irreconcilable camps. Those who followed the eighth-century Iraqi scholar Abu Hanifa in allowing more room for individual interpretation were, and are, considered more "liberal" or "rationalist."[17] Against them were set the followers of the early ninth-century Gaza-born scholar, Shafi'i, who sided with the theological literalists on legal matters.

These differences were the façade behind which lurked a more profound split, setting champions of free will, rationalism, and even Mutazilite theology against traditionalists and theological hard-liners. If the former welcomed the insights of keen intellects, the latter argued that since God created and controls every particle of matter, all talk of rational cause and effect—the basic premise of science—is pointless.

Urban records from Nishapur report that a lecturer in jurisprudence could drew an audience of five hundred in the year 997—clear evidence of the intensity of the feud between the contending parties and of the public's active engagement with it.[18] Many times such assemblies broke down into riots and open conflict, suggesting that the real point at issue had less to do with the minutiae of law or theology than with the question of who would control local politics. The disunity that was reflected in their two hostile factions steadily undermined the power of the patricians as a class. In Bulliet's sorrowful words, theirs was a "long, soft, steadily deepening twilight."[19] In another passage he was more blunt, arguing that the patricians as a class "committed suicide."[20]

While the patricians pummeled each other, Nishapur's tradesmen weighed in against both upper-class parties with their own radically populist brand of traditionalism. Called Kharijites after the group in southern Iraq who had founded the movement in the seventh century, they claimed that religious and civic leadership should go not to the heirs of Ali but to those who were free of sin, and that such good character was as likely to appear in a commoner as in a patrician. According to this argument, a morally pure slave had as much right to lead the faithful as either of the two noisy bands of patricians, or even someone claiming descent from the Prophet. Thousands of Nishapur's commoners professed this openly heretical view and assembled a fighting mob whenever either faction of patricians went too far in pressing its claims to the contrary. One of the Kharijites' leaders was a pious ascetic named Abu Abdallah Karram, whom the Nishapur authorities periodically threw into jail. Far from thwarting him, this attracted followers, called Karramiyya, who appeared in cities across Central Asia.[21]

This three-way struggle persisted through Nishapur's golden era, gradually sapping the energies of the community but at the same time creating an environment in which ideas really counted. In what other city would someone pen eight thick volumes on the lives of local scholars of law and religion and other learned men, as one Hakim al-Bayyi Nasaburi did around the year 1000?[22]

This gave rise to keen competition with other cities. Local representatives of the Tahir dynasty in Merv, wanting to show that their city, too, welcomed jurists and writers, sponsored the issuance of several collections of jurisprudence and poetry to prove it.[23]

FOUNDERS OF AN INTELLECTUAL TRADITION

Nishapur's lively intellectual climate was not solely the product of legal and theological disputes and civic strife. The presence there of articulate Zoroastrians and Christians also played a role, as did, no doubt, the submerged traditions of Buddhism and the ongoing intellectual contacts with India. It is worth noting that when a Nishapur-based scholar from Balkh wanted to study astronomy, he went not to Baghdad but to Varanasi (Benares) in India.[24]

The first philosopher to emerge in Nishapur happened also to have been, in the words of Richard Frye, "the first person anywhere in the Islamic era to have become interested in philosophy," Abul-Abbas Iranshahri.[25] Like many who followed, he was a polymath, interested as much in science as in religion and philosophy. Because he is known to have reported on an eclipse in 839, we know that he worked in the first half of the ninth century. Neither of his two known works on philosophy and astronomy has yet turned up. Barring an unexpected discovery, we must judge Iranshahri from passages from his writings included in the works of other thinkers, and by the fact that great minds like Biruni quoted him with respect.

Iranshahri brought to his philosophizing a deep knowledge of Christianity, Zoroastrianism, and especially Manicheanism.[26] Iranshahri's contention that space is eternal and a manifestation of God's power raises questions, of course, but we do not know how he answered them. Likewise, he is known to have embraced the atomistic views of Aristotle and other Greek thinkers, but again, we don't know how he squared them with the claims of religion, which he accepted. What is clear, however, is that Iranshahri believed that the great questions of existence could and should be addressed by the rational intellect. It is no surprise that Iranshahri produced at least one brilliant student, Muhammad ibn Zakariya al-Razi, the greatest medical clinician of all times, and, as we have seen, in philosophy Razi was a freethinker of stunning boldness and irreverence.

Standing with Iranshahri at the head of the long line of philosophical innovators from the region was Jabir, or "Geber" as he came to be known in the West. We have already noted the pioneering inquiries in which this astrologer-turned–experimental scientist engaged at Baghdad, under the patronage of the Barmaks. Jabir's direct heir in Khurasan, Abu Mashar al-Balkhi, was the alchemist whom we encountered as a key figure in transmitting Aristotle's works to the West. The fifteen European editions of his work before AD 1500 reflect his great impact there.[27] Mashar had a knack for singling out the most provocative lines from the Greek masters, which doubtless played a role in arousing Farabi's interest in Aristotle a century later. Unlike Farabi, Mashar was a cultural dissident who used his Greek sources to discredit Arab thought, which he despised.[28]

Most of the philosophers from Khurasan in the ninth and tenth centuries were Aristotelians. Abu Hasan Amiri, who is known to have left Nishapur only once in his lifetime, defended both rationalist philosophy and Islam, which set him at odds with the local freethinkers.[29] Another Nishapur Aristotelian, Ahmad Sarakhsi, was so highly regarded that he was asked to tutor a young caliph.

Skeptics and Freethinkers of Khurasan

It is probably unfair to judge the intellectual climate of any place on the basis of its more radical voices, yet to ignore them is to distort and diminish the picture as a whole. Khurasan produced more than its share of skeptics and radical freethinkers. Such voices were seen at the time as an inevitable and, to many, unwelcome part of the overall milieu. If one counts Razi as an intellectual son of the region, as surely one must, since he received his education there and found all his top students there as well, then it is fair to say that no region in the young Islamic world surpassed Khurasan for its freethinkers, heretics, and outright atheists. Given the fact that people of the region had been translating and editing the texts of diverse religions for centuries, it is no surprise that these radical thinkers based their attacks on a close reading of the religious texts.

Many prominent skeptics were Muslims, but by no means all of the doubters came from the Muslim side. A man from Balkh named Hiwi launched a blistering assault on the core affirmations of Judaism.[30] Probably himself of Jewish birth, Hiwi's venom against what he considered religious fraud was unrelenting. One of his many diatribes against the holy writ of various faiths focused on the first five books of the Jewish Bible, the Pentateuch. He presented a point-by-point analysis of no fewer than a hundred instances of what he considered inconsistencies, vagueness, absurd details, ridiculous fantasy, or moral obtuseness. Hiwi was a savage critic, and the fact that his preferred target was the Jewish Old Testament did not save him from the wrath of Muslims who, after all, accepted the Pentateuch as divinely inspired.

Other radical freethinkers from Khurasan focused their criticism squarely on Islam. One to do so was Abu Hasan Ahmad Ibn al-Rawandi, who was born around 820 in Lesser Merv (Marv-al-Rud) in what is now northern Afghanistan.[31] His Jewish father had converted to Islam and then turned his back on the faith with a vengeance, actively teaching against it. Young Rawandi began life as a Jew, converted to orthodox Islam and then became a Mutazilite, which he abandoned for Shiism.[32] By the time he finally declared himself an atheist, he had mastered the art of using the Bible against the Bible and the Quran against the Quran to show "The Futility of Divine Wisdom," the title of one his diatribes against all revealed religions.

Rawandi wrote 114 books and treatises on philosophy, politics, music, and grammar, but not one of them survives, nor does any of his poetry. We know Rawandi's *Book of the Emerald* only from a few extracts included in texts by other writers. Unlike the judicious Sijistani, who had constructed an impenetrable wall between religion and secular studies, Rawandi used logic and reason to plumb the nature of religion. Indeed, he was accepted as a master of *kalam*, the much criticized use of argument and debate in Islamic discourse. This was on display as he used the narrative of a conversation between himself and his teacher to mount a blistering attack not only on Islam but on all prophetic religions.

Rawandi began with heartfelt lines about the glory of human reason. He then proceeded to launch a frontal assault on the official faith. What passes for prophesy, he argued, is nothing more than lucky guessing. How is it possible, he asked, for God's various messengers to contradict one another? Convinced that the universe is eternal, he saw no need to invent a creator to bring it into being. As to those angels who intervened on Muhammad's behalf when he met his enemies at the battle of Badr, where were they at the battles he lost? Further, Rawandi proposed that all those miracles performed by Abraham, Moses, Jesus, and Muhammad were simply clever magic tricks designed to dupe the gullible.[33] And, finally, if human beings are able to use God's gift of intelligence to figure out that lying and cheating are bad, then what need have they for revealed religion to tell them the same thing?

Against all this, Rawandi proposed the use of observation and reason, with which all human beings are endowed, and the expression of reason through language, which is also innate in everyone.

Still another man from Nishapur presented himself as an outright materialist in philosophy, committing the extreme heresy in Islam of denying God's attributes. Arrested and jailed, this unnamed figure managed to escape and fled to China, where he is said to have so impressed the emperor that he was named prime minister or vizier. The message this angry Nishapurian imparted to the Chinese court was that Islam was in such a feeble state that even a small detachment of Chinese forces would be able to conquer all Central Asia.[34]

The examples of Razi, Hiwi, Rawandi, and the unknown heretic who denied God's attributes—not to mention Sijistani at his salon in Baghdad— suggest that Khurasan and the adjoining lands of Central Asia had become a very open marketplace of ideas, where free thinking and skepticism abounded and nothing was sacred. But heterodoxy had consequences. The unnamed denier of God's had to leave the country. All three of the other arch-skeptics described above were subjected to relentless criticism and threats from the pious of all faiths. In the case of Razi and Rawandi, that criticism continues among Muslims to our own day. Only the benign and sociable Sijistani managed to escape being attacked for his views.

THE TRADITIONALIST IMPULSE

Many thinkers in Central Asia during the years of Khurasan's rule exhibited a bold spirit of innovation, an eagerness to break boundaries and venture into new areas. Yet there were solid traditionalists as well. Nishapur's community of religious scholars (ulama) included effective partisans of all the more orthodox , and traditionalist views. Among the patrician legal scholars were outspoken champions of the Asharites and other literalist schools. Members of this camp gathered students and followers, lectured publicly, and were a constant presence in the life of the community.

Also part of the scene were several prominent collectors and editors of Hadiths, men like al-Hakim al-Naysaburi (821–875), a contemporary of Rawandi who devoted his life to collecting and evaluating the Sayings of Muhammad. His great enterprise succeeded, and by his death he had issued some two thousand Hadiths, all carefully evaluated with respect to their provenance and the chain of transmission down to the present. Shortly after Naysaburi issued his collection, a contemporary and local rival, Abdul Husain Muslim, issued a competing compendium which gained even greater renown, and is even today considered the second most authoritative collection, after that of his teacher, Bukhari, who also spent many years in Nishapur.

Nishapur may have been home to a nest of traditionalists but it was not a happy one. In his dealings with his peers, Naysaburi showed a quarrelsome and vengeful spirit that was quite out of keeping with the high-minded goals of Hadith collecting. Bukhari finally left Khurasan because of Naysaburi's harassment. Within Nishapur, the Kharijites who dominated the public face of traditionalism, were far more radical than Naysaburi and his colleagues, calling for a popularly elected caliph to reclaim the traditions of the faith which the Abbasid usurpers had corrupted. While Naysaburi's preferred form of combat involved only words, the Kharijites resorted to fistfights.

BALKHI, HASAN NAYSABURI, AND THE IRRATIONAL

To divide Nisahpur's thinkers into innovators and traditionalists does a disservice to both groups, for it implies that the two camps were fixed and stable, whereas in fact they were in constant flux. Worse, it forces important individuals into narrow categories that do not reflect the essence of their thought, which often defies simple categorization.

The Nishapur intellectuals who had rallied under the banner of Aristotle followed very diverse career paths. A good case in point is Abu Zayd al-Balkhi, Amitri's teacher. Raised in a village east of Sistan in present-day Afghanistan, he offered what a German scholar admiringly called a "sober and solid" interpretation of Quranic scriptures. This meant treating the text as Aristotle would have handled any other phenomenon in nature, accepting it as it is and opposing "interpretations" of the kind offered by the Mutazilites and other hair-splitting scholastics.[35] To this point, Balkhi emerges as a peculiar hybrid: an Aristotelian traditionalist. But he was an Aristotelian in his subject matter as well as in his method of study. He was interested in the whole natural world and went so far as to devise a new method of mapping it. He presented the fruit of his research in his *Figures of the Climates*, a series of cartographic exercises that covered the entire Islamic world and presented accurate information (again, Aristotle's influence) on the most distant and remote locales.

Balkhi was as much interested in the study and classification of human beings as of places, and this interest extended to people in all their conditions, including insanity. This curiosity led him down untrodden paths, with the result that he became a pioneer of cognitive psychology and, in a very practical way, a founder of psychotherapy. No one before him, and few for a half-millennium thereafter, wrote more insightfully on what today we call the "mind-body" relationship. His list of symptoms of depression and aggression, anxiety and anger, would be familiar to any urban dweller today. Balkhi then broke each of these down into two subcategories depending on whether each arose from within the body or without.

Having set forth this crucial distinction between neurosis and psychosis, Balkhi then proceeded to consider the best modes of treatment for each. For disorders arising from one's environment, he proposed a combination of "positive thinking" and a kind of talk therapy, detailing each with precision. For mental disorders that have physiological causes, he recognized that healthy thoughts can often alleviate them, even if they don't remove the cause. As to patients suffering from chronic depression, Balkhi did not shy away from treating them with medications.[36]

These are heady thoughts, coming as they do from a man who was born in 850 in what is today the second poorest country on earth and who is known to those few who have heard of him at all as a geographer. They exhibit a serene confidence in reason both as a descriptive and prescriptive tool and a refusal to be constrained by anything but the evidence.

Balkhi embodies the proclivity to break traditional boundaries that we have detected as a characteristic of the Nishapur thinkers as a group. But there is in his trailblazing work on psychology something further, namely, the readiness of a severely rational thinker to accept the irrational and to accord it the same attention and respect that is due any other manifestation of nature. Was this, too, an element of the intellectual tradition of Nishapur, or did it occur to Balkhi during his time in Baghdad? We do not know, but one thing is certain: Balkhi was not the only person in Khurasan to concern himself with mental health and the definition of sanity and insanity. Almost a century later, Hasan al-Naysaburi took up many of the same issues, and with unexpected consequences. As a young man, Naysaburi had supported a proletarian and combative sect, but he then sided with the more patrician but still relatively traditionalist Shafi'i school of jurisprudence. A historian, theologian, and general man of letters, Naysaburi continued to preach to the city's common folk.

We do not know for sure what piqued Naysaburi's interest in insanity, but he made clear that he was bothered by the fact that God, who is responsible for everything, allows some of his people to suffer madness. Like the Job-like hero of Archibald McLeish's play *JB*, Naysaburi confronted the argument that "if God is God He is not good. If God is good He is not God." Such ruminations, which Naysaburi recounted in detail, led him to pen a slim book entitled *Wise Madmen*. A contemporary wrote that Naysaburi's little volume "reached the horizons of knowledge and fame,"[37] which indeed it did.

Naysaburi's first approach to the problem of insanity was to say that we are all a bit crazy, that "no sane man is devoid of a streak of madness." He approvingly quoted a philosopher who answered a question about his condition by saying he was in "a state of soundness stained with disease, and health proceeding towards extinction."[38] A clever quip, but by blurring the distinction between sanity and madness, Naysaburi undermined a core Aristotelian precept and a pillar of rationalism itself, namely, that the observing and thinking mind is not warped by passions. In the end, he judged the mystics right in claiming that the only truly insane person is he "who put his trust in this worldly life, sought it out, and delighted in living."

The effect of such thinkers as Balkhi and Hasan Naysaburi was gradually to erase the line between reason and emotion. We do not know to what extent their fellow scholars were prepared to accept this. But appearing as it did in a city where all three of the main political factions were busy accusing the others of acting out of raw emotion rather than in accordance with the dictates of reason or the strict laws of religion, the philosophers' new insight could not have come as much of a surprise.

During the eleventh century this assumption was to gain broad acceptance, at first in Nishapur and then throughout the entire Muslim world. As that happened, religion itself was to be reshaped to conform more closely to the psyche and emotions of the individual believer and not just to his or her presumed rationality. No longer was true piety to be achieved just by reading scriptures and mastering the arguments of the theologians. Now the action shifted from the outer world to the inner world, specifically, to the soul of the believer and his relationship with God.

The first key stage in this epochal process was the development of a new, quietist form of Islam, Sufism. The second was the appearance of a series of monumental treatises by yet another fiercely argumentative Nishapur philosopher and theologian, Ghazali, who defended and legitimized this shift of focus from the mind to the heart. We shall explore both of these developments in chapter 12, when our attention turns once more to Khurasan, by then under the rule of the Seljuk Turks.

Persian Rumblings from Sistan

Viewed from Baghdad, Central Asia and its capital, Nishapur, was a cantankerous region bent on asserting its political and cultural autonomy. Viewed from the perspective of the lower classes of Central Asia, however, the rulers in Nishapur and their patrician allies had turned their back on their own cultural patrimony to ingratiate themselves with the Arab overlords. Abetting this feeling of alienation on the part of the populace at large was the spectacle of the main patrician parties preoccupied with their conflicts with each other, to the neglect of the public good. During the 850s this unstable situation exploded in a populist uprising. However, the revolt arose neither on the streets of Nishapur nor even in Khurasan but in a remote area to the south and east of Nishapur called Sistan. This "vast but transient" uprising soon overwhelmed Khurasan and adjacent parts of Central Asia and eventually threatened the caliphate itself.[39] Its long-term legacy was to reaffirm the distinctive identity and culture of Persianate Central Asia. Its immediate impact was to set the stage for Abolqasem Ferdowsi to write his epic, the *Shahnameh*.

The territory of medieval Sistan is today a windswept and forlorn region defined by the mountain-ringed Helmand River basin stretching from the eastern border of Iran clear to Kandahar in eastern Afghanistan. Drug-traffickers rather than epic heroes dominate its life today. But a millennium ago Sistan was extremely prosperous, with fertile fields irrigated by canals drawing water from the several large lakes formed by the Helmand River and its tributaries. The local population, of old Iranian stock with almost no admixture of Turks or Arabs, gathered in Sistan's many cities and towns.[40] The future Lord Curzon, passing through the territory in 1889–90, wrote that nowhere else in the world could one see such a proliferation of ancient remains.[41]

To the many branches and groups of Iranian peoples throughout Central Asia and Iran, Sistan was hallowed land. Here were many places named in the Zoroastrians' holy book, the Avesta. Here, too, was the birthplace of the mythic Zal, the progenitor of the ancient Persian dynasties, and of his heroic son, Rostam, who, in a lifetime said to have extended over eight hundred years, gave succor to the kings of Persia in their bloody and endless battles against the eastern realm of "Turan":

in blood and battles was my youth,
And full of blood and battles is my age,
And I shall never end this life of blood.

Rostam's tragedy was to have unknowingly killed his son, Sohrab, in a Central Asian battle, recounted by Matthew Arnold in his dark and moving poem *Sohrab and Rostam*. So great was Rostam's fame a millennium ago that he was memorialized in ancient murals across Central Asia and Iran. Nearly every ruler of Persian stock strained to trace his own lineage to him.

The memory of this heroic age is preserved only in a few place names in Sistan. But the ruins of a grand citadel and palace atop a high basalt outcropping above what was once a shimmering lake hint at the pre-Arab splendor of the region.[42] Here, at Kuh-e-Khwazmi, only a short ride from the Afghan border and southwest of the modern village of Divaneh, were elegant and sprightly murals from the second century AD. Drawing on Buddhist, classical Greek, and Persian sources, they help us understand why the people of Sistan might later have viewed the Arabs and their allies as uncouth imposters.

The capital of Sistan was the moated city of Zaranj, just over the modern border in Afghanistan, some 375 miles southeast of Nishapur. Surrounded by salt marshes formed by the adjacent Helmand River and protected from windblown sand by a huge barrier wall, Zaranj boasted thirteen gates and shops that extended for more than a mile along the roads leading from the center.[43]

During the Arab invasion, this city became a major focus of resistance. Thereafter the locals organized themselves under the populist banner of the Karijiites, the extremist sect from southern Iraq who, as we have seen, claimed that ability rather than genealogy should be the basis for worldly power in Islam. They denied the authority of the caliphs and asserted the right of armed revolt against any Muslim whom they deemed to be deviating from Muhammad's true precepts. Even before the arrival in Zaranj of three brothers from the countryside, one a coppersmith, the second a mule-driver, and the third a carpenter, the Kharijites were on a roll, conquering neighboring regions and scaring the caliphs in Baghdad. The coppersmith, named Yakub ibn Laith (840–879), emerged at the head of this volunteer army, which was Hell-bent on casting off the yoke of Arab rule. Since the Persian word for coppersmith is Saffar, Yakub and his dynasty came to be known as the Saffarids.

Yakub himself had once been a Kharijite but now opposed them. Nonetheless, he managed to attract most of these radical traditionalists into his army.[44] This band of committed populists took control of Sistan in 867 and then headed east to capture Balkh and the two Buddhist centers of Kabul and Bamiyan. They then headed west to break the power of the Abbasids in Khurasan. When the caliph's conceited representative at Nishapur demanded to see Yakub's papers from Baghdad, Yakub brandished his naked sword and declared that it was all the authorization he needed.[45] Soon Yakub's army was at the gates of Baghdad itself.[46] The caliph was flummoxed. Yakub had shipped Buddhist and Hindu "idols" to him from Kabul, but this was merely the prelude to Yakub's demands for power and influence, to which the caliph meekly assented. Yakub seized Khurasan in 873 but died six years later. Troops of the Sistan army then proclaimed his brother Amr, the mule-driver who was known as "The Anvil," as their leader. When Amr again threatened Baghdad, the caliph tried to buy him off by ceding to him control over all Central Asia, at which moment the Coppersmith's dynasty reached its zenith. But this time the caliph's army managed to capture and kill Amr.

The regionwide state that survived for a mere twenty-seven years under Yakub and his brother was no model of progressive rule. Militarized to the core, it threw out what was left of the caliph's administration and relied on a network of spies and Turkish slaves to extract tribute from all and sundry.[47] But this made Sistan far more independent than Tahir's dynasty had been. It also enabled the Saffarids to build palaces in

Sistan and a masterful bridge of twenty-six arches, still standing today, over a tributary of the Helmand River near Herat on the road from Sistan.[48]

None of this would be worth mentioning were it not for the radical cultural agenda of Yakub the Coppersmith and his brother. Both were immensely popular among the common folk and soldiers, and it is not hard to discover why. When an ambassador of the caliph summoned Yakub to Baghdad, Yakub fed him proletarian leeks and barley bread and then issued him instructions:

> "Go and tell the caliph that I was born a coppersmith. . . . The sovereignty and treasure which I enjoy I have acquired by my own enterprise . . . and by my daring; I neither got it as an inheritance from my father nor did I get it from you. I shall not rest until I have sent your head to Mahdiyya [Cairo, the capital of the caliph's rivals, the Fatamids] and have destroyed your house."[49]

Underlying this bravado was a fierce commitment to the cultural inheritance that the Arabs were destroying. A contemporary poet who was evidently close to Yakub put in his mouth the following words:

> the inheritance of the kings of Persia has fallen to my lot.
> I am reviving their glory, which had been lost and effaced by the
>   long passage of time.
> I am openly seeking revenge for them; although [other] men have
>   closed their eyes to recognizing their regal rights, I do not do so.
> So I say to all the [Abbasids], "Abdicate quickly, before you have
>   reason to feel sorry!
> We have gained power over you by force, with our lance-thrusts and
>   cuts from our sharp swords.
> Our forefathers gave you kingly power, but you have never shown
>   proper gratitude for our benefactions.
> Return to your country in the Hijaz [e.g., Arabia], to eat lizards and
>   graze sheep."[50]

It was typical of the Saffarid rulers that their last leader, Amr, demanded for his installation a ceremony based directly on the pre-Muslim court rituals, with barely a trace of Islamic influence.[51]

Most important for the future, Yakub stood forth as a defender of the Persian language. Like all Eastern potentates, Yakub attracted poets who were slavishly eager to eulogize him. Once, while a poet was singing his praises in Arabic, the monolingual Yakub brusquely interrupted him to ask, "Why do you recite something I don't understand?"[52] This momentous gripe was a powerful signal to all potential versifiers. In no time a whole legion of new Persian poets emerged, not in the heartland of what is today Iran but in Sistan and then in Khurasan and other parts of Central Asia. Some of these poets were still beholden to Arabic models, but others declaimed for the first time in the clear voice of the new Persian of the day.[53] First among these was Abolqasem Ferdowsi (ca. 940–1020), from the city of Tus.

TUS: AN AVANT-GARDE BACKWATER

Tus is a mere fifteen miles northeast of Nishapur, around the end of a mountain ridge that separates two valleys. Up to now it has barely rated a mention in our story. Nishapur, with its status as a political center, continental trading power, and its sheer size, seemed to be, and was, the kind of happening place where culture is made and intellectual life flourishes. Tus, by contrast, was desolate, worse than a backwater. Shahid, a poet from Balkh, visited the place about the time of Ferdowsi's birth and penned this quatrain:

> Last night by ruined Tus I chanced to go,
> An owl sat perched where once the cock did crow;
> Quoth I, "What message from this waste bring'st thou?"
> Quoth he, "The message is 'Woe, woe—all's woe.'"[54]

And yet it was this devastated place and not shining Nishapur that gave rise to the greatest figure of the age, the poet Ferdowsi. True, the New Persian language and script owed more than a little to the literary elite of Nishapur. And in the eleventh century it was Nishapur, not Tus, that became the seat of science and letters. Yet the contribution of writers, thinkers, and even statesmen from Tus is so out of proportion to its modest size that it demands an explanation. Or was it mere accident that Ferdowsi and so many other poets and thinkers hailed from there?

We know little about medieval Tus besides the fact that it was divided into two main sectors, that it was a center for the polishing and trading of turquoise and other semiprecious gems, that it exported stoneware, and that it was surrounded by many satellite towns.[55] But all this fades to insignificance compared with one overwhelming political fact: Tus, not Nishapur, had been the capital of Khurasan prior to the Arab invasion.[56] Here is where the old Zoroastrian elite resided and continued to practice their faith down to the time of Ferdowsi and beyond. Here, too, Zoroastrians responded to the Arab invasion by raising a major revolt in the name of their ancient faith.[57] It was here, finally, that loyalty to the memory and traditions of the ancien régime of Sassanian Persia was strongest. The fact that even fewer Arabs had settled in Tus than in Nishapur, where there were only seventy in the early ninth century,[58] assured a high degree of cultural continuity there. Locals also bitterly resented it when the Arabs, after conquering the region, snubbed their city in favor of Nishapur, demoting the former capital to the status of a backward-looking relic.

We can only imagine the sneaking delight with which many in Tus had greeted the rise of Yakub the Coppersmith. Tus had always viewed the Abbasids of Merv with contempt, as déclassé Arabs who were only too glad to use Central Asia for their own purposes. At the same time they looked down on Yakub's proletarian origins. And few, if any, in Tus were prepared to go as far as Yakub and attack Islam. But how could they not admire someone who, for the most part, was singing their song? It is likely that some in Tus privately considered Yakub a kind of modern-day Rostam, wielding his mighty sword not against Turks from the East, as in the epic poems, but against Arabs from the West. And was not Yakub from the same territory as that hero of old?

Whatever their thoughts on Yakub and his brief dynasty, the people of Tus had little time to indulge them, because in 900 Yakub's heirs were driven out of Khurasan and Tus by Ismail Ibn Ahmad Samani, the leader of a new Persianate dynasty based in Bukhara. By midcentury the Samanid rulers appointed a gifted native of Tus, Abu Mansur Abd al-Razzaq, to rule Khurasan as governor. One of his early projects was to assemble all known texts of the national epic, the *Shahnameh*, or *Book of Kings*.

We do not know whether Razzaq launched this great project on his own or on orders from Bukhara. Whatever the case, he assembled a competent team of researchers and charged them not with *creating* an epic but with *assembling* one from the many fragments that had been handed down orally over the centuries, and from the various texts that earlier authors had prepared. At least two Zoroastrian savants from Balkh and one from Merv had preceded Razzaq in this great work of compilation.[59] Conveniently, there were also a number of written histories of Persia under the Sassanians that the team could draw on.

Ferdowsi as a young man almost certainly participated in Razzaq's project. Born in the village of Vazh near Tus, he received a thorough education, thanks to his father's standing as a *dihkan* or patrician of modest means. We don't know the extent or character of his involvement with Razzaq's research team, but it is easy to imagine him as what today we would call a research assistant. At the least, we can be sure that he followed the group's progress with growing interest down to its abrupt end.

The key member of Razzaq's team was one Abu Mansur al-Muameri, whom Razzaq had commissioned to prepare a prose version of the entire epic.[60] Since the extant texts were written in the antique Pahlavi Persian, Muameri had to engage four Zoroastrians to assist in translating them into modern Persian. Their work was completed sometime after 957. This text eventually came into Ferdowsi's hands. Another person who was close to the Razzaq project was the young poet Abu Mansur Muhammad Daqiqi, originally from Balkh, an ardent patriot, champion of the Zoroastrian past, and possibly himself an adherent of that faith. A contemporary of Ferdowsi's, Daqiqi lived life to the full. As he wrote of himself:

> Of all things good and evil in the world,
> Daqiqi's choice is for these four:
> The ruby lip, the lute's sad melody,
> Blood-red wine, and Zoroaster's lore.[61]

Committed to Razzaq's great project, Daqiqi rendered long passages of the vast epic in rhymed couplets, nearly a thousand of which Ferdowsi was to incorporate into his own *Shahnameh*.[62]

FERDOWSI'S SOLO VOCATION

Razzaq did not see his project through to the end. Turkic officers in Bukhara forced the Samanid ruler to fire him and appoint one of their own, the slave Alptegin, in his place. With the sacking of Razzaq, support for his "Epic Project" dried up. Fortunately, Muameri had completed his prose translation. For several years Daqiqi continued to toil away at his verse version, which turned out to be workmanlike but devoid of inspiration. But in 976 Daqiqi's favorite Turkic slave murdered the young poet, bringing the verse edition to an abrupt end.[63] Suddenly Ferdowsi, now in his late thirties, found himself in possession of what must have become a sizable archive on the epic. What person of normal ambitions and endowed with natural poetic talent could have turned his back on this once-in-a-lifetime opportunity?

Several more years were to pass before Ferdowsi committed himself to writing his own *Shahnameh*. Maybe he spent these years gaining the right to use the preparatory drafts his predecessors had left behind. Maybe he tested his abilities by setting a few passages in verse. Or maybe he was in no position financially to drop other activities in order to work on this monumental project. Ferdowsi's decision to compose a verse edition of the entire *Shahnameh* could not have been an easy one. He was now married and the father of two children, yet he had inherited barely enough money to live on. Nonetheless, within a few years after the death of Daqiqi, the poet from Tus commenced what was to become his daily preoccupation for the next three decades.

By the time Ferdowsi completed his herculean labors in the year 1010, he was seventy-one and had accumulated a full measure of resentments. Granted, enlightened officials in Tus had given him food and clothing and even a tax exemption. But other local grandees had appropriated passages of his text without paying him royalties, while still others, their purses closed, lavished on him so many vapid compliments that, as he put, it, "my gall bladder was ready to burst."[64]

That we are still delighting in Ferdowsi's great poem a millennium after he completed it confirms its timelessness. Indeed, in most countries today, the *Shahnameh* is accorded at least the formal respect owed to works of art that have been deemed to be part of the world's cultural heritage. But ritualized clichés about Ferdowsi's "immortal epic" are no substitute for a more substantive appreciation of his masterpiece. Given that even his first English translator expressed doubts about the work,[65] it is worth pausing to identify some of the dimensions of Ferdowsi's achievement (see plate 8).

Acknowledging that sheer size does not make art, what other world epic before Ferdowsi weighs in at fifty thousand carefully composed couplets in verse, more than seven times the length of Homer's *Iliad* or ten times Virgil's *Aeneid*? This actually understates the scale, since each of Ferdowsi's lines is half again longer than those of Homer or Virgil.[66] For sheer productivity, then, no poet beats Ferdowsi.

The length of the *Shahnameh* is the inevitable result of the enormous task Ferdowsi had set for himself. Most epics consist of a recital of legends from the misty past. *Shahnameh* provides this, but nests the early legends within the vast framework of a comprehensive history of the Persian peoples from their first appearance on the world stage down almost to the author's lifetime. This meant that Ferdowsi had to cover the life and times of no fewer than fifty rulers. Half of these dated to the centuries before written records were kept, but the other half were real historical personages. For the early reigns Ferdowsi had to depend on diffuse oral sources or early written accounts based on the oral traditions. For the latter twenty-five reigns, though, he had to mine the historical record, sift evidence, and cut though the hyperbole of official historians.

Like the encyclopedists of his day, Ferdowsi wanted to capture the entire body of information and lay bare its inner meaning. What's more, he strove to present this epic history as he himself perceived it: as a living drama, devoid of dry antiquarianism and hollow moralizing.

A CLEAR-EYED BOOK OF KINGS

Ferdowsi masterpiece is deeply Central Asian in character. This is true in spite of the fact that, with the rise of the Sassanian rulers in AD 224, the Persian capital shifted from the Afghan-Iranian border area first to the circular city of Aredeshir-Khwarrah (modern Firuzabad) in the southwest of Iran, and then to Ctesiphon on the Euphrates downriver from the future site of Baghdad. But the geographical locus of the *Shahnameh* is neither in modern-day Iraq nor in the heartland of what is now Iran. Rather, it is that part of Central Asia lying beyond the Amu Darya. "Turan," as the epic calls it, remained the work's geographical core, for it is there that the great cultural war was played out. The *Shahnameh* mentions several hundred historical places, ranging from Beijing in the East to Rome in the West, but the solid geographical focus of the work is Central Asia, including Khurasan, Sistan, Afghanistan, Samarkand, Khwarazm, and East Turkestan. For Ferdowsi, the fate of Iranian civilization was decided in Central Asia, along the fault-line between the Turkic and Iranian worlds.

The *Shahnameh* is a Book of Kings, and the succession of forty-nine rulers of Persia is its time line. But this was no mindless salute to national glory. On the contrary, the author fearlessly offered up the most scathing descriptions of the howling incompetence of many of Persia's kings, as if to say that the people's troubles arose from man, not God. And while Ferdowsi systematically avoided explicit allusions to the contemporary world, from beginning to end the *Shahnameh* is a searching and profound reflection on the failures of leadership in Ferdowsi's own time. The contemporary references can be veiled or Aesopian, but one is never in doubt that the author fully understood that what he was writing had meaning for his own day.

The sheer passion of some of Ferdowsi's judgments confirms this. He heaped scorn on the legendary King Nowzar, who foolishly alienated his nobility, leaving Iran unable to ward off the attack from the East that led to Nowzar's own death—a parable of the end of the Samanid dynasty. Other rulers also came in for damning criticism, with their transgressions and failures all having a strikingly contemporary ring.

One of the clearest instances of Ferdowsi using history to speak to his own age is his fulsome treatment of the province of Sistan, home of Yakub the Coppersmith and of the legendary hero Sohrab, whose earthly mission was to save the Iranian kings from the consequences of their own folly. Sistan, Ferdowsi reminds us, was also the seat of a dynasty of legendary rulers who were not so much as mentioned in other compilations but who epitomize all the strengths and weaknesses of Persian kingship. It is hard to imagine any of Ferdowsi's contemporaries reading his description of Sistan's mighty king, Faridun, whose mace ridded the world of hordes of evil monsters, without calling to mind the historic figure of Yakub the Coppersmith.

Such a link is even more strongly implied with Rostam himself, who epitomized the kind of elemental force that is needed to extricate kings from ill-advised adventures and the Iranian people from foolish wars. Only Sohrab could save the maladroit ruler King Kavos when he was captured by enemies, or when he built himself a flying machine that plummeted to earth behind enemy lines. Sohrab was not of royal blood (his mother was Afghan), and he could spend many days at a time in drunkenness and carousing. Yet he was the true man of the people without whom the nation would find no salvation (see plate 9).

At times Ferdowsi seems to favor the selection of leaders on the basis of ability rather than of genealogy.[67] But in the end he found this unthinkable, not just because it would have placed him among the Kharijite heretics, but because he believed that the only true leaders are those who are born to rule. Ferdowsi fully embraced the divine right of kings. The challenge for society was to find means of saving itself when leaders succumbed to arrogance, greed, and other forms of hubris, as so often happened.

With stunning candor he repeatedly posed the question of what a good man is to do when his ruler is foolish or malevolent.[68] Rebellion for Ferdowsi was out of the question, for it violated his very conservative views on legitimacy.[69] Nor did he conceive the possibility of an English-type Magna Carta, in which the noble subjects of an unjust king combine to impose limits on monarchical authority. With these paths closed to him, Ferdowsi was left to hope against hope that good advisers would somehow emerge, or that a hero like Sohrab would save the day. With so much to do, it is no wonder that Ferdowsi embraced the folk tradition that claimed that Rostam lived for hundreds of years.

Ferdowsi's ruminations on leadership may seem at first glance to lead to the smug conclusion that civic good is rewarded and evil punished. But he was not so naïve. Directly refuting this view are his tales of Goshtasp and Esfandyar. Both were the sons of kings and therefore legitimate, and both were filled with the good intention of doing that which is right and just. Where the one followed the call of conscience, the other harkened to the call of duty. But in the end both were doomed.

Ferdowsi's world was one in which good intentions were no assurance of positive results, and in which fortune and chance imposed themselves in often harmful ways.

PARADOXES OF PATERNAL RULE

His deep concern for kingship and legitimacy led Ferdowsi to think deeply about the nature of authority itself. Society demands authority, he wrote, but authority is often capricious and unjust. Authority requires loyalty but rarely earns it. In what can only be interpreted as a radical critique of authority as such,[70] Ferdowsi sided squarely with the adviser or assistant who remains loyal even when his leader betrays him.

It is more than likely that his model for such commendable behavior was none other than Razzaq, the able governor of Tus and the initiator of the *Shahnameh* project, whom the Samanid ruler fired and replaced with a Turkic slave. Ferdowsi remained loyal to Razzaq and his vision, and in so doing he included himself in the ranks of those steadfast subordinates who are the saviors of society and culture. Ferdowsi's talented American translator, Dick Davis, is therefore quite right in contrasting the official ethics of the poem, which require the maintenance of authority at all cost, with the poet's own view, which acknowledged the redeeming value of loyalty to higher principles. It is no accident that Khosrow, one of the most noble of all the kings presented in the epic, renounced power for the sake of his soul.

Yet in the end even this solution violated what Ferdowsi sees as the natural order of things. At the back of his mind, royal authority was an extension of a father's authority in his family, a theme that runs like a silver thread through the entirety of Ferdowsi's epic. No work written prior to the *Shahnameh* focused with greater frequency or intensity on the father-son relationship, a fact of which Sigmund Freud was apparently unaware.

A WORLD DEVOID OF SENSE OR WISDOM

Here, again, Ferdowsi avoided simple formulas. Paternal authority may be the pattern for political authority, but it too is flawed by the capriciousness that arises from unlimited power.[71] Worse, paternal authority is also subject to the vagaries of fortune. There is no consolation in the fact that the most dramatic event in the *Shahnameh*, when the hero Rostam kills his son Sohrab in battle, was the result not of evil deeds by either father or son but of accident and sheer chance.

Ferdowsi was preoccupied with the relentless decline of Persian power under the rich and imperious Sassanian empire (224–651) and its final defeat by the invading Arab armies. Though Muslim himself, he exonerated Zoroastrianism for this disaster and in fact treated the great monotheistic faith of the Persian peoples with unalloyed respect. The corruption of power was certainly a factor, but Ferdowsi was more wrought-up over the inadequacy and weakness of leaders than over the mistakes to which they are led by the unrestrained exercise of power. In the end he resignedly highlighted the role of fate, Fortuna, in the affairs of mankind. The royal astrologers foretold the end of Persian self-rule under Yazdegerd, the last emperor, and the leader was powerless to stop it.

After recounting the lives of forty-nine rulers, many of them giants by any measure, Ferdowsi treated the hapless Yazdegerd with touching respect. He followed the king's flight to Merv in Central Asia, where he hoped to regroup his forces. He quoted from Yazdegerd's desperate letter to the evil and duplicitous ruler of Merv, Mahuy, to whom he came "trusting in [his] probity and generosity."[72] After humbly receiving the emperor, Mahuy flattered the ambitions of a prince from Samarkand, whom he persuaded to attack Yazdegerd with an army that came to Merv "glittering as a pheasant's feathers." The governor disguised his own treachery by claiming falsely that the Samarkand army was a band of Turks.

Yazdegerd meanwhile fled to a humble mill where the miller, a decent man, attempted to feed and protect him. But when Mahuy heard that a warrior who "radiated glory" and was "like the springtime itself" was at the mill, he commanded the miller to murder him, warning the miller that if he refused, Mahuy would behead the miller then and there. The hapless miller murdered Yazdegerd. In his dying breaths the last ruler of the Persian peoples declared that

> A man who understands the world soon says
> There is no sense or wisdom in its ways . . .
> The heavens mingle their malevolence

With kindness in ways which make no sense,
And it is best if you can watch them move,
Untouched by indignation and by love.[73]

Were Yazdegerd's words of utter resignation before the powers of fate Ferdowsi's own conclusion on how best to deal with life's utter uncertainties?

One of the most thoughtful scholars of Persian literature, Jan Rypka, who founded the Oriental Institute in Prague, emphatically denied this. On the contrary, Rypka saw the entire *Shahnameh* as a call to action, a paean to the vigorous life. The unrelenting struggles, cascading battles, and constant revolts, he argued, were responsible for Persia's greatness and hence embodied the very essence of Ferdowsi's message in the *Shahnameh*. Going further, Rypka concluded that Ferdowsi's masterpiece is nothing less than a refutation of the quietism and withdrawal from the world that Sufi teachers were preaching among Central Asia's Muslims during Ferdowsi's lifetime.[74]

What cannot be denied is that in the end Ferdowsi said it was all for naught.[75] There was no silver lining. "How did they hold the world in the beginning, that they have left it to us so wretched?"[76] Ferdowsi asked. All the action led nowhere. Iran is no more, and its demise left the present generation in wretchedness.[77] A bleak picture, to be sure, but it is hard to see in it an attack on Sufi passivism, any more than it is a ringing defense of the active life. The reality is far worse. Ferdowsi's celebration of vigorous action may place Persia's epic heroes in the company of Homer's Odysseus and Virgil's Aeneas, but his grim concluding passage tinged them all with the aura of Cervantes's Don Quixote.

Recognizing this, Dick Davis rightly concluded that the *Shahnameh* is a poem at odds with itself.[78] At once a celebration of action and an acknowledgment of its ultimate hopelessness, Ferdowsi's epic presents the reader with conflicting perspectives that remain unresolved. Davis hit the nail on the head when he observed that "at the heart of the enterprise from which the *Shahnameh* arose is silence."[79]

Far from crippling the poem as a whole, this unresolved tension between heroic action and the might of destiny is one of countless elements that give the *Shahnameh* its poignancy, grandeur, and depth. The language is rich, now saturated with drama, now suffused with irony and boisterous humor, and now lyrical and even laconic. For the most part the narrative charges forward with tremendous propulsion, moving exuberantly from one dramatic scene to the next. Ferdowsi's "camera" offers high panoramic vistas, sharp close-ups, and densely packed images from within the action itself. Time and again he stops the frame momentarily to drop in a wry aside or insert a deft image that seems the quintessence of poetry.

All this enables the modern reader to sense the artistic brilliance, psychological depth, and philosophical sophistication that existed at the cultural pinnacles of Tus, Nishapur, Khurasan, and all Central Asia during the tenth and early eleventh centuries. A millennium later these qualities of Ferdowsi's masterpiece are no less arresting than in the poet's own day. As he correctly predicted,

> all the land will fill with talk of me:
> I shall not die, these seeds I've sown will save
> My name and reputation from the grave[80]

CHAPTER 8

❁

A Flowering of Central Asia: The Samanid Dynasty

Around the year 940, when western European writers still aimed for the small circle of readers who knew Latin, and when most Chinese writers wrote for bureaucrats, a poet and singer at Bukhara, Abu Abdullah Rudaki, celebrated his role in society. He did it in three succinct couplets on "The Pen and the Harp":

> Life is a horse, you are the trainer, your choice is to gallop.
> Life is a ball, you are the mallet, your choice to play.
>
> Although the harp player has delicate hands,
> May they be sacrificed to the hand that holds the pen.
>
> There is less oppression, less jealousy because of you.
> There is more justice, more generosity because of you.[1]

The pen, he wrote, is

> a cripple who walks, it has no ears but talks.
> It's an eloquent mute, it sees the world without eyes.
> It is sharp as a sword. It moves like a snake.
> It has a lover's body and a darkened face.[2]

It is easy to imagine an adoring audience responding to him as he sang on timeless subjects like spring, masterfully accompanying himself on the lute (*barbak*) or the harp (*chang*):

> Look at that cloud, how it cries like a grieving man,
> Thunder moans like a lover with a broken heart.
> Now and then the sun peeks from behind the clouds

Like a prisoner hiding from the guard.
The world, which has been in pain for some time,
Has found a cure in the jasmine-scented wind.
A shower of musk streams down in waves,
On leaves, a cover of shiny new silk.

Snow-covered crevices now bear roses.
Streams that had been dry now swell with water.[3]

And, of course, he often sang of love:

For the man who knows love's drunkenness
It's a shame to be sober for a moment.[4]

Among Rudaki's many friends was the aristocratic poetess from Balkh, Rabia, who had written with fierce passion of her unrequited love for a slave:

My prayer to God is this:
That you be bound in love with someone
Unmoving as stick and stone.

For only having suffered love's agony
Of pain and separation
Shall you come to feel and value
My love for you.[5]

Rudaki's heart went out to Rabia, whose own brother murdered her for her indiscretion.

It is no surprise that Rudaki also sang of wine, not as a symbol of mankind drunk with the Divine, as later Sufi poets were to do, but as the drink that lifts one out of melancholia:

Wine brings out the dignity in man,
Separates the free from the man bought with coins.

Wine distinguishes the noble from the base:
Many talents are bottled in this wine.

It's joyous, when you are drinking wine,
Especially when the jasmine is in bloom.

Wine has scaled many fortress walls,
Broken many newly saddled colts.

Many a mean miser, having drunk wine,
Has spread generosity throughout the world.[6]

Rudaki's magically fresh and worldly verse, written in the New Persian that is equivalent to the Tajik tongue of today, exudes the remarkable era in which he lived, when the Samanid family ruled all Central Asia and much of eastern Iran and Afghanistan from its capital at Bukhara. An envious Central Asian writer from the thirteenth century, looking back on this golden age that lasted from approximately AD 800 to 1200, enthused that Bukhara had been

Enlightened with the brightness of the light of doctors and jurists, and its surroundings embellished with the rarest of high attainments.
In every age since ancient times it has been the place of assembly of the great savants from every land.[7]

Born in a village in the north of present-day Tajikistan, Rudaki, while still in his teens, achieved such renown as a singer and instrumentalist that the Samanid ruler of Bukhara invited him to his court. The next emir, Nasr II (914–943), made Rudaki his official poet and boon companion, enriching him to the extent that when the poet moved his home it took two hundred camels to transport his possessions. Rudaki reciprocated with lofty panegyrics and heartfelt poems of praise for the ruler, which he performed at banquets and official occasions. In these set pieces he avoided the usual claptrap about the ruler's prowess in war and in love and dwelled instead on Nasr's active interest in the life of the mind. Never once did he speak of his ruler as fulfilling any religious role in society.[8] Indeed, even though Rudaki expressed his support for the Ismaili cause, his poetry is thoroughly secular in character.[9] In one couplet he opined that "My God created us not for prayers but for the play of one's beloved."[10] He was far more likely to evoke ancient Iranian and Zoroastrian images than Muslim ones.

Rudaki had a deft touch. When his boss lingered for several months on a vacation in Herat in Afghanistan, it was Rudaki who summoned him back to Bukhara and to work with a poem announcing Nasr's return, even before the amir had agreed to go:

> The Amir is a cypress, Bukhara, the garden;
> The cypress is returning to the garden.[11]

After hearing the poet sing these lines, Nasr mounted up without completely dressing and immediately rushed home.

The rise of the Samanid dynasty was greatly facilitated by the waning power of the Baghdad caliphate during the same years. In 861 a Turkish slave murdered the bigoted and boorish caliph Mutawakkil during a drinking bout. This finally put an end to the Mutazilite controversy in Baghdad, but at the price of continued attacks on the city's Christians and Jews, both of whom were required to wear honey-colored garments, ride only on donkeys or mules, and place images of the devil over the doors of their homes.[12] The government was by now completely beholden to its Turkic armed forces, whose officers and rank-and-file had become restive and demanding. Riots were frequent, and in 869 Ethiopian slaves exploded in revolt, destroying whole districts of Baghdad. Such outbursts became increasingly common over the succeeding generations. Finally, on an early summer night of torrential rain, thunder, and terrible lightning in 941, a fearful and awe-inspiring event took place: the towering green-tiled dome of Caliph Mansur's palace, the very symbol of caliphal authority and the focal point of the Round City of Baghdad, collapsed.[13] In an age when even the best educated looked to astrologers for portents of the future, the meaning of this wrenching event was clear to all.

Meanwhile, the sudden rise and precipitous fall of the dynasty of Yakub the Coppersmith in Khurasan had drained the last strength from the Samanis' only potential rival in the region, the Khurasan government in Nishapur. Since Nishapur had received Baghdad's blessing to rule not only Khurasan but most of Central Asia beyond the Amu Darya, the stakes were nothing less than control of all Central Asia. The situation was ripe for a new dynasty, one that would do effectively what the rulers in Nishapur were no longer able to do: provide independence from Baghdad and a stable environment for trade, and keep out the Turkic tribes in the East.

The name of the family that achieved all this and more was Samani. The clan had originated not far from Balkh in Afghanistan and for several generations had produced senior officials who represented the Nishapur government there. Prior to their conversion to Islam, the Samanis had been Buddhists, which may explain why they later allowed coins issued by some of their mints in Afghanistan to feature sun symbols (mandalas) and other Buddhist iconography.[14] This made sense, also, because Buddhism continued to thrive in the Hindu Kush to the end of the Samanid era, around the year 1000.[15]

The Samanis' rise (and later, their fall) was hastened by their fecundity. In the waning days of Nishapur's rule, seven Samani brothers managed to get themselves appointed heads of each of the main cities of Central Asia, including Samarkand, Ferghana, Chach (Tashkent), and Bukhara. Power fell into their hands with only modest help from their armies. But this good fortune left unresolved the question of which of the Samani brothers would actually lead the new dynasty. This was settled by the most gifted of the lot, Ismail Ibn Ahmad Samani (849–907), whose rise began at Balkh, where he destroyed the army of the last heir of Yakub the Coppersmith. He then adroitly marginalized his older brother in Samarkand. In a dramatic break with custom, Ismail did not kill his kin and rival but instead let him stay in that city , grateful to be alive. He did the same with all his other siblings, first stripping them of power and then treating them with dignity.

Ismail Samani's power base was Bukhara, a city with an estimated population ranging upward to many hundreds of thousands.[16] It now became the capital of all Central Asia, including Khurasan, the Ferghana Valley, Khwarazm, and most of what is now Afghanistan. Since the caliphate could no longer hope to exercise any control over the region, the Samanids could afford to show respect toward Baghdad in various ways. They modestly abjured fancy titles, sent gifts to each succeeding caliph, and even recited their name and titles in the Friday prayers. But they refused to pay Baghdad any taxes or tribute. And all the while the Samanid rulers worked diligently to strengthen the sovereignty of their own state.

This meant setting up a functioning system of administration. The Samanid rulers followed the ancient Persian custom, recently copied by the Abbasids in Baghdad, of entrusting the management of official life to competent and loyal chief ministers or viziers. The most effective of these were talented local men who rose through the ranks. One,

Abu Abdallah al-Jayhani, who served as vizier from 914 to 918, was an accomplished geographer who wrote a massive *Book of Roads and Kingdoms*, now lost, that was prized for its detail. Another, Abu 'l-Fadl al-Balami, who served from 918 to 938, was an adept and watchful financial manager whose son came also to hold the same high office.[17] Reporting to the viziers were some ten agencies that functioned like ministries, each with central offices situated around a single square in Bukhara, and each with local representatives in every province.[18] Together these offices managed all aspects of civic life except religion.

Highly literate and given to careful record-keeping, the Samanid bureaucrats took as their symbol the ink bottle. One of these proud professionals even wrote a treatise on management, in which he included details of the various registers in which state salaries and financial transactions should be recorded and a glossary of the most frequently used technical terms in the fields of public administration and finance.[19] That this widely consulted encyclopedia also covered music, poetry, history, logic, medicine, and astronomy added to its popularity. Another contemporary Samanid encyclopedist was Ibn Farighun, who lived and worked at a local court in Chaganiyan, now in Tajikistan. Ibn Farighun's book, too, covers the secretarial sciences (calligraphy, mathematics, and geometry) but also philosophy, astronomy, government, ethics, and alchemy.[20]

The Samanid state, like the earlier regionwide governments based in Merv and Nishapur, was dominated by landed patricians, who controlled agriculture and lived at their rural estates or in town houses. Some of them had been given large feudal holdings in return for collecting the land taxes. The Samanids also respected traders and manufacturers and understood that the state's stability and prosperity depended on the welfare and good will of these groups. In the early years, at least, this translated into moderate taxes on trade that were systematically collected, encouragement for the production of quality goods for export, the pacification of restless Turkic tribes that could threaten caravans, the construction of hotels or caravanserais at convenient intervals along main routes, and constant attention to the stability of the currency. In short, the policy of the Samanid government embodied most of what is today considered sound public policy.

The economic capital of the Samanid state was Samarkand. Traders based there maintained the old Sogdian links with China, Byzantium,

India, and the Middle East. There were even enterprising exporters in Samarkand who delivered goods to the Far East by sending them overland to Basra in Iraq and shipping them from there via the Arabian Sea and Indian Ocean.[21] Balkh, however, preserved its status as the main entrepôt for trade with India, Tibet, and Southeast Asia.

Under the Samanids, long-distance transit trade was paired with the domestic production of high-value portable commodities and finished goods for export. The former included refined tin, lead, copper, and precious metals from locally mined ores. To feed air into the deep mine shafts, Samanid-era technicians developed sophisticated shields to catch and direct the winds, while smelters driven by powerful bellows assured high levels of purity.[22] Even more sophisticated techniques were applied to the production of window glass, the forging of fine knives, the firing of elegant ceramic ware, and the large-scale production of silk, other fine fabrics, and world-class paper. Samanid paper long remained the gold standard in that field globally. These and other domestic manufactures enabled the Samanid state to implement successfully an export-driven development strategy.

Within Central Asia the main coin of trade was either the silver dirham, issued in various forms by mints across the realm, or, at the bazaar level, either a more debased coin of the same size or small coins of bronze. Internationally, though, the handsomely designed and 97 percent pure Samanid gold piece prevailed, becoming the most widely accepted standard of value since the Roman denarius.[23] Marauding Vikings understood this and, like underworld figures today, rushed to convert the fruits of their raids across Europe and the Middle East into this valued currency, which they proceeded to stash away in large hoards in eastern Europe, the Baltic countries, and even on the isle of Gotland in Sweden. A trove uncovered near Stockholm's Arlanda Airport proves that Samanid currency had gained a high international standing by the ninth century.[24] More than a third of all currency in use throughout the Baltic region in early medieval times originated in Baghdad or the Samanid lands, from mints in Tashkent, Bukhara, Samarkand, and Balkh.[25] Simultaneously, Samanid dirhams found their way westward to markets in the Mediterranean and North Africa and eastward to China, India, and even Sri Lanka.[26]

All this produced a thriving commercial life in both Samarkand and Bukhara. Even though it was only the second city in terms of output,

Figure 8.1. Samanid culture rested on solid monetary policy, as reflected in its gold dinars (shown here) and silver dirhams, which served as a reserve currency from India to Scandinavia. ▪ Courtesy of the David Collection, Copenhagen. Ismali I, AD 897, inv. no. C512. Photo by Pernille Klemp.

the capital boasted no fewer than a thousand shops.[27] It goes without saying that this entire system depended on high levels of stability and security. Domestically this meant that the Samanids had to be willing to strike deals with the several local kingdoms they had subordinated but could not directly control. By these means they managed to keep Sistan and parts of eastern Afghanistan safely in their orbit. Internationally it meant that the regime had to maintain a large and effective army, mainly to keep at bay the Turkic tribes that threatened from the east and north. This jihad, which resulted in large numbers of conversions and larger numbers of captured slaves, maintained Samanid security down to the middle of the tenth century. But it meant devoting half the state budget to defense.[28] No wonder that by the end of the reign of Rudaki's patron and friend, Nasr II, in 943, the economic underpinnings of Samanid rule had greatly weakened.

A second and no less serious challenge facing the Samanids was demographic and cultural: it was simply impossible to maintain the required scale of the military by depending solely on the various groups that spoke Iranian languages, whether Sogdians, Khwarazmians, Khurasanis, Pamiris, Baktrians, or Tokharians. Besides, the population that spoke Iranian languages was busy with manufacturing, trade, and urban culture and had lost the martial qualities the situation demanded. And so, like the Abbasids in Baghdad, the Samanids came to rely on converted Turks to fight for them. Over the years, Turkic junior officers who had proven themselves at the platoon and squadron levels rose in the ranks, until by the time of Nasr II they dominated the officer corps. From there it was but a short step to advising the throne and playing a central role in affairs of state.

The fact that these upwardly mobile officers shared a common language and culture with the rank-and-file in the army gave the Turks political leverage in the state. The Turks who rose to senior posts under the Samanids were loyal, not only to the Samanid dynasty but also to the Muslim faith. But their view of Islam tended to be very tradition-bound and rigid, causing them to scorn as heretics anyone tinged with Shiism, let alone freethinkers and secularists. Overall, the Turks' emerging status as power brokers and even de facto rulers could not be denied. Eventually the same process that had led to their rise in the Samanid state enabled them to replace it with a state of their own, in Ghazni, Afghanistan.

In sketching the rich cultural history of the Samanid empire, it would be remiss not to take note of the widespread prevalence there of slavery and the slave trade. Slavery had existed throughout the region since time immemorial and had become a fixture of the economic and military life of the Arab caliphate as well. The Samanids, too, depended on it, and its role in their state did not diminish with time. A single slave-gathering expedition to the Turkic steppes in 893 netted ten to fifteen thousand captives, including the wife of the Turkic chief.[29] Manacles discovered in Samanid mines confirm that many, if not all, of the miners were slaves, probably Turks and Slavs.[30] Members of the elite all held slaves, and even Ibn Sina traveled with two enslaved manservants.[31] More significant, the Samanid state reaped huge profits from selling the slaves it captured on the eastern and northern frontiers, its main customers being Arabs to the west and the caliphate itself.

Notwithstanding this trade in alien Turks and Slavs, the Samanid state represented a landmark in the evolution of Muslim societies away from a narrowly Arabian base to a more ecumenical orientation.[32] This occurred through a process that is highly unusual for what was in reality a postcolonial country: the Samanids elevated the indigenous Persian language and culture and yet simultaneously preserved Arabic, the language of their former colonial overlords, and those aspects of Arab culture that the Bukharans considered positive.

The working language of the Samanid bureaucracy remained Arabic, but the language spoken at the Bukhara court and in most cultural settings other than the mosque was the newly emergent form of Persian that Rudaki did so much to refine. It is too much to say that the Samanis single-handedly ended the "era of silence" that had engulfed Central Asia's diverse Persianate cultures since the Arab conquest.[33] After all, Tahir's successors in Nishapur had launched this process, while Yakub the Coppersmith had also promoted it with passion and militancy. Moreover, Persian-speakers in Baghdad itself used every possibility to advance their native tongue. But it was the Samanids, through their vizier in Tus, Abu Mansur Abd al-Razzaq, who created the environment in which the poet Ferdowsi undertook his *Shahnameh*. Like-minded official patrons in other cities lent support to many other new poets who wrote in Persian, of whom Rudaki was only the most renowned.[34]

Yet none of this occurred at the expense of Arabic. Under Samanid rule, no fewer than 119 Arabic poets from Central Asia appeared on the scene.[35] Some were of diverse Iranian or Turkic stock but more were Arab émigrés, many of whom who used their poetry to vent their spleen against Central Asia and especially Bukhara.[36] Persian-speakers responded with equally vitriolic verses denouncing the difficult life that the Arab conquerors had imposed on their people.[37] Meanwhile, the New Persian verse of Central Asia was enriched by contact with the two-part sonnet-like *ghazals* of the Arab nomads. This gave rise to clever variations in Persian, which in turn influenced Arab poetry. Most Central Asian savants were bi- or trilingual, and the many new terms they introduced into their Arabic writings from Greek and ancient Persian sources did much to further the transformation of what had formerly been a bedouin language into a subtle vehicle for discussing even the most arcane concepts of philosophy and science.

All these developments in Central Asia combined to create a new cultural model that differed radically from what the Arabs had produced on their own. Far more than had occurred in Baghdad, what was understood to be Islamic civilization now ceased to be purely Arabic. Only in the sphere of religion, where the Quran had enshrined the privileged position of Arabic, did the old preeminence remain, and even there it was challenged by new ideas about the faith, many of which arose from Persianate and Turkic Central Asia. Henceforth the hallmark of this new cultural model was a thoroughgoing cosmopolitanism and ecumenism of the sort that had prevailed in Central Asia before the conquest but had not flowered to a like extent in any other center of Muslim life, whether in the Middle East, North Africa, or Spain. To be sure, as the armed bearers of a minority faith, earlier Muslims in Egypt, North Africa, and Spain had ruled over and interacted with many Christians, Jews, and peoples of diverse ethnicities. Baghdad epitomized this pluralism, but there was never any doubt as to which language or culture was the dominant one, at least until the "eastern wind" began to blow from Central Asia. Only in the cosmopolitan world of the Samanids did the interaction of languages and cultures become an active two-way street, with seminal influences flowing comfortably in both directions. Over the following centuries this positive new cultural model was to exert a profound influence on Islamic civilization as a whole.

The Samanid state, like all its predecessors in Central Asia, was in reality a conglomeration of great urban complexes, each with its own local dynasty, traditional elite, and economic and cultural particularities. Yet there was no denying that the city of Bukhara was its political, intellectual, and spiritual heart. Never as large or populous as Merv, Balkh, or Samarkand, this ancient commercial and religious capital was nonetheless perfectly suited to its new role. Long a Zoroastrian center with active fire temples functioning down to the end of the Samanid era,[38] Bukhara at the time of Rudaki was still dominated by the giant tumulus tomb of the pre-Muslim Iranian hero Afrasiab. As we have noted, the city had been so prominent a Buddhist center that it took its name from the Sanskrit word for the Buddhist monastery, or *vihara*, that once existed there.[39]

Built originally on an area of high ground amid salt marshes, Bukhara was distinctive among regional cities in that it could be watered entirely with canals, without resort to the complex underground irrigation systems that were common elsewhere.[40] A century before the rise of the Samanids, the oasis on which the city stood had been surrounded by a 155-mile-long defense wall, the maintenance of which required immense expenditures of labor and money.[41] In a moment of justified boasting, Ismail Samani around the year 900 announced that the wall was no longer necessary, as his new government would secure the entire territory of the state from invasion by nomads.

Even though there were a few broad, paved streets, Bukhara, in the tenth century as today, was a warren of winding lanes and alley. And not a very clean one. Nearly every medieval writer who visited the city commented on the filth and smell that prevailed beyond the zone of official buildings.[42]

This is not to say that Bukhara lacked elegance and refinement. Three large assembly mosques existed there; the Samanid rulers had erected new palaces for themselves and also extensive gardens west of the city's center; and hundreds of patricians had established comfortable suburban residences along the main canals. Unfortunately none of this grandeur has survived. Were it not for two buildings, the so-called Tomb of Ismail Samani and the small Mausoleum of the Arab Father (Mazar Arab-Ata) at the village of Tim, overlooking the Zarafshan Valley near Samarkand, we would scarcely be able to appreciate the refinement of Bukharan architecture at its zenith.

The former, a modest cube crowned by a dome, spans a mere thirty two feet on each side (see plate 10). Dating from the early tenth century, this was probably the garden mausoleum of Nasr II, Rudaki's patron.[43] Its preservation is due to the fact that at some point it was completely buried in sand and remained so for centuries. Three features of this remarkable structure immediately strike the viewer. First, it is a carefully designed jewel box with the geometry and ratios of each element carefully scaled to achieve monumentality within a small compass. To this end the unknown architect slightly tapered the cube, inviting the eyes of the observer to roam upward to the gallery and elongated dome above. Second, the dome, while resting on an innovative form of corner structure or squinch, otherwise differs little from those on Buddhist structures in the region.[44] Third, every part of the cube's exterior and interior, including the unexpectedly classical corner pilasters, is enlivened by baked bricks laid in a profusion of geometric designs. These highly rhythmic patterns create a dramatic counterpoint to the reposeful form of the structure as a whole. And fourth, the symmetry of its four façades and the arched entrances set within recessed rectangular panels looks more to Central Asia's pre-Muslim past than to the Islamic future. No wonder, then, that various writers have traced the prototype for this structure to Manichean temple-observatories, Persian prototypes, and Sogdian memorial structures.[45]

The Uzbek scholar Shamsiddin Kamoliddin has made a convincing case for the Buddhist origins of this design. He notes that patterns in the exterior brickwork repeat earlier designs from Varakhsha and that the superstructure includes a Buddhist mandala. More controversially, he then proposes that the building was an exact reproduction of an earlier tomb nearby that had been constructed to house the remains of a Buddhist princess from China who had rebelled against the Chinese authorities and taken refuge at the *vihara*.[46]

If the Tomb of Ismail looked to the past, the mausoleum at Tim pointed to the future. Dated 977–978, this diminutive structure (a mere 18.5 feet on each side) is at least two generations later than the tomb at Bukhara and suggests how quickly new values were coming into play.[47] Here the blunt goal was not to delight viewers but to awe them. The architect, again anonymous, did not allow visitors to wander around, demanding instead that they plant themselves in wonderment at a spot directly in front of the exaggeratedly tall arched portal or *pishtaq* that he raised like a theater set above the single entrance. This device forcefully concentrates visitors' attention on the interior tomb and on the unknown personage (probably a holy man) buried therein. There is nothing classical about this arrangement. In spite of its extremely modest scale, it provides a pathbreaking foretaste of the increasingly massive and imperious architecture that was to become the norm across Central Asia and, indeed, the Muslim world as a whole.

The Tomb of Ismail treated visitors as an independent and self-respecting patricians, while the mausoleum at Tim reduced them to the status of humble subjects respectfully cringing before authority, in this case religious authority. The Bukhara tomb, linked with the state, celebrated the secular, and its values are intellectual and aesthetic. By contrast, the Tim mausoleum celebrates the piety of the holy man buried there

Figure 8.2. Samani-era mausoleum in Tim, Uzbekistan (977–978). Its theatrical façade (*pishtaq*) anticipated on a small scale the focus on portals that characterizes many Islamic structures down to the present. ▪ Photo by Anvar Iliasov, Aiphoto, Uzbekistan.

and the commanding authority of the traditionalist religious leadership (ulama) of which he was presumably a part. Is it accidental that between the two dates—ca. 944 and 977—the actual power of the Samanid state declined sharply, and with it the intellectual and aesthetic self-confidence of the age of Rudaki, while the voice of the ulama gained strength?

Rudaki's patron, the Samanid ruler Nasr II, was a difficult person. A sickly child, he ascended to the throne at the tender age of nine, after his father was murdered in connection with a corruption scandal.[48] Being so young, the boy-ruler was easily dominated by his vizier, Jayhani, and then spent many years figuring out how to assert himself. This left Nasr so impetuous that his vizier and courtiers adopted the policy of never implementing his demands until at least two days had passed.

Yet this same man became a nearly ideal royal patron of culture. His interest in ideas was genuine, and surviving descriptions of his soirees or *majlises* portray him participating in the discussions with keen interest and considerable intelligence. Once an adult, Nasr was not intimidated by the fact that his vizier could write a commendable work on roads and cities. It was Nasr himself who assigned Jayhani's successor, the savant Balami, the task of translating and abridging the huge and influential *History of Prophets and Kings*, which the Caspian-born historian Tabari had recently issued in Arabic.[49] The ruler clearly understood that this sprawling history of the world, translated, abridged, and with a few timely alterations favorable to his own dynasty, could bolster the Samanis' claim to legitimacy.

It did not hurt that the Samanids were heirs to a rich culture of the written word. Graphomania was so widespread in Bukhara that when Jayhani sent an ambassador, Abu Dulaf, to China, no one was surprised when he returned with a lengthy manuscript describing all he had seen there. The many bookshops that filled a plaza near one of the city's three main mosques offered this and countless other laboriously copied volumes in many languages. Competition among the booksellers drove down prices and turned the dealers into hard-driving salesmen. One shopper, the young Ibn Sina, had had no intention of buying a book he chanced upon at one such stall, since the tome purported to explain what the rising young intellectual considered the "incomprehensible" metaphysics of Aristotle. But when the dealer chased him down the street offering a further price cut, Ibn Sina relented. This chance purchase, at the price of three gold pieces, turned out to be Farabi's guide to Aristotle's *Metaphysics*. Reading it proved to be a life-changing experience for the aspiring medical doctor and philosopher.[50]

In about 996 Ibn Sina, now sixteen, was summoned to the palace to consult with court physicians who had been unsuccessfully treating the ailing ruler. When his advice saved the ruler's life, the young doctor boldly requested permission to use the royal library. The grateful regent assented, thanks to which we have from Ibn Sina's pen the following description of Bukhara's own "Storehouse of Wisdom":

> I was admitted to a building which had many rooms; in each room there were chests of books piled one on top of the other. In one of the rooms were books on the Arabic language and poetry, in another on jurisprudence, and likewise in each room [were books on] a single science. So I looked through the catalog of books by the ancients and asked for whichever one I needed. I saw books whose names had not reached very many people and which I had not seen before that time, nor have I seen since. I read these books and mastered what was useful in them and discovered the status of each man in his science.[51]

As with Caliph Mamun's court library in Baghdad, the Bukhara library was a powerful magnet for talent, drawing philosophers and scientists from throughout the realm and abroad. Among the many luminaries who graced Samanid Bukhara was the Nishapur-born specialist on Greek thought, Abu Hasan Amiri, who crossed swords with local philosophers; the astronomer Ibn Amajur al-Turki; and the mathematicians Abul-Wafa Buzjani and Abul-Hasan al-Nazami; not to mention many philologists. The names of such medical scholars as Rabi' ibn Ahmad al-Akhawayni al-Bukhari, Hakim Maysari, and Abu Sahl Masihi may be unknown today, but a millennium ago these figures stood at the forefront of the field.[52] Meanwhile, historians scattered across the Samanid realm issued volumes describing individual cities and towns.[53]

As a group, the savants of Bukhara were saturated with classical Greek learning but eager also to make contact with the Indian sciences. They shared a passion for new knowledge and a culture of humanism that enthroned reason as the best tool for unlocking the secrets of nature and for explaining humankind's place in the cosmic order. An aspiring philologist and lexicographer from Nishapur, Abu Mansur al-Thaalibi, was taken to Bukhara by his father and had the good fortune to attend a discussion among the city's savants. His father rhapsodized: "Oh, my son, this is a memorable day; an epochal moment as regards the assembling of talent and the most incomparable scholars of the age. Remember it

Figure 8.3. Doctors and experimentalists had access to finely produced equipment, such as this eleventh-century scientific glassware. ▪ Photo by Anvar Iliasov, Aiphoto, Uzbekistan.

when I am gone. . . . For I scarcely think that in the lapse of years you will see the likes of these meeting together again."[54] Samanid Bukhara, Thaalibi concluded, was "the home of glory, the Kaaba of sovereignty, the meeting place of the most eminent people of the age, the horizon of the literary stars of the world, and the fair of the greatest scholars of the period."

The Shiite Challenge, from Without and Within

Thaalibi, for all his gushing, was silent on one matter: religion. Maybe he assumed that his rubric "greatest scholars" included learned Islamic jurists and other members of the caste of the religiously learned. More likely, he understood the secular tone of the prevailing high culture. For the freethinking leaders of Samanid high culture tended to view religion as at best a symbolic representation of the truths of reason that had been devised for the benefit of the common people. Thaabili may also have wished to sidestep the deep division between Shiite and Sunni Muslims that was opening across the Muslim world and, not least, within the Samanid empire.

Strange to say, this growing fractiousness among Muslims arose just as Islam was becoming the majority faith throughout the region. Many adherents of other religions in territories conquered by the Arabs had converted or emigrated. Others had been successfully marginalized, while still others had made their peace with the Muslim order. If Muslims had earlier identified the "other" as Zoroastrians, Buddhists, Christians, or Jews, they now assigned that role to fellow Muslims with whom they disagreed.

The waning power and spiritual decay of the Baghdad caliphate had brought to the surface old resentments and new ambitious among those whom it ruled. Iranians, for one, had never accepted the diminished role that came with the Arabs' destruction of the Sassanian empire in the seventh century. Three brothers from the Buyah family of northern Iran saw their opening and rushed in. After first defeating the caliph's Turkish forces in 934, they took control of most of Iran, parts of Oman, and the caliphate itself. Their conquests effectively transformed the caliphate into an Iranian empire. Their capital at Rayy (just east of modern Tehran) became a center for the recovery of Persia's political identity, which had been smothered since the rise of Islam, and also the scene of serious intellectual activity. All this was occurring in the very years of the Samanid ascent.

Significantly, the Buyids, as they are known, were Shiites, while the caliphate remained Sunni.[55] But the Buyids had good reason to respect the caliph's titular authority, even if he was Sunni: they now faced a rival caliphate, this one unabashedly Shiite.[56] The Fatimid state, with its capital at Cairo, was the creation of the Ismaili branch of the Shiites. It already ruled Egypt and northern Africa, but its leaders had begun casting their eyes eastward. This emerging confrontation underscored the fact that two of the biggest power centers in the Islamic world were now in Shiite hands. Many concluded that the Islamic world had entered an age of Shiite hegemony. Scholars like to speak of a "Shiite century," an era when Shiites took Islam's political and intellectual helm, which they date to roughly the years 950–1050.[57]

The reality of this supposed new religious hegemony was less impressive than the phrase "Shiite century" implies. To their credit, the Fatimids fostered trade and also exhibited a commendable tolerance of other faiths. Like the Buyids and Samanids, they were open to the world of ideas and generously supported a range of religious scholars. A madrasa they founded at the Al-Azhar mosque in Cairo later branded itself as "the world's first university," even though it taught only religious subjects until 1961.

But there were problems. Few understood the Ismailis' form of spirituality, which blended Islam with both rationalism and Neoplatonist mysticism. Most Muslims, who were Sunnis, were suspicious of this arcane branch of Shiism and considered the Cairo caliphate illegitimate, a view the Shiite Buyids in Iran shared. At the same time, the Fatimids were totally dependent for their security on mercenary soldiers, first Berber tribesmen from North Africa and then Cherkess (Circassian) Turks from the North Caucasus. This posed a constant threat of civil war.

The Buyids' vulnerabilities were even greater. They were not really a state at all but a loose, dynastic confederation that was unable to control its own territories, let alone those of others. As a result, most of their energies, and those of the Baghdad caliphate they controlled, were directed inward, or westward toward the competing caliphate in Cairo.

The fragmentation of Buyid rule and the problems of the competing caliphate in Cairo created a political opportunity for anyone with a strong will and an army to enforce it. The Samanids, who were already in power when these developments occurred, were up to the challenge. Their ascent had marked the emergence of a powerful and self-confident Sunni dynasty in the Persianate world. Now they could serve as a counterweight to the emerging Shiites in Rayy and Cairo, while at the same time being relatively free of the dogmatism that so easily comes with politicized religion. Indeed, tenth-century Bukhara was in most respects a very tolerant place. It viewed Mutazilites and other fringe groups with concern but did not purge their books from libraries, as happened in purportedly tolerant Spain.[58] In short, the Samanid rulers successfully resisted Shiite pressures from without, all the while retaining a relatively open society. But it was less successful in its dealings with Shiism from within.

At first the Ismaili missionaries who trickled into Bukhara from North Africa caused no particular stir. Among them were highly sophisticated Central Asians, like Abu Yaqub al-Sijistani from eastern Afghanistan, who was an accomplished Neoplatonist philosopher.[59] Contributing to the initial absence of conflict was the fact that the missionaries focused their attention on the upper classes and tended to work in secret. However, these early Ismaili missionaries won over the Samanid governor of Nishapur and leaders of many cities of western Afghanistan before moving on to Bukhara, where they converted the amir's secretary and then the regent himself, Nasr II.

The sub rosa character of this process suggests that conversion occurred in the name of the secretive Brethren of Purity, a diverse group with many Ismaili members. Its adepts swore to "shun no science, scorn no book, and not cling fanatically to any single creed."[60] After all, they reasoned, they all "derive from a single principle, a single cause, a single world, and a single Soul." Claiming not to adhere to any particular religion, the Brethren considered themselves totally independent and free to go wherever the exercise of reason led them. This opened them to mathematics, the sciences, logic, and philosophy.[61] In their quest for higher knowledge, members of this unabashedly elitist movement also called on esoteric knowledge, including astrology, mysticism, numerology, and symbolic interpretations of the Quran. This potent combination of reason and mysticism was to exercise a powerful influence on all Ibn Sina's thought.

Leaders of the Sunni religious establishment in Bukhara were shocked that some members of the government appeared to be colluding with these radical Shiites. Working hand-in-hand with officers from the Turkic guard, the ulama raised a revolt in 943, overthrew Nasr, beheaded the Ismaili missionaries and the vizier who had backed them, and suppressed their remaining followers. Rudaki was blinded and expelled from Bukhara.

The story did not end there. Several provincial governors, including the learned Razzaq, Ferdowsi's protector at Tus, continued as active Ismailis, while a fresh group of Ismailis at the court in Bukhara and others in Herat, in present-day Afghanistan, actively promoted their version of Islam. Two generations after the purge of Ismailis from the government, the young Ibn Sina grew up in a staunchly Ismaili household, was taught by Ismaili scholars, and went on to carry with him throughout his life their blend of rationalism and mysticism.

TRADITIONALISM'S NOT SO SILENT MAJORITY

If the upper ranks of Samanid Bukhara were open to rationalism, free thought, and Ismailism, the city as a whole was a bastion of traditionalist Sunni Islam. Even before the rise of the Buyids in Iran and the Fatimids in Egypt, members of the Bukharan ulama and other scholars elsewhere in Central Asia had cast themselves as the protectors of the true (Sunni) faith.

The great question at issue among people across the ideological spectrum concerned the source of authority and law in the Islamic world. Over the first decades after Muhammad's death in 632, there had been a direct laying on of hands, with each caliph named on the basis of genealogical succession. But within a generation this process was hotly contested, resulting in the murder in 661 of the fourth caliph, Muhammad's cousin and son-in-law, Ali. We have already noted how Ali's supporters, later known as Shiites, looked to a hereditary imam to define and interpret the law. Against them the proletarian Kharijites who had taken root in Khurasan argued that the leader of the faithful should be chosen on the basis not of heredity but of moral purity. Finally, there were a few intellectuals who, taking inspiration from the Greek classics, proposed that wisdom and a capacity for critical thought should be the basis of civil authority.

Those who came to be called Traditionalists responded to all these currents and movements with a starkly simple formula. The only acceptable sources of law and civic authority, they argued, were the Quran and the Hadiths of the Prophet. The Quran, of course, was God's word. But his Messenger, in conversations with the Companions, left a wealth of practical and philosophical guidance. At first, the Companions transmitted these utterances or Hadiths by word of mouth. Over the succeeding centuries, people began collecting them and writing them down. Most of the earliest collectors of Hadiths were from Central Asia.[62] Like all those who followed, they cleaved to the traditionalists' faith that if all the words of the Prophet could only be assembled, they could provide authoritative guidance on all questions of law and life.

During his celebrated facedown with Caliph Mamun, Ahmad ibn Hanbal of Merv had stood forth as a rock-solid defender of this new traditionalism. In time he became the patriarch of what Patricia Crone has termed the "Hadith Party," an expanding band of deep-dyed traditionalists and literalists who opposed both the Mutazilites and the rationalists.[63] Within a generation the caliphs were using their office to impose this dogma on society.

Some years after Mamun's inquisition, when Ibn Hanbal had outlived his last official persecutor, a young friend from Bukhara, Muhammad al-Bukhari (810–870), presented the pious old veteran with the manuscript of a compendium of sayings and traditions of the Prophet that he had been gathering for sixteen years. Bukhari offered it as a meticulously selected and rigorously edited text. He then proceeded to issue his compendium for the public and then to defend it before audiences of thousands across the length and breadth of Central Asia.

By this process, Bukhari's opus, entitled *An Abridged Collection of Authentic Hadiths with Connected Chains [of Transmission] Regarding Matters Pertaining to the Prophet, His Practices, and His Times*, became universally accepted as a core element of the Islamic canon and Islam's second most holy book after the Quran, a position it still holds today. Bukhari's close ties with Bukhara established that city as "The Dome of Islam." Ismail Samani was twenty-one years old when Bukhari died in 870. Bukhari's tomb outside the city became a major pilgrimage point, visited by multitudes of the faithful from across the Islamic world.[64]

Since Muhammad lived twenty years after he first began disseminating the revelations he had received, the number of Hadiths attributed to him was formidable. Moreover, the process of telling and retelling them caused their number to rise exponentially. But serious problems arose concerning their authenticity and accuracy. Many narrators recalled individual Hadiths quite differently. As in the childhood game of telephone, the very process of retelling the Hadiths caused the message to shift and then to change again, often to the point of becoming unintelligible. Nor were people immune from the temptation to twist Hadiths to make them better serve their own needs. Worse still, the mere fact that so many people believed that arguments could be settled by citing the Prophet's words became an invitation to fraud. Thousands of bogus "Sayings of Muhammad" circulated, offering the Prophet's purported views on everything from the attractiveness of one city as opposed to another to the degree to which specific ethnic groups could, or could not, be trusted.

Many of the forgeries were concocted and disseminated to support particular theological or political positions.[65] Especially popular were those that could be interpreted as hostile to the authority of the caliphs and those that appeared to place limits on the exercise of reason. By the ninth century the forging of Hadiths had become something of an industry and the stage on which many battles, both religious and secular, were waged.

It was not until a century and a half after Muhammad's death in 630 that the process of assembling Hadiths and traditions really gained momentum. The hope of finding certainty in an uncertain world drove thousands of professionals and amateurs into what we would now call the field of oral history. Joseph Schacht, the German-born expert on Islamic law, has argued that most, if not all, of the Hadiths can be traced back to no earlier than the eighth century.[66] Whether or not this is the case, none of the actual collectors doubted the authenticity of the oral tradition. We have noted that this project particularly attracted Central Asians, with the city of Merv producing several of the earliest compilers. Scholars from Balkh, Tirmidh, and other centers were no less active.[67] The names of no fewer than four hundred Hadith collectors from the region are known,[68] far outstripping the rest of the Persian world and all the Arab lands as well. A single town, Amul on the Amu Darya in present-day Turkmenistan, claimed nearly a dozen Hadith specialists.[69]

Is this activity a reflection of the fact that for nearly a millennium prior to this, Zoroastrian divines, Buddhist monks, and Christian, Jewish, and Manichean scholars from the region had been assembling, codifying, and translating the holy books of their respective faiths?

MUHAMMAD AL-BUKHARI: AN ORPHAN'S QUEST

By the early ninth century, the eastern Muslim world was awash in Hadiths, both real and fraudulent, purportedly tracing to Muhammad. Ibn Hanbal alone assembled a collection of twenty-nine thousand, while

Bukhari claimed to have heard and memorized more than seventy thousand.[70] Similarly, some seven thousand "narrators" had been identified and, where possible, debriefed. This mother lode of conflicting evidence created an urgent need for some kind of vetting process. Again, this task fell overhwhelmingly to scholars from Central Asia. By the time it was over, Sunni Muslims throughout the world accepted only six major collections as authoritative and canonic. Of these, five were the work of Central Asian experts,[71] among whom Muhammad al-Bukhari was universally acknowledged as preeminent.

Muhammad of Bukhara was orphaned early. One can speculate that his lifelong search for authority and guidance in life traces in part to this fact. The canonic version of his life has him memorizing the entire Quran by age seven or eight and then beginning to collect Hadiths a year later. His father left sufficient wealth that at age sixteen Bukhari, his mother, and his brother could afford to perform the Hajj and then spend several more years in Arabia visiting the shrines of Islam and meeting Hadith collectors in Medina and Mecca. Over the next sixteen years, Bukhari traveled, taught Hadiths, interviewed several thousand narrators, and compiled his *Collection of Authentic Hadiths*. The final two decades of his life he spent with students in Nishapur and Bukhara, with frequent side trips for what was in effect a sustained book tour. A medieval biographer estimated that the total attendance at his various presentational meetings was upwards of ninety thousand.[72]

Aside from a passion for archery, Bukhari focused entirely on his work, never marrying and avoiding all contacts with officialdom. When a ruler asked Bukhari to deliver him a copy of his *Collection of Authentic Hadiths*, Bukhari declined, on the grounds that he was not answerable to secular authorities.[73] Judging by the very restrained language he uses to criticize narrators whom he considered to be "weak," Bukhari must have been of mild temperament. Given this, it is all the more surprising that in both Nishapur and Bukhara he found himself locked in rancorous debate with other Hadith scholars, who challenged many of his judgments, faulting him both for what he had included and what he had not. In light of the later canonization of Bukhari's work, the challenges that arose on every side from his own contemporaries, many of them scarcely less expert than he, must strike us today as strange indeed.[74]

Among Bukhari's challengers was al-Hakim al-Naysaburi, the quar-relsome Hadith collector whom we met in Nishapur. Naysaburi accused Bukhari of having accepted Hadiths from questionable narrators. For good measure he extended this accusation to Bukhari's local student and disciple Abdul Husain Muslim, later the author of one of the other six authoritative Hadith collections. Ominously, Naysaburi also accused Bukhari of being wobbly on the crucial question of the createdness of the Quran, a charge that led to Bukhari's departure from Nishapur in disgrace. After a period spent at his home in Bukhara, he then traveled to Samarkand, where several hundred experts confronted him for a solid week over what they considered the many shortcomings of his work.[75]

With 7,250 Hadiths in Bukhari's collection, not to mention the tens of thousands he had decided to exclude, critics had more than enough to chew on. Since some of their concerns pertained to widely accepted Hadiths that Bukhari had rejected, his protégés Abdul Husain Muslim from Nishapur and Hakim Tirmidhi from Tirmidh (Termez) rushed to defend their master by issuing their own compendiums. In both cases they made clear that Bukhari might have included the several thousand additional Hadiths they were offering had he not chosen to abridge his collection.[76]

THE SCIENCE OF ORAL HISTORY

More serious questions regarding Bukhari's method arose in his own lifetime and have lingered through the centuries. Hadith scholars, for example, differed on their evaluation of the reliability of many of his informants and also disagreed on the process for resolving such differ-ences. Thus the people of Medina claimed that any Hadith tracing to their city was ipso facto valid because of Medina's pedigree as the burial place of the Prophet.[77] Others disagreed. To take another example, all Hadith scholars agreed on the validity of any Hadith that could be traced through a valid chain of narration to one of the Companions of the Prophet. But this did not allow for the possibility that a given narra-tor may have had a faulty memory or was in deep old age when he or she transmitted what had been heard. Even the Companions' authority was questioned. The Prophet's favorite wife, Aisha, for example, was nine or ten when Muhammad married her and nineteen when he died. Yet she was still bringing forth new recollections of her husband's words a half century later. Were all her memories unerringly accurate? Finally, Naysaburi himself acknowledged this general problem of human memory when he pointed out that professional Hadith scholars themselves could not be relied on to recall accurately everything they had recorded.[78]

Against such criticism, medieval scholars sometimes argued that if a Hadith was "widely accepted" it was likely true.[79] However, this standard would appear to conflict with the principle of a valid chain of transmission through authoritative narrators on which the entire enterprise was built. This in turn raised again the question of the signs by which one could know that a given narrator was authoritative. The decisive criterion, it turns out, was that he or she had to have been a pious Muslim and of good moral character. Well and good, but a skeptic like Razi or Ibn al-Rawandi, as well as modern scholars,[80] could not help but inquire into the definition of a "good Muslim" and suggest that it might exclude anyone whose views, and the Hadiths they transmitted, were not safely orthodox. Finally, what excludes the possibility that a given Hadith was wholly fabricated, along with multiple "reliable" chains of transmission? Such doubts led a contemporary Baghdad scholar to observe that the whole matter of historical traditions turned on highly subjective opinion.[81]

LAWS, ON EVERYTHING

The issuance of collated collections of Hadiths marked Islam's point of transition from oral to written culture and a basic shift in the hierarchy of media that had heretofore favored memory over the written word.[82] Central Asians were already adept at the written word and plowed forward in their work, even though they knew full well that their compendiums would henceforth cut the reader off from what a recent analyst called the "authorizing chain of transmission."[83] They, the religious jurists themselves, had long overseen this process and now stood to lose if it ended. They accepted this because their compendiums advanced a much more important goal, namely, to provide a concrete and detailed body of religiously sanctioned dos and don'ts that could become the basis of law.

It is true that Islamic law, or Sharia, has always allowed ample room for interpretation. But this was not what the compilers of Hadiths sought. Henceforth, written religious law, rather than caliphs, imams, opaque philosophers, or heretics, would regulate personal and social behavior. This would finally put an end to the confusion and uncertainty that had reigned since the death of the Prophet two centuries earlier. Thanks to the recovery of the putatively authentic voice of the Prophet, the words of the Quran would finally be implemented and its vision for humankind fulfilled. Bukhari and the other leading collectors and collators of Hadiths were quite literally canonized,[84] and their work placed beyond the reach of criticism.

Nothing was too trivial to fall within the law. So that it would better serve as a guide to action, Bukhari organized his compendium by subject. Included among the main headings are such seemingly random topics as bathing, invoking God for rain, dreams, menstrual periods, mortgages, the transference of debts, manners and dress, and the meaning of eclipses. Specific Hadiths cover the treatment of head lice, how to trim one's beard and clean one's teeth, Peeping Toms, eating while standing, and even whether men should stand or squat to urinate. But also present are headings on revelation, knowledge, the creation, marriage and divorce, dealing with apostates, the manumission of slaves, and the oneness of God.

Viewed as a whole, Bukhari's grand compendium provided a bridge between the world of early Islamic Arabia and daily life. And even though it offered countless instances of the harshest punishments, in other respects it softened the severe call to obedience that permeates the Quran. In an echo of the Golden Rule, the Prophet asserted in one Hadith that "none of you truly has faith if he does not desire for his brother what he desires for himself."[85] Time and again the *Collection* exhorts the reader to consider the intention underlying deeds and to recognize that all human actions, including those that appear most malevolent, are ordained by God.[86] Life is a covenant between mankind and the deity. In the words of Lenn E. Goodman, "God [was] now a party to every undertaking, and good faith means purity of intent, a conscience, and consciousness open to His scrutiny."[87]

In the end, Bukhari's *Collection of Authentic Hadiths* severely narrowed the sphere of choice in human affairs. It implied that divine will, as known through revelation and interpreted in the Hadiths, can and must guide every human action, and it sharply reduced the need for interpretation. Islamic jurisprudence had always had a "window" that allowed the ulama to exercise independent judgment when applying Sharia. Bukhari did not close that window, but he began the process by which it could be closed.

By virtue of its infinite detail, the result of the great Hadith project by Bukhari and his five contemporary scholars was to regularize and systematize what had heretofore been a bewildering maze of orally transmitted regulations. From start to finish the *Collection* is filled with the language of imperatives.[88] Even before the end of the Samanid era, many Muslims began to feel the negative effects of what they perceived as an overdose of legalism and formalism. Over time this was to emerge as one factor among many that caused large numbers of the devout to recoil against the increasingly rule-bound world of Muslim literalism and traditionalism and to embrace the new mysticism championed by Sufi visionaries.

Traditionalism Triumphant

In the short run, however, Bukhari's achievement was greatly to strengthen and embolden the traditionalists at the expense of the party of reason and philosophical inquiry. The fact that the entire Hadith project, with its problematic assumptions on evidence, occurred at the same time that Farabi and others were analyzing and setting down the austere rules of logic shows how far apart the two—philosophers and traditionalists—had grown. The traditionalists understood this and went on the offensive. Just as the civilization of the Samanids reached its zenith in the second quarter of the tenth century, a sympathizer of Bukhari's from Samarkand named Muhammad Abu Mansur al-Maturidi (853–944) issued a bold series of "Refutations" denouncing Mutazilism, Shiite heretics, the Ismailis, and philosophy generally.[89] An authority on all the religions of the region and a formidable jurist, his pronouncements attracted readers from East Asia to North Africa. Maturidi, whom we encountered earlier on these pages, was someone to be taken seriously.

Gradually during the final generations of Samanid rule, the traditionalists and literalists gained the upper hand in Bukhara. And not just Bukhara. Proud of their collectors of Hadiths, people throughout Central Asia adopted a proprietary attitude toward the faith. No less than the Arabs, Central Asians now saw themselves as the chief bearers of Islam and arbiters of things Islamic. They also took measures that would make their region the heartland for all the Religions of the Book. The pious of all faiths could now make pilgrimages to what were believed to be the remains of Job near the village of Jalalabad in the Ferghana Valley, of Solomon at the mount near the city of Osh in what is now southern Kyrgyzstan, of Daniel near Samarkand, and of the prophet Ezekiel at Balkh. By some unknown process, the bones of St. Matthew found their way to a monastery at Lake Issyk Kul, in modern Kyrgyzstan, which became a pilgrimage site for Christians and Muslims and became known as far afield as Spain.[90] The proliferation of mausoleums of Islamic saints and holy men meant that Muslim pilgrims could find shrines in practically every town and village. A single shrine at the village of Ispid-Bulan in Tajikistan commemorated no fewer than 2,700 Companions of the Prophet.

While supportive of the faith, the champions of reason and logic in the Samanid flowering found the spreading literalism and flight to tradition deeply unsettling. To be sure, there had long been a note of world-weariness and melancholy in Samanid poetry, as, for instance, when a friend of Rudaki's from Balkh observed:

Search the world through and through; you will not see
One man of wit who's not by grief oppressed.[91]

But now the rationalists and freethinkers were being shown the door by an unlikely alliance of traditionalist members of the ulama and officers of the palace guard. The latter were known to be lax in their Islamic observance and all too devoted to drinking wine, but they could nonetheless be mobilized in the name of true belief, especially when their own power was at issue. Now both groups were convinced that irreligion and heresy were rampant in Bukhara.

More alarming were the new Turkic armies that appeared on the eastern and southern borders, posing threats that demanded the government's full attention, and all the resources it could muster. Taxes rose precipitously to pay for the army, trade was in free fall, and dissident members of the still-prolific Samani family began once more to challenge their kinsman in Bukhara. By 980 the Samanid state and the rich culture it had sustained for more than a century were in crisis.

ABU ALI AL-HUSAYN IBN SINA

Even as these events were unfolding, the greatest mind of the Samanid era and one of the most seminal thinkers of the Middle Ages anywhere was coming of age in Bukhara. The golden age of Rudaki was long in the past, but even as the sun of Samanid culture was setting, it produced a man whose contributions to medicine and philosophy were to have the most profound impact on the entire Islamic world and on Europe and Asia as well.

Ibn Sina is best known in the history of world science for his magisterial *Canon of Medicine*, which became the foundation stone for European medicine and a transformative force in Indian medicine. So lasting was Ibn Sina's impact in the West that as late as the seventeenth century, when Robert Boyle was striving to refound medical science, he took as his first task to challenge and finally move beyond the heritage of this Bukharan from six centuries earlier. Even then Boyle remained indebted on many points to Ibn Sina.

No less noteworthy were Ibn Sina's contributions to philosophy. As if mixing oil and water, he managed to integrate the doctrines on being of Aristotle, grounded on common sense, with the more esoteric and mystical notions of the Neoplatonists, at the same time preserving a role for God in human affairs. Two centuries after Ibn Sina's death, Thomas Aquinas seized on his writings to justify the doctrines that earned the Italian the title of "Doctor Angelicus," which the Catholic Church bestowed on him. Many Muslim divines today call Ibn Sina their "Third Teacher," after Aristotle and Farabi.

Ibn Sina was born at a town seven miles northwest of Bukhara in 980. For the first thirty-two of his fifty-seven years, he lived in Central Asia, and then he spent the rest of his life in constant flight from city to city in western Persia. He was astonishingly precocious, to the extent that by the time he left Bukhara at age twenty-two he had written his first two books, sparred with the other greatest mind of the age, established the patterns of thought and work that were to endure throughout his life, and plotted out the main directions of his subsequent writings.

Credit this to inborn talent, if you will. But in addition to that, Ibn Sina benefited from an education that was extraordinary in both its breadth and depth, exposing him to every important intellectual and cultural current of the era. Detailed information on Ibn Sina's early training has come down to us—more, in fact, than for any other cultural figure of the Islamic Middle Ages. Because his education epitomized the best of the era, it is fitting to conclude our survey of the Samanid world with a closer look at the process by which the young man became the "Third Teacher."

Ibn Sina, like the Samanids themselves, descended from a family of senior officials from Balkh in Afghanistan. His father had headed the Samanid government's tax department in a town just north of the old Bactrian capital before being assigned to similar posts at Kharmaythan and then Afshana, not far from Bukhara.[92] His work involved far more than shuffling papers. It demanded a solid command of the newly imported Indian system of numbers, and of mathematics, statistics, and Khwarazmi's practical algebra. After the birth of al-Husayn Ibn Sina, his second son, the father managed to get himself transferred to the city of Bukhara, where he could more systematically advance his sons' education. Whether at this point or earlier, the father had come to the favorable attention of the ruler, Nuh. The father's new duties required him to travel, and he sometimes took his sons with him. Four decades later Ibn Sina recalled his being amazed, at the age of five, at the speed with which everyday objects became petrified in the muds along the Amu Darya.[93]

Down to his tenth year, Ibn Sina studied the Quran, memorizing long passages and mastering the art of eloquent reading from the scriptures. As happened to every other precocious Muslim child, his proud elders credited him with having memorized the entire book. Neighbors were so impressed that they called him the sheikh, an honorific title usually reserved for religious scholars five times his age.[94] Ibn Sina emerged from these studies with a solid knowledge of holy writ and a fluent command of Arabic, in addition to his native Persian.

The next step was to master the practical skills that had proven so beneficial to his father's administrative career. Rather than have the lad sit in a classroom, his father sent him to work for a local vegetable trader, often said to have been an Indian, whom he commissioned to teach Ibn Sina the Indian number system, mathematics, accounting, and algebra. Clearly the plan was to prepare the young man for a career as a senior official. In the business-oriented Samanid state, this meant hands-on training in the world of commerce.

The next phase of Ibn Sina's education entailed the systematic study of "philosophy," which included logic, theology, and jurisprudence. To guide the eleven-year-old and his brother through these challenging fields, the father engaged a philosopher from Khwarazm named Abu Abdallah al-Natili, who moved into the family home.[95] Natili was no ordinary pedant but an accomplished scholar who had written an Arabic translation of a work by the first-century Greek botanist, pharmacologist, and medical expert, Dioscorides. Later Ibn Sina was grandly dismissive of Natili but also acknowledged that his interest in medicine arose in the course of his studies with this man of letters. As to Natili's translation of Dioscorides, Ibn Sina did not forget it and included large sections of it in the fourth book of his *Canon*.

Of equally lasting importance was the fact that Natili was an adherent of the Brethren of Purity, the secret society in which many Ismaili Shiites participated, and close to Ismaili missionaries in Bukhara. Among those attracted to the the Ismailis' doctrines of rationalism, humanism, and mysticism was Ibn Sina's father. His Shiism was reflected in the names he chose for his sons, Husayn and Ali, after the first two Shiite martyrs. Thanks to the tolerance that continued to prevail under Samanid rule, the father's career did not suffer from this, even though Bukhara's powerful traditionalists vehemently opposed Shiites in general and Ismaili Shiites in particular.

The young Ibn Sina spent long evenings listening to discussions about the "World Spirit," the "World Reason," and other arcane topics drawn from the *Epistles* of the Brethren of Purity. While his brother Ali joined the Ismailis, Ibn Sina himself did not.[96] Yet throughout his life he remained closer to the Ismailis than to any other branch of Islam. Evidence of this is to be found less in the answers he propounded than in the questions he asked. To the end of his days, he was inquiring, as did Natili and his father, into what could be known through the rigorous exercise of reason illuminated by "the sanctity of divine power."[97]

The oft-told story of Ibn Sina's instant mastery of the works of Ptolemy, Euclid, and other classics of Greek science and philosophy is a core piece of Ibn Sina lore. Basically, Natili—or possibly Ibn Sina's next teacher, an Ismaili savant named Abu Mansur al-Qumri—told the young man to work ahead on his own, as he had nothing more to teach him. Natili returned to Khwarazm, and Qumri, a medical doctor and freethinker,[98] at once began building on the twelve-year-old's interest in medicine. When stumped by a problem, Ibn Sina rarely turned to Qumri, whom he claimed to have taught more than he learned from the man, and instead went to the mosque to pray. In a revealing foretaste of his later views on the workings of the unconscious, Ibn Sina reported that the answer would usually come to him spontaneously, even during sleep.[99] Significantly, he concluded that prayer was fundamentally a process of intuition.[100]

The decision by Ibn Sina's father to allow the young student to spend the next four years studying on his own is testimony not only to paternal wisdom but also to the stimulating intellectual climate in late Samanid Bukhara. This was the period when Ibn Sina chanced upon Farabi's book on the metaphysics of Aristotle in a Bukhara book store. He also engaged in legal disputations and continued his study of medicine. When not studying, the young Ibn Sina spent time with the "golden youth" of the city's elite. He developed a love of music that later found expression in a seminal work on the theory of music, and he also gained an appreciation of poetry. Only fragments of his many verses in Persian survive,[101] along with a handful of Arabic rhymed *qasidas*, the two-part, sonnet-like Arabic verse form usually used for praising rulers. Ibn Sina, however, used *qasidas* to expound on logic, medicine (including everyday tips for good health), and the soul.[102] He also acquired a taste for wine, which he drank throughout his life to aid his thinking and writing, and for convivial evenings with students. With scant concern for Muslim orthodoxy, his *Canon* would enumerate both the negative and positive effects of wine, encouraging moderation but by no means banning it.[103]

By age sixteen Ibn Sina was treating patients on his own, in the process learning firsthand what could not be gained from books. The prevailing standard of medical knowledge throughout Khurasan and the Bukhara region was very high. Even before Razi issued his thirty-volume *Comprehensive Work on Medicine*, a chemist from Nishapur had circulated throughout Central Asia his own *Book on the Foundations and True Essence of Drugs and Medicines*, with detailed descriptions of 585 medicines.[104] With so much material readily at hand, it is no wonder that Ibn Sina could confidently announce that "medicine is not one of the difficult sciences."[105]

Soon senior doctors sought his counsel. Just at this time Amir Nuh took ill and his doctors, frustrated at the failure of their own remedies, consulted with Ibn Sina on how best to treat the ruler. They followed his advice and the amir promptly recovered, making the young doctor a hero. For nearly two years the young scholar then pored through the books in the royal library, taking notes and making files. At the ruler's request, he also translated a book on Zoroastrianism from Middle Persian, which included medical material that later found its way into the *Canon*. It has already been noted that two of Ibn Sina's future students were Zoroastrians.[106]

Two events in 997 ended this phase of Ibn Sina's life: the urban fire that destroyed the royal library, and the death of Ibn Sina's champion, the regent Nuh. They also marked the end of an education that was remarkable for both its breadth and its depth, and which left the student, Ibn Sina, fully conversant with the most noble achievements, ancient and modern, in the sciences and the humanities.

Ibn Sina, Biruni, and the Cosmos

The one great shortcoming of Ibn Sina's education was in the area of character. The death of the Amir Nuh in 997 had left Ibn Sina's father without a patron, but the court stepped in with various gifts and grants to his talented son. This left Ibn Sina, now eighteen, on top of the world. Pampered and flattered since childhood, he fell prey to what proved to be a lifelong arrogance. Who but Ibn Sina would dictate a self-congratulatory autobiography at the age of thirty-seven and use it to settle scores dating back to his childhood?[107] Later in life he would quote the following verse, indicating that over time he had gained in perspective but not in modesty:

When I became great, no country could hold me:
When my price went up, I lacked a buyer.[108]

His conviction of his own invincibility left Ibn Sina short-tempered, impatient with anyone he judged to be mediocre, and quick to resort to outright nastiness.

These traits all flared up during a youthful exchange of letters with a fellow student of Natili's named Abu 'l-Faraj ibn al-Tayyib. Fluent in Latin and Greek, the author of a book entitled *The Cause of Things*, and also a medical doctor, Abu 'l-Faraj was secretary to a Christian bishop in Baghdad. At first Ibn Sina acknowledged Abu 'l-Faraj's work in medicine while mercilessly attacking his philosophical ideas. Abu 'l-Faraj responded with criticism of his own. Then Ibn Sina descended to across-the-board abuse, going so far as to pen an essay "On the Reprimands of a Feces-Eater." An otherwise admiring medieval biographer characterized Ibn Sina as "poisonous, and foul of speech, with an evil tongue."[109]

A much more formidable challenge soon emerged in the person of Abu Rayhan al-Biruni, eleven years older than Ibn Sina and the only man in Central Asia, the Muslim lands, and, arguably, the whole world who was Ibn Sina's intellectual equal. Both were young, both had been praised to the skies, and both considered themselves the masters of all knowledge.

Biruni had just moved to Gurganj, the new capital of Khwarazm, some 250 miles northwest of Bukhara. He, too, had heard about the whiz-kid from Bukhara from Ibn Sina's proud teacher, Natili, who now also resided in Gurganj. Biruni was also in touch with Abu 'l-Faraj, who doubtless had nothing good to say about the brash and abusive upstart. Naturally Biruni was eager to test the knowledge of this young rival. He did so by addressing an initial set of eighteen questions to Ibn Sina. These questions, Ibn Sina's answers to them, Biruni's rejoinders to Ibn Sina's responses, and his several further sets of questions, along with Ibn Sina's final responses and rebuttals, take us deep into the inner workings of Central Asian intellectual life.[110] Having noted this correspondence in chapter 1, it is now necessary to examine it in greater detail.

There is nothing quite like the Biruni–Ibn Sina exchange. Stretching over the two years 998–999, the letters reveal two young titans of thought locking horns over some of the most basic questions of philosophy and science.[111] They also attest to the vitality of intellectual life in Bukhara and Central Asia during the last days of Samanid rule. Whether or not this was part of a tradition of sharp exchanges of ideas in the Islamic world, as has been claimed, it was among the first known example and the most pointed.[112] Since it included more than fifty questions and answers on discrete topics, it will be necessary to limit our coverage to a few highlights.

Biruni fired the opening shot with blunt questions on Aristotle's *Physics* and on his newly translated cosmological treatise, *On the Heavens*. Biruni questioned Aristotle on point after point, forcing Ibn Sina either to defend his master or abandon him. On most issues Ibn Sina cleaved loyally to Aristotle. Thus, he defended the Greek thinker's view that the heavenly bodies have neither mass nor weight, and that Earth alone has gravity. Biruni thought otherwise. Ibn Sina's universe was static, while Biruni's, with his affirmation that "all things strive toward the center," opened the path to a more dynamic universe driven by interacting gravitational forces. This query came to focus on what both saw as an anomaly, namely, that if heat rises, how does the heat of the sun reach Earth?

Ibn Sina focused in on Biruni's assertion that vacuums can exist in nature. Basing his argument on "observable evidence," Biruni had argued that this was possible, while Ibn Sina, whose more logical and metaphysical mind could not accept the notion of a vacuum, protested. Yet at a later stage of the debate, Biruni asked, if heat causes things to expand and cold to shrink them, why does a sealed bottle of water break when it is *either* heated *or* frozen? Ibn Sina this time offered the dubious argument that frozen water indeed shrinks, but the "impossibility" of a vacuum causes the bottle to break. Similarly, he claimed that water enters a bottle when you suck out the air precisely because a vacuum is impossible.

Biruni also challenged Ibn Sina on whether all motion of celestial bodies must be linear or circular, as Aristotle held, or whether elliptical motion was also possible. On this and other points Biruni emerged as the radical innovator, challenging Ibn Sina's resort to philosophy and the authority of the ancients. Without a touch of subtlety he declared

that philosophers like Aristotle and Ibn Sina have no business addressing problems that can only be solved by mathematical astronomers like
himself. By considering even the possibility of elliptical motion in the
universe and elliptical orbits, Biruni opened the door that Johannes Ke-
pler (1571–1630) strode through in the seventeenth century. Ibn Sina,
by contrast, on this point played the part of a medieval schoolman cleav-
ing to the authority of his ancient master.

On several other questions, though, it was the Bukharan who blazed
the path forward and Biruni who dragged his feet. Ibn Sina explicitly rejected Plato's notion that human sight arises from rays emanating from
the eye and asserted instead that vision works in the opposite direction.
This in turn enabled Ibn Sina to explain why objects that are farther
away appear smaller, and to elucidate the process by which the mind
adjusts visual data on size to bridge the gap between impression and
reality. On both points Ibn Sina was centuries in advance of both Byzantine and European thought, which were based on the ideas of Plato.
Whereas the Platonic idea of visual rays emanating from the eye underlay centuries of reverse perspective in Byzantine and Western painting,
Ibn Sina's clear exposition anticipated and explained what five centuries
later came to be called "Renaissance perspective."

A dramatic moment in the exchange occurred when Biruni asked
whether we are alone in the universe or other worlds also exist. On this
matter Aristotle was unequivocal. Basing his claim on the view that
all matter was concentrated on Earth, he argued in his treatise *On the
Heavens* that "there is not now a plurality of worlds, nor has there been,
nor could there be." A half millennium later, most European thinkers
still clung to this notion and vigorously defended it against heretics like
Giordano Bruno (1548–1600), the Italian Dominican friar who was
burned at the stake for arguing that the universe contains an infinite
number of inhabitable worlds. But neither of our Central Asian correspondents found any rational grounds for believing that we are alone in
the universe, although neither Biruni nor Ibn Sina went on to revise the
geocentric structure of the universe propounded by the ancient masters.

After his bold initial sally, Ibn Sina, aware that his position flatly
contradicted the views of Islam and the other Abrahamic religions, introduced a fine distinction that kept him within the bounds of faith.
There *can* be other worlds, he argued, but they must differ from ours

in significant but undefined ways. This is so because God's creation of Earth and of human life was unique. Biruni flatly rejected this intrusion of theology into the argument, adding that problems in the physical world can never be solved with metaphysics.

Throughout his career Biruni was to base his thinking on the firm evidence of observation. In the exchange with Ibn Sina, he brought forth such perplexing facts as a ray of light refracted off a bottle of cold water being able to burn but the same light reflecting off an empty bottle would not burn. And why does ice float, even when we know it to be composed of the same elements as water? Biruni later became the first thinker to develop the concept of specific gravity, which he proceeded to apply to hot and cold water, among other materials.

Ibn Sina was no less insistent on the importance of observation and was to take it as the basis of his *Canon of Medicine* and other scientific works. But he took more seriously than Biruni the truths to be attained by the exercise of logic alone and from revelation. Biruni acknowledged the realm of metaphysics but never allowed it to impinge on the truths gained through observation and quantitative analysis.

Biruni's criticism of metaphysics led to an even more fundamental exchange over the origins of the universe. The doctrine of God's creation was deeply rooted in Islam, as it is in Judaism and Christianity, through the book of Genesis. The slightest whiff of deviation on this point could lead to charges of heresy, as had happened even to the certifiably orthodox Bukhari at Nishapur. Biruni's blunt questioning about the "createdness" or eternity of the universe allowed Ibn Sina little room to wriggle. But he did so nonetheless, and most adroitly. He agreed with Biruni's shocking assertion that Earth and universe were eternal but then invoked Neoplatonic concepts to argue that what was eternal was the *concept* or *idea* of Earth and the universe and not their *materiality*. God's initiative had therefore been required to transform idea into full being.

This struck the skeptical Biruni as empty philosophizing, and he said so. He knew from observing fossils and geological deposits that the world had *not* been created whole and complete, and he knew that Ibn Sina understood this as well. In this sense they were both sharply heretical, their arguments having more in common with evolutionary geology and even elements of Darwinism than with the Quran or Bible. Yet in the end Ibn Sina partially pulled back, contriving a complex metaphysical explanation that he hoped would square the circle of science and revelation.

Neither Biruni nor Ibn Sina set out to upset the prevailing Aristotelian view of nature. Even Biruni, whose observations made him more prone to skepticism, raised questions not about Aristotle's views per se but about the reasoning and evidence with which Aristotle supported them. In terms of Thomas Kuhn's *Theory of Scientific Revolutions*, they were both focusing squarely on *anomalies*, not on the grand cosmological or scientific systems themselves. Thus the inquiring Biruni did not challenge Aristotle's notion that all celestial motion must be circular. But since he knew from observation that a vacuum *could* exist, he queried Aristotle's claim that elliptical motion would create a vacuum and was therefore impossible. This in effect shifted the burden of proof on celestial circular motion from the skeptics to the defenders of circularity.

Viewing the Ibn Sina–Biruni debate as a whole, it is clear that from the outset Biruni managed to throw Ibn Sina on the defensive. None of Biruni's questions, for example, dealt with medicine, where Ibn Sina had a home-court advantage. Both debaters were astute observers of nature, but Ibn Sina perceived the world to some extent through the colored lens of metaphysics, while Biruni did not. At bottom, the result of the debate was to begin to create (or recognize) a "fault line between philosophy and science."[113] Biruni was not indifferent to philosophy and metaphysics, but he had come to them late, and they remained peripheral to his main concerns.[114] He flatly rejected Ibn Sina's assumption, drawn from Plato rather than from Aristotle, that the abstract realm of the ideal is superior to the realms of either mathematics or materiality.

Rather than accept that their differences were fundamental and hence irreconcilable, each of the young polemicists chose to ridicule the arguments of the other. What started as an earnest intellectual exchange quickly descended to a vulgar quarrel. The tone of Ibn Sina's responses from the outset was arrogant and condescending. At one point he charged that "there is nothing more absurd" than Biruni's questions, and at another he waived off Biruni's objections by saying "it is inappropriate for you to pursue what your intelligence prevents you from pursuing" and "you have a poor command of logic." On his side, Biruni cast his questions almost as if he was asking, "Can you possibly believe that . . . ?" The proper responses, he assumed, were self-evident to "anyone who does not insist on falsehood." Eventually Ibn Sina withdrew from the fray, claiming he had more important things to do. He handed his side of the correspondence to a bright but arrogant student, who defended Aristotle against Biruni's searching questions with a quarrelsome and supercilious tone that would have stood out even in a modern faculty lounge.

We do not know if it was Ibn Sina or Biruni who made the exchange of letters public, but this occurred almost immediately. The fact that the letters put Ibn Sina in a particularly bad light makes one suspect Biruni as the culprit. No sooner had the public read them than an uproar ensued, amid which "it became impossible to discuss philosophy in Bukhara."[115] Word of the philosophical jousting also reached distant Baghdad, where Abu 'l-Faraj, whom Ibn Sina had treated so rudely, exalted that "he who kicks others gets kicked in return." Exulting in Ibn Sina's discomfort, he confessed that he himself had prompted Biruni to challenge Ibn Sina in the first place. "Biruni," he exulted, "did this for me!"[116]

The affair revealed both of these young geniuses as difficult and even harsh personalities. Biruni later acknowledged that he had been rude, which was more than Ibn Sina ever did.[117] The exchange involved living issues that were of importance to many beyond the narrow circle of scientists and philosophers. At bottom the questions and responses juxtaposed two radically different intellectual worlds. The first, represented by Ibn Sina, assigned reason a crucial role but not to the point that it would fundamentally challenge revelation. This approach welcomed research and experimentation but in the end turned above all on logic and language. The second, represented by Biruni, relied on observation, experimentation, and mathematics to answer life's great questions, rather than on intuition or revelation. Like the Afghan-born philosopher Sijistani, who had insisted that scientists and philosophers should have nothing to do with religion and vice versa, Biruni sharply divided the realms of reason and faith and rarely concerned himself with the latter.

In the year 1956 the British scientist and novelist C. P. Snow delivered a lecture at Cambridge entitled "The Two Cultures," which he subsequently issued as a book.[118] In it he argued that the worlds of science (including mathematics) and the humanities (including philosophy and metaphysics) were diverging, and that this growing intellectual and moral rift posed serious dangers to modern civilization. The Ibn

Sina–Biruni correspondence laid bare the outlines of this problem a millennium ago. This emerged with striking clarity when Biruni, amid the argument over whether other worlds exist, archly informed Ibn Sina that "the metaphysical axioms on which philosophers build their physical theories do not constitute valid evidence for the mathematical astronomer." Here, for the first time, is a clear juxtaposition of Snow's two cultures: the philosopher who bases his case on metaphysics versus the scientist who relies on mathematics.

Their exchange of letters having ended, the two protagonists went their separate ways. Biruni continued his research at Gurganj, while Ibn Sina, at the request of erudite neighbors, tossed off two books, the first a concise encyclopedia covering all the sciences except mathematics, and the other a treatise on ethics, grandly titled *The Allowed and the Forbidden*.[119] Over the coming years the two moved farther and farther from each other intellectually, with each focusing on his own greatest strengths: medicine and metaphysics for Ibn Sina, and astronomy and the physical and social sciences for Biruni.

TWILIGHT IN BUKHARA

Curiously, the last Samanid ruler, Abu Ibrahim Ismail, known as Muntasir, was himself a poet. Responding to those who beseeched him to restore the happiness and glory of past times, he wrote:

> They ask me "Why not adopt a face of merry cheer,
> A house adorned with carpets rare, with many hues bedecked?"
> Can I, amidst the warriors' shouts and cries, the voice of minstrels
>   hear?
> Can I, amidst charging steeds in fight, the rose-bower sweet elect?[120]

In 999 Muntasir was driven from Bukhara, and, after struggling to mount a last-ditch rally, he suffered a final defeat in 1004. Meanwhile, just as the Ibn Sina–Biruni correspondence was drawing to a close, Ibn Sina's father died, leaving his sons without a source of support. Forced to make a living, Ibn Sina accepted an administrative post under the new Karakhanid rulers who had just assumed power in Bukhara.[121] His new duties required that he move to Gurganj, which he did in 1005. His departure from Bukhara marks the symbolic end of the golden age of the Samanids.

Let us return in closing to Rudaki, the joyful and wise singer with lute and goblet. Blinded and driven from Bukhara in the anti-Ismaili purge,[122] he returned impoverished to his native village and died there. This was several generations before the end of the Samanid dynasty. Yet one of his final songs stands as an elegy for the era as a whole:

> My dark-haired beauty, you can't possibly know,
> What shape I was in a long time ago!
>
> You can stroke your lover with your curls,
> But never saw him with locks of his own . . .
>
> Gone are the days when he was happy,
> When joy was plentiful and sorrow slight . . .
>
> You never saw him when he used to tell tales,
> And sang songs that rivaled the nightingales'.
>
> I was happy, my soul was a meadow
> Filled with joy, never having known sorrow . . .
>
> But I'm no longer the friend of nobles. The days
> When princes favored me are gone.
>
> Times have changed, and so have I. Bring me my staff.
> It's time for the cane and the beggar's purse.[123]

❁

CHAPTER 9

A Moment in the Desert: Gurganj under the Mamuns

KHWARAZM: A TECHNOLOGY-BASED SOCIETY

We do not know how Ibn Sina traveled to his new job in Gurganj, the capital of Khwarazm, more than twelve days by camel from Bukhara. He could have joined a caravan for the trip across the Black Sands (Kara-kum) Desert. More likely he went due west and then took a boat down the Amu Darya, a branch of which passed Gurganj before flowing an-other 370 miles northwest to the Aral Sea and the Caspian. Today we view Khwarazm and the Aral Sea region as a remote and forbidding desert. In 1004, though, it was known as one of the most developed and best-connected regions in all of inner Asia. The fact that the Amu Darya flowed on to the Caspian down to the late medieval period assured that the region along its lower reaches was connected with the larger world in ways that are scarcely imaginable today.[1]

Gurganj stood at the crossroads of four of Central Asia's main trans-port routes. The road east passed through the earlier capital, Kath, and eventually reached Farabi's hometown of Otrar, and beyond that, Chach (Tashkent), Taraz (Talas), and Balasagun, before passing into East Turkestan (Xinjiang) to reach Kashgar and, eventually, China itself. The southern road went to Merv, Bukhara, Balkh, Kabul, and eventually India, while the caravan route north led up the Volga Valley to the land of the Rus and then to Scandinavia.[2]

Their location at the crossroads of continental trade assured the pros-perity of Khwarazm's cities—provided they could channel water from the fast-flowing Amu Darya onto their agricultural land. As early as the sixth century BC, the people of Khwarazm had become masters of

Figure 9.1. Biruni's Khwarazm was a land of fortified towns and castles belonging to rulers and rich landlords. The restored walls at Khiva in western Uzbekistan typify these protective bastions. ▪ From published report *Khorezm Ma'mun Academy*, n.a., issued by the Academy of Sciences of Uzbekistan, under Cabinet of Ministers decree no. 532 (Tashkent, 2006), 18.

hydraulic engineering, diverting whole rivers into freshly dug channels to serve major centers tens of miles away, and dividing them again into canals to provide water to more remote towns. Nowhere on earth were irrigation technologies more highly developed than here.[3] Even today, enormous beds of ancient canals can be traced for many miles through the desert. Thanks to these skills, Khwarazm for a thousand years before Ibn Sina's arrival teemed with large, prosperous cities and the walled castles of patricians (*dihkans*).[4] In a single oasis belt less than fifteen miles across, more than a hundred fortified castles of various sizes have been identified.[5] Since Khwarazm, like much of Central Asia, was, and is, a seismically active zone, builders invented subtle construction techniques to make them earthquake resistant.[6]

These urban centers were in direct contact by land and sea with all Eurasia, sending locally manufactured swords and other products to India[7] and the Middle East, and transshipping forest products from northern Europe and silk and other goods from China. Money generated political power. German and Russian scholars have speculated on the existence before 500 BC of a "Great Khwarazm Empire" extending from the Black Sea eastward to the Tian Shan and south to Herat in Afghanistan.[8] Whether or not such a state existed, when Alexander the Great swept through Central Asia he greedily eyed Khwarazm until the local ruler arrived with a grand entourage to consult with him. Meeting as equals, they struck a deal that caused Alexander to withdraw and head toward India.

All but unknown today, the civilization of Khwarazm between AD 100 and 600 was a highly sophisticated society. Its capital and main religious center, Topraq Qala, in what is now western Uzbekistan, featured a walled, 1100-by-1600-foot rectilinear compound. Elegantly built, it consisted of grand three-storied palaces and temples with formal chambers ornamented with frescoes depicting fifty male and female figures vigorously dancing in couples or alone. Both in subject and in form, these exuberant dancers reflected the influence of India. Other halls were devoted to commemorating past heroes and to martial themes. The walls of the main temple sanctuary also featured built-in benches, on each of which sat a sculpted male figure one and a half times larger than life-size, flanked by standing life-sized statues of females and males. Each of these somewhat eerie groups represented the spirit of a specific day of the month.[9]

The Khwarazmian solar calendar, related to the Zoroastrian system, is known to us thanks to Biruni, who argued that it was in advance of most other ancient systems for measuring time.[10] Religious life centered around local cults linked with Zoroastrianism, traces of which have been detected among rural Uzbeks even today.[11] Fire temples have been found all over the region.[12] Also enriching the religious mix were Iranian-speaking Manichaeans, and Melkite Christians from the ancient

Greco-Roman centers of Antioch, Alexandria, and Jerusalem. The language of Khwarazm, which has only recently been deciphered, was yet another member of the highly diverse Iranian family of languages and came to be written in the Aramaic script.[13]

We know from deciphered documents found at Topraq Qala that slaves formed a significant part of the Khwarazmian labor force.[14] But the economic system as a whole was sufficiently complex as to demand skills of a high order. The technologies of water management, agriculture based on seasonal variations in the water supply, record-keeping for a sophisticated tax system,[15] and the management of a stable national currency all placed a premium on mathematical, scientific, and technical knowledge.[16] Khwarazmians developed all these, and especially mathematics and astronomy, to an exceptionally high degree. We have seen how al-Khwarazmi drew on local traditions in these fields for his pathbreaking work. Others were to build on his achievements, most notably the great Khwarazm polymath Abu Raihan al-Biruni.

The Arabs' Invasion and Its Aftermath

Nowhere was the Arab jihad of the seventh and eighth centuries more destructive of local culture than in Khwarazm. While it is true that the Khwarazmian state showed signs of decay prior to this, the invasion inflicted a near-fatal blow. The conquerors' wrath was sharpened by the fact that it had been in Khwarazm that Central Asian leaders convened to plan their resistance to the Arab onslaught. Libraries, archives, and an entire literature in the Khwarazmian language were put to the torch, and scientists and other bearers of the local civilization systematically killed. As Biruni wrote, the Arab general Qutayba "extinguished and ruined in every possible way all those who knew how to write and read the Khwarazmian writing, who knew the history of the country, and who studied the sciences."[17] When Biruni later compiled his lengthy *History of Khwarazm*, now lost, he had to cull his evidence from the few scraps that survived the conquest.[18]

It took the two centuries from 750 to 950 for Khwarazm to recover from the Arabs' blow. But Khwarazmian civilization had deep roots and it did so magnificently, to the point that well before the year 1000 it had again become a major intellectual center. Indeed, between 980 and 1017 there was no city anywhere that surpassed Gurganj, the new capital of Khwarazm, in the distinction of its scientists and writers (see plate 11). Yes, the two great giants of the age, Biruni and Ibn Sina, were both there. But so too were lesser-known luminaries like the mathematician Abu Nasr Mansur Iraq, the astronomer Abul-Wafa Buzjani, and the medical scholar Abu Sahl Masihi. In the one figure of Abu Khayr Khammar, a Nestorian Christian from Baghdad, were combined great skill as a logician and as translator of ancient Greek philosophical works (Theophrastus, Porphyry) from Syriac to Arabic, and also expertise in medicine, which found expression in his studies on gerontology, pregnant women, and the treatment of diabetes and epilepsy, from which he himself apparently suffered.[19] Even though this effervescence was to be cut tragically short, such a stable of talent gives Gurganj a distinguished place in the history of civilization. It stands with Merv, Nishapur, and Bukhara as a focal point of the early phase of Central Asia's Age of Enlightenment.

Since the fourth century, the shahs of Khwarazm (Khwarazmshahs) had ruled the region. Through careful management they had transformed formal vassalage into de facto self-governance, preserving their independence against the Sassanian Persian empire, the Arab caliphate, and the Samanids. Distance and adroit politics enabled them also to preserve an open intellectual climate. Nominally Sunni but sympathetic to Shiism, the Khwarazmians avoided both the hard religious orthodoxy and radical heresies that prevailed elsewhere.

By the tenth century the shahs of Khwarazm had moved their capital from Topraq Qala to Kath, a swaggering commercial center built near a branch of the Amu Darya near the modern city of Nukus in Uzbekistan. Even in the tenth century the river was eating away at the walled city and would eventually swallow it. But for the time being it thrived in a chaotic way. Here is how the Jerusalem-born Arab geographer Maqdisi (ca. 954–1000) described this city of contrasts:

> The town is indeed magnificent, having scholars, litterateurs, prosperity, agricultural products, commerce. The builders here are skillful, and the like of their reciters of the Quran are not to be found in Iraq. Here is perfection in chanting, excellence in reciting, similarly in appearance and reputation. . . . [But] here are numerous channels along the streets into which they publicly defecate. . . . It is not safe

for strangers to appear abroad until daybreak because of the extent of the excrement; the natives walk in it and carry it on their feet into the communities. The people have a course nature, an ugly disposition; their food is bad, their city vile.[20]

BUZJANI, POLYMATH AND PEDAGOGUE

Amid this blend of magnificence and squalor lived the great mathematician Abul-Wafa Buzjani (940–998).[21] A native of Afghanistan, Buzjani had enjoyed the caliph's support in Baghdad, but after his patron's death he accepted an invitation from the shah of Khwarazm to relocate there. He proved a good investment. Following directly the path of practical mathematics blazed by Khwarazmi a century earlier, Buzjani penned a useful book on mathematics for merchants and civil servants, with concrete advice on calculating land taxes and the management of investments. He also wrote a book on geometry for craftspeople and another on practical mathematics. Since businesses had yet to adopt the Indian number system, he presented his advice in plain prose, or in a form requiring only ruler and compass.

Buzjani is remembered today for his pioneering research on geometry, trigonometry, mathematics, and astronomy. In geometry he advanced new means of solving equations with compass and ruler and figured out how to construct parabola from points and also certain types of hectagons and polyhedra. In trigonometry he is credited with having been one of three researchers to prove the law of sines, along with Khujandi and his student, Abu Nasr Mansur Iraq. Buzjani's method of developing sine and tangent tables produced results accurate to the eighth decimal point, as opposed to Ptolemy's three. And by applying sine theorems to spherical triangles, Buzjani opened the way to the development of new methods of navigating on open water. Besides developing many of the fundamentals of modern trigonometry, he was also the first mathematician writing in Arabic to use negative numbers, a concept known to the Indians and Chinese but rejected by the Greeks. Interestingly, this appeared not in some theoretical work but in his primer on *What Is*

Necessary from the Science of Arithmetic for Scribes and Businessmen, where it was included in a passage on debt accounting.

An accomplished astronomer, Buzjani in 988 had built an observatory at the palace gardens in Baghdad with a twenty-foot quadrant and a fifty-nine-foot sextant, and he probably constructed another such instrument at Kath. His particular interest was the moon, a natural concern since the Muslim calendar is based on lunar months. It was in this connection that he devised his sine and cosine tables, both of which he used to pin down the moon's orbit, and also to chart the declination of certain stars. He gathered much of this research into a pioneering work on mathematical astronomy.

During his later years Buzjani coached an equally talented local astronomer, mathematician, and expert on trigonometry who happened also to be a prince of the royal house of Khwarazm. Abu Nasr Mansur Iraq (960–1036) was anything but an avocational scientist. Working along the same lines in astronomy and trigonometry as his mentor, Iraq took up the knotty problems of spherical astronomy.[22] He delved into the *Spherics* of Menelaus of Alexandria (70–140) and then advanced to Ptolemy's trigonometric calculations with chords, in the process moving trigonometric functions close to those familiar to us today. He also followed Buzjani in developing astronomical and trigonometric tables, focusing especially on improving the data compiled by the Central Asian astronomer Habash Marwazi. His mathematical innovations greatly simplified the task of finding quantitative solutions to problems in spherical astronomy. In addition to writing several treatises on the astrolabe, he compiled a major study on the spherical nature of the heavens that was far in advance of the work of Arab astronomers.[23] Such thinking fed directly into the development and elaboration of the spherical astrolabe or armillary sphere, the most advanced astronomical instrument until modern times.

A Coup in Khwarazm

The decline of Samanid rule tempted the rulers of Khwarazm to reestablish the ancient grandeur of their dynasty and region. But rivalries among potential heirs to the throne in Kath so weakened the state that a competing dynasty based in the city of Gurganj, ninety miles to the northwest, arose to challenge the royal house. Simultaneously two other aspirants emerged on the scene, both of them Turkic, and both of them burning to fill what appeared to be a power vacuum in Khwarazm: the Karakhan clan from present-day Kyrgyzstan and the ruthless new dynasty from Afghanistan led by Mahmud of Ghazni. In what would in the end prove to be a foolhardy show of bravado, the pretender from Gurganj, Abu Ali Mamun, took on them all, and also the Samanids for good measure. The civil war that erupted in 992 ended three years later with the utter rout of the army of the shah of Khwarazms and the collapse of his government.[24] In a few swift moves, Abu Ali Mamun succeeded in marginalizing both Turkic powers as well, but at a fearful price: Mahmud of Ghazni, unaccustomed to defeat, vowed that he would one day crush Mamun's new dynasty and bring Khwarazm under his control.

The Youth of Muhammad Biruni

Back in 973 a comfortable family from Kath had a son, Abu Rayhan Muhammad al-Biruni. Soon thereafter the child was orphaned. Thanks to his family connections, young Biruni was adopted by a prince of the Khwarazm royal house, who raised him with his own son, Abu Nasr Mansur Iraq, who was about thirteen at the time.[25] The two became as brothers, sharing an upbringing in the Khwarazmian language. Iraq was already studying astronomy and mathematics with Buzjani, so it was natural for Biruni to follow in his footsteps, the more so because the orphaned Biruni at an early age showed strong potential in those fields. The close personal and intellectual bonds that arose between Biruni and Iraq were to last throughout their lives. Over the following half century, each was to dedicate no fewer than twelve works to the other.

By age sixteen Biruni had read the *Geography* by the Samanid vizier Jayhuni and set out on his own to calculate the latitude of his hometown of Kath, which he did by using the maximum height of the sun. He also conceived a bold plan to construct a globe that would include natural features, an idea that had occurred to Claudius Ptolemy around AD 150 but was otherwise undeveloped until Martin von Behaim, geographer to the king of Portugal, took it up in 1492.[26] The sixteen-foot globe Biruni constructed was destroyed during Khwarazm's civil war.[27]

Both the globe and the measurements of Kath reflect Biruni's almost tactile interest in what could be seen and measured (see plate 12). More than any other thinker of the Middle Ages—East or West—Biruni embraced and acted on the maxim of Pythagoras, the Greek mathematician of the sixth century BC, that "things are numbers."[28] He was also a tinkerer, inventing what might be called an "observation tube" for viewing heavenly bodies. Though it lacked a lens, it nonetheless focused the eye on a given object and excluded peripheral light. The addition of a lens to Biruni's contraption in the seventeenth century created the modern telescope. He also invented an eight-geared machine for computing lunar and solar calendars. It is revealing that Biruni's education did not touch on philosophy and metaphysics until he was twenty.[29] Rather than dwell on abstractions, Biruni instead began work on what was to be one his major books, *Geodesics*.

The civil war in Khwarazm destroyed more than Biruni's globe. Iraq managed to stay aloof from the struggles and probably remained in Khwarazm, but Biruni, now twenty-three, fled to the Buyid capital at Rayy in Persia. With no money, he took a room with a merchant family from Isfahan and worked at whatever jobs he could find. Fortunately for Biruni, the great Central Asian astronomer and mathematician Khujandi had just built his famed observatory at Rayy, and he now accepted the eager young refugee as his student. In addition to writing an innovative paper on wall sextants, Biruni traced the source of errors in Khujandi's own calculations to the fact that one wall of the structures that housed his sextant had settled.[30]

After two years in Rayy, Biruni learned that it was safe to return to Kath. Back home, he carried out studies demonstrating for the first time how geographical distance and time could both be calculated on the basis of a lunar eclipse, and he used the method to correct measurements of the temporal distance between Kath and Baghdad.[31] Since the capital of Khwarazm had been moved to Gurganj, and since his friend Iraq had made his peace with the new regime and moved to the new capital, Biruni moved there as well. It was then that he initiated the exchange of letters with the "youth," Ibn Sina, described earlier. But there was no work for Biruni in Gurganj, so he once more abandoned

Khwarazm for the city of Gorgan near the Caspian Sea, where he was to spend five years at the court of the local ruler, Qabus. Having only recently returned from eighteen years in exile, Qabus was eager to re-establish his status as a regional power and patron of culture. A willful tyrant, he was also a fine poet and stylist and was content to let Biruni work on his own projects. Thanks to such support, Biruni was able to complete an ambitious volume that had been his main preoccupation in Kath and Gurganj, *The Chronology of Ancient Nations*.

Biruni Invents History

Biruni's *Chronology* is one of the most astonishing works of the Middle Ages and also, for the modern reader, one of the most perplexing. Bulging with chapters on everyone from the Egyptians, Greeks, Jews, Persians, Muslims, pre-Muslim Arabs, Zoroastrians, Khwarazmians, and "Pseudo-Prophets," the *Chronology* appears at first a comprehensive history of the known world, with all the main events in each nation's history carefully enumerated. But then it plunges for more than half its length into excruciatingly detailed discussions of each people's calendar system and even its method of counting the years, months, days, and hours. It is tempting to conclude that Biruni the historian and Biruni the astronomer were two different people who had been forced into a bizarre collaboration.

The *Chronology* undeniably reveals the author's immense erudition. On the historical side, Biruni seems to have pored over the chronicles of all cultures and the holy books of all religions. He delved into their holidays and feasts and recounted how each is celebrated and then drew comparisons across cultural lines and chronological eras. He even reported on what each religion and culture forbade, and he attempted to explain the rational bases for such prohibitions. Frequent references to oral traditions, as well as insights gained from interviews, enriched the written record.

Biruni treated all these diverse sources with a skeptical eye, arguing that people everywhere are prone to falsify their histories and pedigrees, foreshortening and lengthening time to suit their prejudices. "Lies," he wrote, "are mixed up with all historical records and traditions."[32] In the absence of a scientific basis for their edicts, why do different religions defend their positions so vehemently, especially when they are subject to change over time? Muslims at first directed their prayers not toward Mecca but Jerusalem, while Manicheans meanwhile orient their prayers toward the North Pole. Frustrated, Biruni suggested that the evidence "seems to indicate that a man who prays does not need any *Kibla* [the architectural niche that indicates the Muslim direction of prayer] at all."[33]

On the astronomical side, Biruni presented every group's views and assumptions regarding the heavenly bodies and demonstrated how these affected their concepts of time as embodied in their calendar system. In carrying out this analysis, he delved into heretofore unknown or unstudied cosmological constructs, drawing material equally from the scientific worldviews and astrological systems of every culture. Why did pre-Muslim Arabs hold that the day begins at sunrise while other cultures claim that it begins at midnight or noon? The only method that is "compatible with science," he argued, is to measure from the meridian, whether midnight or noon.[34] Why did the Arabs come up with their names of stars, some of which are patently inaccurate? He criticized Muslims for taking over some of the desert Arabs' most unenlightened views on cosmology, fulminating that "it is astonishing that our masters, the family of the Prophet, listened to such doctrines."[35] Biruni was no kinder to Jews and Christians, whose cosmologies on important points were "obscurity itself."[36]

The astronomical issue that most absorbed Biruni's attention was the manner in which each religion or nation adjusted the length of the year to round out the annual cycle. Without it, New Year's Day and all other holidays would gradually migrate through the year. Called "intercalation," this became a kind of litmus test by which Biruni measured the intellectual seriousness of cultures. He praised the Egyptians for the precision of their intercalations, which extend down to seconds. He was less generous toward the Jews and Nestorian Christians, even though their systems of intercalation were widely copied. He noted that the pre-Muslim Arabs had a primitive system of intercalation, and he considered it simply a mistake for the Prophet Muhammad to have explicitly rejected the adjustment of the year to reflect astronomical reality (see plate 13).[37]

Carefully hiding behind the words of another author, Biruni concluded that this decision by Muhammad "did much harm to the people."[38]

To this point the reader may be impressed with the range of Biruni's knowledge and his extraordinary candor but might wonder what grand purpose it all served. In several long passages on the Creation according to the various religions, Biruni finally showed his hand. After carefully reviewing the claims of all the holy books on the date of the Creation and of Adam and Eve, Biruni mischievously summarized them in a single table. He went on to present a series of further tables covering the dates of major historical event in each tradition and, finally, drew these together in various comparative tables. The results were ridiculous and devastating. Not only were the various chronologies incompatible with one another, but in most cases they were simply absurd. It is impossible, he concluded, to date human events from the Creation, Adam and Eve, the Exodus, or any other such event in prehistory.[39] Worse, the various systems of dating more recent events like the birth of Alexander the Great, the life of Zoroaster, or the fall of Rome were a hopeless muddle. Simply stated, history as a rational chronology of events did not exist.

Once he had rejected the pretensions to chronological accuracy of all religions and national mythologies, on what possible basis could Biruni ground historical events? On this core point he was absolutely clear: history must be based on reason, which in this case means the truths of astronomy. Without a rational system of counting time, chronology cannot exist, and without chronology there can be no rational understanding of the past. The only people to have clearly grasped this were the ancient Greeks. Unlike all the religions and national mythologies, the Greeks had been "so deeply imbued with, and so clever in geometry and astronomy, and they adhered so strictly to logical arguments, that they were far from having recourse to divine inspiration."[40]

At this point in his argument Biruni moved from description to prescription, proposing steps by which the age-old mess created by religion and national mythologies could be corrected, or at least alleviated. His method was to create a system for converting dates from one system to another. Today his system would be embodied in a simple computer program. A millennium ago Biruni presented it in the form of a chessboard including eras, dates, and intervals. Anyone who was "more than a beginner in mathematics"[41] could manipulate the chessboard so as to translate from one system to another. The method, he boasted, would be equally useful to historians and astronomers.[42]

The immediate purpose of Biruni's chessboard was historical, namely, "to fix the durations [of the reigns of kings] by the most correct and perspicacious method."[43] He was well aware that commercial interchange among diverse peoples requires a common system of dating events,[44] and that all interactions among peoples require some common system of reckoning the passage of time.

Prior to the appearance of Biruni's *Chronology*, there had been no universal history, nor could universal history have been written, because there existed no unified matrix for measuring time that extended across religions and civilizations.[45] Biruni's was the first global calendar system and hence the essential tool for the construction of an integrated global history. To be sure, he failed to include the Chinese, Indians, and other more distant peoples in his sample of civilizations and calendar systems. But his reason for excluding them was as simple as it was rational: "I have not met with anybody who had an accurate knowledge of this subject. Therefore I turned away from what I cannot know for certain."[46] Biruni, by grounding his concept of human history on the solid firmament of astronomy and reason, gave all peoples of the world a simple method for fixing dates on a single calendar system. Not until recent centuries have thinkers fully applied the concept of a universal history to which Biruni's *Chronology of Ancient Nations* opened the path.

The massive *Chronology* was but one of fifteen books and papers Biruni turned out at Gorgan. These included a weighty *Keys to Astronomy*, a book on the astrolabe, four tracts against astrology, the "art of deception," and a pair of historical studies. One of his astrological studies contained a stunning apologia for interdisciplinary research and for science as such. It has often been claimed that in Islam the concept of "science for the sake of science" or "pure science" does not exist.[47] Yet this is precisely what Biruni championed:

> The duty of the servant of science is not to differentiate among its kinds, even though it may not be easy for him to [master] all its branches. Rather, he should know that in an absolute sense science is good in itself, apart from its [content of] knowledge, [and] that its lure is everlasting and unbroken. . . . He [the servant of science]

should also praise the assiduous [ones] whenever their effort [arises from] delight [in science itself] rather than from [the hope of achieving] victory in argument.[48]

For all these works, as for the *Chronology*, Biruni drew on massive materials he brought from Khwarazm or had sent from there by his friends Iraq and Masihi.[49] But as Biruni's star rose, his royal patron grew more demanding, to the point that he was monopolizing Biruni's time. When Biruni resisted, Qabus turned down his request for money to fund an expedition to measure by land Earth's meridian angle.[50] Frustrated in Gorgan, and attracted by what he heard had become a very hospitable intellectual climate at Gurganj, Biruni returned to his native Khwarazm.

Gurganj of the Mamuns

Khwarazm's new capital was teeming with activity when Biruni arrived there in 1004.[51] A huge new wickerwork dam diverted the entire Amu Darya past the city walls, rendering the city comfortable and feeding both mills and cool gardens in the suburbs. To celebrate his new dynasty, the founder, Abu Ali Mamun, had embarked on a building spree that his successors continued and expanded.[52] Over the following two decades the new rulers constructed for themselves a large palace near the northern gate, a great new mosque, and, nearby, a giant "victory" minaret.[53] Although it was his younger son, Abu Abbas Mamun who erected the minaret, Abu Ali Mamun already had good reason to celebrate, for he had claimed the lands and title of the shahs of Khwarazm and fended off foreign invaders, including the most serious threat to the regime, Mahmud of Ghazni. His elder son, Abu Hassan Mamun, continued to keep Mahmud at bay by accepting Mahmud of Ghazni's sister as a wife, while Abu Hassan's brother and successor, Abu Abbas Mamun, did the same with another of Mahmud's sisters. All this was done in hopes of sustaining the peace. Only too late did the rulers of Khwarazm realize that Mahmud had been playing a cat-and-mouse game with them. But by then he was poised to strike.

The founder of the new dynasty, Abu Ali Mamun, set about to revive regional culture and intellectual life. He began by assembling writers and thinkers. A contemporary wrote that he was so enthusiastic a champion of the world of learning and especially of its new rationalist currents that he sometimes breached the Muslim faith.[54] This may have been a factor leading to his assassination in 997. But both his sons worked with equal vigor to revive and enhance Khwarazm's standing as a center for science and literature. Their first priority was to complete the reconstruction and expansion of the library of the shahs of Khwarazm, the great collection that the Arabs had destroyed 250 years earlier. Thanks to the Mamun family, a medieval writer was able to record that the library at Gurganj "had no equal, either before or afterwards."[55] Because of its large holdings and the fact that it sponsored the work of translators, the library became a magnet for writers and scholars from many fields, effectively filling the void left by the fire at the Samanid library at Bukhara.

For all the interest of the Mamun brothers, Hasan and Abbas, the cultural effervescence in Gurganj would not have reached its great heights without the enthusiastic leadership of the vizier, Abu Husain al-Sakhli. A wealthy local patrician or *dihkan* and avid promoter of literature and the arts, Sakhli wrote poetry in both Arabic and Persian, including a quatrain on *The Stars*. Ibn Sina, who in this period was careful with his dedications, inscribed a number of treatises in Sakhli's honor.[56] It was Sakhli who came up with the idea of holding scientific evenings at Mamun's palace on the north of town. These assemblies, which have anachronistically been referred to as "the Mamun Academy," were in fact much like the Barmaks' gatherings in Baghdad and intellectual evenings in Merv, Balkh, Nishapur, or Bukhara. Along with recitations and discussions, they featured debates and contests of various sorts, all designed to showcase the luminaries present. The one difference is that none of the others could have claimed to include the two greatest minds of the Middle Ages, Biruni and Ibn Sina, not to mention a legion of other major poets, writers, and scientists. During its brief existence Mamun's academy was the intellectual center of the world.

Thanks to the diligence of scholars in Uzbekistan, the roster of thinkers and writers at Mamun's Gurganj is being rescued from the shadows of time.[57] Abu Sahl Masihi, who had spent time in Bukhara and who was said to have first aroused Ibn Sina's interest in medicine, was among the many thinkers resident there. He was working at the time on his *Book of the Hundred*, a massive and carefully organized compendium of medical knowledge that doubtless prompted Ibn Sina to write his own compendium, the *Canon of Medicine*.[58] Masihi's enormous work was heavy on theory but short on therapeutics, which may explain why it was never translated into Latin. But in his passion for organizing and simplifying, Masihi opened the way for Ibn Sina's more accessible work. Also participating actively was Biruni's friend, the mathematician and astronomer Abu Nasr Mansur Iraq. The sciences were represented also by chemists,[59] several astronomers besides Biruni and Iraq, and a cadre of mathematicians. Among the latter was the elderly Abu Khayr Khammar, whom we have met as a medical researcher and translator, but who doubled as a mathematician who was known for his work on the spherical sine law and for coming up with the theorem on the properties of isosceles triangles.

There were also amateurs among the learned thinkers who assembled at the Mamun court. Among these were the diplomat Abu Abdullah Muhammad ibn Hamid, who doubled as a poet, and Abu Ali Hasan al-Khwarazmi, a judge by day and by night an expert on the algebra needed to calculate inheritances. Most participants, including the large group of poets at Gurganj, were drawn from Central Asia.[60] But so great was the attraction of Mamun's Gurganj that top scholars came there from elsewhere as well. This welter of activity in diverse spheres of knowledge invited attempts to make sense of it all. One to do so was Abu Abdullah al-Khwarazmi, who authored a major work that organized and classified all the various branches of knowledge according to an overarching rational system.[61]

Were the thinkers at Mamun's academy chosen on the basis of their adherence to some common religious or philosophical stance? It is true that arch-rationalist Mutazilites were a powerful force in Khwarazm, so much so that it was assumed that any Khwarazmian was also a Mutazilite,[62] a view that persisted there down to the Mongol invasion. It was in Khwarazm that Mahmud al-Zamakhshari (d. 1144), author of a famous Mutazilite commentary on the Quran, was born and flourished. In spite of this, there was no "party line" to which thinkers at Mamun's academy had to adhere. The luminaries included Mutazilites and anti-Mutazilites, Platonists and Aristotelians, pious believers and freethinkers. Two of the major participants, Abu Sahl Masihi and Abu Khayr Khammar, were Christians. In short, the intellectual climate at Mamun's Gurganj was notably ecumenical and open, with none of the narrow traditionalism and mainstream orthodoxy that were even then on the rise elsewhere in the Islamic world. The one common element among these diverse figures was a commitment to the unconstrained exercise of reason. It was reason that had proven the world to be round, and it was reason that would now bring enlightenment through all spheres of knowledge.

SUPERSTARS IN SEPARATE ORBITS 1004–1010

This, then, was the world into which Biruni plunged after his return from abroad in 1004. Now thirty-one years old and hailed everywhere for his *Chronology*, he entered on a period of peaceful but intense work in many fields, including hydrology, geodesics, and mineralogy. His systematic approach was evident in mineralogy, where findings from this period ended up several decades later in his seminal book on the field. In astronomy, Biruni's work was facilitated by his large new ring-shaped instrument for measuring the meridian of the solar transit and a hemisphere sixteen feet in diameter, which he used for plotting graphical solutions to geodetic problems.[63] Amid all these activities, he also found time to pen a defense of Khwarazmi's algebra and a critique of the mathematician Habash, along with proposing an original method of determining the chord of arcs.[64]

Less than a year after Biruni's return to Gurganj, Ibn Sina arrived there as well. He had been driven there by necessity and had important administrative duties to fulfill. As an agent of the Karakhanid Turks who now ruled Bukhara, he was bound to be an object of suspicion, at least as first.[65] But his fame had preceded him, and after a short time he, too, was part of Mamun's academy, participating in the soirees organized by the vizier, Sakhli. There is no evidence of contacts between Ibn Sina and Biruni. But since they were both in a city that one could walk across in a half hour, they doubtless met. The absence of reports on their interaction suggests that their relations were unremarkable. More important, Ibn Sina once again could treat patients and forge ahead in his medical research. So productive was this respite for Ibn Sina that after a half dozen years he had assembled all the materials needed to write his masterpiece, the *Canon of Medicine*.

This peaceful interlude was not fated to last. The death of Abu Hasan Mamun in 1009 set in motion a whirlwind of events that would eventually destroy the learned world the Khwarazm shahs had so carefully constructed, bring down their dynasty, cause the sacking of Gurganj, and throw the lives of both Ibn Sina and Biruni into turmoil. The driving force behind this drama was the wily and ruthless Mahmud of Ghazni, the sharp-clawed cat who had been biding his time while the mice played their intellectual games at Mamun's court.

Abu Abbas Mamun took over as shah of Khwarazm after the death of his brother in 1009. Like both his father and brother, Abu Abbas was well educated and a patron of culture. He saw to it that the academy continued almost unchanged. Biruni praised him as "learned, brilliant and steady in affairs of state" but went on to charge him with pursuing naïve and self-destructive policies.[66] In fact Abu Abbas did little to bolster the economy or security of his realm and was utterly profligate with money. His main strategy was to play for time.

This became evident as early as his coronation. From far-off Ghazni in Afghanistan, Mahmud sent his congratulations and also the offer of another of his sisters as a wife to the new ruler of Khwarazm. Abu Abbas pocketed the proposal for five years, by which time it was obvious that his delayed acceptance was a sign of growing weakness. Also at the coronation, the caliph in Baghdad conferred on Abbas the honorific title of "Eye of the State and Ornament of the Faithful."[67] Fearful that Mahmud would interpret his acceptance of this honor as an assertion of sovereignty, Abu Abbas Mamun sent an envoy to intercept the caliph's delegation en route and quietly bring the document and robes back to Gurganj. For this delicate diplomatic mission he selected Biruni. Mahmud must have smiled at Abu Abbas's display of weakness.

During or shortly after the coronation, Mahmud made a third and far more ominous move. In a letter oozing with venomous sweetness, Mahmud wrote Abbas that he had heard of the glorious scholars and scientists at the Gurganj court and requested that Abu Abbas share their genius with him by dispatching Ibn Sina and a group of others to Ghazni.[68] It was clear that Mahmud did not have in mind a brief social visit. In effect he demanded that Mamun close his academy and send its luminaries to Mahmud's court in Afghanistan.

According to a near-contemporary historian from Samarkand, Nizami Arudi, Abu Abbas called the scholars together and told them, "Mahmud is all-powerful. He has many troops and has already taken Khorasan and India and is sniffing at Iraq. I cannot turn a deaf ear to his order. . . . What do you say to that?"[69] Ibn Sina had already been driven from his home once by Turkic invaders, and he now foresaw a lifetime in the jaws of the enemy. He and Masihi refused outright. Wasting no time, they immediately fled Gurganj, striking out on their own onto the desert west of Gurganj, without a caravan. Biruni and the others were no more enthusiastic but did not flee, preferring instead to stay put in Gurganj. The shah of Khwarazm tried to buy off Mahmud by sending some minor figures to Afghanistan, but Mahmud was unrelenting.

When Mahmud learned of Ibn Sina's flight, he sent letters to rulers throughout the region ordering his arrest and even demanding his head.[70] But with a long head start, Ibn Sina managed to elude his pursuers. After a series of harrowing adventures that are recorded differently by every medieval writer on the subject, Ibn Sina reached the city of Gorgan near the Caspian, the very place where Biruni had spent six productive years. Fortunately, Ibn Sina's voluminous notes and papers also reached Gorgan, and he was soon able to turn his full attention to writing his magisterial work on medicine. His colleague Masihi did not survive the rigors of the trip, however, and died in a sandstorm in the Dekhistan Desert.

THE CANON OF MEDICINE

Ibn Sina's masterpiece, the *Canon of Medicine*, did not appear out of thin air. Even though it came to be acknowledged as the peak of Islamic medicine, it had important predecessors that shaped both its form and its contents. We have seen how the Barmak family supported medical research in Baghdad and also noted the strong medical centers at Merv and Nishapur, and the presence in Bukhara of a whole band of medical practitioners during the years Ibn Sina was growing up there.[71] Many other leading medical experts were working at Balkh, Samarkand, and Gurganj.

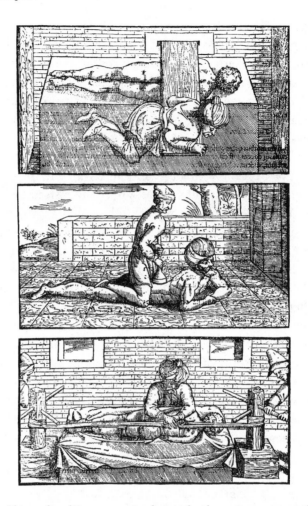

Figure 9.2. This medieval European print depicts the alternative treatments for spinal disorders detailed by Ibn Sina in his *Canon of Medicine*. ▪ Dr. Jeremy Burgess/Science Source/Photo Researchers.

By Ibn Sina's generation the focus shifted from individual diagnoses and cures to systematizing the medical enterprise as a whole. The pivotal figure in this enterprise was Muhammad ibn Zakariya al-Razi, the Merv-trained doctor from Rayy, the city whose ruins adjoin modern Tehran. The feisty Razi was said by contemporaries to have taken Socrates as his imam.[72] He had no doubt that reason was the best tool for attaining knowledge, which is why he recommended a "spiritual physic" to purge the mind of all irrational passions, including faith.[73] He also bombarded the public with tracts on such topics as "On the Fact That Even Skilful Physicians Cannot Cure All Diseases," and "Why People Prefer Quacks and Charlatans to Skilled Physicians."[74] Lest anyone miss his point, Razi opened his greatest work with a passage on "The Necessity of Death."

Razi was a fiercely focused clinician. His study of smallpox and measles is a classic, undergoing no fewer than forty editions in Europe alone between 1475 and 1866. He was also a dedicated experimentalist, and the first scientist anywhere to organize research projects with systematic controls of the sort we consider normal today. Razi's hard-headed rationalism was particularly evident in his comments on fraudulent treatments: "Since many wicked people tell lies with regard to the properties [of medicines], it would be useful not to leave these claims scattered but to collect them and write them up. . . . We shall not accept any property [of a medicine] as authentic unless it has been examined and tried."[75] He even used himself as a laboratory, preparing careful notes on his self-diagnosis and treatment of a bad fever or a swollen testicle.[76]

When Razi died, his students organized his mass of notes into an encyclopedia, called *The Comprehensive*. Included in this immense compendium were entries on virtually every known medical problem, including the views of Galen and other ancients and insights gained from Razi's own clinical practice and research. Only a half-millennium later (in 1486) did part of *The Comprehensive* reach Western readers. Yet even after such a long interval, European medical professionals seized on it as essential reading. But at twenty volumes (only ten of which survive), Razi's giant compendium was simply too expensive to duplicate, and few copies circulated even in the Muslim world.[77] For all practical purposes, Razi's compendium was inaccessible, bibliographically dead.

This opened great prospects for Ibn Sina. The fact that medicine was not a field in which his rival, Biruni, claimed any special expertise doubtless made it all the more attractive. And so Ibn Sina began work on his *Canon* in about 1012, completing it a decade later. There are obvious similarities between Ibn Sina's *Canon* and Razi's *The Comprehensive*: the summary in each entry of the views of Galen and other ancient sources; the inclusion of the author's own clinical observations; the definition of medicine, in Ibn Sina's oft-repeated phrase, as "the art of removing impediments to the normal functioning of nature"; and the prevailing faith in rationalism. Like Razi, too, Ibn Sina at the same time penned related studies on specific problems, among them his *Ten Treatises on the Eye*. This, the first systematic textbook on ophthalmology, was also the first work to present the physiology of eye movement, the first to point out that the lens is flat, not curved, and the first to describe the functioning of the optic nerve. He also wrote an important separate treatise on medicines he deemed beneficial to cardiac health.

It is tempting to enumerate the various ways the *Canon* surprises with its modernity. Dr. Ibn Sina emerges from his pages as a champion of good health through exercise and sleep, the medicinal value of mineral water, the positive influence on health of a good marriage, and the impact on health of climate and the environment. He also called for the early surgical treatment of cancer, the testing of new drugs on animals, and the use of alcohol as an antiseptic.[78] Other passages speak directly to issues of public health arising from residence in humid or maritime conditions, high altitude, and so forth.[79] Separate sections address problems of sexually transmitted diseases, while others discuss factors contributing to high blood pressure and means of reducing it. He also cautioned about the contagious nature of tuberculosis and called for the boiling of water to prevent the spread of disease.

A particular strength of the *Canon* is its extensive treatment of pharmacology. No fewer than seven hundred drugs and medicines, most of them natural products, are discussed on its pages and their effectiveness evaluated on the basis of actual experience or clinical trials. Much of Ibn Sina's material in this section was original. Along with local drugs and preparations known to both the ancient and modern experts in the Mediterranean world, he also considered drugs from India and China. Going beyond the simple question "Does it work?" Ibn Sina demanded to know for each drug how fast it works, under what conditions it works, at what strength it is effective, and for which categories of medical problems it is appropriate.

Ibn Sina was not thwarted by the religious and moral concerns that prevented dissection in the West down to the seventeenth century.[80] He considered it essential to an understanding of the functions of the various organs. And while he seems not to have engaged in much dissection himself, he drew on his practice of surgery for many entries in the *Canon*, including his discovery of the *cerebellar vermis* in the brain, which controls locomotion, and his conclusion that the frontal lobe of the brain might control the exercise of reason.

Ibn Sina's *Canon* was immediately recognized as a very useful book, not least because its moderate size—one thick volume in five parts—rendered it manageable. Any practicing doctor could find in it enlightening analyses and useful advice. Arabic readers ordered their own copies and quickly assimilated its rational and analytic method. Those in the region who could not read the Arabic original soon had access to a Persian translation. Within a generation the *Canon* was solidly established as the most important text for the teaching and practice of medicine throughout the Eastern world. Other worthy treatises, such as the concise guide written by Ali Ibn Hindu from Rayy, simply went by the board. Well trained by Abu Hassan Amiri from Nishapur and Ibn Sina's colleague from Gurganj, Khammar, Ali Ibn Hindu penned a solid textbook,[81] but it disappeared under the tide of interest in the *Canon*.

A few medical experts found fault with Ibn Sina's work. One Arab doctor in Spain was so incensed by the second book that he cut off the margins of his copy and used the paper to write prescriptions.[82] Other scientists in the Arabic world defined their research on the basis of gaps in the *Canon*. One to do so was Abu Hassan al-Nafis from Damascus, who, inspired by Ibn Sina, pinpointed the arteries to the heart and described for the first time the process of pulmonary circulation. He published his findings in a book respectfully entitled *Commentary on Anatomy in Avicenna's Canon*.[83]

It took a century and a half for the *Canon* to reached European readers. But after the appearance around 1180 of a Latin translation by Gerard of Cremona, the indefatigable translator of Khwarazmi and Farabi, universities across the continent adopted it as their basic medical text. Some continental institutions continued to teach from it down to the seventeenth century. The Doctour of Phisik in Chaucer's *Canterbury Tales* boasted of having read it, presumably as a student. Dante, in his *Inferno*, placed Ibn Sina in "limbo," the benign waiting room outside Hell proper that was reserved for the most noble non-Christians deemed worthy of salvation; also there were Aristotle, Socrates, and Euclid—as well as the Amazon queen Pentheselia for good measure. The first of several dozen printed editions of the *Canon* appeared in Latin in 1473.

Figure 9.3. In his engravings, the eighteenth-century German Georg Paul Busch celebrated leaders of the early European Enlightenment. However, he also reached back seven hundred years to salute Avicenna (Ibn Sina) as a kindred Enlightenment spirit.
■ Courtesy of U.S. National Library of Medicine, History of Medicine Collection.

The impact of the *Canon of Medicine* reached far beyond the Middle East and Europe. A Chinese translation appeared in the fourteenth century, by which time Indians had already been reading it for two hundred years. Under Ibn Sina's influence, Indian doctors devised an entire system of "Greek" healing, called "Unani" from the Hindi word for Greece, with the *Canon* as its cornerstone. In due course Unani medicine was enriched by direct contact with the works of Galen and Hippocrates, and also Razi and al-Nafis, all the while preserving the system of humors that Ibn Sina had adapted from the Greeks. Even today there exist in India a number of colleges and hospitals specializing in Unani medicine, in addition to a government-sponsored National Institute of Unani Medicine at Bangalore.

Returning to the *Canon* itself, one further characteristic of Ibn Sina's great project must be noted: its clear overall structure, which derived from the author's conception of medicine as a whole and of its place in the larger structure of human knowledge. Here the modern reader is in for a surprise. Whereas Razi considered medicine as a noble and self-standing body of knowledge, Ibn Sina viewed it merely as a "derivative and practical science." And if Razi saw medicine as the summation and pinnacle of human learning, Ibn Sina considered it "not worthy of inclusion in the roster of theoretical sciences."[84] The one point where medical science reached higher, claimed Ibn Sina, was in its theory of humors and temperaments, in other words, in the one part of the *Canon* that strikes readers today as the most "medieval."

Such views would have struck Razi as obscurantist nonsense. Unlike Ibn Sina, he saw medicine as an end in itself, a noble calling that owed its place in the order of things solely to its own achievements and not to its position with respect to some larger and purely hypothetical structure of knowledge (see plate 14).

Both Ibn Sina and Razi accepted only those medical truths that were verifiable on the basis of observation and experiment. Both would have agreed that diagnostic work—as opposed to Ibn Sina's talk of humors and temperaments—relied mainly on inductive logic. But where Razi triumphantly embraced inductive reasoning, Ibn Sina at times seemed almost to apologize for it. At one point in the *Canon* he contrasted the views of the "physicians" to those of the "philosophers,"[85] with the clear implication that the latter were working on a much higher plane due to their more complex systems of logic and argumentation. For all Ibn Sina's preoccupation with the cases before him, his deepest interest was not in specific diseases and cures but in the broad categories of humors and temperaments that, he argued, were the essential armature on which all medical knowledge was built.

Ibn Sina's whole edifice of knowledge pointed upward, first to the all-encompassing category of "physics" or the natural sciences and, beyond that, to metaphysics and religion. Razi, the religious skeptic and foe of Platonic mumbo-jumbo, reveled in the uniqueness of every case before him. Ibn Sina, by contrast, acknowledged that single cases may be instructive but considered them ultimately ephemeral, as opposed to the all-encompassing truths that lie above and beyond individual phenomena. Razi was gloriously content to be a medical man. Ibn Sina yearned to be something more and believed that higher realms of thought beckoned him. This nobler realm, however, was concerned less with the paradigms of science than with existence itself. Philosopher Dimitri Gutas sees in Ibn Sina's stance "the evil legacy of Neoplatonism," with its neglect of the specific and its endless quest for the eternal and abstract forms that supposedly underlie reality and our thinking about it.[86] This was not Ibn Sina's view in the *Canon*, of course, but it became so as he gradually shifted his main focus away from medicine and toward philosophy, metaphysics, and religion.

The Fall of Gurganj

While Ibn Sina was quietly working in Gorgan, the situation in Gurganj grew steadily more ominous. The vizier, Abu Husain al-Sakhli, who apparently favored a more active response to Mahmud, had a falling out with his ruler and fled Khwarazm for Baghdad in 1013.[87] The following year yet another letter arrived in Gurganj from Mahmud in Ghazni. Sensing the moment to assert his control over Khwarazm, Mahmud demanded that his name, rather than that of Abu Abbas Mamun or the caliph, be read each Friday at every mosque in Khwarazm.

Abu Abbas knew full well that this ritual blessing of the secular authority, called the *khutba* in Arabic, had immense symbolic meaning for the populace, as well as for Mahmud. Fearing the distant enemy more than his own people, the shah of Khwarazmi announced that he intended to honor Mahmud's request.[88] This proved a near-fatal mistake, for the entire patrician class, as well as the leadership of his army, erupted in a fury of rage. Catching wind of this impending rebellion, Mahmud shot off yet another missive, this one in the form of an ultimatum, informing Abbas that if he could not keep his own patricians and officers in line, he, Mahmud, would do so for him.

Plate 1. The walls of Balkh today. Mighty Balkh, known as the "mother of cities," gained immense wealth and cosmopolitan sophistication by trading with India, China, and the Middle East. ▪ Kenneth Garrett/ National Geographic Stock.

Plate 2. Tenth-century residence at Sayod, Tajkistan. Both before and after the Arab invasion, wealthier urban residents of Central Asia lived in comfortable homes with brightly painted walls, running water, and systems for cooling and heating. ▪ Interior decor, Sayod, Hamadoni district, Khatlon region, 9th–11th centuries. From Hamrokhon Zarifi and D. Nazriev, eds., *Tajik Arts and Crafts: Through Centuries*, 2nd ed. (Dushanbe, 2010), 190. Courtesy of His Excellency Hamrokhon Zarifi.

Plate 3. Papermaking in Khwarazm. Central Asians invented usable paper by making it from cotton rags rather than overly absorbent and stiff mulberry or bamboo fibers, as the Chinese had done. It was the Central Asian's high-end product that spread westward. ▪ From published report *Khorezm Ma'mun Academy*, n.a., issued by the Academy of Sciences of Uzbekistan, under Cabinet of Ministers decree no. 532 (Tashkent, 2006), 67.

Plate 4. Fresco from sixth-century Afrasiab, now Samarkand. The worlds of statecraft, diplomacy, national epics, and religion intermingle on the lively murals that adorned the palaces of local rulers. ▪ Photo by S. F. Starr.

Plate 5. Fragment of a mural from a house at Balalyk-Tepe, Tajikistan. In pre-Islamic Central Asia the life of the mind flourished amid conviviality and sociability. ▪ From Guitty Azarpay, *Sogdian Painting: The Pictorial Epic in Oriental Art* (Berkeley, 1981), plate 3.

Plate 6. A page from a fifteenth-century Egyptian copy of an astrological treatise by Abu Mashar of Balkh, an astrologer known in the West as Albumashar, who did more than anyone else to introduce, or reintroduce, Aristotle's works to Europe.

Plate 7. Quranic prohibitions against depicting the human form were frequently ignored, as in this alluring painting of a woman, one of many from a bathhouse at Nishapur. ▪ From Charles K. Wilkinson, *Nishapur: Some Early Islamic Buildings and Their Decoration* (New York, 1986), fig. 33, The Metropolitan Museum of Art. Image © The Metropolitan Museum of Art.

Plate 8. The Persian world had long celebrated its ancient heroes, but Ferdowsi's *Shahnameh* greatly stimulated that activity, as in this depiction of an early seventh-century siege of Aleppo in Syria. ▪ "Anushirvan besieges the Rumis in Aleppo," 942/1536, BL, Add. 15531, 460b. © The British Library Board.

Plate 9. Miniature Central Asian painting depicting a central moment of Ferdowsi's epic *Shahnameh*, when the hero Rostam unknowingly killed his brave son Sohrab. ▪ Courtesy of the Bodleian Library, University of Oxford. Ms. Ouseley Add. 176, fol. 92a.

Plate 10. Tenth-century Tomb of Ismail Samani at Bukhara, with both Zoroastrian and Buddhist design elements, as well as hints of classicism and nomad ornamentation. ▪ Photo by Alastair Rae.

Plate 11. Thinkers who gathered at Gurganj in Khwarazm around AD 1000 included some of the greatest minds between antiquity and the Renaissance. This contemporary miniature shows some of them in earnest discussion. ▪ From published report *Khorezm Ma'mun Açademy*, n.a., issued by the Academy of Sciences of Uzbekistan, under Cabinet of Ministers decree no. 532 (Tashkent, 2006), 64.

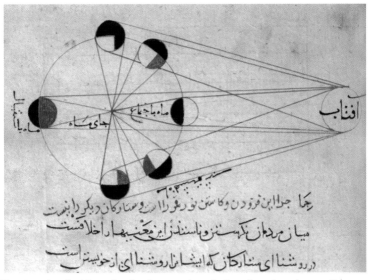

Plate 12. Biruni was barely thirty when he discovered a technique for using a lunar eclipse to calculate both geographical distance and time. ▪ From a manuscript, *Kitab al-Tafhim*, setting forth this process. ▪ Reproduction from Seyyed Hossein Nasr, *Islamic Science: An Illustrated Study* (World of Islam Festival Publ. Co., 1976).

Plate 13. The Prophet Muhammad preaching. Of relevance to science was his attack on intercalation—the annual addition of extra hours or days to assure that the calendar precisely adheres to solar time. Biruni severely criticized this stance, claiming it had caused much harm. ▪ Courtesy of the Bibliothèque nationale de France. Or. ms. Arabe 1489, fol. 5v.

Plate 14. Ibn Sina dismissed Razi by saying he should have stuck to analyzing urine samples, but he himself considered urine analysis to be a valuable diagnostic tool, as shown by this illustration from the *Canon*, depicting patients lining up to present their glass beakers to the great physician for his diagnosis. ▪ Courtesy of the U.S. National Library of Medicine, History of Medicine Collection.

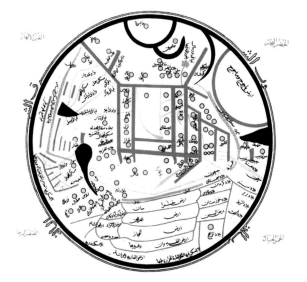

Plate 15. The circular format of Mahmud of Kashgar's ethno-linguistic map of the Turkic peoples was outmoded by 1074, but he broke with convention by placing its center at Balasagun rather than Mecca, and by including the first known map of Japan. ▪ From Serik Primbetov, *Atlas Turan* (Almaty, 2008). Courtesy of Serik Primbetov.

Plate 16. Even though tower minarets were a Turkic invention, architects from the various Persian groups were soon building them as well, as in this minaret from 1108–1109 in Jarkurgan, Uzbekistan, designed by architect Ali bin Muhammad of Sarakhs. ▪ Courtesy of OrexCA, Tashkent, Uzbekistan, www.OrexCA.com. Photo by V. Jirnov.

Plate 17. A diminutive Mahmud of Ghazni receives an honorific robe from the caliph, whose authority and throne he later came to covet. ▪ From Rashid al-Din, *History of the World*. Or. ms. 20. fol. 121r, Edinburgh University Library, Special Collections Department.

Plate 18. The murder of Nizam al-Mulk on October 14, 1092, by Ismaili assassins (shown here in an early miniature painting) triggered the chain of events that led to Ghazali's crisis. ▪ Courtesy of the Topkapi Palace Museum. TSMK. H. 1653, fol. 360b.

Plate 19. Sultan Sanjar's piety, reflected in this picture of him giving alms, could not save the final Seljuk ruler from a humiliating and bitter fate. ▪ Attributed to Bihzad, "The Old Woman Petitioning Sultan Sanjar," from *Khamsah of Nizami*, c. 1490, fol. 18r, BL, Add. 25900. © The British Library Board.

Plate 20. Sanjar's vast mausoleum at Merv. Its ingeniously engineered double dome, visible at a full day's journey from the capital, launched a movement that created prototypes for such domes in the West. ▪ Peretz Partensky/Langton Labs.

Plate 21. A Mongol siege in progress. Mongol attackers employed carefully constructed siege engines and pots of burning naphtha to break the defenders' will. ▪ From Rashid al-Din, *History of the World*. Or ms. 20 fol. 124v. Edinburgh University Library, Special Collections Department.

Plate 22. Chinggis Khan as portrayed by a Chinese artist in his own century, suggesting more the strategist and builder than the bloodthirsty field commander. ▪ Reproduced by permission of National Palace Museum, Taiwan, Republic of China.

Plate 23. Nasir al-Din al-Tusi (1201–1274), Khurasan-born polymath and founder of the Maragha observatory under the Mongols, at work with his multinational team of scientific colleagues. ▪ Persian miniature. Folio number 1418, Rare Works of Art Library, Istanbul University Library, Instanbul. Courtesy of the Istanbul University Library.

Plate 24. Timur, known in the West as Tamerlane, as depicted by later court painter Kamoliddin Bihzad (1450–1537) from Herat, Afghanistan. ▪ "Timur Receiving the Subject Princes at his Ascension," from *Zafarnama*, c. 1485, ff82v–83r. The John Work Garrett Library, The Sheridan Libraries, Johns Hopkins University.

Plate 25. Timur covered the entire outer and inner walls of his buildings with brightly colored tiles, a practice that soon spread throughout the Muslim world. ▪ Photo by Brian J. McMorrow.

Plate 26. This mega-mosque, named for Timur's Uyghur wife, Bibi Khanym, had barely been completed when its giant arches began to collapse. This photograph shows the sprawling structure after it was reconstructed during the Soviet era. ▪ Photo by Brian J. McMorrow.

Plate 27. Timur was often his own architect and construction manager, as seen from this contemporary painting of him driving the workmen at the Bibi Khanym mosque, Samarkand. ▪ Attributed to Bihzad, "The Construction of the Masjid-i Jami in Samarkand," from *Zafarnama*, c. 1485, ff359v–360r. The John Work Garrett Library, The Sheridan Libraries, Johns Hopkins University.

Plate 28. The career at Herat of the artist Bihzad coincided with the lives of Leonardo da Vinci, Michelangelo, and Bellini. Later Islamic rulers collected Bihzad's works the way European monarchs collected Renaissance masterpieces. ▪ Detail from "Two Seated Men (drawing attributed to Bihzad), with Verses about the Inconstancy of Prophetic Vision by Sa'di; folio from an album." Harvard Art Museums/Arthur M. Sackler Museum, Gift of Philip Hofer in honor of Stuart Cary Welch, 1972.299. Photo by Katya Kallsen. © President and Fellows of Harvard College.

Plate 29. Attempts to conduct astronomical research in the Ottoman empire, like that depicted here, fell prey to opposition and indifference, as happened also among the Saffavids in Iran and Mughals in India. ▪ From *Sehinsename* (Book of the King of Kings). Folio number 1404, fol. 57a, Rare Works of Art Library, Istanbul University Library, Istanbul. Courtesy of the Istanbul University Library.

Facing rebellion and worse, Abu Abbas again turned to Biruni to mollify the dissidents. Entering the lions' den, Biruni confronted the enraged rebels with what a contemporary called "a tongue of silver and gold."[89] The compromise he proposed was to include Mahmud's name in the *khutba* in all cities of Khwarazm *except* Gurganj and Kath. Knowing Mahmud's reputation for greed, he also proposed to send eighty thousand gold pieces to Ghazni as a gesture of goodwill.

Biruni was now acting as de facto vizier, with his sovereign deeply dependent on him as the one man who could help him wriggle out of Mahmud's trap. Biruni's solution was to involve a third power in the face-off, the Karakhan clan from the Tian Shan region to the east. The Turkic leader of this rising power had maintained cordial relations with Mahmud and sealed them by sending his daughter to be one of Mahmud's wives. Just beneath the surface, however, he, Mahmud, and the shah of Khwarazm were locked in the three-way struggle over the remains of the Samanid state. This competition was in turn complicated by the fact that the Karakhanids were themselves divided into competing clans, each of which was tempted to cut a secret deal with Mahmud against the other. This had prevented any kind of common front against Mahmud by the Karakhanids and the Khwarazmians.

Biruni, ever the mathematician and chess player, devised a strategy to solve what he perceived to be an equation with at least four variables. His solution was, first, to help the Karakhanids settle their internal problems and then, second, to sign a peace treaty with the Karakhanids as a whole, bringing in Mahmud as well. By this means the Karakhanids and Khwarazmians would form a common front but without challenging Mahmud, who would then be free to focus on his campaigns in India and Persia. But the bold gambit proved too clever by half. Biruni got his agreement, but there the matter ended. Mahmud correctly perceived that the deal had exposed the utter weakness of both his partners. This freed him to concentrate his full attention not on India or Persia but on Khwarazm. Early in 1017 he advanced his entire army from Ghazni to Balkh, whence he could quickly descend the Amu Darya clear to Gurganj. The Khwarazm shah once more began maneuvering desperately to gain time.

By now Abu Abbas's officers and patricians were in full rebellion against his refusal to stand up to Mahmud. In their eyes, Biruni's schemes amounted to capitulation and had opened the way for Mahmud to advance to their doorstep. Taking matters into their own hands, on March 17, 1017, they mounted a coup against Abu Abbas Mamun, killing him in the process. They then prepared for a last-ditch defense against Mahmud, who promptly attacked Gurganj to "save" the government of his brother-in-law from the rebels and to rescue his own sister. Mahmud quickly rolled over all resistance. Once in control, he rounded up all the rebellious patricians and officers in a public square and gave orders for his military elephants to crush them underfoot. He then turned loose his troops to sack the city.

Within days Mahmud's victorious forces had taken prisoner thousands of Khwarazmians and packed up their booty on thirty thousand horses. These he dispatched under guard to far-off Ghazni. One writer recorded the total number of captives as five thousand,[90] while Mahmud's court historian claimed the column of prisoners sent off with halters around their necks stretched from Balkh to Lahore in present-day Pakistan. Most of the captured Khwarazmians ended up fighting with Mahmud's army in India.

Biruni had already fled Gurganj to escape the wrath of the patricians and officers, but he could not elude Mahmud, who had never abandoned his plan to lure scientists from Khwarazm to his court in Afghanistan. Biruni realized he could no longer escape this fate. But he also saw in it a silver lining. Like Khwarazmi before him, his fascination with Indian science and culture was of long standing. In his *Chronology* he had rued the fact that he had found no reliable source of information on Indian astronomy and history. By 1017 Mahmud ruled the entire Indus Valley and was the most powerful ruler in India. Now, if Biruni played his cards right, he could have full access to this treasure house of knowledge. And so, with Gurganj obliterated as a scientific and cultural center and all escape routes blocked, Biruni bowed to the inevitable and set off for a new life in Afghanistan. As an honored trophy of conquest, he was allowed to bring with him his entire archive of books and papers. Accompanying him were the Nestorian Christian mathematician and physician Khammar and his dear friend, the mathematician Abu Nasr Iraq.

Thus ended the remarkable moment in the desert that was Gurganj under the Mamun rulers of Khwarazm. No one knew at the time that this was to be the last great flowering of Iranian culture in Central Asia.

Talented members of the various groups that spoke the diverse eastern Iranian languages continued to do important work, but henceforth they were to carry out their research under the rule of Turkic overlords. The new Turkic hegemons had no connection with the deep roots of religion and science in Central Asia and were largely ignorant of the heritage of ancient Mediterranean civilization that had inspired the flowering of science in the Islamic world. In time many Turkic people were to master this knowledge and contribute to it as scientists of the first rank. Several enlightened Turkic leaders were also to lend their support to very significant research. Yet in a larger sense, the fall of Gurganj marked a turning point in the civilization of Central Asia.

POSTLUDE: IBN SINA SYNTHESIZES ALL KNOWLEDGE

The story of Gurganj's brief flowering cannot end without a further word on Ibn Sina. From the time of his abrupt flight in 1010 to his death in 1037, the philosopher was in constant motion, his shifts from one city to the next prompted by the many political upheavals that infected the region like a virus. The main driver of this turmoil was the inexorable pressure exerted on all the Buyid rulers by Mahmud of Ghazni and his army. Ibn Sina realized full well that Mahmud had wanted—more accurately, demanded—that he join his court at Ghazni. He knew that Mahmud would never forgive him for eluding his grasp, and he could therefore expect no mercy if caught. Ibn Sina's restless wanderings covered the length and breadth of Iran, with stops in Gorgan, Rayy, Qazvin, Hamadan, and Isfahan. The only common element among these diverse city-states is that they were all ruled by Buyid families that adhered to the Shiite faith.

Besides his fear of Mahmud, the driving force for Ibn Sina's moves was the incessant political instability of the various Buyid courts at which he sought refuge. Ibn Sina's reputation as an able administrator earned him high posts in several governments, including the rank of vizier, which he held at Hamadan. The death of a ruler, a bout of illness on Ibn Sina's part, the outbreak of domestic insurrection, and military threats from neighboring rulers all played a role in forcing him to flee from place to place. At one point he was incarcerated in a fortress, and at another a popular uprising against policies he had instituted as a senior official caused him to flee incognito into the desert, where thieves made off with a number of important manuscripts he had stored in his saddlebag. Only in Isfahan did he find a peaceful refuge, but by then he had already entered old age.

It is a miracle that Ibn Sina found time amid this unfolding chaos to dictate the texts of a score of important articles and all the major books that earned him a central place in the worlds of philosophy and metaphysics. He accomplished this mainly during evening sessions with a handful of students and scribes, after a full day spent negotiating the petty intrigues and conflicts that afflicted every minor court. No ascetic, he regularly accompanied these sessions with wine and, if we are to believe one of his students, an impressively active sexual life.[91]

The main products of these years were entitled *Book of Healing*, *Book of Deliverance*, and *Book of Remarks and Admonitions*, all three of which, translated into Latin, elicited a powerful and positive response in the West.[92] Both the *Healing* and the *Deliverance* wandered freely through the realms of science and philosophy, with passages on physical mechanics or geology standing close by other passages on the soul. Some of these short sections are remarkable in themselves. For example, in a famous passage from the *Book of Healing* on the formation of mountains, Ibn Sina laid out the principles of evolutionary geology:

> Either they are the effects of upheavals of the crust of the earth, such as might occur during a violent earthquake, or they are the effect of water, which, cutting itself a new route, has denuded the valleys, the strata being of different kinds, some soft, some hard. . . . It would require a long period of time for all such changes to be accomplished, during which the mountains themselves might be somewhat diminished in size.[93]

But in comparison with his earlier work, it is clear that Ibn Sina's focus had shifted to the determination of "first principles," that is, to what humans can learn about ultimate realities through the exercise of logic and reason. In these seminal works Ibn Sina largely set aside the practical concerns that informed his *Canon* and focused instead on the core issues of philosophy, metaphysics, and religion that were generating deep divisions among the Muslim faithful and that were to explode in frontal ideological conflict within a generation.

It is impossible to summarize briefly the dense and complex arguments that Ibn Sina presented in these later works. Scores of learned tomes have been devoted to each of his major themes. Suffice it, then, to point out some of the signposts along the way, and the final positions to which these led him.

After their initial dispute, the intellectual lives of Ibn Sina and Biruni had followed radically different paths. Both began with Aristotle's assertion, in the first line of his *Metaphysics*, that "all people by nature desire to know." But what Ibn Sina and Biruni sought to know differed sharply. Biruni was fascinated by the stuff of observable nature, including the movement of the planets, the qualities of minerals, and the varieties of human cultures and their diverse ways of explaining the world around them. He especially thrilled at the possibility of describing the phenomena he observed in mathematical terms. Ibn Sina, in his medical studies, had started out in the same direction and delved deeper into the specifics of diseases and cures than anyone before him. But these were the concerns of his youth and early adulthood. Increasingly he wanted to understand the fundamental nature of being itself. This was, and is, the sphere of cosmology and metaphysics.

In the tenth century the question of being and all it entailed—the extent of our understanding, the origins of the visible world, the role of some sort of divine spirit in that process, and the nature of being itself—had an urgency that is hard for a less philosophical age to appreciate. Two great intellectual forces were at their zenith at the moment Ibn Sina began to address these issues. On the one hand, a formidable body of ancient Greek learning, especially the works of Plato, Aristotle, and their diverse disciples, had burst on the scene, thanks to translations made in Gundeshapur, Baghdad, Merv, and other centers. For all their differences, Greek thinkers had exercised their rational intellects to forge answers to these questions, confident that they had the essential tool for unlocking the universe. Some, like the later followers of Plato, ended up in a kind of mysticism, but even they used reason to chart out the human condition.

On the other side were ranged the Abrahamic religions—Judaism, Christianity, and now Islam—which offered sharply different answers that were based ultimately on the workings of a divine spirit that was knowable only through revelation and faith. Among the Abrahamic religions, Islam was at its triumphant zenith, bearing specific answers to questions regarding the origins of the universe and man's place in it. All these answers traced directly to the initiative of God, as revealed through revelation and apprehended through faith.

Sijistani had attempted to build a wall between reason and revelation. Kindi and Farabi took the opposition approach, attempting to square the circle of reason and revelation. Farabi came closest, but his argument strongly favored the rationalists' approach. In the end, the effect of these early Muslim thinkers was mainly to sharpen the debate and render it yet more urgent. The English philosopher Bertrand Russell was right to call this set of issues a philosophical "No Man's Land."[94] But then Ibn Sina charged in with his boundless confidence in his own powers. In his effort to resolve the tension between reason and revelation, he deployed all the intellectual weapons he had drawn from the Neoplatonists of late antiquity, the Mutazilites, the Ismaili Shiites, and mainstream Muslim theologians, heaping criticism on each of them as he did so.

In the end Ibn Sina came as close to reconciling science and revealed religion as anyone in the Middle Ages. Muslims and Christians alike believed he had developed the grand synthesis that both had yearned for. In the West Ibn Sina's formulation inspired not only Thomas Aquinas but generations of his followers. But in Central Asia a devastating counterattack was to be mounted within a generation.

Launching into his argument, Ibn Sina reminded his reader that reason itself takes several different forms, which can be ranked hierarchically in terms of their ability to address ultimate questions. At the most mundane level there is the visible world and all those practical sciences devoted to studying it. But he followed Plato in arguing that the reality of a triangle is more than any specific triangular object that we may perceive and study with our senses. The tool with which the human mind deals with this *concept* or *essence* of triangularity is mathematics, which therefore ranks above practical reason as a tool for attaining truth. But where does the concept of a triangle, or of anything else in our world, come from? Were these concepts always there, or did something bring them into being? Neither practical reason nor mathematics is able to address such questions, which are the proper realm of the science of metaphysics. Like Farabi before him, and like Farabi's ancient master, Aristotle, Ibn Sina held that the sole form of reason that can grapple effectively with problems of metaphysics is logic. By "logic" he did not mean "being logical" as we use the phrase in daily life. Rather, he had in mind the rigorous processes defined by Aristotle and expounded by Farabi, which involved defining premises and reasoning deductively (or, less frequently, inductively) to arrive at conclusions. The core of this process was the syllogism—the process of inferring a conclusion from two prior propositions. Such logical reasoning, Ibn Sina argued, can lead us to truths that neither science nor math can attain.

By these neat steps, Ibn Sina, like the followers of Plato before him, transformed the old question of "Where did the world come from?" Religious prophets had answered by declaring that the world is the product of God's creation, while most philosophers argued instead that the world was eternal. Using the tools of logic, Ibn Sina brought forth a version of the same argument he had employed as an eighteen-year-old against Biruni, namely, that the *material world* came into existence at a specific time, but the deeper *concept of a world* is eternal. The act of creation was not the process of replacing nothing with something, as the theologians maintained, but rather, the transformation of the preexisting reality of "existence as *concept*" into material existence. Practical scientists who could not grasp the idea of concepts, or mathematical scientists who did not go beyond the mere recognition of concepts or essences, were as confused as the theologians. Ibn Sina maintained that at the deepest level their views were in fact compatible.

If this line of reasoning pointed toward a reconciliation between religion and science, it was only a first step in that direction. Left open was the question of the human soul and of God. Ibn Sina advanced into this territory by inquiring first into the origins of concepts, that is, of the underlying realities of all existence. The soul, he wrote, is that something within us which becomes conscious of *being*.[95] By this he did not mean consciousness of our specific existence in space and physical time but of being as such. The body can perceive material *things*, but it is the soul alone that can grasp immaterial realities. As to the nature of the soul, Ibn Sina drew a distinction between the animal soul, which perishes with the body, and what he called the *rational soul*, which is eternal.

He then employed logic and syllogisms to pick his way down an intricate path that led to his rational demonstration of the existence of God. Causation, he reasoned, cannot be retraced ad infinitum. Eventually, one arrives at the notion of a "Necessary Being," "First Being," or "Giver of Forms," from which emanate all forms of existence, both in essence and materiality.[96]

Logic, the queen of the sciences, thus brought Ibn Sina to a concept of God that was compatible with the views of all contemporary scientists other than those few who were outright materialists and atheists, and also with the truths announced by the prophets.[97] To this point Ibn Sina was simply repeating the line of argument that Farabi had advanced a century earlier. But then he laid aside logic and introduced a higher means of arriving at truth, namely, through the exercise of *intuition*. To Ibn Sina, the case for intuition as a special way of reaching truth seemed obvious. He reported that when stumped by a problem he would often go to the mosque, where the solution came to him *intuitively*.

In a particularly bold passage, Ibn Sina went so far as to suggest that a certain few people are prophets because they possess exceptional powers of intuition. With their keen intuition, prophets grasp and present in accessible language the same great truths to which both science and theology aspire. Prayer itself, for Ibn Sina, was the exercise of intuition.

This remarkable proposition was eventually to bring down the most withering assault on Ibn Sina's legacy—Ibn Sina himself had been dead sixty years by the time the attack occurred. To many of the pious, the assertion that revelation and intuition are identical, or nearly so, was outrageous. Surely, they claimed, this shifted the entire focus of revelation from God to man, diminishing both divine revelation and faith. Never mind that Ibn Sina recognized Muhammad's role as a Messenger of God, founder of Sharia, and promoter of justice. Like the ancient Greek philosopher Protagoras (490–420 BC), Ibn Sina seemed to many to have asserted that man (thanks to both reason and intuition) is the measure of all things.

Ibn Sina fully understood that in writing his *Book of Healing* and other philosophical works he had entered onto extremely controversial territory. Both the questions he asked and the answers he reached were fraught with danger. Thus Aristotle and the philosophers held that the earth was eternal, while Islam traced its existence directly to the actions of the Creator. Ibn Sina could have sought out some bland middle ground between the two positions on this and other key points. But instead of merging them he bridged them, and without compromising either. He did the same with the problem of evil, seeming fully to recognize its reality yet in the end affirming, in a rather Panglossian moment, that good outweighs evil in the world, and that much that appears evil is only so in relation to some good that is being denied, and does not signify that human beings are inherently corrupted.[98]

It was this Great Reconciliation that Western Scholastics found so compelling and which caused the young Thomas Aquinas to orient his entire metaphysics around the Central Asian's arguments.[99] In the West only St. Augustine in the fifth century, Aquinas in the thirteenth century, and the Berlin philosopher Hegel in the nineteenth century came so close to addressing all the main problems of philosophy and religion within the framework of a single system of thought.

Not everyone in the Muslim world saw the Bukharan's magnum opus in so positive a light. Within two generations there appeared another philosopher who frontally challenged Ibn Sina on just about every point. This man was Abu Hamid Muhammad al-Ghazali (1058–1111), known in the West as Algazel. Ghazali, too, was a Central Asian, from Tus in Khurasan. His attack on Ibn Sina was comprehensive, subtle, and, in many respects, devastating. Since it arose amid dramatically changed political and cultural conditions, we will hold off examining it until after we have reviewed the developments that came to shape that later era.

Meanwhile, there was no further contact between Ibn Sina and Biruni. After the productive years they shared at the Mamuns' court in Gurganj, the arch-rivals went in opposite directions, both literally and figuratively, Ibn Sina to Iran and Biruni to Afghanistan and beyond. During the difficult years in which Ibn Sina composed his great philosophical treatises, his competitor Biruni was quietly toiling away in what is now Pakistan and Afghanistan, preparing his own magisterial *Canon* devoted to astronomical and mathematical knowledge. Notwithstanding their nomadic existences, both reached their peak of creativity while in exile. Ibn Sina eventually contracted an intestinal illness, probably colon cancer, which a student blamed on his hyperactive sex life. His self-treatment involved a course of constant enemas, but these failed and Ibn Sina died in his fifty-seventh year, in 1037.[100] Biruni lived on another eleven years and died peacefully in 1048.

The upheaval that had driven these two great minds from their Central Asian homelands was the explosive rise of powerful new Turkic states: the Karakhanid rulers in Balasagun, Kyrgyzstan, and Mahmud of Ghazni in Afghanistan. Since the rising Turkic powers henceforth defined the political and cultural life of the entire region, it is to them we must now turn.

❀

CHAPTER 10

Turks Take the Stage: Mahmud of Kashgar and Yusuf of Balasagun

Mahmud al-Kashgari, or Mahmud of Kashgar, was a man on a mission—several missions, in fact. He had been raised in Kashgar in what is now China's westernmost province of Xinjiang. Through most of the twentieth century we thought of this region along the Tian Shan or Celestial Mountains as the fault line between the Chinese and Soviet Russian land empires. In the eleventh century it was the place where Islam, Buddhism, and the animism of the nomads all collided. Kashgari was a good Muslim, which is why in 1072 he found himself in Baghdad, home of the severely weakened caliphate. But he was also a Turk, and a rather thinskinned one at that. This gave rise to his first mission.

Kashgari had come to Baghdad several years earlier. Everyone knew, of course, that whatever limited powers the caliph still wielded were due largely to the Turks around him. Turkic people had been evident at the political seat of Islam at least from 719, when the new Abbasid rulers swept in from Central Asia to take over the moribund caliphate of the Umayyads. Many of the conquering soldiers, and even some of their officers, had been Turks from the countryside around Merv and Nishapur.[1] Every year thereafter the caliphs purchased Turkish slaves in the markets of Bukhara and Nishapur to man their army and, most important, staff the elite guard that protected them from the palace plots, uprisings, and insurrections that were constantly boiling up. And more than a few Turks, slave and free, found their way into the caliph's administration.[2] By the 900s it was clear to everyone that while the caliphs themselves may have been Arabs and the culture of Baghdad was increasingly shaped by Persianate people from Central Asia and Persia proper, it was the Turks from Central Asia who wielded the real power.

This was true practically everywhere in the Muslim world. Far to the west, the Fatimid government in Egypt came increasingly under the control of its Turkic slaves, the Mamelukes, who eventually seized power for themselves. Even the independence-minded rulers in Bukhara had come to depend on enslaved Turks to man their army. In fact, Samanid bureaucrats had worked out a career ladder by which a lowly Turkic slave could rise through the army from the humble status of groom to the most senior positions at the court of what was a proudly Persian-speaking state.[3] It was like the ancient Romans' *cursus honorum*, except this social escalator was for slaves, not aristocrats. When the Samanid dynasty fell in 1002–1004, it was due to internal opposition mounted by such men, as well as to external opposition raised by an emerging Turkic state to the east.

Given the immense power of the various Turkic tribes and clans across Central Asia and the Muslim world, it must have been all the more humiliating that Turks were treated so shabbily by the very people who should have welcomed them. In Baghdad most Turks lived in segregated neighborhoods. The city's majority population of Persians, Arabs, and others viewed the Turks as conniving and dangerous, and whenever a disturbance broke out, which was often, it was assumed to be the work of Turkic malcontents and gangsters. Many in Baghdad considered mobs and Turks to be synonymous.

Most humiliating of all, nobody bothered to learn Turkish. True, fruit sellers in the bazaar may have picked up a few words, but certainly not members of the elite. And why should they? Both Persian and Arabic speakers assumed that the Turkic languages were for rough-hewn foot soldiers and the urban poor but offered nothing for the scientist, poet, or seeker after wisdom. They grudgingly accepted the Turks as political heavies but denied them the slightest civilizational role. Kashgari recognized in all this the worst kind of cultural prejudice and took it as his personal mission to change it. To this end he mobilized an astonishingly innovative and bold set of skills for which he is still remembered today.

Behind this probably lay a second mission, "probably," because he left us only one hint that he embraced it. Back in 1055, only seventeen years before Kashgari arrived in Baghdad, a new mass Turkic army had swept to power in all Central Asia and then gone on to conquer the caliphate itself, reducing the caliph, God's Captain on Earth, to the status of a dependent vassal. The Seljuks and their confederation presented themselves as good Muslims, which they had been for several generations, and the caliph could only applaud their declared intention to purge the caliphate of Shiite sympathizers. But like it or not, the caliphs now had to pay homage to their new Seljuk Turkish overlords. The caliph meekly presented his daughter in marriage to the Turkish sultan and sealed the deal by grandly declaring the Turk "king of the East and West." With these lofty words, the heir to Muhammad's earthly power ceded temporal authority to someone whom he, the caliph, must have viewed as an upstart and a ruffian.[4]

This marked a critical stage in the ongoing decline of the caliphate but opened splendid opportunities for a talented and widely traveled middle-aged Turk like Kashgari, one who was fluent in both Arabic and Persian as well as the main Turkic languages. Kashgari hinted that he was mindful of these possibilities when he dedicated his work to the caliph, who may have remained in his own mind the Commander of the Faithful but was in fact reduced to the status of "Deputy of the Lord of Worlds," that is, the vassal of a Turk like himself.

Kashgari's plan was to force the caliph, and beyond him all Arabs and Persians, to acknowledge that the time had come for them to study the Turkic languages and come to grips with Turkic culture. Even though doubtless prompted by the sting of prejudice directed against Turks over the years, Kashgari's plan was nonetheless unapologetically triumphalist. But he did not merely hector his readers. Instead, he offered them a practical plan for learning Turkic languages and acquainting themselves with Turkic culture. His method was to write a kind of Turkish-Arabic dictionary, one that included not only words and phrases but also proverbs, sayings, poetry, and pithy nuggets of folk wisdom from across the Turkic world. All were to be offered in the original Turkic languages in Arabic script and then translated into accessible Arabic. In addition, Kashgari proposed to present thumbnail sketches of the various Turkic tribes and their customs, and even a map so that readers could look up where each group lived.

THE MAKING OF AN ETHNIC PROPAGANDIST

This was a tall order, made the more challenging because there was no obvious precedent for it, no model that Kashgari could emulate. True, there was a so-called classical school of geographers who wrote in Arabic. The best of these had lived a century and a half before Kashgari and was a fellow Central Asian, Abu Zayd al-Balkhi from Balkh.[5] His geographical book, now lost, was widely copied by Arabs who wanted to depict the various political regimes of the Muslim world. There were also dictionaries of various sorts, although never one for Turkish.

One area where Kashgari did not innovate was his choice of a wheel-shaped form for the map that he appended to his text. Though still in common use in Europe, the Arab world, and China, these geometrical schematizations were in fact old hat, having been superseded by maps showing longitudes and latitudes that Khwarazmi and his contemporaries had developed two centuries earlier.[6] The form of Kashgari's map may have been outmoded, but he was on new ground when he used it to show the relative locations of specific ethnic and linguistic groups, append to it a dictionary of their languages and dialects, and then presented examples of their best poetry and wisdom—all in one book.

Andreas Kaplony, a specialist on the *Compendium of the Turkic Dialects* from Zurich, points out that Kashgari's purpose went far beyond the mere showcasing of Turkic dialects: in his manual he wanted to provide the key to learning them all. It was as if a single author today would purport to provide in a single textbook everything needed to master not just French or Spanish but *all* Romance languages. He began with his own native tongue, Khaqani. As Kaplony explains, "To switch to one of the other languages, he gives his reader phonetic and morphological rules to apply and quotes the exceptions to these rules, namely, the words used in this or that tribe. He emphasizes that the reader who memorizes all his examples and applies all his rules can understand any Turkic language."[7]

Kashgari launched into his book by declaring with stunning immodesty that he and he alone was the right man for the job:

> I have travelled through [the Turks'] cities and steppes and have learned their dialects and their rhymes. . . . Also, I am one of the most elegant among them in language and the most eloquent in speech; one of the best educated, most deep-rooted in lineage, and the most penetrating in throwing the lance. Thus, I have acquired perfectly the dialect of each one of their groups and I have set it down in the accompanying book in a well ordered system.[8]

He then confided that God himself had been his research assistant on what will doubtless become an "everlasting memorial and an eternal treasure." No blurb by a modern publisher could possibly top this one, penned by a self-publishing author a millennium ago.

Kashgari came by his healthy self-image naturally. His family was from the town of Barskhan, on the barren southern shore of Lake Issyk-Kul in what is now Kyrgyzstan.[9] Situated astride an important route between the Seven Rivers region and Kashgar, Barskhan prospered, and Kashgari's father, a city official who was related to the Karakhan clan that had recently taken power in the region, was promoted to a senior post in the Karakhanid capital at Kashgar. It was there that Mahmud al-Kashgari was born and raised.

The ancient city of Kashgar is the heart of one of the most lovely and bountiful oases in all Central Asia. Here the two great trade routes from China to the West come back together, after dividing in order to skirt the fearful Taklamakan Desert to the north and south. Winding lanes lined with willow, white poplar, and apricot trees connect the old city with the satellite towns that are a feature of all ancient Central Asian cities. In one of these, now called Wupar, Mahmud of Kashgar was eventually to be buried. His tomb is still visited today.

The reason Mahmud's father moved to Kashgar is that this ancient center of Buddhism, Nestorian Christianity, Manichaeism, and Judaism had become one of the capitals of a new confederation of tribes that were both Muslim and Turkic. Strange to say, we don't know the contemporary name of this state. But nineteenth-century European historians named its rulers the Karakhanids (or "Black [i.e. 'noble'] Princes") and the name stuck.[10]

THE ASTONISHING KARAKHANIDS

Hundreds of years earlier the Karakhans and several related Turkic tribes had been pushed out of the Mongolia-China border area and moved west into what is now the central part of China's autonomous province of Xinjiang.[11] A main branch, the Uyghurs, settled there and developed a written language based on the old Aramaic alphabet, which they had learned from the Nestorian Christian merchants who had long been established in the area. But the Karakhan clans kept moving and by the mid-800s had reached the eastern fringes of the Samanid state. Once more the old line of demarcation between Buddhism and Islam—the line established at the Battle of Taraz (Talas) in 751—made its presence felt, with the Muslim Samanids to its west and the Buddhist Karakhanids to the east, with the actual line running along the western side of the Tian Shan. Would the Karakhanid army maintain its westward momentum, or would the Samanids summon enough resolve to stop them?

Then, in the 950s the new rulers of Kashgar proclaimed their conversion to Islam. After their official conversion, the rulers immediately suppressed the Buddhists as idol-worshippers, but other religions continued to be practiced there for several hundred years. Since other Turkic groups had been in the region for nearly a millennium and not converted to Islam, it is worth asking why this one so abruptly abandoned its traditional faith. It may have been because the Karakhanid people had taken up a settled life and become city-dwellers. Perhaps they looked on Islam as the most suitable, if not inevitable, religion for urbanites. But the Uyghurs further east were thoroughly urbanized and for the time being remained Buddhists. It is also true that the Samanids had sent missionaries to the Karakhanid leaders. Thanks to the indefatigable Russian scholar Bartold, we even know one of these Muslim proselytizers by name.[12]

Whatever their motivation, the Karakhanids' decision to convert was an adroit move diplomatically. In their preconversion identity they were bound to have impressed the tottering Samanids as pagan intruders with strange and forbidding names like Falcon, Wolf, or Camel. Now they could present themselves instead as fellow Muslims and even protectors of the faith.[13] When in the 990s their forces appeared at the gates of Bukhara, their leader, Ilek Nasr, who still used the honorific title Arslan, "The Lion," among his Turkic colleagues, could announce that he came only as the Samanis' friend, coreligionist, and protector. On October 23, 999, a Karakhanid army entered Bukhara without opposition, took control of the Samanid treasury, rounded up the remaining Samanis, and settled into their palace.[14]

The Karakhanid dynasty filled the vacuum caused by the erosion and then implosion of the Samanid state. Early in this latest Turko-Persian confrontation, the Samanids had counterattacked by seizing the city of Isfijab (Sayram) in what is now south-central Kazakhstan. But the merchants of that important mercantile center ignored Bukhara's demand that they cease trading with the enemy and the move failed.[15] Now such defections by traders and also regional noblemen grew increasingly common. By the start of the new millennium most of the eastern lands of the former empire were in the Turks' hands.

Even before their triumph, the Karakhanids found themselves beset by the same centrifugal tendencies that had so weakened the Samanids, but in the case of the Turks these forces of division arose from within the ruling family itself. Typical of nomadic families, all the sons of the founder demanded their share of his patrimony. After they settled into urban life, this meant that each one appropriated for himself a capital city and a territory to go with it. By the time Mahmud Kashgari set out for Baghdad, there were no fewer than four Karakhanid capitals: the oldest at Kashgar; a second at the ancient city of Samarkand; and two others at Uzgend and Balasagun, both in present-day Kyrgyzstan. Over time these last two emerged as the most important, at least until the Balasagun khanate moved its headquarters back across the mountains to Kashgar during Mahmud Kashgari's lifetime.

The persistent tendency toward fragmentation within the Karakhanid clan was the Achilles heel of this first Turkic Muslim state, as it was to be for the many other Turkic dynasties that followed. In truth, it was no state at all but a loose confederation of appendages, the ruling houses of which were linked by blood ties. Had this been the whole story, they would hardly have deserved Kashgari's massive effort in their behalf, nor would the other new Turkic dynasty, the Seljuks. But there was more to the picture, specifically the economy, concerning which the Karakhanid rulers showed themselves far more united and more effective than they did in the political sphere.

One must stand astonished that these nomadic khans, who had earlier focused all their energies on maintaining the kind of army that would enable them to survive on the embattled steppe and who, even after adopting a settled way of life, were content to leave the tedious work of commerce to others, could have emerged as astute managers of their combined economies. In this sphere they showed themselves to be fast learners indeed. This is why many Muslim communities opted to trade with the Karakhanids and even to align themselves with them politically.

The Karakhanids built caravanserais to foster trade along heretofore neglected routes. Notable among these was the monumental

Ribat-i-Malik on the road between Bukhara and Samarkand and Bukhara. Built in 1078 by the Karakhanid shah Shams al-Mulk Nasr, this resplendent building, with its multiple domes and fluted adobe walls, looks more like a country palace than a hotel for traveling businessmen. Another, the enigmatic Tash Rabat, still stands at a lonely bend in the road near the top of a difficult pass through the Tian Shan on the road connecting Uzgend and Kashgar. This structure long mystified archaeologists, some of whom concluded that it had first been a Christian monastery and only later used for commercial travelers.[16] Recent findings support its commercial role and date it squarely to the Karakhanid era.

Karakhan leaders also excelled at the minting of money and fiscal policy generally. Numismatists have identified no fewer than thirty Karakhanid mints, most of which produced the lower-valued leaded copper coins used by bazaar traders.[17] For silver and gold coins they prudently relied on the Samanid dirhams that were known and trusted throughout the known world, or on their own copies of them.[18] A study of nearly two hundred coin hordes confirms a brisk commerce throughout the politically divided realm, conducted mainly with coins of lower denominations. The rulers' populist economic policies facilitated this trade. When butchers petitioned a Karakhanid ruler to allow them to raise their prices, he reluctantly consented but promptly banned the consumption of meat. When he finally yielded to the butchers' pleadings, the ruler forced them to pay a penalty in order to get their old price back.[19]

Mahmud Kashgari Takes Up the Banner

Whatever specific policies the Karakhan rulers might pursue, Kashgari had not the slightest doubt that the culture and values of Turkic nomads were destined to redefine the "civilized" world. As he wrote in his introduction, "[Turks] are the 'Kings of the Age,' appointed to rule over mankind. . . . [God] strengthens those who are affiliated with them and those who work in their behalf." By Turks, of course, Kashgari meant the entire population, including the ordinary men and women who create and transmit proverbs, folk poetry, and, above all, the Turkic languages. This was the natural populism of the horizontally organized nomadic peoples.

Figure 10.1. Tash Rabat, an austere but durable Karakhanid-era caravanserai high on the mountain pass between Kyrgyzstan and Xinjiang. ▪ © Hermine Dreyfuss 2005. All rights reserved.

To nail his argument in behalf of the Turkish language and the Turks, Kashgari invoked divine will. Even though he was writing two centuries after Imam Bukhari released his authoritative collection of Hadiths, Kashgari solemnly declared that imams in both Bukhara and Nishapur had confided to him authentic Sayings of the Messenger of God that exhorted humankind to "learn the language of the Turks, for their reign will be long."[20] In other words, learning Turkish and studying Turkic culture was a religious duty. He then cited another purported Hadith of Muhammad affirming that "Turks are superior to all other beings."[21] Karakhanids and Seljuks alike must have reveled in this public relations masterstroke.

With his forward written, Kashgari turned to the body of his *Compendium of the Turkic Dialects*. He carefully noted the date on which he sat down to write: January 25, 1072. Counting the editing process, he was to devote five years to the effort, completing the work on or about January 9, 1077.[22] From the first page to the last he stuck to his populist and egalitarian message, avoiding all temptations to impose a hierarchy on the many Turkic languages and cultures. His reports on such ethnographic and folkloric elements as cuisine, kinship, and folk medicine were relentlessly enthusiastic and resolutely nonjudgmental.

This was the easier because Mahmud Kashgari excluded from his picture nearly all exogenous influences on the Turks. True, he dwelt with fascination on the so-called Alexander saga tradition in Central Asia, but by the time of his writing these events had occurred more than a thousand years earlier, long enough for them to have become fully absorbed into the warp and weft of Turkic life. Otherwise Kashgari presented his Turkic tribes as being free of external influences, as wonderful a view as it was inaccurate.

Kashgari's readiness to plunge into the timeless and earthy core of Turkic culture was one of his most important innovations. He included extended poems and proverbs in the original Turkic, but, in consideration for his intended audience, the body of the text was in Arabic. The style is systematic and sober, as befits a lexicon and language primer. Yet it reads like the report of an excited anthropologist who has just returned from exotic places and is reporting for the first time to an audience of wide-eyed but educated laypeople. Kashgari modestly avoided using the first person. But he never allowed readers to forget that all this intriguing information was coming to them thanks solely to this enquiring social scientist and linguist, who had trekked to strange places and whiled away months with unfamiliar people.

Kashgari did not equate "exotic" with "alien." His decision to include large numbers of proverbs and folk poems in his compendium was made precisely to show his "globalized" readers that the received insights of the Turks need not take second place to any other culture in terms of wisdom. Using proverbs to buttress his case, he painted a picture of a shrewd people who harbored no illusions about human perfectibility: "Throwing a harness over an ass's head does not make it a horse." Yet "He who does evil to others does it to himself." One need not shine at court to lead a fulfilled life, for the world offers many possibilities: "Better to be the head of a calf than the foot of an ox."

Kashgari's Turks understood that nothing is achieved without struggle: "He who would gather honey must bear the sting of bees." They were fighters but understood the cost of conflict: "Two camels fight and the fly between them dies." In the end, everyone is responsible for his or her actions, for "Every sheep is hung [in the butcher's shop] by his own feet." Modesty, then, was highly valued, and also realism about one's importance: "The hare was angry with the mountain, but the mountain was unaware of it."

Though pastoral nomads, the Turks believed that knowledge itself was important: "He who knows and he who does not know are not the same." Taking a long view, they did not yield to pessimism, for "One crow does not make winter." And in the end, hard work pays off: "He who marries early enlarges his family; he who gets up early goes a long way."

Even in Kashgari's day the folksy tone of such nuggets would have conjured up a timeless steppe world that is far removed from the turmoil of modern urban life and its incessant changes. Then, as now, such a vision would have been attractive to harried urbanites, and especially to those in Baghdad who now found themselves subject to the supreme authority of precisely the bearers of this hearty folk culture.

In this context the "wheel map" of peoples that Kashgari appended to his volume carried a message of its own: "You may not know who these Turks are or where they came from, but rest assured that they have long been masters of a huge part of the lands defined by the World-Encircling Ocean." This belief that the inhabited world was surrounded by water traces straight to Aristotle; in his mention of Gog and Magog, the mythic land lying beyond impenetrable mountains, Kashgari drew on the Bible. The color-coded map that conveyed this message took note of Egypt, India, Russia, China, and most lands in between. Most astonishing is its portrayal of Japan, which was prominently highlighted by a green semicircle (see plate 15). Aside from being the first known Turkic map, Kashgari's schema is renowned today as the world's oldest map depicting Japan.[23]

The peculiar format that Mahmud chose had one enormous advantage over a grid based on longitude and latitude: it had a center. Arab cartographers loved wheel maps because they could be drawn to show Mecca at dead center of the circle. In a telling detail, Kashgari shifted the map's focus. His subject is not the world of Islam—neither Mecca nor Baghdad is even shown—nor just the world of the Turks. Instead, Kashgari depicts almost the entirety of the known world and shows all of it revolving around Balasagun, the Karakhanids' main capital at the time. This was politicized cultural revisionism with a vengeance.

For all Kashgari's practicality and astuteness, his picture of the Turks includes a strong element of romanticism bordering on the kind of self-deception to which all nationalists are prone. Let us note again that Kashgari systematically neglected to acknowledge external influences on the Turkic languages and cultures. Indeed, he goes so far as to say that whenever a Turk gets too close to an alien (in this case Persian) world, he begins to lose touch with his own linguistic and cultural identity.[24] Yet wasn't Kashgari's own life a manifestation of precisely such cosmopolitanism? He no longer prayed the ancient Turkic prayers, he had fully absorbed the Arabic language and its associated culture, and he was linked by family ties to a dynasty that was busily assimilating every aspect of the rich Samanid culture. Yet this otherwise practical man seemed also to yearn for some lost Turkic Eden, that place where words exuded an elemental vitality, all was poetry, and timeless wisdom prevailed.

For all his boosterism, Kashgari was at bottom something of a pessimist. Like Ferdowsi, he seemed intent on setting down important elements of his pre-Islamic heritage, fearful that they might soon be lost forever. To this end, he included a lengthy discourse on the Turks' twelve-year calendar based on a cycle of animals. The Muslim calendar that Karakhanids and Seljuks had recently adopted had rendered the old system obsolete, yet it continued to thrive side-by-side with the twelve-month calendar and was still being used by the Mongols two centuries later.

Kashgari struggled to present himself as both an insider and an outsider. Yes, he was himself a proud Turk and never let his reader forget it. But his picture of the Turks exuded the globalist perspective of his day. He did not hesitate to employ techniques he had learned from the Arab lexicographers and Persian folklorists and antiquarians, including Ferdowsi himself.

Viewed thus, Kashgari's great book emerges, first, as a politically motivated celebration of everything Turkic written for an audience of anxious Persians and Arabs who were just beginning to come to grips with the fact that they were now ruled by Turks. Kashgari, drawing on extensive field research and his own heritage, assured them that all would be well, and that new opportunities would open before them. But they should be clear about the new realities. The Prophet himself had indicated the superiority of Turkish culture and of Turks, or so Kashgari claimed. Anyone who was so foolish as to contest this would "expose himself to [the Turks'] falling arrows."[25] In other words, anyone rejecting the message of this book will die!

Second, Kashgari's *Compendium* is an endearing tribute to timeless traditions of Turkic culture by a man whose own life had taught him that only change is permanent. The tension between these two aspects of the book is palpable throughout. Far from being a source of discord, the tension gives the book a soul, lifting it high above the level even of a ground-breaking language text, dictionary, lexicon, and pioneering work of cultural anthropology.

AN AGE OF "CULTURAL RETROGRESSION"?

In his magisterial works on Central Asia, Vasilii Bartold wrote dismissively about the Karakhanids and what he considered their negative impact on culture. Indeed, Bartold believed emphatically that the source of all the great civilizations of Central Asia was overwhelmingly the urbanized Persianate population and that the mainly nomadic Turks, whether the Karakhanids, Mahmud of Ghazni, or Seljuks, brought nothing more to the region than militarism, neglect of the trading economy, and overall cultural decline. There is some justification in this charge, for it was the Karakhanid army that overran much of East Turkestan and wantonly destroyed a thriving and ancient Buddhist culture there. When a great Buddhist monastic library was threatened by Karakhanid forces, its staff sealed it up in a cave to protect it.[26] But Bartold did not cite this in support of his thesis, instead focusing on the Turks' impact on the Persian-speaking Muslim oasis cultures.

In the introduction to his study of Turkestan, Bartold decried the fact that the Karakhanid era left posterity only one scholarly work, *Examples of Diplomacy in the Aims of Government* by a Samarkand writer named Muhammad bin Ali al-Katib. Bartold noted in passing the same author's well-known poem *The Great Book of Sindbad* but obviously did not consider it either high scholarship or high art. For good measure he acknowledged another writer from Kashgar, this one a historian, but noted disdainfully that his book on his hometown was riddled with unsubstantiated legends and outright mistakes.[27] He did not bother to take note of such other Turkic writers as Ahmed Yugnaki, who penned a lengthy work in verse entitled *Gifts of Virtue*.[28]

On this basis, Bartold concluded that "The period of the rule of the Turkish Kara-Khanid dynasty was without doubt a period of cultural retrogression. . . . In spite of the good intentions of individual rulers, the view that the kingdom formed the personal property of the Khan's family . . . [led inevitably] to the decay of agriculture, commerce, and industry no less than of intellectual culture."[29] But then Bartold proceeded to qualify his own conclusion. Maybe no historical masterpieces were written during the Karakhanid years, but Bartold himself mentions a *History of Turkestan* and a volume entitled *Turkish Peoples and the Marvels of Turkestan*. And what about Mahmud Kashgari's magnum opus, let alone the book by Yusuf of Balasagun, to which we will turn shortly? It turns out that neither was known in Bartold's day. The former was preserved in only one manuscript copy, which was found in an Istanbul archive and published only in 1917–1919, just after Bartold's volume went to press. Three copies of the latter existed, but basic scholarly work on them awaited a later generation. One can only speculate whether knowledge of these literary treasures would have caused Bartold to alter his sour judgment.

AVENUES OF TURKIC CREATIVITY

Whether or not one views the Karakhanid era as a time of cultural retrogression, it is worth inquiring into the character and texture of creative life in those years. Clearly no great astronomer, mathematician, chemist, or doctor appeared in the Karakhanid lands or found support from their rulers. But that does not mean that major achievements did not occur in other areas.

It is surprising that the formerly nomadic Karakhanids should have emerged as enthusiastic patrons of architecture and the building arts, but this is precisely what occurred. Thus, as soon as they seized the citadel at Samarkand, the Karakhanids erected several lightly built pavilions that were adorned with the most intricate and sophisticated painted vegetable-like patterns in tan, blue, white, and green. The chance discovery of these in 2002 makes one wonder how many other Karakhanid era frescoes have been lost.

More substantial structures followed over the next century. Karakhanid-era folk ballads referred to their capital at Uzgend in the eastern Ferghana Valley as "the city of our souls" and "our own special city."[30] There, on the high bluffs above the Kara Darya, the rulers created a complex of three separate but interlinked mausoleums, each with elegantly inscribed inscriptions in Arabic and Persian. These were for Karakhanid rulers, not saints or holy men.[31] All three are rectangular blocks with very highly decorated formal portals harkening back to the Ismail Samani mausoleum in Bukhara. And like the Samani mausoleum, their surfaces are ornamented with wonderfully intricate geometric patterns of fired brick. Careful study has revealed the precise geometric matrix for the elevations of these commanding structures, and for the glorious ornamentations on their façades.[32]

There is every reason to believe that similarly sophisticated buildings once adorned the main Karakhanid capital at Balasagun. However, that question cannot even be raised without first acknowledging what proved for a century to be the vexing question of where the city actually stood. The two main candidates are less than four miles from each other, but one, Ak Beshim, falls within the borders of Kazakhstan while the other, Burana, is on the territory of Kyrgyzstan. Thanks to this, a good-natured intra-Turkic rivalry came into play, with the Kyrgyz advancing arguments in favor of their site in the most earnest technical reports while the Kazakhs make their case in deluxe volumes, the latest of which disquietingly features the author on the back cover, dressed in full military uniform and sporting a chest-full of medals.[33]

Figure 10.2. Three Karakhanid mausoleums, Uzgend, eleventh century. The Karakhanids, who had only recently abandoned nomadism, quickly embraced and mastered the architecture and urban high culture of the Samanids, whom they had conquered.

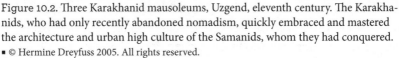

Summarizing the findings of several dozen archaeologists and philologists, it is now clear that this important Karakhanid center was in Kyrgyzstan, at a site called Burana in the broad Chu River Valley, some fifty miles east of the Kyrgyz capital of Bishkek. The other possibility, Ak Beshim, lies 3.7 miles to the northwest. In 1961 it was identified as the ancient city of Suyab, a very large (the outer walls embrace eight square miles) center of manufacturing and trade that had boomed in the sixth through ninth centuries but faded just before the dawn of the Karakhanid era.[34] The Karakhanid princes clearly understood that this point of the Chu Valley was the perfect site for the capital of a new state that embraced both sides of the Tian Shan. In light of the depleted and probably run-down condition of Suyab, they launched their new city at the best alternative site nearby.

And what a site they chose! Nowhere in Central Asia does one more vividly sense that one is on a major east-west thoroughfare than in this broad valley flanked by snow-capped mountains to the north and south. As one arrives from the west, the topography seems to pull one eastward toward China, as the Chu Valley narrows and eventually turns into a gorge. Reaching Balasagun from the east, however, medieval travelers would have felt that all Central Asia lay before them.

Both Balasagun and Suyab were what we would now call multicultural centers. Like most medieval cities in the densely populated Seven Rivers region that lies just west of the Tian Shan and Alatau Mountains in Kyrgyzstan and Kazakhstan, they had congregations of Syrian Christians (two churches in Suyab) and in Suyab two Buddhist shrines and a Zoroastrian temple. The Karakhanid rulers suppressed Buddhism but did not touch the Christians or Manicheans, both of whom, unlike the Turks themselves, were active in commerce. This may explain the establishment and existence over several centuries of the Holy Trinity Monastery on the picturesque northeastern shore of Lake Issyk-Kul. Because it contained what were believed to be the remains of St. Matthew the Evangelist, this ancient establishment was an important pilgrimage site over several centuries before it was lost under the rising waters of the lake.[35]

Balasagun had a densely built-up urban core (*shahristan*) with high walls that encompassed a rectangular area of fifty acres and were fully sixty-five feet thick at the base. Unusual for Central Asia, the capital had no citadel. Was this because Turks were more accustomed to the role of besieger than of besieged? A mile and a half farther out was a second wall that protected the commercial area or *rabat*, with probably a third wall beyond that. In the area of this outer ring stood at least five semi-urban estates, large walled compounds with dozens of rooms and broad central corridors up to a hundred feet in length. Running water, baths, and under-the-floor heating systems rendered these multistoried estates very comfortable, even by modern standards.

Much, including the main mosque, remains to be discovered and excavated at Balasagun, although, being a flat site that has long been plowed, much has also been destroyed. What is known for certain is that at the heart of the urban core or *shahristan* stood two mausoleums as at Uzgend, in all probability also memorials to Karakhanid rulers. At Balasagun they were round structures, however, with domes or tent roofs. Immediately adjoining them still stands a cylindrical tower or minaret, known today as the Burana Tower.[36] A similar tower stands at Uzgend. Both of them are now greatly reduced from their original height of 131 and 147 feet, respectively. Their outer walls are still divided by horizontal bands that recall weavings, and these in turn define sections of complex ornamentation in the brickwork. These minarets appear to be part of an important manifestation of creativity in the era of the Karakhanids.

Turkic Victory Towers

Specialists have long known that such tall, cylindrical towers or minarets originated in Central Asia and Afghanistan, and that they spread from there to India in the East and to the later Persian and Ottoman empires in the West. Earlier, mullahs issued the call to prayer either from the rooftops (as in Arabia in the early days); from towers based on Roman or Byzantine prototypes (as the Umayyad rulers did in Damascus); from prayer towers built into the outer walls of mosques (again, as did the Umayyads and North Africans); or, in the unique case of the Abbasid capital of Samarra north of Baghdad, from a weird spiral structure copied from ancient Persian ziggurats.

Beginning just before the rise of the Karakhanids and flowering under their rule, Central Asians followed a bolder course, constructing free-standing and slender columnar minarets. Scholars have catalogued no fewer than sixty such edifices in the region. By far the best known of them is the proud and sinister Kalyan minaret (built 1127) in Bukhara, known also as the "Minaret of Death" because the local authorities hurled condemned criminals to their deaths from its pinnacle.

What has not been adequately acknowledged is that most of the earliest of these appeared under the Karakhanids, and that nearly all the others were sponsored either by the Karakhanids or by other Turkic regimes, whether the Ghazni rulers in Afghanistan and India or the Seljuks in the western part of Central Asia and Iran.[37] The Karakhanid origins of the towers in Bukhara, Vabkent, Uzbend and Balasagun are certain, as are many others. The other Turkic dynasties that appeared soon after the Karakhanids seem to have zealously taken up the construction of such towers. One of the few major cylindrical towers of the century not to have been constructed under a Turkic regime is the amazing the 213 foot tall Jam minaret in Afghanistan, which somehow escaped the notice of the outside world until 1886. In this case, though,

Figure 10.3. The Kalyan minaret, Bukhara, 1127. After conquering the former Samanid capital of Central Asia, the Turkic Karkhanids placed their victory monument in the very center of Bukhara. ▪ © Hermine Dreyfuss 2005. All rights reserved.

Figure 10.4. The 65-meter-high minaret from the 1190s at Jam in central Afghanistan celebrated the Ghorid dynasty's victory over what remained of Mahmud's Ghazni.
■ Photo by David Adamec.

the local Persianate Ghorids who constructed it had just defeated a Turkic army from Ghazni, Afghanistan. They celebrated their victory over the minaret-building Ghaznis by erecting a giant one of their own.[38]

Before concluding that these were a peculiar phenomenon of the Karakhanids and the other Turkic rulers who appeared on the scene slightly later, it is important to ask whether these minarets might have been conceived and executed locally, rather than by the Turkic over-lords. That local architects were involved and that some of them came from the settled Persian (Tajik) population is proven by the extraordinary Jarkurgan tower at Chamangan, Uzbekistan. With its columnar form arising from the integration of eight profusely ornamented sub-columns, it stands as one of the most resplendent buildings remaining from Central Asia's Age of Enlightenment (see plate 16). The architect, Ali bin Muhammad of Sarakhs (now just across the border from Turkmenistan in Iran), was so pleased with his work that he signed the tower in the baked bricks. But assuming that at least some of the architects and builders of these monumental towers were local and possibly Persian (Tajik), who was it who conceived the idea of erecting them, and who paid for their construction?

A clever researcher from Kyrgyzstan, D. Imankulov, has recently come up with a highly innovative insight on this issue. Using modern technologies, he measured the height of a number of major minarets of the Karakhanid lands from the eleventh and twelfth centuries. He then measured the diameter at their bases and arrived at the ratio of 1:2,666.[39] This ratio turns out to have been remarkably constant, to the point that it defines the canonic form of nearly all Central Asian towers of the period. Imankulov then applied it retroactively to the now-truncated minarets at Uzgend and Balasagun and arrived at their original heights.

It is possible that architects and builders wandered from place to place in the eleventh century proposing and building such similar towers, the way European master stonemasons were to do with Gothic cathedrals a century later. But the power and resources needed to initiate such grandiose projects in the very heart of the great urban centers was far more centralized in Central Asia than in Europe.

The path leads back to the basic reality of life in Central Asia, the complex interaction of settled oases people and the nomadic tribes that came in successive waves from the East. In good times this relationship was symbiotic and mutually beneficial, with each side deriving economic benefits from the other. But whenever a new nomadic army emerged on the doorsteps of the settled cities, a new modus vivendi had to be worked out. The immigrant Karakhanid population was not large, its leadership was divided, and the Karakhanids' control always tenuous. They therefore had to take special measures to "plant their flag."

We have seen that one way they did so was to erect handsome mausoleums for their deceased rulers in the urban centers. Far more evocative, and hence more effective, would have been the erection of minarets that could dominate the cities and surrounding countryside, impressing all with the Karakhanids' presence and might. This, one might hypothesize, was the main reason the Karakhanid rulers worked with such zeal to construct minarets. The noted art historians Richard Ettinghausen and Igor Grabar were therefore almost certainly correct when they characterized these so-called minarets as "victory towers."[40] The canonic ratio of the resulting structures reflects their common genealogy.

Over time the columnar victory towers acquired other civic functions. The German historian Ernst Dietz documented their role as guideposts for travelers, pulpits from which official decrees were announced, and, in Bukhara and elsewhere, places for administering death penalties.[41] This is not to deny the core religious function of minarets in general and of some of these in particular. But it is worth noting that neither at Balasagun, at Uzgend, nor elsewhere did the Karakhanids erect major mosques, and at Balasagun the structures that stood nearest to the minaret were royal tombs, not places of worship. V. D. Goriacheva of the Kyrgyz Academy of Sciences speculated that the mosques may have been built of wood or that they were of the simple open-air type common in early Arabia, neither of which would have left strong traces in the ground. Neither theory seems likely, however, for why would anyone go to the immense trouble of erecting a huge tower and then link it with a diminutive or impermanent mosque?

It remains only to inquire into the origins of the cylindrical form of the Karakhan victory towers. Several intriguing hypotheses have been advanced. Thus it has been suggested that they trace back to the stone memorial piers or *balbals* that early Turks erected over their leaders' graves.[42] Others have found their progenitors in the towers on which Zoroastrians exposed their dead, while still others have sought a possible source in the Buddhists' columnar memorial *stambkhas* that were once common in Afghanistan and Central Asia.[43] Noting the star-shaped plan of Mahmud of Ghazni's victory tower in Afghanistan, scholars have pointed to a general pre-Islamic origin for them.[44] One further possibility might be suggested. Because the steppe-dwelling nomads lived in a horizontal world, they had always been fascinated by height. They built tumuli over their dead that were as tall as possible, far taller in their original form than in their eroded state today. How could such people not be inspired by the prospect of building brick towers that reach to the skies?

Whatever the origins of the victory tower form, the exteriors of these columnar structures were invariably ornamented with bricks laid in a profusion of geometric designs. As noted in chapter 1, an unknown medieval Italian traveler was so impressed with the patterns on the Kalyan Minaret in Bukhara that he copied them on the exterior brickwork of the Doge's Palace in Venice. Some of these patterns were drawn from nomadic fabric designs, but others trace to the ornamentation of the great Samanid buildings of the preceding century. As such, these ornaments attest to the process of assimilation to urban life that began the moment the Karakhans dismounted from their horses and settled down in their four capitals.

THE KARAKHANIDS ASSIMILATE

In many spheres the Turkic Karakhanids showed themselves to be eager students of their Persian/Tajik subjects. Everything from the design of their houses, the cut of their clothes, and the pottery on their tables reflected the influence of the established oasis dwellers of Central Asia. Thus, even while Turkification advanced at an unprecedented pace on the oases, an overarching process that was the Karakhanids' chief legacy, the Karakhanid elite simultaneously embraced many aspects of the prevailing oasis culture and of the larger Islamic world of which they were now a part.

In addition to building mosques, mausoleums, and various other monuments, the Karakhanids set up typically Islamic foundations or *waqfs* for charitable purposes.[45] One Karakhanid ruler went so far as to build a major new hospital in Samarkand, permanently endowing not only the salaries of the doctors and auxiliary staff but even the budget for the kitchen, heating, and lighting.[46] He also constructed, staffed, and permanently endowed a religious school or madrasa, where pious youths could be instructed in the traditionalist and orthodox truths of the Sunni faith.

It is hard to imagine another case of newcomers accepting and mastering the intricacies of a complex, settled urban culture as quickly as did the Karakhanids. It was all well and good that the dynastic head preferred to live like his forefathers in tents at the army encampment outside the walls of Balasagun rather than in the well-appointed city.[47] Most of the formerly nomadic elite did otherwise. They were by no means deaf to Mahmud Kashgari's call for them to remain true to their steppe culture, but they did this even while assimilating much of the new world in which they found themselves. Thus, even while the Turks were transforming Central Asia, they were themselves being transformed by the civilization of the region they now called home.

No one more fully embodied this two-way process of cultural assimilation than the Karakhanid philosopher, poet, and statesman Yusuf Balasaguni, or Yusuf of Balasagun, the author of a didactic volume for rulers entitled *Wisdom of Royal Glory*.[48] He completed his book in about 1069, only three years before Mahmud Kashgari sat down to write his *Compendium*. Yusuf Balasaguni's great achievement was to have shaped and disciplined the vivid but unruly spoken Turkic into a flexible literary language, and then to use that language to compose over two hundred pages of rhymed verse on a subject that would have been familiar to any educated Persian or Arab: advice for princes. It represented a near-perfect cultural merger.

THE WISDOM OF ROYAL GLORY

Balasaguni's plot can be easily summarized. An ambitious youth arrives at the court of the prince or khan just when the latter begins searching for a new chancellor, or vizier. The young man gets the job, does brilliantly, starts a family, and spends long hours engaged in philosophical discussions with his boss. But he suddenly dies. The prince assumes responsibility for the vizier's young son, who eventually grows into his father's old post and renews the discussions, offering advice to his sovereign on both practical and philosophical topics. The prince is delighted and delegates extensive powers to his new vizier. Realizing that his young deputy needs a backup, the prince asks the vizier to recruit a second adviser. The vizier turns to his brother, who has abandoned the city for the life of a religious recluse—nepotism seems not to have been a concern. But the hermit rebuffs two letters from the ruler and consents to a meeting only when it is agreed that as soon as it is over he will return to his mountain retreat. Shortly thereafter the hermit dies. The vizier considers following his relative and adopting the life of a religious ascetic. But on his deathbed the hermit reminded his brother of his duty to his prince, and the vizier therefore assumes once more his official responsibilities, now seeking more than ever to do good to others.

So far so good: Balasaguni's volume is a fairly typical tale in the tradition of Persian and Arabic "mirrors for princes." But the story is merely the armature on which the author hangs his much richer allegorical, ethical, political, and religious subject matter. Thus the prince, called "Rising Sun," symbolizes justice because the sun is constant and steadfast; the first vizier, called "Full Moon," represents capricious fortune, which can quickly wax and wane, as actually happens with the vizier's sudden ascent to power and early death; the second vizier is "Highly Praised," the wise and practical leader who employs intellect to dispense detailed (if often banal) advice on how the prince should treat everyone down to cup bearers and beggars; and the hermit is "Wide Awake," because, like a Muslim Sufi or a Buddhist monk, he has awakened his soul to mortality and eternity.

Balasaguni recounts the complex interplay of these human and allegorical personages in verse, using rhymed couplets of equal numbers of syllables, a meter he borrowed from Ferdowsi's *Shahnameh* and applied for the first time to a Turkic tongue. Balasaguni's role as innovator anticipates what Chaucer was to do with English, Dante with Italian, and Luther with German.

A quick reading reveals that the book is built around a single polarity, but what are its opposing elements? One possibility is that the prince stands on one side and that all the other figures, whom he either hires or tries to hire, are on the other. Viewed thus, the work emerges as an eleventh-century guide to human resource management. But the mere juxtaposition is clearly between just two of the actors: the vizier and the hermit. Robert Dankoff, who masterfully translated both Balasaguni and Kashgari, sees it as a clash between civic duty and individual conscience.[49]

Balasaguni Embraces the World and Rejects the Sufi Alternative

Blasaguni presents the vizier as the sole support of the prince, and hence of society. Yes, there are nobles, an army, and religious law, but none of these can assure that society will live in peace and security. Only the presence of a competent and energetic minister or vizier can enable the prince to rule effectively and society to thrive. The vizier may be called "Highly Praised," but he is the chief pillar of humankind's communal existence.

In sharp contrast is the hermit (Wide Awake), whose life is lived outside society. Yusuf Balasaguni bends over backward to present this other-worldly and difficult figure in a positive light. The hermit is a Sufi, seeking to purify his soul in order to be able to perceive the divine light. But far from being an ecstatic Sufi who strains to the light through poetry and song, Wide Awake is a grim nihilist, for whom all earthly joy is sham. "Children are enemies," he intones, "a father is to be pitied, for his food is poison." "When a friend's heart is broken he becomes a secret enemy." "When Adam sinned, God made this world a prison as the punishment." Against all this his brother advances perfectly sensible arguments. The prince goes further, counterattacking with the accusation that those pursuing a life of renunciation—Sufis—seek praise for doing so and thereby commit the sin of pride.[50]

In the end, the role of Wide Awake is less to defend an alternative vision of life than to serve as a foil to his brother the vizier and the prince. Goaded by Wide Awake, the vizier comes back again and again, championing active engagement with the messy world, affirming the possibility of doing good for others, and, in Balasaguni's words, "telling Wide Awake that the next world is won through this world."[51] In the end the appealing figure of Highly Praised, who acknowledges his own delight in the very things his brother so bitterly denounces, wins the day. At the very time when Sufism was beginning to sweep the Turkic world as a reaction against a decaying civic life, *Wisdom of Royal Glory* offers a civic manifesto, a ringing affirmation of the city against the lures of the hermit's cave.[52] In spite of this, none of the other disputants comes close to the hermit Wide Awake in sheer passion and intensity. The others speak from the mind, but the hermit's voice arises from his very guts. This fact introduces a curious element of ambiguity into *Wisdom of Royal Glory*.

The Hermit Who Lost the Battle but Won the War

At two points in the book Yusuf steps from behind the curtain and addresses the reader directly. In both cases his voice is decidedly grim.

First, in the introduction he speaks at length about the onslaught of old age: "O greybeard! Prepare now for death and mourn your years gone by."[53] Second, Yusuf Balasaguni appended to his book two short odes. The first is a further rumination "On Old Age and the Loss of Youth." The second, however, "On the Corruption of Time and the Treachery of Friends," is a bitter denunciation of the times in which he lived, when "hearts and tongues are two-faced," there are no "true neighbors," "kin act like strangers" and "whoever I held close like my own eye turned out to be an enemy like the devil himself."

This is the bitter voice of Wide Awake himself. Like the hermit, Balasaguni directs his attack squarely against the civic realm and the society of his fellows. In that same voice Balasaguni concludes,

> Here I have found nothing better for myself than to abandon society and live far removed from men. Let them not hear my name or see my form, and if they seek me let them not find me. They string like scorpions, suck blood like flies, and yelp like dogs—which shall I strike first? Now I writhe and groan, having fallen among shameless folk. May treachery's oppression not reach me, may I be delivered from the wicked and the faithless. Give me sustenance, my God! Let me witness the face of our beloved Prophet, and the faces of his four companions.[54]

Against this background, how are we to view this book by Yusuf of Balasagun? On the one hand, it is a passionate salute to civic life and all that can be achieved through positive endeavor guided by intellect and wisdom. On the other hand, it proclaims the very opposite, a life of renunciation and passive withdrawal fed by a realization of the utter folly and emptiness of society. Balasaguni makes clear that in his personal case the shift from the former to the latter arose from the disenchantments that come with age. But he was also reporting that society no longer shone with the luster of hope and philanthropy, and that serious people were beginning to doubt whether the civic realm could ever bring fulfillment in life. Was the Age of Enlightenment beginning to fade even as Turkic peoples were participating for the first time in the high culture of urban Central Asia? To the extent this is so, it strikes an ominous note for the future.

A KASHGAR AFTERGLOW

In 1070 or shortly thereafter, Yusuf Balasaguni presented his manuscript to the reigning Karakhanid prince at his palace in Kashgar. He read it to the monarch, who rewarded him with the title of privy chamberlain and the pension that came with high posts. Seven years later Mahmud Kashgari returned to the capital after completing his *Compendium*. Since Yusuf was by then between fifty-seven and sixty-six years old, and since Mahmud died only in 1102 (at the age of ninety-seven!), it is more than likely that the two actually met. Whether or not they knew one another, however, both Kashgari and Balasaguni were honored and respected. When they died, their tombs in villages south and southwest of Kashgar became points of visitation and pilgrimage. Mahmud's still stands in the village of Upad, while Yusuf's, in the village of Tainap, was destroyed by Red Guards during the Chinese Cultural Revolution.

Unfortunately the flowering of Turkic languages and culture that these great scholars and poets initiated could not be sustained. The local khan who had patronized them both was named Ibrahim Tamghach Bughra ("Abraham Tamghach the Camel"). A tough leader, he was once shown graffiti that local gangsters had scrawled on the wall of Samarkand. To their boast "We are like an onion: the more we are cut, the more we grow," he responded with his own graffiti: "I stand here like a gardener. However much you grow I will uproot you."[55]

The presence of a resolute leader did not prevent dark clouds from gathering over Kashgar, Balasagun, and Uzgend at the very time the two authors had returned to Kashgar. The Seljuk Turks had already attacked the Karakhanids from the west, forcing them to acknowledge them as suzerains. Then Ibrahim Tamghach died in 1072, just two years after bestowing honors on Yusuf Balasaguni. His successors were weak and ineffective.

Over the next generation a new threat appeared in the form of the Karakhitai tribes. Buddhists and shamanists, they were pushed out of China by the Turkic Tunguz and began working their way westward after 1125.[56] A Karakhanid army from Kashgar defeated them in an initial confrontation, but within a decade they were back, ten thousand strong, sweeping through Kashgar and Balasagun into the heart of Central Asia. The Karakhanids' days were numbered.

In one of his more optimistic passages, Yusuf of Balasagun had announced that states are formed by the sword but maintained by wisdom. A few pages later he offered a very different judgment: "Two things, gold and the sword, hold the realm together."[57] Within a generation of the deaths of Yusuf Balasaguni and Mahmud Kashgari, neither gold, the sword, nor wisdom sufficed to save the Karakhanids from their fate.

CHAPTER 11

❖

Culture under a Turkic Marauder: Mahmud's Ghazni

A New Type of Ruler

We left Abu Rayhan al-Biruni at a perilous moment. He had loyally served his besieged regent in Gurganj and then, after Mamun was murdered, negotiated with the enemy generals to secure the best possible deal for his homeland. He had not forgotten Sultan Mahmud's letter to the shah of Khwarazm, Abu Abbas Mamun, demanding that he ship off the stars of Mamun's academy to adorn his court at Ghazni in Afghanistan. He had put off Mahmud before but now realized that he was in no position to negotiate. In short order Biruni, then forty-four, and several other leading lights of the now defunct scientific center in Khwarazm set out across the Karakum Desert for Afghanistan. Biruni was to spend the rest of his life in the service of Mahmud of Ghazni and was eventually buried at Mahmud's capital, where his tomb can still be seen today.

Who was this Sultan Mahmud of Ghazni, who had married off his sister to the ruler in Gurganj, then incited a lynch mob against his new brother-in-law and crowned the sordid affair in 1017 by kidnapping the men of science who had been the glory of Mamun's court?[1]

Now forty-six years old, Mahmud epitomized a new type of Turkic ruler who emerged in Central Asia in the eleventh century: the head of a permanent slave army who ceaselessly pursued wars of conquest throughout his thirty-three-year reign (997–1030). Slave armies were not new, of course. The conquering Arabs, the caliphate itself, and its two wayward offspring, the Samanids of Bukhara and the Fatimids of Cairo, had all relied on slave soldiers to maintain themselves in power. But all these forces had included at least a core of nonslaves, whether

Figure 11.1. Ruins of Mahmud's capital at Ghazni, Afghanistan, as depicted in a mid-nineteenth-century steel engraving by Albert Henry Payne, possibly after a painting by Lt. James Rathway, *Town and Citadel of Ghaznee*, 1848, now in the British Library. ■ Courtesy of Frances Pritchett.

Arab tribesmen or Persians.[2] Most contemporary Turkic rulers, including the Karakhanids, preferred to build their military around free members of their tribe. Mahmud was the first Muslim ruler in the region to staff his war machine almost entirely with slave officers and soldiers of the line, and to make the maintenance of that force the very core of his rule.[3] While it is true that this marks a step forward in the development of a permanent state,[4] as opposed to the personal patrimony of single leader, the creation of this militarized conquest-state carried a heavy price.

Mahmud's highly mobile army rarely fell below the force of 100,000 that he amassed to attack Balkh in 999.[5] In recruiting and deploying his slave soldiers, Mahmud was blind to color, ethnicity, and religion. He did not hesitate, for example, to send Hindu forces against the Turkic, Persian, or Indian armies that were defending Muslim cities. Even his own household consisted mainly of slaves. Far from being constrained by his Muslim faith, Mahmud believed that the highest religious authority, the caliph, had validated his actions and confirmed all the dubious privileges he so freely exercised.

Figure 11.2. Mahmud of Ghazni pillaged Hindu shrines and enslaved tens of thousands of Indians but gladly issued dual language coins for his Indian subjects, in Arabic and Sanskrit. ▪ Courtesy of the David Collection, Copenhagen. AD 1027, inv. no. C367. Photo by Pernille Klemp.

It was fitting that Mahmud was himself born a slave. His maternal grandfather, Alptegin, was a Turk who had risen to prominence in Bukhara as a member of Ismail Samani's bodyguard. In due course he advanced to head the army,[6] making him the most powerful man in the state and also one of the richest, with five hundred villages paying him levies.[7] After serving as head of the Samanid garrison in Nishapur, Alptegin in 962 was named governor of Ghazni in Afghanistan. He promptly claimed for himself the title of king of Ghazni, all the while proclaiming his undying loyalty to his Samanid overlords (see plate 17). Mahmud's father, Subuktegin, was also a slave, from the town of Barskhan on the shores of Issyk-Kul, the same town where Mahmud Kashgari was born. Subuktegin had the good fortune to marry Alptegin's daughter.

The Ghazni kings never tired of reminding the world that their rule was based on the sophisticated and cultured model of the Samanids. While true with respect to both the language and culture that Mahmud supported, it does not reflect either the structure or functioning of the state itself. The claim of kinship to the Samanids is no more credible than its variant, that Mahmud's was the last state to be patterned after the caliphate, with its Persianized elite backed by a Turkic army composed mainly of slaves. Both claims are only half true. Far more than either of these precedents, the state that Mahmud brought to full flower was an unalloyed despotism.

Staffed at the top mainly by Turkic allies of the sultan and backed by a slave army of Turks and Hindus, Mahmud's imperial state, during its brief flowering, was able to control and demand payments from its subordinate provinces to a degree that would have exceeded the caliphs' wildest dreams. When Mahmud's vizier failed to confiscate enough wealth from the city of Herat, Mahmud first demanded that the key official make up the difference from his own pocket, and when that did not suffice, he tortured the man to death.[8] No wonder that Mahmud's gold coins kept their value!

Mahmud's state was not the kind of "oriental despotism" depicted by Karl Wittfogel,[9] grounded in the complex organizational networks needed to manage large hydraulic systems. Nor was it the kind of trade-based autocracy that had been the norm in Greater Central Asia. The latter, after all, had included important elements of pluralism and in-clusiveness, for the simple reason that their economies depended on the productivity of manufacturers, farmers, and traders, and not on the state alone. By contrast, Mahmud's entire system was based on the primitive extraction of wealth from conquest and from the peoples it ruled. This focus caused him to neglect or, worse, destroy the very social groups that might have sustained it.[10] This thwarted social development, isolated the Ghazni state from normal commerce and cultural contacts, and rendered it unsustainable.[11]

Contemporaries from the western border of Persia to the Ganges had good reason to think of Mahmud of Ghazni solely in terms of the de-struction wreaked by his army during its endless campaigns. Following his brutal conquest of Gurganj, Mahmud went on to attack the city of Rayy near modern Tehran. Here the Shiite Buyid dynasty had gathered a large group of eminent thinkers, including the Neoplatonist Miskawayh, a leading ethicist and historian.[12] All were driven out and the large li-brary destroyed. Earlier in his career Mahmud had battered Balkh and oppressed its population with punitive taxes. He had similarly brutal-ized Merv, Sistan, and other cities of Central Asia. In Nishapur in 1011 a famine induced by Mahmud's onerous taxes led to the death by star-vation of 100,000 people and widespread cannibalism, which began by the eating of exhumed corpses and ended by citizens of Nishapur killing one another on the streets for food.[13] All this took place during intervals between no fewer than twenty savage raids on India. Gandhara, Multan,

Lahore, Kannauj, Ajmer, Gwalior, Peshawar, and Shimnagar were all devastated by Mahmud's armies.

Even by the bloody standards of the day, Mahmud stands out for his relentless savagery. In his far-flung wars of conquest, Mahmud established a new model that was followed subsequently by Muhammad of Ghor, Tamerlane (Timur), and such Mughal rulers as Babur, Akbar, and Aurangzeb. Could it be an accident that the trajectory of militarized regimes from Mahmud forward parallels the decline of Muslim cultural life across Central Asia and northwestern India?

Even in his own day writers worried that Mahmud's reign might mark the beginning of a downward spiral for the region's civilization. Anyone speculating on this possibility soon comes up against the inconvenient fact that Mahmud, even as he pursued his savage wars of conquest, also patronized some of the greatest minds of his age, notably Biruni and Ferdowsi. He constructed monuments of architecture that overwhelmed their beholders, and he lent his support to a phalanx of writers, including several whose Persian poetry still delights and enchants readers.

The reason we find this a conundrum is that it challenges our modern assumption that learning and the arts arise naturally from just political and social orders and, conversely, that bad politics somehow poisons the life of the mind and spirit. The easy way out is to argue that culture under the Ghazni kings flourished in spite of their leadership and not because of it. But as we shall see, this misses an essential truth concerning Mahmud of Ghazni, namely, that he was in some respects a true cultural heir of the Samanid and Khurasan monarchs, even as he obliterated much of what they created.

Mahmud's Rise and the Government That Supported Him

Mahmud's rise was greatly facilitated by his father, Subuktegin, who successfully transformed the modest Samanid satrapy of Ghazni into a rising empire. It was Subuktegin who, in the course of military campaigns in Afghanistan, Khurasan, and India, introduced his two teenage sons to the art of war. The pivotal moment in Mahmud's rise occurred when his father bequeathed his expanding empire not to the twenty-three-year-old Mahmud but to his elder brother, Ismail. Mahmud promptly tracked down Ismail and murdered him.[14]

No sooner had Mahmud gained control over Subuktegin's legacy than a Karakhanid army attacked him from the north. The prize over which the two Turkic armies fought was what remained of the dying Samanid empire and its capital, Bukhara. Mahmud succeeded in overwhelming Bukhara and crowned himself sultan, but neither force could defeat the other. And so Mahmud and the Karakhanid leader struck a deal, dividing the spoils of the Samanid state and establishing a border along the Amu Darya.

Mahmud then launched the first of his twenty campaigns in India. But while he was away, the Karakhanids in 1006 attacked Ghazni territories in Khurasan and Afghanistan. Confronted by this act of treachery, Mahmud rushed home and, with the help of his weapon of choice, a corps of five hundred mounted and heavily armed fighting elephants, annihilated the invading Karakhanid army.[15] After this resounding victory, Mahmud recaptured Nishapur from Samanid diehards who had seized it in his absence.

By 1006 Mahmud was firmly committed to a life of ceaseless warfare. Down to his death in 1030 he campaigned constantly, pausing only long enough to embellish his summer capital with grand buildings and to erect a winter capital and resort for himself and his inner circle. To pull this off he needed a reliable system of governance and an administrative structure that could manage his huge wealth, provision his army, and nip rebellions in the bud.

The standard view is that Mahmud achieved this by relying on two management models: first, that of the caliphate in Baghdad, where top-down control was exercised by an all-powerful vizier, and, second, the more systematic administrative model that had been developed by the Samanids in Bukhara, which employed effective agents at the local level and, with respect to their overlord in Baghdad, made lavish shows of loyalty to mask their de facto autonomy from the caliph.[16] This picture is only partly true. For while Mahmud indeed had a vizier, or *wazir*, he invariably hated these officials from the moment he appointed them.[17] And while Mahmud kept the old Samanid administration in place, he wholeheartedly courted the caliphs' support down to the time when he decided that he himself, rather than the caliph, should rule the Islamic world. This campaign for total control began as early as the year 998, when Mahmud anointed himself with the heretofore unknown title of "sultan."[18]

The sole element of continuity in Mahmud's system, besides the sultan himself, was a council made up of his most trusted acolytes. This was a consultative body of sorts, but it was also the main executive authority, implementing Mahmud's will through deals with subordinate leaders throughout the realm. Backing up the council was an effective postal system that was closed to all but official correspondence, and an extensive network of spies, who checked to see that central decisions were faithfully implemented and punished all lapses. Finally, to assure that his subjects appreciated his tireless efforts, Mahmud had an official news writer, a chief propagandist, who bombarded the public with the sultan's version of the events of the day[19]

Far more than the caliphate or the Samanids, Mahmud's system existed to maintain the army. Not only did this require the constant provisioning, feeding, and rewarding of tens of thousands of troops in the field, it meant the feeding and care of Mahmud's most awesome weapon, his corps of fighting elephants. These mighty pachyderms were equipped with elaborate seats and attack spears and were overseen by a legion of handlers, most of whom were Hindus from India. We will turn shortly to the system that Mahmud devised for extracting money from the subject peoples. Suffice it for now to note that the price of this system was to suppress transport and trade and to put vast quantities of arable land out of cultivation. Even beyond the devastation of home territories caused by Mahmud's army, this led to the steady decline of the economy wherever Mahmud and his successors held sway.

"God's Shadow on Earth"

What was the great purpose that Mahmud's military machine served? The sultan had no doubts on this score. He saw himself as "God's shadow on earth"[20] and cultivated an image of strict Sunni orthodoxy by claiming to having memorized the Quran and by writing (or commissioning someone else to write) a book on Muslim jurisprudence (*fiqh*).[21] Contemporary writers shared this image of the man[22] and wrote that his earthly vocation was to advance the cause of Sunni Islam at home and abroad and to destroy whatever stood in its way. Some have seen this behavior as typical of the newly converted and a manifestation of a restless militarism of the Turkic steppes, as opposed to the more serene cultures of the Persianate oases, which were often self-assured enough to be venturesome in matters of faith. But it might be noted that the Turkic Karakhanids were also orthodox Sunni Muslims and even more recent converts, yet their faith was cool where Mahmud's was hot, broad and relatively tolerant where Mahmud's was narrow. Also, it was precisely during Mahmud's lifetime that many Muslim thinkers began to have serious qualms about the moral validity of holy war, or jihad, posing a choice between Islam and death.[23]

Even though most of the writers who extolled Mahmud's exertions for the faith were on his payroll, their claims about his religious motives are not unfounded. He was the first to take Islam into the heart of India, and when he sacked one Hindu temple after another he did so in the name of Islam. Deeply loyal to the caliph and his Sunni creed, Mahmud drew the sword against perceived heretics of all sorts, and especially Shiites. He hounded down and murdered Ismaili Shiites in Iran and Pakistan[24] and obliterated the power of the Shiite Buyid court at Rayy, which had controlled the caliphate for a century. In pursuing these campaigns, Mahmud was ruthless, deploying his elephants to trample religious dissidents and committing both schismatics and their libraries to the flames.[25] Even Biruni was not above suspicion, with courtiers questioning his orthodoxy, which was admittedly shaky.

Acknowledging all this, it is important to note the countless instances in which Mahmud's actions sprang from quite different motives. In Nishapur he began by supporting the orthodox Sunnis but then found it convenient in Khurasan to back the schismatic Karramiyya, who had declared total war on the caliphs and their riches.[26] Mahmud's loyal secretary used a poem to present this cynical maneuver as an act of statesmanship, but everyone understood it for what it was.[27] Meanwhile, Mahmud's Karramiyya puppet carried out a reign of terror that involved extortion, show trials, purges, forced confessions, and poisonings, to the extent that "people saw that his [Mahmud's] saliva was deadly poison and his exhalation meant ruin."[28] Then Mahmud abruptly shifted ground and again supported the mainstream Sunnis.

Mahmud shed as much blood of fellow Muslims as of pagans. And while he systematically sought out and destroyed Hindu temples in India, he did not hesitate to deploy unconverted Hindu troops against Muslim forces. Having bought the loyalty of these and other conquered Hindus, he gave them an entire quarter in his capital and granted them a tolerance he would never have accorded to deviant Muslims.[29] Meanwhile, he did not hesitate to turn conquered Indian provinces back to their unconverted Hindu rulers as long as they paid him tribute. And for all his professed loyalty to the religious authority of the caliph, Mahmud by the end of his reign was in the process of subjugating the caliphate itself.[30]

Was Mahmud a hypocrite? Probably not, if one assumes that a hypocrite is conscious of the gap between principles and practice. Yet Mahmud's record shows that Islam served neither as a consistent guide for, nor constraint on, his actions. When it was convenient he wrapped himself in the robes of piety, but just as often he acted out of sheer lust for conquest, even as he remained convinced that he was a true jihadist.

PAYING FOR IT ALL

This brings us once more to the question of how Mahmud financed his sultanate. A typical Central Asian leader would have used every chance to encourage transport, trade, and commerce and then tax it. This was not Mahmud's approach. In fact, no other ruler of Central Asia showed such total indifference to trade and commerce as Mahmud. Instead, to fund his troops and elephant brigades he relied, first, on the traditional Muslim tax on the agricultural lands of nonbeliever subjects, the *kharaj*. Totally lacking in scriptural justification, this tax was nonetheless highly lucrative. But it did not suffice for Mahmud's gargantuan needs. To fill the gap he relied on war booty, mainly from India.

From the time of his first raid on the Indus Valley in 986, Mahmud focused single-mindedly not on converting Hindus but on plundering their riches. In this he followed in the footsteps of his own father, Yakub the Coppersmith, and, still earlier, of the Sassanian Persians. But if for these rulers pillaging was a welcome by-product of conquest, for Mahmud it was the shining goal. In the course of his campaigns he besieged and captured some of India's greatest fortresses.[31] Contemporary biographers invariably state that fifty thousand civilians died in each battle, which is quite possible.[32] And in describing the loot he sent back to Afghanistan, they run out of superlatives.

The canonic fifty thousand deaths is also the toll cited for Mahmud's desecration and destruction in 1025 of one of Hinduism's most holy shrines, the temple to the moon god Mahadeva at Somnath on the southern coast of Gujarat.[33] To get there Mahmud drove an army across the bleak expanses of the Tar Desert at immense cost in lives. The Hindus valiantly defended their shrine, and when it finally fell Mahmud packed whole caravans with gold and the temple's holy lingam, which he installed at the Friday mosque in Ghazni. No less valuable were the tens of thousands of Hindus he sent to Ghazni as slaves. His staff biographer reported that there were so many slaves from India that "the drinking places and streets of Ghazni could not hold them all," and that "a white freeman was lost among them." He goes on to note dryly that Mahmud's actions let to the glutting of slave markets across Central Asia, driving down the price of chattel labor.[34] Nothing reveals more clearly Mahmud's priorities than the fact that as soon as he finished pillaging Somnath he turned the site back to the Hindus and departed.

CHARACTER AND CULTURE IN THE WORLD OF MAHMUD

For all his talk of being the sword of Islam, the driving force in Mahmud's restless life was pure avarice. When his court biographers listed his greatest victories, they invariably were those that yielded the most plunder. His unalloyed love of money surfaced in even his most trivial dealings, For example, after his conquest of Nishapur, he learned of an extremely rich man of that city. Summoning him to Ghazni, Mahmud accused him of being a heretic, for which there was absolutely no evidence. The poor man vehemently denied the charge and indicated he would pay whatever was necessary to clear his name. Mahmud took all his money and left him a document absolving him of all charges.[35]

In his personal life the Turkic sultan was austere, except for the fact that he drank wine and had a long-term homosexual relationship with a Turkmen slave whom his courtiers cordially hated.[36] Not surprisingly, he was headstrong, impatient, and utterly intolerant of anyone who contradicted him. Otherwise his demeanor was reserved, and he was known as a hard worker. He professed a stifling form of traditionalist Sunni Islam, against which even his own son rebelled—by building for himself a palace, the walls of which were adorned in the Indian style with paintings of naked men.[37]

Mahmud was deeply insecure and at the same time pathetically vain.[38] Why else would he have concocted the title sultan for himself and then gone on to invent dozens of other titles with which to embellish his resume?[39] Why, too, would he have assembled up to four thousand elegantly attired warriors to greet him in two long files whenever he returned in triumph to Ghazni?[40] His goal may have been to outrank the caliph,[41] but in pursuing it Mahmud showed himself to be painfully aware of his lowly slave origins.

Mahmud was a man of contradictions. Raised by his father in a scholarly environment, he was literate in Persian and Arabic, as well as Turkic.[42] Like his father, he aped the elegant and artistic court life of the Persianate Samanids,[43] even to the point of commissioning some of the most splendid, and also eccentric, ornamented books ever created.[44] But then he adopted Arabic when he began his campaign to dethrone the caliph. We know he had a major library but know nothing of its contents beyond the fact that it included a copy of *The Easterners*, an important book by Ibn Sina that is now lost.[45] Even as he enriched his own library with books plundered from throughout the region, he wantonly destroyed the libraries of others, like the great repository at Rayy, if he suspected they contained unorthodox writings.

The one point of consistency in Mahmud's approach to culture was the avarice with which he collected those whom he considered talented. We have seen this in his virtual kidnapping of Biruni and the other great scientists whom Mamun had assembled at his court in Gurganj.[46] In the end Biruni was joined on the road to Ghazni by his close friend and former teacher, the mathematician Abu Nasr Iraq, and by the Nestorian Christian physician Khammar. Khammar, already in advanced years, converted to Islam in Ghazni, briefly practiced medicine there, but died soon thereafter.[47] Meanwhile Mahmud issued an arrest warrant for Ibn Sina.[48] Upon learning that Abu Nasr Iraq was also a good artist, Mahmud had him paint a precise likeness of Ibn Sina, which he sent off to Nisa with a demand for Ibn Sina's head.[49] But Ibn Sina had long since eluded Mahmud's net.

As for Biruni, he settled for a while in Ghazni, where he appears to have been assigned the role of court astrologer. Biting his tongue, but not forgetting his earlier tracts on the subject, he began assembling material for a weighty tome, *Astrology*, which he managed to complete only a decade later. Superficially addressing all the main themes of astrology, Biruni's detailed study brims with material drawn from his scientific work. It touches on everything from astronomy and mathematics to history and his beloved topic—calendar systems.[50] It is clear that Biruni had been assigned to write on astrology, and that he snuck in as much material as possible from his personal scientific research. Curiously, he dedicated his *Astrology* to a woman, Rayhanah, whom he did not identify and who is otherwise unknown.

No sooner had Biruni arrived in Ghazni than he began at once to carry out astronomical observations with improvised instruments. Within three years Mahmud had supported his construction of a new "Yamani Ring," an astronomical device named after Mahmud's title as "Right Hand of the State."[51] Along with his other inventions—a planisphere, an orthographic astrolabe, and his precursor of Galileo's telescope without a lens—it signified Biruni's return to his usual intensive work schedule. Soon he was to decamp for Lahore in what is now the Pakistani sector of Punjab.

That Mahmud pursued Ibn Sina, Biruni, and the other scientists from Gurganj so relentlessly attests more to his collecting instinct than to any serious interest in science. A story still being retold several generations later has Mahmud demanding that Biruni foretell the future, including the sultan's impending actions. When Biruni's prognostications turned out to be accurate, he almost paid with his life. Mahmud considered science a form of wizardry.[52]

In Ghazni the involuntary immigrants from Khwarazm joined a veritable army of writers, painters, architects, copyists, gilders, historians, painters, and goldsmith from throughout Mahmud's realm. Some had been forced there, others had been lured by handsome salaries and the promise of high positions at court, and still others had gladly entered the service of a regent who ruled so much of the known world and who could back up his professed love of culture with unimaginable resources. Thus, the same all-consuming avarice that flooded Ghazni with gold, artistic treasures, and slaves also transformed the bleak capital into a cultural center that was unrivaled in its day.

"THE GLITTER OF THE EARTHLY WORLD": MAHMUD AS BUILDER

No art surpasses architecture in its appeal to dictators. Whether in triumphal arches erected by Roman emperors, Darius's palace at Persepolis, where sculpted friezes depict humbled foreign grandees bearing tribute, or the palace of the Parthian kings at Nisa, with its receiving hall dominated by a throne of ivory, architecture proclaims power and conquest. The world's first sultan may have been born a slave, but he understood this truth as well as any leader before him.[53] Not that he neglected the other arts: suffice it to say that books, metalwork, and ceramics from Mahmud's reign all rank among the finest achievements of Central Asian art. But with virtually unlimited resources in the form of plunder from India, Mahmud could indulge his passion for architecture to the fullest. No wonder that he boasted (in lines inscribed on his victory tower in Gurganj) that he had created "the glitter of the earthly world."[54] Mahmud began building early and continued until his death.

He typically laid the expense of maintaining his grand edifices on the reluctant local public, so the process of decay began immediately. The cost to the citizens of Balkh for the upkeep of his formal gardens in that city was so burdensome that he first tried to foist it onto local Jews and then lost interest in the site.[55] Many of Mahmud's projects, including his summer capital at Ghazni and his winter capital at Lashkari Bazar, were at strategic points that were to become war zones over subsequent centuries. With few buildings surviving from Mahmud's day, we must reconstruct his activity as builder from the literary and archaeological evidence.

When Mahmud moved there, Ghazni was still a Zoroastrian center where Buddhists in an earlier era had built numerous monasteries and stupas.[56] Mahmud began by building a great dam on the Jikhai River that still stands fifteen miles north of the city. He then proceeded to erect a sumptuous new palace, residences for his principal aides, an ornate Friday mosque for six thousand worshippers, a hippodrome inspired by stories of the one in Constantinople, a 144-foot-high and exquisitely ornamented victory tower or minaret,[57] a madrasa or college with his state library, and a walled complex to lodge his fighting elephants and their army of grooms. A mile distant he erected for himself a massive two-part tomb and tower consisting of a shaft made up of vertical panels, a dramatic flange, and a cylindrical shaft, now destroyed, above. The massive doors bore elegant inscriptions in Kufic script that are in advance of anything done at the time in the better-known architecture of Fatimid Cairo.[58]

Typical of Mahmud, this monument was covered with no fewer than eight large inscriptions. Here and elsewhere the sultan exploited architecture to extol his own greatness, in the process turning buildings into what historian Robert Hillenbrand calls "huge billboards proclaiming various messages to those who enter them."[59] Lest anyone miss the point, Mahmud's secretary and publicist, Utbi, assured his readers that Mahmud's mosque at Ghazni far outshone the Grand Mosque in Damascus with respect to plan, scale, and sheer magnificence, and that Ghazni's other monuments similarly outdid their rivals elsewhere in the Islamic world.[60]

Entering the palace through doors sheathed with gold melted down from Hindu statues, one passed along corridors ornamented with mosaics of gold and lapis lazuli and paved with square and hexagonal slabs of white marble from India. Huge trees hauled from the Indus basin were also used for roofs and columns. Eventually one reached the sultan's domed audience chamber, which featured a gilded throne that "far outshone" those of Persia and Egypt. Excavations have confirmed that both the palace and mosque were heavily ornamented with sculpted terracotta, an ancient Iranian art that reached unprecedented heights under Mahmud and his successors. They also included ample space for the exhibition of trophies that Mahmud had laboriously hauled back to Afghanistan from Hindu temples he had sacked in India.[61] Biruni, with undisguised disgust, reported that Mahmud had placed a fragment of the great statue from Somanath at the doorway of his mosque in Ghazni so people could wipe their feet on it before entering.[62]

The large construction crew that created all this apparently consisted of both free and enslaved craftspeople. Utbi, in a unique moment of levity, or at least sardonic irony, reported that all of the craftspeople received "copious wages and complete reward" for their work. But while some were paid with money from the sultan's treasury, others—the slaves—got only "promissory notes from the Treasury of Heaven."[63]

Whether or not Ghazni surpassed all other cities in size, spacious plazas, and grand edifices, as the hyperbole-prone Utbi claimed, it briefly became a major international emporium. But its design and character were defined not by commerce, which Mahmud largely ignored, but by the sultan's passion for asserting his power through awe-inspiring buildings and grand ceremonies. The same motive underlay his decision to build the great victory tower at newly conquered Gurganj. Because the shahs of the Mamun dynasty had erected their own victory tower there only a few years before their fall, Mahmud set about constructing nearby his own 203-foot-tall Kutlug Timur minaret, which stills stands today.[64]

Besides Ghazni, Mahmud's grandest architectural indulgence was undoubtedly the palace complex he erected at Lashkari Bazar in the Helmand Valley of Afghanistan. Even though the mighty eighty-five-foot-tall ceremonial arch that remains in the town of Bost near the entrance to the site was built after Mahmud's death, its existence confirms that everything connected with Mahmud's winter capital was built to an awesome scale. The architectural heart of Mahmud's state was outside Bost at a place known locally as "Mahmud's City" (Shahr-i-Mahmud). The well-watered and comfortable site stretches for more than four miles along the elevated plain adjoining the Helmand River, offering ample room for Mahmud and his elite to build their palaces, and also for the many official buildings, mosques, and the barracks needed to house the sultan's guard and elephant corps.

The principal palace stretches for a third of a mile above a bend in the river. Excavations by French archaeologists between 1929 and 1952 revealed a heavily walled but spacious compound of several score rooms built around a vast central courtyard flanked by four arched portals or *iwans*.[65] As one of the earliest courtyards defined by four great arches, it set a model that Muslim architects emulated throughout the world over the next half millennium.[66]

Figure 11.3. A photograph from the 1930s of Mahmud's winter capital at Lashkari Bazar along the Helmand River in Afghanistan, his opulent celebration of the Persianate cultures he had conquered. ▪ From *The Illustrated London News*, March 25, 1950.

The central audience hall must have made a stunning impression, with its intricate ornamental panels in delicately molded brick, ornately sculpted stucco, and a central basin in the form of rose petals. Brilliant colors were everywhere, with the ceiling painted in bold primary colors and the floors set with brightly colored marble. The jewel of the audience hall was the large and richly colored murals depicting rural scenes of blooming flowers, cavorting gazelles, fruit trees, and birds. Amid this bucolic splendor were depictions of Mahmud's royal bodyguards. Dressed in long, heavily brocaded and belted kaftans and wearing boots with pointed toes and tops of embroidered felt, these painted figures reminded visitors of Mahmud's power even in the absence of the sultan himself.

The fragments of these wonderful paintings that are now in the National Museum in Kabul conjure up a court culture devoted to the leisurely and risk-free thrills of the hunt. The court historian Beyhaqi, writing about Mahmud's son Masud, described the sultan wielding his bow from the wall of his pleasure pavilion as six hundred gazelles were driven before him. The same author described the sultan on a leisurely outing on the Helmand River. Seated on a barge festooned with banners of silk and accompanied by a retinue of falconers, Masud whiled away the hours hunting, as musicians serenaded him.[67]

Considered purely as architecture, Lashkari Bazar stands as one of the most significant achievements of Central Asia's golden age. On this site selected by Mahmud himself arose a series of monuments that anticipated later architecture by several centuries. The expansive spaces,

the great portals defining a stately central square, the magnificent ornamentation in brick and stucco, and the brilliant colors all spread from
Lashkari Bazar eastward deep into India and westward across Persia to
the Middle East. Nearly three centuries later the monuments were to
exercise a powerful influence on the successors of Tamerlane (Timur).

HISTORIANS UNDER GHAZNI RULE

That Mahmud was a generous patron of Sunni orthodoxy cannot be
doubted. His official biographer, Utbi, outdid himself in describing the
rare and valuable books on theology that his master pillaged for his library at Ghazni, the size of the endowment that Mahmud set up to support learned divines at his college there, and the generous salaries the
faculty received. But this did not make him a serious patron of letters,
for there is no hint that the institution supported anything but narrowly
orthodox religious studies and what Utbi referred to as "chanting."[68]

To appreciate Mahmud's contribution to the written word we have
to look elsewhere, to historical writing and poetry. In both spheres
Mahmud emerged as an active, if hopelessly narcissistic, patron. For
Mahmud and for his son, Masud, history was the verbal equivalent of
architecture, an ideal instrument for glorifying one's own achievements.
Thanks to this, a bevy of talented historians left us an unusually rich
body of writing on the life and times of these two rulers.

The story began with Mahmud's secretary, the hard-working and
loyal Utbi, and his great memoir on the reigns of Mahmud and his father, Subuktegin.[69] Of Arab background from the Persian city of Rayy,
Utbi candidly acknowledged that he wrote to extol his boss.[70] He followed Mahmud through all his early campaigns and described them
with the directness of a firsthand observer. His account of the ghastly
famine that befell Nishapur as a result of Mahmud's taxes is the more
credible because of its candor. Elsewhere, though, Utbi often paused to
settle scores not only with Mahmud's critics but with his own enemies
as well.

Utbi wrote his history in an informal epistolary style, replete with
quick asides, riveting details, and entertaining one-liners that have little
or no bearing on his formal duty to glorify the Ghazni throne. He thus

reveals himself as a first-class writer, a stylist who wanted to reach an intelligent readership. Clearly, being in Mahmud's pay did not overly inhibit him, nor did it prevent several other historians from producing works of lasting value.[71]

Of all the contemporary historians of the Ghazni dynasty, by far the most distinguished was Abolfazi Beyhaqi (995–1077) from a town near Nishapur, who first found employment in the Correspondence Department under Mahmud's son Masud and eventually rose to head that office.[72] From this position Beyhaqi was privy to the inner workings of the dynasty. Neither a toady nor a renegade, he clearly found his task difficult at times, which led him to pen a monograph on the tribulations of being an official secretary. Like many independent-minded civil servants, Beyhaqi kept detailed notes on all that passed before him and probably squirreled away some official documents besides. Unfortunately most of these were taken from him when he retired and he had to work from memory and oral sources, bitterly complaining all the while.[73] He managed to produce a gigantic work of some thirty volumes. While most of the text focused on the reign of Mahmud's son Masud (Beyhaqi titled it the *History of Masud*), several volumes were devoted to Mahmud as well. Unfortunately only three volumes and a few additional fragments of this historical and literary masterpiece survive.[74]

Beyhaqi was a senior official of the Ghazni state, but he did not write a word of his history until he was nearly fifty and safely in retirement. From the outset he realized that he was engaged in the writing of what he unabashedly termed a "mighty monument."[75] His *History* is indeed the most credible source on the era, and for good reason. Beyhaqi had a lofty view of the historian's calling. He rued that "most people are so constituted that they prefer the absurd and the impossible,"[76] and hence "the number of people who can distinguish truth and reject falsehood is very small."[77] It thus falls to the historian to ferret out the reality of each situation, no matter how difficult that may be. For example, when Beyhaqi described Mahmud's wanton destruction of Gurganj and its academy in Khwarazm, he candidly avowed that "it comes hard for me to set down such words, but what can I do? One must not show partiality in writing history"[78] With impressive dispassion he detailed the assassination of a renowned vizier and all the self-indulgent vainglory and sybaritic folly into which the Ghazni rulers sank.

Then there is Beyhaqi's wonderful prose style. His thumbnail sketches of leading officials are worthy of a great novelist, as are his concise yet definitive judgments on the motives of leaders when acting out of arrogance or fear. He deftly indicated how the most trivial concerns gave rise to great historical shifts, while his dispassionate reporting permitted readers to draw conclusions on their own, without his preaching to them. Beyhaqi's language is rich and expressive and his writing fluent. At times gossipy and informal, Beyhaqi has been called by Bosworth "the Persian Pepys." Unlike Pepys, however, Beyhaqi revealed little about himself. Yet the appellation is amply warranted by his mischievous accounts of writers and artists at the Ghazni court, and by his ruthless judgments on the actions of leaders. Indeed, Beyhaqi was far more than a Ghaznavid Pepys. Here, from the heart of a forgotten empire that existed a millennium ago in Afghanistan, is the authentic voice of a keen student of human folly, an independent writer and thinker who fully grasped the timeless predicament of moral men in immoral societies.

THE COURT POETS

Nizami Arudi, a twelfth-century writer from Samarkand, observed that "a king cannot do without a good poet, who will make his name immortal and record his fame in verse and books."[79] Mahmud collected poets the same way he collected Hindu temple idols. And he did so out of the same motives, namely, greed and the desire to show his own greatness on the battlefield and as a man. Mahmud's "guild of poets" is commonly said to have numbered four hundred versifiers.[80] Even if this is quadruple the actual number, it still attests to a level of literary lion-hunting that knows few equals. Nearly all these poets wrote in Persian, although when Mahmud began to cast a covetous eye on the caliphate he hired some Arabic writers as well.[81] Even though he himself was a Turk, Mahmud showed not the slightest interest in Turkic letters.

The recruitment of this stable of poets was anything but systematic. One of them, Farukhi from Sistan, rose to eminence when he happened to observe one of Mahmud's local chiefs branding a horse and penned some laudatory verse describing the event. The official hired him on the spot and soon passed him up to Mahmud.[82] Others were dragooned from conquered territories, mainly from the cities of Balkh, Sistan, and nearby parts of Central Asia. Above all it was Mahmud's good fortune that several of the region's most resplendent courts, beginning with the Samanids at Bukhara, closed their doors just as Mahmud rose to power, throwing scores of poets and singers onto the market. Most found their way to Ghazni. The original model for all this activity, as for so much in Ghazni, was the Samanid court at Bukhara, which Mahmud's predecessors had so loyally served. If the Samanid ruler Nasr II enjoyed the company of a single Rudaki, why should Mahmud not have four hundred Rudakis?

The life of a court poet in Ghazni must have involved a lot of sitting around. We know this from written accounts of leisurely afternoons spent improvising intricate riddles, to be answered in verse.[83] The one product that the poets were expected to turn out were poems in praise of their Ghazni patrons. Written to be proclaimed orally, these verses, called panegyrics, followed strict conventions.[84] Their profusion of wooden formulas and tiresome superlatives convey a deep and, to modern ears, loathsome obsequiousness.

Take, for example, the poet Abu Ishak of Merv (known as "Kisai"). Not without talent, Kisai was capable of describing an azure-blue water lily thus:

> Color like heaven, and like the heavens, radiantly bright,
> But cup all yellow, as in the light of the fortnight-old moon;
> Yet your azure hue is like a monk grown sallow from a full year's
> fast,
> Who from the merit thus amassed is garbed all in blue.[85]

But when it came time for Kisai to heap praise on Mahmud, he did it with the tongue of a seasoned flatterer:

> O Shah, we well may call thy hand our jewel,
> For from it comes a never ceasing shower of gems;
> Though God hath made thy soul of bounty and noblesse,
> How, when that soul is fatigued, has it yet the power to breath?[86]

Such banality was nothing new for Kisai, who had made a good living writing similar verse for the rulers in Bukhara and then in Baghdad. He was one of many who turned sycophancy into a profession. Could he have done otherwise? Poets at the time could neither imagine nor be attracted by the ideal of the impoverished but soulful artist ensconced in his garret. Most of the four hundred poets at Ghazni had voted with their feet.

But might it have been possible for even one poet to write his views more candidly? Back in Samanid Bukhara, a writer whose patron, the vizier Balami, failed to pay him penned the following angry screed:

> Belami's vizierate is a complete mess,
> A lock attached to a pile of ruins.
> He respects neither savants, notables, nor scribes.
> No one more deserves a beheading.[87]

In the more relaxed atmosphere of Bukhara, the author, a satirist named Lahham, lived to see another day. The only poet in Mahmud's realm to criticize the regent was Ferdowsi, but he did so from a safe distance and in veiled language.

If mass-produced panegyrics had been the only poetry that sprang from Ghazni, Mahmud could be dismissed as a trivial patron. But the ruler himself is said to have penned verses, and even if the two surviving examples may be the work of others, he certainly understood poetry. The man who enjoyed the title of "King of Poets" (*Malik-us Shu-ara*) at Mahmud's court, Abul Qasim Unsuri, wrote some thirty thousand lines of verse, of which only a few thousand have survived. Unsuri came from Balkh and was among those artists whom Mahmud claimed for himself after capturing that city.[88] Unsuri's success at Ghazni owed much to his tactfulness. During a drunken party Mahmud had ordered his male companion to cut off his flowing locks. The next morning he realized he had done something stupid but was at a loss as to what to do. Unsuri saved the day with the following lines:

> Though wrong that your Idol's hair was shorn,
> Why wake and sit in grief forlorn?
> It's time for mirth and glee—to call for wine!
> Trimming the cypress' locks served but to adorn.[89]

Another court poet, Abu Nazar Abdul Asjadi from Merv, demonstrated a similar response to remorse, writing from the throes of a hangover that

> Of wine and praise of wine I do repent,
> Of lovely maids, fair chins with silver blent.

Lip words! My heart lusts still for sin.
Oh God! Such penitence Thou dost resent.[90]

In addition to the obligatory panegyrics, which intermingled syco-phantic tributes to his patron with highly refined descriptions of nature, Mahmud's poet laureate authored numerous graceful and affecting ro-mances.[91] In this he was joined by several other of Mahmud's poets, who sang of the problems that arise when young people from differing social worlds fall in love. Asjadi, his hangover passed, left these touching and musical words from a weeping lover:

Tear drops dripping from my eyes I shed,
Like the cloud, or murmuring, murmuring stream;
These drops the dripping rain have far outsped,
These murmurs like my sad heart's murmurings seem.[92]

Through extended works in this spirit, Mahmud's court poets did more than anyone before them to develop the romantic epic, the most impor-tant literary innovation of the age, and to this day a vital genre in many languages.[93]

At least one more of Mahmud's panegyrists, Farukhi, attained a very high literary standard. An accomplished musician and singer, Farukhi accompanied the ruler on his military campaigns and on the royal hunts. In works like his "Hunting Scene," he interspersed praise for Mahmud with timeless maxims and elegant notes on the beauties of nature and the joys of wine and love.[94] Because he served not only Mahmud but his two successors, Farukhi had to negotiate the obvious decline that oc-curred, which he did through lucid but complex poems built around the symbolic image of the garden. His verse on the death of Mahmud is one of the finest elegies in Persian.[95]

FERDOWSI: A SLAVE OF TURAN

If Unsuri, Asjadi, and Farukhi rose like promontories above the flat ter-rain of court poetry in Ghazni, Ferdowsi was a mountain, albeit a distant one. We left Ferdowsi at his family home in Tus, where he was already hard at work on his *Shahnameh*. When the Persian-speaking Samanids of Bukhara took control of Khurasan, he briefly benefited from their patronage of his work on Persia's great epic. But in 993 the Samanids were driven out of Khurasan by one of their own Turkic slave soldiers, whose grandson, Mahmud, declared himself ruler of all Central Asia. Mahmud's rise presented Ferdowsi with a conundrum.[96] On the one hand, the self-proclaimed sultan was a Turk and thus represented the forces of Turan against whom the Persian heroes of the *Shanameh* had struggled. On the other hand, Mahmud admired the Persian culture of the Samanids and used Persian rather than Arabic or Turkic at his court. Ferdowsi may even have looked to Mahmud to continue the Samanids' declared mission to revive the culture of old Persia.

Impoverished and with a daughter and granddaughter to support, Ferdowsi swallowed his pride and set off for Ghazni. Semilegendary accounts of his visit have him showing up at a garden where Unsuri and other poets had gathered. Suspicious of this new arrival, Unsuri challenged him to complete the second lines of several diptychs, or two-line verses. Ferdowsi passed the test masterfully but concluded that he would never be welcome among these toadies, who passed their days in riddles and poetic games. With promises of generous support from Mahmud, he returned to Tus, taking care thereafter to forward sections of the emerging *Shahnameh* manuscript back to Ghazni, along with some panegyric verses of his own.

We know from the following that Mahmud read these. In his role of champion of Sunni orthodoxy, the sultan besieged the Shiite city of Rayy, where Ibn Sina had spent time in his ceaseless effort to escape Mahmud's grasp. The ruler complained of Mahmud's troops but otherwise spent his days in chess and debauchery until Mahmud had him captured and brought before him. "Have you not read the *Shahnameh*'s history of the Persians and the chronicles of Tabari, with their Muslim history?" "Yes," the captive replied. Mahmud retorted, "Your actions contradict your words. And [by the way,] don't you see that one king has checkmated the other?"[97]

Exhausted by his labor of more than three decades, Ferdowsi penned the final lines of his epic on March 8, 1010.[98] Once more he made the arduous journey across Afghanistan to present his completed work to Mahmud in person. We do not know exactly what transpired during this visit, but the results were disastrous. Ferdowsi, now over seventy years old, set out for home empty-handed, with none of the sixty thousand gold pieces he thought Mahmud had promised him. What should have been the celebration of a lifetime's work had turned into a tragedy, not only for Ferdowsi but for his family.

Why did Ferdowsi leave Ghazni empty-handed? At least six possibilities have been proposed. First, Mahmud, who was notoriously stingy, may simply have refused to honor his earlier commitment. Second, Unsuri and the other Ghazni poets may have prevailed on the sultan out of sheer jealousy, a charge that was levied at the time.[99] Third, Unsuri and Farukhi may have made the case to Mahmud that Ferdowsi devoted far too much space to idealizing heroes from the distant past when contemporary heroes (read Mahmud) were so readily at hand.[100] A fourth explanation focuses on the unfounded claim that Ferdowsi, having given up on Mahmud, had presented a copy of the *Shahnameh* to a Shiite prince in Iran, making himself persona non grata in Sunni Ghazni. Equally far-fetched is the fifth argument, which has Mahmud displeased with Ferdowsi for painting the Zoroastrians of old in so positive a light.

Finally, it is argued that Ferdowsi was simply too pro-Persian. Back when Mahmud promised to support Ferdowsi, his main enemy had been the Turkic Karakhanids, whom he was trying to drive out of the Samanid lands. Ferdowsi's project reinforced Mahmud's claim to champion Central Asia's Persianate population against the Karakhanid clans. Now, however, Mahmud saw the possibility of destroying the despised Shiite Buyid regimes in Iran and then moving on to liberate Baghdad itself from Shiite control and placing himself at the head of the caliphate.[101] As part of this drive, he demoted Persian as a court language and strove instead to present his pro-Arab credentials. He rightly feared that the Persians could seize on Ferdowsi's *Shahnameh* to strengthen their resolve to resist Mahmud. The argument is logical, but concrete evidence for it is lacking, as, indeed, it is for all the other possible explanations for the disaster that befell Ferdowsi.

Ferdowsi's tragic end is wrapped in legend. We know that he left Ghazni at once for Tus, stopping one night at an inn near Herat. Mahmud, having heard that Ferdowsi had left Ghazni embittered, belatedly sent an elephant loaded with silver to pay the poet. But Ferdowsi had been promised sixty thousand pieces of gold, not silver. The elephant-load of money caught up with the poet just as he was buying a sherbet to enjoy after his bath. Infuriated, Ferdowsi disdainfully divided the money between the bath attendant, the sherbet seller, and the elephant driver. In his fury he is said to have appended to the *Shahnameh* a vitriolic passage in which he denounced Mahmud's duplicity, stinginess, and lack of honor.[102]

It is now accepted that this anti-Mahmud diatribe came from someone other than Ferdowsi. But even without it, Ferdowsi's had clearly stated in the *Shahnameh* his utter contempt for people like Mahmud. Indeed, several passages are fairly seen as coded attacks on the sultan who had loomed like a specter over the poet's life. For example, when a tyrannical and corrupt usurper of the waning Persian throne boasted that "I am the world's king," Ferdowsi deferred, angrily reminding his reader that "he sent armies out in every direction. He promoted evil men, as might be expected of a scoundrel of his character, appointing criminals as governors everywhere, and wise men had to bow and obey them. On all sides, truth was humiliated and lies flourished."[103] Ferdowsi left no doubt that he was speaking of the present, and that the real target of his wrath was Mahmud.

Even then, the story did not end. An ancient legend relates how Mahmud, stung by Ferdowsi's rebuff, finally sent the full shipment of sixty thousand gold pieces to the poet at Tus. But just as the caravan bearing the money arrived, a funeral cortege emerged from the city gate. It was Ferdowsi's.

The full story of Ferdowsi's relation to his patron, Mahmud of Ghazni, will never be known. What is clear is that Mahmud seized the opportunity to bring one of the world's great writers under his wing. Then, whether because of changed political or religious circumstances, envy on the part of court poets, or sheer stinginess, he had a complete falling out with the poet. One could never imagine such a thing occurring at the court of the Samanid kings in Bukhara, at Nishapur, or among the Turkic Karakhanids in Balasagun and Kashgar. The episode casts a shadow over Mahmud's already dark character. Even more, it speaks eloquently of the new model of personal, militarized, and conquest-driven governance that Mahmud introduced. Himself a slave, he created a slave state and then proceeded to inflict that system on the arts and letters. Ferdowsi finally recoiled from it all, but by then it was too late.

Natural and Social Sciences under Mahmud

Sometimes we come to understand things better by their absence. Mathematics, astronomy, medicine, and the natural sciences had aroused the curiosity of Central Asians since ancient times. Cross-fertilized by constant interaction with both India and the West, the best minds of the region could not help but compare and contrast these diverse traditions and seek the deeper truths that each brought partially to light. Enlightened officials burnished their reputations by patronizing such thinkers.

Against this background, it is all the more striking that Mahmud of Ghazni all but ignored these fields. True, he tried to corral the scientific innovators at Mamun's court in Khwarazm. Most of them escaped his grasp, and two of the three who came to Ghazni were never heard from again. Only the great polymath Abu Rayhan al-Biruni continued to labor in the sciences.

Under Mahmud the intellectual range of Central Asia's Age of Enlightenment began to narrow. Besides the trio from Khwarazm, he is not known to have patronized any other natural scientists, medical experts, or social scientists. But so brilliant was Biruni's work during his thirty-one years in Mahmud's realm that he made up for all those who were absent. As a friend wrote, "His hand was almost never without a pen, his eye without an observation, and his mind without a thought."[104] Biruni was Mahmud's one-man Academy of Sciences, and most of his 180 known works were produced while he was in Mahmud's domains.

Biruni went to Ghazni on his own free will. The main lure for him was India, which Mahmud had spent a generation conquering and pillaging. Back in Gurganj, Biruni had long admired the works of Indian astronomers and mathematicians, which he had studied in Arabic translations prepared in Baghdad or Merv.[105] This respect was fostered by his qualified regard for the ancient Greek scientists whom Ibn Sina had defended with such ardor.[106] In his *Chronology*, Biruni expressed regret at having found no one who could inform him on Indian calendar systems and history. Mahmud offered a timely window on this rich scientific heritage and Biruni eagerly opened it. Passing through Ghazni, he pushed on to India, where he remained for thirteen years.[107]

From his base at Lahore, Biruni appears to have made two extended research trips to Sindh, as well as shorter expeditions to areas in the north.[108] He also came to know Kashmir well and even wrote a book for scientific colleagues there. During all these travels Biruni showed himself to be the intellectual omnivore he had always been. One imagines him traveling with a whole wagonload of notes and scientific equipment. Few of the 160 papers and 20 books that resulted from his field research have survived, but we know from their titles that they embraced fields as diverse as hydrology, mathematics, astronomy, geography, medicine, theology, geometry, geology, mineralogy, anthropology, geodesy, zoology, and pharmacology. He carried out pioneering research in each of these fields, in the process advancing innovative theories on such varied subjects as the medicinal treatments of deafness, both the radius and circumference of Earth, and the hydrostatics of artesian wells.

In all his work Biruni, like modern scientists, was eager to apply number and measurement to the diverse phenomena of nature. Who else but Biruni would note that

> Among the peculiarities of flowers there is one really astonishing fact: viz., the number of their leaves, the tops of which begin to form a circle when they begin to open, is in most cases conformable with the laws of geometry. In most cases, [furthermore,] they agree with the chords that have been found by the laws of geometry, not with conic sections. You scarcely ever find a flower of seven or nine petals, for you cannot construct them according to the laws of geometry in a circle, as isosceles [triangles].[109]

And who but Biruni would throw out the line that "if you would count the number of seeds of one of the many pomegranates of a tree you would find that all the pomegranates contain the same number of seeds as that one, the seeds of which you had counted first"?[110]

It is impossible to summarize Biruni's many findings, but a few examples are revealing. During his travels Biruni paused to pin down the precise latitude and longitude of many of the cities he visited, beginning with a fort in present-day Pakistan. He also calculated and published the precise coordinates for Ghazni, literally putting Mahmud's capital on the map.[111] He carried out these calculations using a variant of Khwarazmi's old method and also a newer system that employed the tangent and sine functions he had learned from the Indians and from those Arabic mathematicians like Nasr Mansur who had taken an interest in Indian mathematics. In his travels Biruni also observed geological stratification and fossils, which led him to conclude that much of India had once been a sea.[112] While this was not the first time that his work led him to question whether God had created Earth whole and complete as it is today, it confirmed his conviction that Earth itself was undergoing constant evolution. Of greater consequence were Biruni's studies on planetary cycles, and also the observations that enabled him to reach impressively accurate estimates of the size of the moon and known planets and their distance from Earth.[113]

Seemingly as an afterthought, when Biruni found himself detained for a few days at Nandana Fort near the small Punjab town of Pindh Dadan Khan in what is now Pakistan, he once again calculated Earth's diameter and circumference, this time surpassing in accuracy even his own earlier findings from Khwarazm.[114] His calculation for the diameter was a mere 10.44 miles less than the modern measurement. He achieved this by sighting the dip angle from the top of the mountain and applying the law of sines formula to it and the height of the mountain, which he had also calculated with the help of an astrolabe. Even though other astronomers had applied the law of sines to their field, no one before him had applied the law of sines to the solution of so practical problem, let alone to measuring the size of Earth. Besides being far simpler than using two distant points on flat land, this method produced a degree of accuracy that was not equaled for six hundred years.[115] In this same project Biruni's measurement of one degree of meridian differs by only 2,034 feet from our modern figure, an astonishing achievement.[116] All this study culminated in his masterwork on astronomy, the *Canon* (*Canon Masudicus*), which he later drafted in Ghazni.

Such research alone would have earned Biruni and, by extension, his patron a significant place in the annals of astronomy. But they were equaled and surpassed by the massive and pioneering research that Biruni conducted on India itself. In an intellectual tour de force that has no antecedent in the ancient or medieval worlds, and which was equaled only in recent centuries, Biruni offered a comprehensive intellectual, economic, and cultural analysis of the wellsprings of Indian civilization.

Figure 11.4. Defying its title, Biruni's *Elements of Astrology* offers questions and answers on mathematics and astronomy for the new learner. This fourteenth-century copy is one of only 22 of his 180 known works to have survived. ■ Abu Rayhan al-Biruni, *Kitab al-Tafhim*, ms. no. 445, al-Biruni Institute of Oriental Studies. From published report *Khorezm Ma'mun Academy*, n.a., issued by the Academy of Sciences of Uzbekistan, under Cabinet of Ministers decree no. 532 (Tashkent, 2006), 115.

The resulting book, entitled simply *India* (*al-Hind* in Arabic), stands as one of the greatest achievements of all time in the area of interdisciplinary social science, international studies, theology, intellectual history, and the history of science.

BIRUNI'S INDIA

Given Biruni's chronic graphomania, it was probably inevitable that he would write a book on the country in which he had spent so many years. By the eleventh century, books of travel in foreign climes were already popular among Arabic readers. The genre culminated 350 years after Biruni when Ibn Battuta (1304–1368), an Arab from Morocco, published an account of his travels across parts of Europe and Africa and eastward to Central Asia, China, and Southeast Asia. Marco Polo's account of his travels appeared about the same time.

Biruni's study of India was not only much earlier but immeasurably deeper than the books by Battuta, Polo, or any other writer on international themes before modern times. Drawing on the author's travels, it presented much information of interest to the curious. And along the way it passed on Biruni's insights on the topics that to this day prevail in serious books on foreign countries, such as politics, economics, society, international relations, culture, and religion. But for Biruni, these subjects were means to an end, not the end itself.

Biruni had long since concluded that Indian learning was on the same high plateau as that of ancient Greece, with no rival elsewhere. The great purpose behind his *India* was to identify the highest achievements of Indian thought in the various sciences and to trace them to their deepest sources within the culture. This quest led him eventually to Hinduism, which he analyzed with penetration and dispassion—at the very time that his patron, Mahmud, was winning plaudits throughout the Islamic world for his savage jihad against Hindu temples and shrines.

This great undertaking required years of careful preparation, beginning with intensive language study. Biruni confessed to "a great liking" for his classes in Hindi and Sanskrit and noted that in this respect "I stand quite alone in my times."[117] Yet it was tough going for a man who was nearing fifty. What particularly bothered him was the existence in Sanskrit of many words for the same thing. In the end he learned the languages well enough to sprinkle through *India* more than a thousand Sanskrit terms in transcription.[118] At the same time he regretted that he was unable to translate the Sanskrit holy books into Arabic.

As Biruni got down to work, he realized he had to defend his decision to write a book on India and Indians before skeptical readers of Arabic. Anticipating his critics, he accepted not only that the Hindu Indians "are our enemies" but also that "they differ from us in everything." Indeed, he conceded, "if ever a custom of theirs resembles one of ours, it has the opposite meaning."[119] Besides, was not the arrogance of the Hindus the stuff of legend throughout Central Asia and the Middle East?

Biruni carefully responded to each of these objections. Yes, Hindus are our enemies. But they have good reason to fear us and to frighten their children with stories about us. And besides, it is still useful to understand them. And yes, too, many Indians are arrogant and overweening, but some of their best thinkers generously shared with the author valuable texts and other materials. As to all the information on Indian culture, is it not important for Muslims to be able to converse intelligently on religion, science, or literature?

Having thus prepared the ground, Biruni dealt briefly with prior works in Arabic on India. He graciously acknowledged the best of them but then tore into the rest for their shallow research, trivial explanations, and endless display of their authors' prejudices.[120] "Everything which exists in our literature on this subject is second-hand information which one has copies from the other, a farrago of materials never sifted by the sieve of critical examination."[121] Biruni, by contrast, would avoid polemics, seek to free himself of the prejudices "which are liable to make people blind to the truth," and instead simply "relate without criticizing."[122] True to his vow, he more than once cautioned his readers that "we may not like it, but . . ." and then proceeded to relate something that was bound to shock Muslim readers.

Biruni introduced his book with a remarkable statement about methodology: "To execute our project it has not been possible to follow the geometric method." In other words, the scientific method that worked for astronomy or mathematics did not work for society. Instead, he proposed and defended a *comparative* method that identifies specific social functions and moves on to consider how specific societies address them. This marked the start of rigorous comparative studies in sociology, religion, the history of science, and so forth, and an important step toward the definition of a social science methodology that is truly scientific.

From the outset Biruni warned that he would be very demanding with respect to evidence. Naturally he would present his own observations from the field, but he would also draw on the primary written sources to verify them. Finally, he announced that he would be particularly cautious in using the evidence of Indian traditions. He rejected the notion that "sound" oral traditions can somehow be distinguished from unsound ones, disdainfully dismissing them all as "reports from the inventors of lies."[123] Even the oldest written traditions, he asserted, are merely a branch of hearsay.

This line of argument must have struck many Muslim readers as sheer blasphemy. The absolute authenticity of the Hadiths of the Prophet had been established solely on the basis of oral history. Bukhari and the other compilers had presented their method of analyzing the oral record as "scientific." Muslim clerics and scholars had not only accepted this claim but placed it beyond challenge. Now came Biruni, denouncing nearly all orally transmitted evidence on the history of religion as hearsay and the work of liars, and rejecting wholesale all conclusions derived from it.

He proposed to rely instead on the close analysis of actual texts and on the comparative method. Here, too, he was rigorous, to the point of analyzing the various factors that could cause authors to falsify their findings.[124] The many citations in the text from the Bhagavad Gita, or simply Gita (Song of God), and the countless comparisons with Greek, Christian, and Muslim practice show that Biruni was as good as his word.

The rigorously scientific cast of Biruni's mind is vividly evident in his discussion of Indian literature. Not content merely to list the various forms of Sanskrit and Hindu poetry, for example, he produced a strict typology based on meter and then presented his findings in the form of a chart.[125] He realized that such an approach may not engage his reader, let alone address his skepticism about Indian achievements in literature or other fields. So he repeatedly paused to highlight the positive. At one point, for example, Biruni interrupted a discussion of the stone cisterns that Indians construct to store water to note that Muslims are unable to produce such masterpieces of cut stonework.[126] He was cautious with such praise, however, and respectful of his readers' sensitivities. Thus he praised Buddha but not Buddhists; Hindu philosophy but not the Hindu philosophers. In this way he enlivened analysis with sympathy and then tempered sympathy with prudence.

BIRUNI ON INDIAN SCIENCE

Since Biruni's interest in India arose out of his study of Indian science, he launched his book with a rigorous *tour d'horizon* of Indian learning in that sphere. While he dutifully listed prior studies on Indian science, he made it clear that all were quite inadequate. Clearly aware that some Indian savants were not above using their knowledge to dupe the gullible, he also offered some sharp criticism of scientists who prey on popular ignorance. Then he got down to business, turning first to astronomy and summarizing Indian work in that field by topic.[127] He treated such issues as equinoxes and the "cupola of the earth" in detail and without compromise, as if he were writing for an audience made up exclusively of astronomers. Here, and throughout the book, Biruni assumed an exceptionally sophisticated readership.

No subject engaged Biruni's attention more consistently that the development of an astronomically exact calendar. After thoroughly discussing Indian notions of duration and time, he dissected the various Indian calendar systems, comparing each of them with Central Asian and Western alternatives and finding the best of the Indian systems to be impressively precise. The cornerstone of his case was provided by a series of long and detailed tables in which he presented the results of his own reworking of Indian calendars, with data carried out to a staggering twelfth digit.[128] In a world without computers, this was no mean achievement.

This led naturally to what can only be described as an exhaustingly detailed disquisition on Indian weights and measures.[129] In these chapters Biruni stepped forth once more as the encyclopedist he had always been. It was but a step from this to a further analysis of Indian mathematics, including a separate article on Indian calculations of the value of pi.

TIME AND HUMAN HISTORY

Biruni's passion for the precise measurement of time is strikingly modern. This was one of the core issues around which the immensely sophisticated corpus of his scientific thought arose. Within the larger astronomical issue was the question of time in human history. On this issue we see Biruni fully engaged, sitting at his desk and fuming, even raging, against what he considered the Indians' wanton sloppiness with respect to the chronology of human affairs. "Unfortunately," he groused, "the Hindus do not pay much attention to the historical order of things . . . and when they are pressed for information and are at a loss, they begin telling tales." The Hindus, he concluded, were "useless" on dates.[130] The Gita is loaded with dates, but not one of them could be readily translated into any universal system of time. Puzzled by the indifference to what Biruni considered a natural human desire, he set for himself the immense task of reworking the entire corpus of Indian history, both divine and human, in terms of the best Central Asian, Greek, and Arabic understandings of chronology. His analysis, festooned with tables worked out to fourteen decimal points, must have been the work of years, or of a large team of researchers.

RELIGION AS CULTURE

The deeper Biruni plunged into Indian science, the more he realized that he was dealing with something utterly different from the Greco-Roman worldview that had initially defined his own understanding of reality. The central task of his masterpiece was to pinpoint that difference and trace the Indians' approach to its roots. The defining reality of Indian civilization, he concluded, was Hinduism, and to this he devoted the largest section of his book.

Few writers before him had attempted to ferret out the deepest well-springs of another culture, let alone one that was saturated with a religion that was so radically at odds with his own. None did so with such success. The task was extraordinarily sensitive. Unlike the case of astronomy or mathematics, Biruni could not assume that members of his audience had the slightest idea of what he was offering them. Worse, he knew full well that the very subject might be anathema to those of his readers who believed that theirs was the only true faith and that all others represented varying degrees of ignorance, error, or apostasy.

This presented a formidable pedagogical challenge. Somehow he had to anchor his presentation in what his readers already knew—what they *actually* knew, as opposed to what they *should* know or *thought* they knew. Biruni met this challenge with immense subtlety. His key tool was the comparative method, which he applied with greater sophistication than any would-be cultural anthropologist or sociologist before him. Over and over again he drew parallels between Hindu beliefs and practices and those of other religions known to his readers. How, for example, could he convey the Hindu notion of multiple levels of being? Biruni calmly explained to his reader that "if you compare these traditions with those of the Greeks regarding their own religion, you will cease to find the Hindu system strange."[131]

Over the course of his volume he drew often striking parallels between Hinduism and the beliefs of pre-Muslim Arabs, Christians,

Muslim Sufis, and Jews, as well as orthodox Muslims. Only rarely did he give up and simply report that a given doctrine may seem strange but is at least rationalized by a body of serious thought. He does this with Hindu views on astrology and cosmic phenomena, which, he avowed, may seem peculiar to foreign readers but at least are supported by theories that are "very lengthy and at the same time very subtle."[132]

It was also necessary to address directly some legitimate prejudices against his subject that prevailed at the time. Most of these arose from exotic tales of snake charmers, idol worship, and dark superstitions that earlier writers had disseminated to Central Asia and the Arab world. Biruni, speaking from the lofty pinnacle of Central Asia's intelligentsia, rejected this criticism out of hand. Of course such practices exist, he acknowledged, but in India as elsewhere one must separate the religious beliefs of the intellectuals from those of the uneducated masses.[133]

Having conceded this point, Biruni proceeded to explain rationally many of the very superstitions that had thwarted understanding in the first place. For example, to explain the phenomenon of Hindu snake charmers, he compared their peculiar music to the strange singing that preceded ibex hunts in Central Asia. Both are understandable, he said, when one remembers that they arose from people who were accustomed to hearing the sounds produced by birds and animals.[134]

Biruni adopted a similar line of analysis in considering idol worship. Given Islam's proscription against human images, he was on sensitive ground, but he forged ahead anyway, tracing its origin to the desire of the living to recall those who have passed from life.[135] This process of honoring, he suggested, was then extended to these who build the idol and then, in a kind of shorthand, to the image itself. Finally, he suggested that Hindus honor their idols on account of those who had them built, not on account of the materials of which they were constructed.[136]

Biruni took a different tack in addressing the multiplicity of Hindu divinities. Methodically reviewing the use of the word "god" in Hebrew, Greek, and Arabic, he found telling similarities among them. On this basis he suggested that Hindus were really monotheists of sorts, and that their crowded pantheon was merely a way of representing the diverse manifestations of the divine that all religions recognize.[137]

The Hindu concepts of reincarnation and the transmigration of souls posed a thornier problem, for they frontally contradict the common sense of practitioners of all three of the Abrahamic religions. The effort to come to grips with them sent Biruni on a quest that culminated in some of the most daring thoughts about humankind and nature before modern times. His initial question was implied but not stated: what happens to all those immortal souls as their numbers mount with the increase of population and the passage of time? Muslims or Christians assumed that the process must lead to their exponential proliferation. Biruni had his doubts. He had observed that all living organisms seek to reproduce themselves to the greatest extent possible. But, he reasoned, nothing in nature can proliferate forever. Thus "the life of the world depends upon sowing and procreating. Both processes increase in the course of time, and the increase is unlimited, whilst the world is limited."[138]

With these brief lines Biruni posed the same question that the English pastor and scholar Thomas Malthus (1766–1834) was to raise eight hundred years later in his epochal *An Essay on the Principle of Population*. Biruni offered a concise and bold answer. The ever-growing population would eventually become unsustainable unless some process of selection were to occur. Elsewhere in his book he had considered the demographic impact of epidemics and natural disasters. He knew that these can kill off large numbers of people and thus regulate the size of the population. Yet because such disasters affect all equally, they can also have the cumulative effect of weakening the species as a whole. Left to itself, nature draws few distinctions when deciding who will survive: "Nature . . . does not distinguish [among living organisms], for, its action is under all circumstances one and the same. It allows the leaves and fruit of the trees to perish, thus preventing them from realizing that result which they are intended to produce in the economy of nature. It removes them so as to make room for others."

Like Malthus, Biruni in this passage saw nature as blind, a stern master who, for the sake of the survival of the species, punishes equally the fit and the unfit. Up to this point in his exposition, Biruni did not present a world in which the fittest necessarily survived. Acknowledging this hard truth, he then brought forth striking counterexamples, where some form of rational selection appeared to be taking place: "The agriculturalist selects his corn, letting grow as much as he desires, and tearing out the remainder. The forester leaves those branches which he perceives to be excellent, whilst he cuts away all the others."

These are easily dismissed as the work of rational human beings. But nature, too, draws distinctions when deciding who should survive: "The bees kill those of their kind who only eat but do not work in their bee-hive." In other words, selection is not as random as it may seem at first. It intervenes to prevent Malthusian debacles by indiscriminately thinning out the population of whichever species of plant or animal threatens to dominate. Yet both human beings and other forms of life base their selection on the fitness or unfitness of each organism to survive: "If the earth is ruined, or is near to be ruined, by having too many inhabitants, its ruler—for it has a ruler, and his all-embracing care is apparent in every single particle of it—sends in a messenger for the purpose of re-ducing the too great number and of cutting away all that is evil."[139]

Natural selection is thus at work, thanks to the constant care of a benevolent and rational God. Now, since for Biruni "all things are di-vine,"[140] Nature in this instance is God, who assures the maintenance of all forms of life through the exercise of natural selection. Indeed, whether or not we understand all its workings, natural selection is the basic principle that regulates everything that lives. In reaching this point, Biruni, writing in Afghanistan a millennium ago, took a long step toward Charles Darwin's conclusions regarding the same processes. The difference is that for Biruni, God selects individuals and species on the basis of some predetermined set of criteria formed by his overall plan, much like a farmer, whereas in Darwin's case the criteria are determined by the selection process itself.[141]

There remains another crucial difference between Biruni and his Vic-torian successor: Biruni's species are static. His concern is to explain the preservation and healthy maintenance of living organisms, not their evolution through time. Biruni was not blind to evolution, having ob-served fossils that have no counterpart among living species. In fact he had long since proposed a theory of geological evolution to explain pro-found changes over time. Yet to the extent that he did not apply the same reasoning to the evolution of living organisms, he remained a man of his times, or of all times down to the mid-nineteenth century.

Having presented his readers with these revolutionary notions, Biruni then concluded that the concepts of reincarnation and the trans-migration of souls are by no means nonsense, for they represent a kind of conservationist ecology applied to the human soul. He added one further flourish by noting that not only were these notions to be found in Hinduism, but they arose also in the thought of Muslim Sufis, who would have been well-known to his readers.

Hindu rituals did not escape Biruni's attention, and in his exposition on them he called on neither the holy texts nor the works of subsequent Hindu thinkers. Instead, he rendered them intelligible by pointing to parallels in the religion of classical Greece and by showing their practicality. Thus he found little difference between the burning of the dead by Hindus and by Homeric Greeks and traced both to a belief in the unity of creation.[142] As for the Hindu abstention from eating beef, is this not simply a matter of rational economics in light of the fact that "the cow serves man by carrying his load, by plowing and sowing his fields, and by providing milk"?[143]

Because the Hindu injunctions to fast, give alms, and undertake pilgrimages to holy sites all parallel Muslim practice, Biruni felt no need to explain them theologically and passed instead to a more general evaluation of how they worked in practice. All three are obviously beneficial to health and communal welfare. But Hinduism is content to leave them as voluntary duties rather than treat them as divine commands.[144] In other words, the Hindu religion trusts human beings to behave rationally. Unstated by Biruni is the obvious comparison to Islam, which considers all three duties to be obligatory and imposes them with the force of law.

BIRUNI ON HINDUISM, ISLAM, AND MAHMUD

The picture of Hinduism and Indian culture that emerges from Biruni's *India* is complex, nuanced, and sober. He found much to admire in his subject, yet he steered clear of advocacy. Nor did he peer at India through an Orientalizing lens, finding everywhere a romantic "otherness." His goal was not to appear evenhanded, which would have meant skewing the evidence to give the appearance of balance, but to be rigorously accurate in his analyses. In *India* we see the work of a remarkably inquisitive polymath, whose approach marked a significant advance in the scientific study of culture and society.

Fair-minded throughout, Biruni left the reader with the picture of India as a highly creative civilization that has addressed many of the same problems of philosophy and science that preoccupied ancient Greeks, Arabs, and Central Asians, albeit in its own distinctive way. Never stated but implied on every page was Biruni's assumption that humankind is one. Except for its use as a foil to his own views, there was no "we" and "they" in Biruni's world, only different cultural solutions to the same challenges that face all human beings. In this he echoed Aristotle but was at sharp odds with many mainstream Muslims. Biruni's critics lend support to those who have questioned the existence of true humanism in the so-called Islamic renaissance.[145] But Biruni himself, on the pages of his *India*, stood forth as just that, a Muslim humanist. He told his reader that Indian culture and its thinkers were different but related to Muslim readers through the problems they sought to address. In the end, they were equals.

Precisely because of this, he could not avoid the question of why Hindu Indians hated Muslims with such ardor.[146] In answering this, Biruni reminded his readers of the wars of religious jihad and conquest that Muslims had waged against Hindus. He provided unvarnished details on the sheer brutality of it all, and on the greedy pillaging that resulted. The Quran may say that "there is no force in Islam," but the facts known to Hindus belied this.

Acknowledging this, Biruni asked why had the caste-bound Indian masses refused to embrace the egalitarianism that is supposedly the hallmark of Islam. First, he posited that this may arise from the fear that conversion to Islam could be punishable with death, as would happen if a Muslim converted to another faith. But he pointed out the Hindu holy books are far milder on apostasy than the Quran is. Having laid aside this argument, he then cited overwhelming evidence that in those parts of India where Islam prevailed, the caste system had actually been strengthened and hardened.[147] Nor, he added, did Islam bring enlightenment in its wake. On the contrary, he argued, Muslims in India had allowed science to be perverted by religion. Realizing this, many of India's greatest intellectuals had fled from those areas that came under Muslim control.[148]

This discussion brought Biruni face to face with the role of his own patron, Mahmud of Ghazni. Mahmud's court historians, Beyhaqi excepted, presented the story of Mahmud's assaults on India as a heroic tale of selfless struggle and sacrifice in the name of Islam. Biruni knew better and scarcely bothered to cloak his account in Aesopian language. With unvarnished bluntness he reported that the recent Muslim (read Mahmud's) wars of conquest had everywhere wrought devastation.[149] "Mahmud utterly ruined the prosperity of the country," he wrote, "and performed there wonderful exploits by which the Hindus became like atoms of dust scattered in all directions. . . . The scattered remains, of course, cherish the most inveterate aversion towards all Muslims."[150] With their prosperity destroyed, their caste system reinforced, and their best minds driven into exile, how could Hindus not hate Muslims?

In light of Mahmud's status as Biruni's patron, the ferocity with which the scientist attacked Mahmud and his jihad in India is truly astonishing. Even though he had earlier opined that the Indian masses were ignorant and certain of their leaders even worse, he did not hesitate at this point to praise several Hindu leaders who fell before Mahmud's sword, even setting them up as models of decency and bravery.[151]

But the question persisted: why did the Hindus lose? Here Biruni offered an insight that any modern political scientist would understand: Mahmud won because he had a totally centralized command structure while the Indians' polity was characterized by a high degree of decentralization and self-government.[152] What had been their virtue in normal times became in wartime a liability.

Taking Biruni's *India* as a whole, the work more than fulfilled the expectations that the author raised in his introduction, namely, that he would enumerate some of the greatest achievements of Indian science and learning, that he would consider the central contribution of Hinduism to Indian intellectual life and civilization, and that he would do this as a scientist and "not as a polemicist."[153] When the author confessed his "perfect uncertainty" on a given issue, as he often did, or when he excoriated those who refuse to avow their ignorance by a frank "I don't know,"[154] readers understand that they are dealing with a rigorous analyst, a discerning clinician of culture, or, as we can confidently say today, a great figure of the social sciences. In both its comparative method and its international perspective, *India* was centuries ahead of its time, a true breakthrough. One of Biruni's biographers only slightly exaggerated when he called it the "finest monument" of Islamic learning.[155]

CODA

Shortly before Mahmud's death one of his stable of loyal poets, Umarah of Merv, wrote this grave stanza:

> Be thee not proud, even though the world has chanced to make you great.
> Many are the great whom the world brings swift to low estate.
> This world's a snake—a charmer who seeks all within his power to bring;
> But the charmer oftimes from the snake receives a mortal sting.[156]

In the year 1030 Mahmud of Ghazni died of malaria. Avaricious to the end, one of his last acts was to demand that his jewels be exhibited before him in glittering rows.[157] Did he also pen verses on the transitory nature of human striving and the vanity of life? A contemporary biographer made this intriguing claim,[158] but no such death-bed confession has ever come to light. What we do know is that within seven years of Mahmud's death, much of his kingdom had fallen away, thanks to the sudden rise of yet another Turkic people, the Seljuks. Mahmud's son Masud perpetuated the Ghazni line and used his inherited riches to maintain a splendid court life at Ghazni and Lashkari Bazar and to support the historians Utbi and Beyhaqi, and the poets Unsuri and Sistani.

As to Biruni, Mahmud's death did not interrupt what turned out to be an immensely productive period. The hard-drinking and sybaritic Masud, who had made a career of frustrating his father's expectations of him, cast himself as a keen supporter of science and of Biruni personally. Biruni obliged his new master, filling the role of court astrologer and writing a whole treatise on the orientation of the *qibla*, or direction of prayer, so that a prayer in Ghazni would truly be aimed toward Mecca.[159] Amid these tasks, Biruni completed his lifelong project on *Geodesy*, which includes a pathbreaking analysis of gravity and sly critiques of scientific concepts and contradictions in both the Bible and the Quran.[160]

He also drafted a study, based on research dating back to his youth, on the physical properties of shadows and their relationship to light, which led him to advance innovative proposals regarding their use in geodesic and astronomical studies.[161] Here, in his typical way, Biruni casually dropped momentous points in the midst of seemingly unrelated discussions. For instance, he directly presaged the use of polar coordinates and also suggested that acceleration involves nonuniform motion. It is curious that shadows became a subject of rigorous scientific study long before they appeared in Central Asian or Persian painting. Finally, amid all this activity, Biruni somehow found time to pen a long work on the history of Mahmud's reign, of which no copy has survived.[162]

Of seminal importance to the history of science was the vast compendium of astronomical and mathematical knowledge that Biruni wrote between 1031 and 1037. Known as the *Canon Masudicus* because of its dedication to the new sovereign, this magisterial work treated astronomy and related fields of mathematics and trigonometry with the same comprehensive thoroughness as Ibn Sina's *Canon* treated medicine. Indeed, Biruni chose the title "Canon" as a response to his old rival's *Canon of Medicine*, completed less than a decade earlier (1025). For this monumental work Masud rewarded the author with an elephant-load of silver, which Biruni politely refused.[163]

Both an encyclopedia of astronomy and related areas of mathematics and a summary of his own research over a lifetime, Biruni's *Canon Masudicus* brought medieval astronomy in the Arabic language to its highest pinnacle. Whether it began its long decline then as well, or sustained significant work for several more centuries, as George Saliba maintains, is less important than the fact that it was Biruni, working alone in Afghanistan, who brought astronomy to its medieval peak.[164]

The *Canon Masudicus* comprises eleven books, which deal respectively with the calendars of different peoples and their astronomical bases; properties of the circle; the mathematical astronomy of the fixed stars and the sun regarded as a point on the ecliptic; a mathematical treatment of terrestrial latitude and longitude; the motion of the sun and moon and eclipses of both; and the movements of the five planets.[165] Throughout his massive text Biruni maintained his focus on a single theme: the numerous anomalies he observed in areas that Ptolemy and his system had supposedly settled for all time. He had assembled these over an entire lifetime from his own observations (often repeating studies of the same subject that revealed unexpected changes), from the works of other Central Asian and Arab scientists, and from the evidence assembled by other scientists, mainly Indians. In subject after subject, we see Biruni deep in what Thomas Kuhn called "normal science," identifying and coldly analyzing the "anomalies" that are the essential drivers of scientific revolutions.

This process is evident in a score of areas, but Biruni's exposition of the geocentric system of Ptolemy is particularly striking. Biruni began by arguing that there is no compelling evidence that Earth is stationary, nor, for that matter, that it is in motion: "The rotation of the earth does in no way impair the value of astronomy, for all appearances of an astronomical character can quite as well be explained in accordance with this theory as the other" (i.e., an immovable Earth).[166] The occasion for these daring speculations was his contact with an astronomer, mathematician, tinkerer, and astrologer from what is now the Iran-Afghanistan border region, Abu Said al-Sizji (ca. 945–1020).[167] Sizji had wandered wide but dedicated much of his writings to the prince of Balkh. Biruni entered into correspondence with him and eventually was able to examine a bold new astrolabe that Sizji designed on the hypothesis that Earth revolves:

> I have seen a kind of simple astrolabe constructed by Abu Said al-Sizji . . . known as the boat-like astrolabe. I liked it immensely, for he invented it on the basis of an independent principle that is extracted from what some people hold to be true—namely, that the absolute eastward visible motion is that of the earth rather than that of the celestial sphere [i.e., sun]. I swear that this is an uncertainty that is difficult to analyze or resolve. The geometricians and astronomers who rely on lines and planes have no way of contradicting this [theory.] However, their craft will not be compromised, irrespective of whether the resulting motion is assigned to the earth or to the [sun]."[168]

He concluded by calling on physicists to either disprove this thesis or accept it. In passing, Biruni brushed aside the notion of the sun as a metaphysical or mythological entity and affirmed that it is in fact a fiery material substance with a specific diameter and distance from Earth.

He then turned to the question of Earth's movement in space. Unlike Razi, who rejected the geocentric principle outright, but with scant evidence, Biruni tentatively accepted it as a relevant hypothesis while noting that a heliocentric system was just as fully possible. Needless to say, there were few people anywhere in 1037 who would so casually have placed a heliocentric polar system on the same level of probability as a geocentric one; then declared that he, as an astronomer, could live with either view; and finally called on physicists to either validate or disprove it.[169] Again, it is worth stressing that Biruni's main concern was not to prove that Earth was moving or stationary but rather to show that from a mathematical perspective it makes no difference. This had been asserted since ancient times. But in his *Codex Masudicus*, Biruni addressed the point in such a way as to nudge physicists to address the larger question.

Another spin-off from this general discussion was Biruni's bold note concerning ballistics:

> When a thing falls from height, it does not coincide with the perpendicular line of its descent but inclines a little. . . . [A falling object] has two kinds of motion: one is the circular motion it receives from the rotation of the earth, and the other is straight, which it acquires in falling directly [toward] the center of the earth.[170]

After setting forth the fundamentals of plane geometry and trigonometry needed for the remaining books, Biruni introduced several innovations and refinements in spherical geometry that were immediately applicable to spherical astronomy. For example, he was the first Islamic scientist to introduce second-order interpolations of trigonometric tables.[171] He also proposed a new mathematical method for analyzing the acceleration of the planets.

Biruni then set forth once more his new method of using a gnomon's shadow and time in a place of known longitude to determine latitudes. This innovation, valuable in itself, became the first step toward a much bigger discovery.

BIRUNI DISCOVERS AMERICA

In his *Codex* Biruni gathered together the conclusions from his research on Earth's circumference that he had carried out at Nandana. He then set about fixing all known geographical locations onto this accurately sized terrestrial sphere. It is not known whether Biruni constructed another model of the globe to do this or carried it out paper. He certainly did not need the model, however, for by now he was an expert in the fast-developing field of spherical geometry and could easily make the necessary adjustments between a plane and spherical surface. His list of longitudes and latitudes had grown substantially since his earliest collection and now included more than seventy sites in India alone, as well as thousands of other locations stretching across the entire Eurasian land mass.[172]

When Biruni transferred these data onto his more precisely measured circumferential map of Earth, he noticed at once that the entire breadth of Eurasia from the westernmost tip of Africa to the easternmost shore of China spanned less than about two-fifths of the globe. This left three-fifths of Earth's surface unaccounted for.

The most obvious way to account for this enormous gap was to invoke the explanation that all geographers from antiquity down to Biruni's day had accepted, namely, that the Eurasian land mass was surrounded by a "World Ocean." Alexander the Great, who had been personally tutored by Aristotle, had had this facile hypothesis in mind when he and his army emerged from the mountains of Afghanistan onto the western side of the Indus Valley. From this vantage point he fully expected to gaze out at the World Ocean. But instead he saw only more land.

Was more than three-fifths of Earth's circumference really nothing but water? Biruni considered this possibility but rejected it on the grounds of both logic and observation. Why, he mused, would the forces and processes that had given rise to land on two-fifths of Earth's belt not also have made themselves felt in the other three-fifths as well? Reasoning thus, Biruni concluded that somewhere in the vast expanses of ocean between Europe and Asia there must be one or more heretofore unknown land masses, or continents.

Proceeding by logical steps, he then asked if these unknown continents were empty wildernesses or were instead inhabited by human beings. To this point he had relied on his research at Nandana on Earth's circumference, on his voluminous data on the longitudes of the world's cities and known geographical features, and on simple logic. To advance further he now turned to his data on longitudes. He noted that human beings inhabited a broad north-south band stretching from what is now

Russia to southern India and the heart of Africa. Assuming that this band represented Earth's habitable zone, he asked if the unknown continent or continents were situated only in latitudes lying north and south of this band.[173]

In answering this, there were no further field observations to which he could turn, but he did have the tools of logic. Noting that the Eurasian land mass stretched roughly around Earth's belt, and that it covered a broad north-south band, he hypothesized that this was the result of powerful processes that would surely have obtained elsewhere. Known evidence of Earth gave him no grounds for believing that the unknown continents would be squashed into the northernmost and southernmost latitudes. Reasoning by anology to Eurasia, he concluded that these unknown land masses would have to be inhabitable, and in fact that they were inhabited. As he stated in the *Codex Masudicus*, "There is nothing to prohibit the existence of *inhabited lands* in the Eastern and Western parts. Neither extreme heat nor extreme cold stand in the way . . . it is therefore necessary that some supposed regions do exist beyond the [known] remaining regions of the world surrounded by water on all the sides."[174]

Did Abu Rayhan Muhammad Biruni discover America in the first third of the eleventh century? In one sense, definitely not. He never laid eyes on the new continent or continents about which he wrote. By contrast, the Norsemen had actually touched land in North America shortly before AD 1000—briefly, to be sure, and without really understanding what they had found. Leif Ericson was so uninterested in the forested shore that he did not bother to returned later, nor did any of those who heard oral reports of his travels or read about them in later Norse documents. Still, if "discovery" includes the groping and unreflective processes of Norse seafaring, then the prize that partisans of Christopher Columbus, John Cabot, or unknown seafarers from Bristol in England have claimed for their heroes must definitely go to the Vikings.

But Biruni should also wear the crown of discovery, and the cumulative and analytic process by which he reached his conclusions is at least as deserving of honor as are the Norse traders, if not more so. His tools were not wooden boats powered by sail and muscular oarsmen but an adroit combination of carefully controlled observation, meticulously assembled quantitative data, and rigorous logic. Not for another half millennium did anyone else in Europe or Asia apply such rigorous analytic tools to global exploration.

Viking explorations were also cumulative, with a few bold adventurers using ad hoc reasoning to build on what they had heard about earlier travels. But Biruni went far beyond this, beginning with his assembling of all known human knowledge of the subject at the time. It drew on the wisdom of the ancient Greeks and Indians, and also on the work of medieval Arabs and fellow Central Asians. Moreover, Biruni devised absolutely new methods and technologies to generate his voluminous and precise data and then processed it with the latest tools of mathematics, trigonometry and spherical geometry, as well as the austere methods of classical logic. And he was careful to present his conclusions in the form of hypotheses, on the understanding that other researchers would test and refine his findings. This did not occur for another five hundred years, but in the end the European explorers confirmed his hypotheses and vindicated his bold proposals.

Solar Insights and Specific Gravity

Following this exposition, Biruni devoted a long passage to the movement of the sun, in the course of which he offered a simple formula for relating its zenith at a given point to latitude. Several further chapters called on measurements taken over the past three centuries to take issue with Ptolemy on important points, including the ancient claim that the sun's apogee is unchanging. After presenting measurements on the size of the moon and its distance from Earth, he proceeded to a detailed analysis of the movements of Mercury and Venus. The *Canon* ended with a discussion of both Western and Indian astrology, tracing the more credible claims to solid astronomy and dismissing the rest as fantasy.

Here, as elsewhere in the *Canon*, Biruni for the first time assembled the insights of classical Greek, Indian, and Islamic astronomy and subjected them to rigorous mathematical analysis; where he judged the classics to be faulty, he advanced bold new insights of his own. With respect to the problems it posed, the analysis it pursued, and the conclusions it offered, Biruni's *Canon* reflected his penchant for rigorous and dispassionate analysis and astonishingly bold hypothesizing. As arguably the single greatest medieval work on astronomy, it is all the more surprising that the full Arabic text of Biruni's *Canon Masudicus* was not published until 1954–1956, and then not in a critical edition. Even today there exists no authoritative and reliable English translation, and none at all in French or German.[175]

In the same years Biruni completed what is recognized as the first scientific treatment of mineralogy, *A Collection of Information for Recognizing Gems*, known as the *Mineralogy*. Extending Archimedes's principle, this work is less technical than most of Biruni's writings and features long digressions on everything from the lexicography of terms for minerals in Eastern and ancient languages to comments on the location of major deposits and methods used to mine them, and even citations of poetry devoted to each gem or precious metal. Amid it all he paused to discuss crystallography and the origins of minerals, with discussion of both igneous and sedimentary processes.

In this volume Biruni for the first time introduced the concept of specific gravity. Having discovered the concept, he then proceeded to apply it to scores of minerals and metals. His measurements were so extraordinarily precise—accurate to three decimal points—that Europeans were unable to equal them until the eighteenth century.[176] First translated from Arabic to a Western language only in 1963, Biruni's *Mineralogy* has yet to find its place in the history of science.[177]

The Fall of Ghazni

Eventually Mahmud's son and heir, Masud, was driven from his throne by the rising tide of the Seljuk Turks. His successors withdrew from most of Central Asia and focused their attention instead on Afghanistan and the Indus Valley. But by the twelfth century the Ghazni state was under assault from all sides. A new dynasty from the remote Afghan town of Ghor had carefully studied Mahmud's style of conquest. In 1146–1147 the king of Ghor took control of Mahmud's Indian possessions and celebrated by erecting a splendid triumphal tower at his capital on the Hari River at Jam (now Ghor province, Afghanistan) which, at 213 feet, dwarfed even Mahmud's victory towers at Ghazni and Gurganj.

In a fitting finale, the ruler of Ghor thoroughly plundered Mahmud's Ghazni, the "Bride of Cities."[178] This long overdue act gave rise to a mordent epitaph penned by Nizami Arudi, the twelfth-century poet from Samarkand:

> How many a palace did great Mahmud raise
> At whose tall towers the Moon did stand agaze
> Whereof one brick remaineth not in place.[179]

The last Ghazni ruler died in near poverty in 1186.

CHAPTER 12

❁

Tremors under the Dome of Seljuk Rule

GHAZALI'S CRISIS

August 1095 was a painful month for the most influential, most power-ful, and richest intellectual in the Muslim world. Abu Hamid Muham-mad al-Ghazali (1058–1111) was the author of several widely admired books attacking all the freethinking Muslim philosophers who were oriented toward Athens and Alexandria rather than toward Mecca. He was easily the most prominent faculty member at Baghdad's renowned Nizamiyya School (madrasa), only recently set up to prepare a new gen-eration of Islamic jurists who would adhere strictly to orthodox Sunni principles. Ghazali was himself a highly competent practitioner of the formal logic that he was to criticize in others as dangerous and hereti-cal. Three hundred admiring students packed his lectures, soaking up his ferocious yet cogent critiques of all the intellectual icons of the age. Moreover, high government officials regularly consulted his views and paid him handsomely to deliver just the right oration at ceremonial events.

But now, suddenly, the vain Ghazali could not digest food, his throat was dry and choked up, he was unable to form words, and he lost all desire to mount the lectern.[1] This most self-confident young academic suddenly was struck dumb. Ghazali's painful crisis—which he himself termed an "illness"—dragged on for two months. Given the symptoms he himself cites, it is likely that what began as a professional or spiri-tual crisis ended as a nervous breakdown. When eventually it passed, Ghazali gave away all his property beyond what was needed to support his wife and young children, donned the humble robes of a pilgrim, and departed Baghdad on foot, alone.

A deep-dyed Central Asian, Ghazali had been born in Tus, where Ferdowsi had lived and worked. He had studied under the best minds of Nishapur and had taught there until a rising vizier from his own home-town, Nizam al-Mulk, arranged for his move to Baghdad. Yet now, when his career was in crisis, Ghazali set his sights not on a homecoming to Khurasan and Central Asia but on Jerusalem, Mecca, and Damascus, where he was to spend whole days in isolation atop a minaret. When he returned ten years later he claimed to be a new man, a "Sufi" who ignored the dogmas of the faith in favor of direct, mystical communion with God.

Even in the earlier phase of his life, Ghazali had always been search-ing for certainty. As Farid Jabre wrote in his classic study of the thinker, it was this drive that informed both Ghazali's attacks on the rationalists and the quest on which he embarked after his crisis.[2] Now he had found it. The path to truth, he averred, was through self-purification, a disci-plined process of stripping away the extraneous and the false that would leave the believer able to grasp God's truth directly. Avoiding Baghdad, he now headed back home to Nishapur, a new man. There he put the finishing touches on a new approach to religion and the life of the mind that would profoundly change not only the world of Islam but of Chris-tianity as well.

It is too much to say that Ghazali's severe demotion of the rational intellect was solely responsible for the waning of the free intellectual enquiry that had been the ornament and glory of Central Asian civili-zation for three hundred years. Ghazali, in the end, was quite nuanced in his argument, and many other factors besides Ghazali helped bring about this change. Moreover, he was careful to qualify his critique of reason in important ways. On the questionable grounds that mathemat-ics and certain other fields were irrelevant to the great questions of God and salvation, he exempted them from his general condemnation of rational inquiry. Yet this reprieve was only partial, for Ghazali argued that science itself bred a kind of rationalism that leads to skepticism and atheism. Neither reason nor logic was relevant to the real purposes of humankind, and the entire scientific and philosophic quest—indeed, the very idea of a society committed to the pursuit of new knowledge— struck Ghazali as nothing but a hollow delusion. As to Ibn Sina's "intu-ition," this, Ghazali felt, was all about Ibn Sina and not about God.

By contrast, the "way of mysticism" represented a superior form of cognition, a step above reason in the same way that maturity as a stage of life is superior to childhood. Whatever his intentions, Ghazali's writings provided a potent intellectual tool for anyone itching to silence those who dared champion rationalism and enlightenment in the name of Islam.

At the time of Ghazali's birth in 1058, Ibn Sina had only recently completed the philosophical masterpieces that stand among the greatest achievements of the medieval mind. Farabi's great works on philosophy, now two centuries old, were everywhere revered as classics. But Ghazali branded both thinkers posthumously as heretics and therefore, according to the Quran, punishable by death. As to Biruni and his followers, they were utterly beyond the pale. Biruni, who had actively championed the freethinker Razi, had written that "the sciences arise from the needs of humans and are essential to their life."[3] This is precisely what Ghazali most vehemently denied. Fifty years after Ghazali's own death in 1111, the champions of free and uninhibited inquiry were in full retreat, under withering attack from a new orthodox mainstream whose dogmatic partisans marched under Ghazali's banner.

THE ARC OF CHANGE IN EUROPE AND CENTRAL ASIA

Were Ghazali's diatribes against the use of the rational intellect to attain knowledge of God the cause or culmination of the great cultural shift that occurred in the late eleventh and twelfth centuries in Central Asia and the Islamic world? The question calls for an inquiry into the context of Ghazali's life, and specifically into the great political and cultural upheaval that was taking place around him. This exploration must deal with the rise of a great new Turkic empire—the Seljuk state—in Central Asia and the other lands of the caliphate; the flowering of economic and cultural life under that empire; the culmination of the long-festering schism between Shiites and Sunnis; the hardening of doctrine into dogma that accompanied that split; and the process by which millions sought refuge in new forms of personal religion. Only then can we hope to gain some understanding of the nature of Ghazali's self-appointed mission and of its place in the intellectual life of his age.

The larger context is also important because it was precisely in these years—roughly 1050 to 1160—that other great shifts were occurring elsewhere in the Eurasian world. In the West the early twelfth century was a period of awakening. The weakening of feudal bonds and revival of Mediterranean trade generated new prosperity. The rise of monastic intellectual centers at Cluny and elsewhere and the translation into Latin of key works by Khwarazmi, Farabi, Ibn Sina, and others served as yeasts that soon affected broader sectors of literate society.

To be sure, there were still Viking raids, pressure from Saracens, the uncertain new Magyar and Slavic monarchies, and mass illiteracy, all of which held back Europe's advance. And only 135 years after Ghazali's attacks on "heretics," Pope Gregory IX launched his own "inquisition of heretical depravity" in 1231.[4] Yet in spite of all this, a fresh momentum was palpable in Europe during Ghazali's lifetime.

Central Asia, as well as certain centers in the Middle East, still boasted formidable economic, cultural, and intellectual resources. As we shall see, Seljuk rule unleashed an economic boom that dotted the Central Asian countryside with monuments to prosperity that can still be seen today. But what was happening under the surface? Did Seljuk rule foster the headlong intellectual momentum of previous centuries, or did it instead cause it to flag?

SELJUK AND TURKMEN TRIBES ON THE MOVE

The Seljuk empire eventually embraced all of what is now Iran and extended to modern day Turkey and the Caucasus. But it was born in Central Asia, it was ruled at its zenith by a Central Asian, its last capital was in Central Asia, and it was there that it met its end.[5] Its greatest statesmen, scientists, thinkers, architects, and poets were nearly all Central Asians. Down to the tenth century the Oghuz Turks—roughly the same people as the Turkmen, Azerbaijanis, and the Turkic raiders who moved onto the territory of modern Turkey—had lived a nomadic existence on the steppes north of the Aral Sea, hiring themselves out from time to time to the Samanid governors as soldiers.[6] Eventually they were pushed into Khurasan, where they entered Central Asian history. Over time the descendants of one of their leaders, Seljuk, came to dominate the larger tribal confederation. He accomplished this by defeating petty rulers on the battlefield and then demanding tribute from them.

Richard W. Bulliet has argued that the movement of the Oghuz into Khurasan from the northern steppes and then from Central Asia into northwestern Iran was precipitated by a prolonged period of bitter cold winters.[7] Drawing on contemporary memoirists and analyses of Asian tree-rings carried out by the Lamont-Doherty Laboratory in New York, he posits the intriguing thesis that Central Asia experienced a prolonged cooling during the eleventh and early twelfth centuries. This resulted, for example, in the deaths of some 100,000 inhabitants of Nishapur in the year 1011.[8]

What is not clear is why this "big chill" would have driven Seljuk tribes either to the south or the west, where presumably the cold was equally severe. Clearly these are issues that warrant further study. For now, a more political explanation seems in order, namely, that the Seljuk tribes rushed in to fill the power vacuum created by the declining Ghazni state.[9] The Seljuks had long posed a direct challenge to Masud, the son of Mahmud of Ghazni, who was trying desperately to hold onto Khurasan. The two armies squared off in May 1040 at Dandanqan, near Balkh.

Masud arrived on the battlefield astride an immense female elephant and with fifty thousand heavily armed foot soldiers. The Seljuks had left their baggage two days' march away and showed up with only sixteen thousand lightly armed but highly mobile cavalrymen. The Ghazni forces were decimated and Masud fled to India, abandoning on the battlefield the portable throne he had hauled with him.[10] The Seljuk chief, Toghril, immediately sat on the throne and declared himself amir of all Khurasan. Within months the rich and powerful cities of Nishapur and Merv opened their gates to Toghril's forces, to be followed by Balkh and Herat.[11] The practical rulers of all these cities had observed Toghril and his Seljuk army and noted that he showed little or none of the rapaciousness and greed that were the hallmark of the rulers from Ghazni. In the end the Seljuks triumphed by not being Mahmud.

The caliph in Baghdad also watched the rise of this new power in Central Asia and saw in it a great opportunity. He had long been beholden to Persia's Shiite rulers, the enlightened but doctrinaire Buyids, who controlled Baghdad. Mahmud of Ghazni had proposed to "liberate" the caliphate, but at the price of subjecting Baghdad to his savagery.

Now the caliph saw a better alternative in Toghril, whom he invited to conquer Baghdad and return it to the fold of solid Sunni orthodoxy. The Seljuk leader, only too glad to assist the desperate caliph (who had even offered the Turk a daughter in marriage), marched into Baghdad in 1055.[12] Meanwhile tens of thousands of Oghuz tribesmen poured out of Central Asia into Persia, taking Isfahan and Hamadan and then moving on to the Christian kingdoms of the Caucasus. They were simple marauders, more interested in booty than empire, but the Seljuk chief, Toghril, could scarcely refuse the rich cities and kingdoms they handed over to him.[13]

This haphazard process of expansion reached its peak in 1071, during the reign of Toghril's nephew, who bore the grandiose Turkic military title of "Valiant Lion" (Alp Arslan). Masses of Oghuz marauders had crossed into eastern Anatolia and gradually expanded their presence into Byzantine territory. Finally, Alp Arslan resolved to consolidate this northwestern frontier so he could concentrate his army against the Shiite Fatimids in Egypt.

On August 25, 1071, Alp Arslan defeated the Byzantine army at Manzikert in eastern Anatolia, but he quickly struck a deal with the Greek emperor Romanus in favor of the status quo. Had it not been for events back in Constantinople, the Byzantines could easily have mounted a rematch. But an insurrection in the capital overthrew Romanus and quashed all thought of a return engagement.[14] Thus, thanks more to internal dissension within Byzantium than to Seljuk military prowess, all Anatolia lay open to the Oghuz tribesmen.[15] The expansion of Turkic and Muslim power at Manzikert led directly to the later European Crusades.

What accounts for the Seljuks' success? In part it was due to the relentless activity of Oghuz freelancers. Whether or not they were driven by the search for warmer climates, they were venturesome and effective in bringing territories under their control. But more benign factors also came into play. All those who had previously been ruled by the rapacious and brutal Ghazni state welcomed the new conquerors as a more magnanimous alternative.[16] Sunni leaders in Central Asia and Iran greeted the Seljuks, who were safely mainstream Sunnis, as the power that would drive the Shiite dynasties into oblivion and stand up to radical sectarians like the Ismailis. No less important, most of the Oghuz tribesmen who swept westward out of Central Asia continued their nomadic life in the countryside they conquered, leaving cities and the urban economies in peace and in the hands of the local elites.[17]

The Seljuk empire reached its zenith under Toghril's very capable nephew Alp Arslan (r. 1063–1073) and, after the murder of Alp Arslan in Khwarazm, under the latter's son, Malikshah (r. 1073–1093).[18] Leaving only a high commissioner to keep check on the caliph in Baghdad, they established their capitals first at Hamadan and then at Isfahan in western Iran.[19] Down to the death of Malikshah, all Khurasan was ruled from these distant places. The rest of Central Asia was divided among the Karakhanids in Balasagun, Bukhara, and Samarkand; the new line of shahs of Khwarazm, whom the Seljuks had installed in Gurganj; and the rumps of Mahmud of Ghazni's state in eastern Afghanistan and of the Karakhanids in Kyrgyzstan.

Alien rule had never suited Central Asians, who for millenniums had managed either to wriggle out from under taxes imposed by foreign rulers or to gain for themselves a decisive voice in their rulers' affairs. Under the Arabs they had succeeded in applying both tactics simultaneously, thanks to the combined influence of Abu Muslim and the Abbasids, with their strong Central Asian orientation, Caliph Mamun, and the Barmaks. Now, under the Seljuks, another Central Asian—the adroit and powerful vizier, Abu Ali al-Hasan (1018–1092) from Tus— dominated and controlled the entire government for three decades down to 1092. In effect he *was* the ruler: the honorific title by which he became known—"Nizam al-Mulk"— means "Order of the Realm."

NIZAM AL-MULK: A MACHIAVELLI FROM TUS

Nizam al-Mulk's task was not easy, but he was fortunate to serve rulers who were either capable or malleable. The first Seljuk ruler for whom he worked, Alp Arslan, was intelligent and strong but gladly listened to Nizam al-Mulk's advice. His successor, Malikshah, came to power as a mere boy, and Nizam al-Mulk could easily dominate him throughout the nineteen years of his rule. The greatest challenge facing Nizam al-Mulk arose not from the personality of the rulers but from the nature of Seljuk rule. Like the Karakhanids, the Seljuks were a clan of brothers and cousins, each of whom felt himself sovereign in his own territory.

This made for a loose confederation rather than a unified state. Unlike Mahmud of Ghazni, whose family had gained top-level administrative experience under the Samanids, the Seljuks came to power directly from nomadism. And unlike both Karakhanid and Ghazni rulers, Abu Ali al-Hasan inherited no functioning central bureaucracy.

To these challenges the future Nizam al-Mulk brought a very specific body of personal experience. His father, a native of Beyhaq near Nishapur, had served as senior official in the revenue department in a village near Tus.[20] He sent his son to study religious law with a well-known jurist at Nishapur and then to Merv.[21] After Masud's humiliating defeat at Dandanqan in 1040, the family fled to Afghanistan, where the twenty-two-year-old son spent several years working in the administration. Once Khurasan settled down under Seljuk rule, Abu Ali al-Hasan headed back to his native Khurasan, which the future sultan Alp Arslan had been sent as governor. Under this scion of the Seljuks, he rose quickly to the rank of senior administrator at Merv. Abu Ali al-Hasan's successful performance in that role earned him the viziership and eventually the title by which he is still known.

No practitioner of the art of government in Central Asia, Persia, or the Arab lands proved as effective as Nizam al-Mulk. Suffice it to say that he was able to finance the sultans' military campaigns while at the same time successfully promoting the private economy and maintaining a stable currency. The gold dinars issued at Nishapur, Merv, and other Central Asian mints became standard instruments of trade across Eurasia, staving off for a time the rising inflation that was later to be reflected in the issuance of degraded silver coinage.[22]

The key to Nizam al-Mulk's success was his use of the Arabs' traditional land grants (*iqta*) as a means of paying army officers. By focusing the grants in border areas, he not only assured stability but placed the heaviest tax burden on peripheral territories rather than the Seljuks' economic heartland in Persia and Central Asia.[23] Indeed, taxes in the great centers like Merv, Nishapur, and Isfahan actually diminished under Nizam al-Mulk's regime. Thanks to this savvy policy, manufacturing flourished, international and regional trade boomed, and cities expanded, especially in Central Asia.[24] Typical of the aggressively mercantile spirit that defined the age, "Chinese" export pottery manufactured at Nisa in present-day Turkmenistan soon drove the Chinese originals from Western markets.[25]

Toward the end of the reign of Malikshah, the sultan asked his vizier, Nizam al-Mulk, to prepare a strategic assessment of the entire Seljuk government, pointing out areas of deficiency and recommending measures to address them. The result, known as the *Book of Government*,[26] stands today as the most comprehensive exposition of the practical arts of governance from the Islamic Middle Ages.

The *Book of Government* is blunt to a fault. Nizam al-Mulk subjected even the most petty flaws to merciless criticism. Slaves and servants should stand at attention while on duty; officials should not be given multiple assignments; public and private audiences, once announced, should never be canceled; and posting stations must be unfailingly supplied with fodder. With similar forthrightness he waded into more sensitive territory: women must be strictly excluded from matters of state; army units should be drawn from diverse races; decisions should not be taken in haste; drinking parties are essential for the ruler's relaxation but must be arranged so as to preserve a public face of decorum and to prevent underlings from putting on airs; and on matters of state, the ruler should consult only with his vizier—that is, Nizam al-Mulk. The key word in these and other pieces of advice is "should." The clear implication is that the ruling sultans had failed to abide by these rules.

It is often said that the *Book of Government* is saturated with Nizam al-Mulk's admiration for the ancient Sassanian state and his desire to revive the Persian practices of governance that had worked so successfully for half a millennium prior to the Arab invasion.[27] That Nizam al-Mulk was born near Ferdowsi's hometown of Tus two years before the poet's death helps explain how the great vizier might have come to admire the government described in the *Shahnameh*. But another model was both more immediately at hand and more significant: the government of Mahmud of Ghazni. Besides having actually worked in Ghazni for several years, Nizam al-Mulk, in his early years, was surrounded by officials whose sole work experience was under the Ghaznis' system, and it was from these men that he drew the large staff that he later brought with him to Isfahan. Central Asians may have loathed Mahmud of Ghazni, but his was now the system of rule they knew best.

The clearest evidence that Nizam al-Mulk sought to emulate Mahmud's administrative practice was in his approach to information. Over and over again he complained that he had insufficient intelligence on the practice of tax collectors, judges, prefects of police, mayors, and even matters of religion and religious law. His solution? To import Mahmud's system of spies: "Spies must constantly go out to the limits of the kingdom in the guise of merchants, travelers, Sufis, peddlers of medicines, and mendicants, and report back everything they hear, so that no matter of any kind remains concealed." Only with the help of an active network of professional spies could the ruler discover if army officers were plotting against him, if the rural peasantry was restive, or if a foreign ruler was preparing to attack. Thus informed, the ruler could then "set out immediately with all speed and, coming upon them unawares, strike them down and frustrate their plans."[28] Such a system, reinforced by an efficient postal system serving only the ruler, had been the very heart of Mahmud's state. Now Nizam al-Mulk proposed to emulate it.

Nizam al-Mulk failed in his effort to import Mahmud's system, which is why he devoted so much favorable attention to it in his *Book of Government*. But by analyzing it so carefully and championing it so systematically, he gave it an enduring standing in the political science of the entire Islamic world. Four centuries before Niccolo Machiavelli penned his *The Prince* in Florence, the *Book of Government* enshrined a harsh and systematic Machiavellianism in the political thought of Central Asia, Persia, and the Arab world.

Fortunately Nizam al-Mulk's actual practice was very different, especially with respect to the economy. His background in Merv and Nishapur had given him a keen appreciation for the role of commerce and trade in the workings of the economy, and of fiscal policy as a tool to support them. By contrast, Mahmud and his son Masud had systematically beggared the economy, undercut trade, and impoverished urban life. Nizam al-Mulk's positive legacy lay in his failure to implement the kind of police state he advocated in his great treatise on government, and his success in the economic sphere, which the *Book of Government* barely touched on. Thanks to this paradox, agriculture, manufacturing, and commerce thrived under Seljuk rule for a century and a half, and the expanding urban centers gave rise to a last great flowering of Central Asian culture.

Seljuk Architecture and Manufactures

The Russian historian Bartold dismissed the Seljuks as "strangers to all culture."[29] It would have been more accurate to say that when they swept out of Central Asia, they were still a nomadic people whose culture arose from centuries on the steppes rather than from urban life. Many of the Oghuz tribesmen who moved westward continued their nomadic existence. Their leaders, however, like the Karakhanid and Ghazni rulers before them, quickly mastered the skills of city life. Rather than dismiss the Seljuks as "strangers to all culture," we might more productively ask how they managed so quickly to master these new skills, and whether they brought to their new urban existence any traces of their rich nomadic culture.

One answer to the latter question is clear: for all the diversity of architecture, the arts, and crafts in the Seljuk era, they were all defined by rich surface decoration. This is manifest in the increasingly elaborate sculpted and painted plaster on the walls of nearly all buildings, in the intricate patterns on the surface of bronze ware, and in the rich patterns and colors of the "engobbed" pottery of the era. None of this was completely new: recall the heavily ornamented exterior of the tomb of Ismail in Bukhara. But surface decoration was so pervasive in this period that it became a Seljuk signature in the arts. Its origins can be traced directly to the textile arts of the Seljuks' nomadic forebears.[30]

The greatest monuments of Seljuk architecture have more to do with commerce and religion than with government. True, one can visit the ruins of Seljuk palaces in Anatolia and Isfahan, as well as the equally impressive palaces of local rulers, such as that at Khuttal in Tajikistan. But far more typical of the era were the many caravanserais constructed to facilitate trade, and the new types of mosques and commemorative tombs that sprang up everywhere. Both of these arose under the direct patronage of Nizam al-Mulk and reflected his priorities. Like the Barmaks in tenth-century Baghdad, the great vizier from Central Asia virtually defined the building arts in the Seljuk age.

To gauge the might of Seljuk architecture, one need only visit any of the dozen caravanserais on the ancient routes between Merv, Sarakhs, Balkh, and Nishapur.[31] The Seljuks constructed them to accommodate

Figure 12.1. In spite of their nomadic background, Seljuk rulers eagerly embraced commerce, as is reflected in this enormous caravanserai they erected on the Nishapur–Merv road. ▪ Photo by Harry Reid, London.

the surge of caravan traffic at the most heavily traveled knot of the Silk Road. Some of these travel lodges, like the formidable Rabat-i-Malik, were unapologetically fortresses, while others were so large that they could be mistaken for walled cities. All attest to the comfortable merging of commerce and art that occured under Nizam al-Mulk.

Nizam al-Mulk was also directly responsible for introducing a new type of mosque featuring a large domed and open-sided hall standing directly in front of the semicircular niche that indicated the direction of prayer. This innovation, which Nizam al-Mulk constructed at the Friday mosque at Isfahan to give shelter to his sultan at prayer, was subsequently copied throughout the Islamic world.[32] That this new construction at Isfahan was part of his broader support for the conservative Shafi'i school of law gave rise to rioting, which forced him to concede a section of the mosque to the more moderate Hanafis and their architects.[33]

Nizam al-Mulk's reign also gave rise to the craze for constructing monumental tombs over the remains of notables and holy men. This practice traced to the seventh century BC, when the Achaemenid empire erected a massive tomb for Cyrus the Great. The Tomb of Ismail in Bukhara relaunched the custom in Muslim times, as did the small mausoleum at Tim and many more shrines built to house the remains of holy men. Under Seljuk rule, however, every city and town competed to have the largest or most resplendent monument. The proliferation

Figure 12.2. Civil engineering in Central Asia thrived under the Seljuks, as evident in the Malan bridge near Herat and in the scores of other bridges, aqueducts, markets, and caravanserais they erected. ▪ UN Photo/Basir Seerat.

of Sufi holy men or *pirs* assured an endless supply of worthies to be commemorated architecturally. Commercial wealth fueled this mania, as it did at the same time in western Europe, where local burghers paid for the construction of chapels and shrines to commemorate their local saints. However, no region in the Christian West or Islamic Middle East outdid Central Asia in its enthusiasm for building tombs or *mazars*, with the result that from the twelfth century on they constituted the most widespread and enduring Muslim building type.[34] Particularly notable examples from the Seljuk period include the Khudai-Nazar-Ovliya mausoleum, the Muhammad ibn Zayda mausoleum, and especially the tomb of the last Seljuk ruler, Sanjar, all at Merv. All were constructed according to a geometry as strict as that observed by the architects of ancient Rome.[35] Many Seljuk buildings also featured pointed or "equilateral" arches. Whether or not these becamne a source for the Gothic arches that were appearing at the same time in Europe is a subject much debated by specialists.

Central Asia and the Seljuk realms were not the only places where architectural innovation was going forward in this period. In western Europe, following the resolution of the church-state controversy,

Romanesque buildings were increasingly defined not by their walls but by their ever-expanding arches. This led in time to the rise of a daring new Gothic architecture, in which walls virtually disappeared in a flood of windows and light. At precisely the same time when Sanjar's massive tomb was rising above the sands at Merv, Abbot Suger was building his pathbreaking basilica at the monastery of St. Denis north of Paris, blazing a path to further innovations in engineering and aesthetics. Seljuk architecture was a great achievement, but in the end it represented more a culmination than the launching of a new direction.

Can the same be said of Seljuk civilization as a whole? The prosperity engendered by Nizam al-Mulk's prudent economic policies gave rise to flourishing arts and craft industries. The few fragments of textile to have survived reveal a continued development of this ancient craft. On a similarly high level were the metalwork and ceramics. Finely crafted bronze vessels, oil lamps and sculptures inlaid with silver, fine tools and weapons of tempered steel with brass and silver inlays, elegant vases and platters exquisitely painted with human figures or enlivened by a host of brilliantly colored new finishes all stood at the pinnacle of their art. All these techniques were invented in Central Asia. It is no wonder that the proud craftspeople regularly signed their products. Thus a Khurasan craftsman engraved on his bronze cooking pots his name, "Khojo from Tus." Another, Abd al-Razzaq Naysaburi from Nishapur, signed his perfume sprinklers and boldly emblazoned on them the word "Happiness."[36] Such joie de vivre doubtless boosted sales: an unsigned bronze Seljuk pot from Merv bore the cheery inscription "Enjoy Your Meal."[37]

Both Khojo and Abd al-Razzaq produced for export. In fact, several of the latter's fine pieces have turned up thousands of miles distant from their place of manufacture.[38] In nearly all the major crafts, the artists and artisans from cities like Herat, Nishapur, and Merv set the pace for the entire Islamic world during these years.[39]

The World of Learning under the Seljuks

Both architecture and the crafts can flourish in an intellectual void. What, then, was the state of scientific research and philosophy under Seljuk rule? Were there any Seljuk patrons who rivaled the earlier

Barmaks, Caliph Mamun, or the the Khwarazm dynasty in Gurganj? The only figures from the Seljuk era who met this high standard were Sultan Malikshah in Isfahan and his vizier, Nizam al-Mulk. It was they who intiated the research that led to the introduction of a new solar calendar in 1079. Several later medieval writers claimed that these two officials also funded an observatory at Isfahan, and that the great mathematician Omar Khayyam was brought from Central Asia to head that facility. More certain is the central role of Nizam al-Mulk, who conveniently arranged that the solar cycle began with the national festival of Novruz on the vernal equinox.[40] But aside from Malikshah and Nizam al-Mulk, official patronage for learning in this period was modest at best. The only other Seljuk leader to take a serious interest in learning was one al-Tughrai, a vizier from Merv, who penned a book on astrology.[41]

State patronage of thinkers and writers had been the norm in Central Asia for centuries, but its decline under the Seljuks did not mean that intellectual life languished. In Europe at this time, learning began to move beyond the monasteries to the royal courts. Beyond this, the rising cities and the founding of universities in this period prepared the way for entirely new forms of cultural patronage that would blossom later.[42] Central Asian cities also boomed during Seljuk times, and their courts provided an environment for learning that was no less stimulating than what prevailed among their counterparts in the West. Two in particular stood out: Nishapur and Merv.

Proud Nishapur experienced so vibrant an economic and cultural effervescence during the Seljuk era that it has been called the intellectual capital of the Seljuk world.[43] Had this accolade been based only on the mathematician and poet Omar Khayyam, the philosopher Ghazali, and the vizier Nizam al-Mulk, Nishapur would have deserved it. But a host of lesser lights worked there as well—men like the mathematician al-Nasawi, who picked up the notion of decimal fractions from a Syrian thinker and gave it wide currency.[44] In addition to a deep tradition in the sciences, Nishapur continued to boast the most authoritative teachers of law. It was also home to a deep tradition of nonconformist thought and outright skepticism, which added an intellectual frisson to the overall climate. This combination of strengths is why Khayyam, Ghazali, and Nizam al-Mulk all headed there to complete their education. No city in the Islamic world boasted a more yeasty intellectual environment at the time.

Perhaps too yeasty, for during the entire Seljuk era Nishapur was once more riven by fierce debates among the competing schools of Islamic law and over contending strands of religious thought. Hell-fire literalist sects also flourished, and one of them briefly seized power in the city, instituting a reign of terror.[45] As was the case earlier, all this contentiousness spread throughout the populace.[46] By the eleventh century Nishapur was once more the scene of perpetual strife, with street fights among the contending parties a constant occurrence. A yeasty environment was one thing, but daily brawls were quite another.

Urban violence, along with the clashes between the Ghazni army and the rising Seljuks at the outset of the era, gave Nishapur and Khurasan an aura of instability that belied their commercial wealth. A number of intellectuals chose to leave. Among them was the splendid poet Abu Mansur Ali Asadi from Tus, known as Asadi. Once he completed his education, he departed for the mountains of Azerbaijan, where he attached himself to a local court and versified *The Epic of Garshasp* (*Garshaspnameh*), which ranks second only to Ferdowsi's *Shahnameh* among Persian epic poems.[47] When the philosopher Ghazali returned from his extended travels, he too avoided the prevailing turmoil in Nishapur and retreated into a life of seclusion.

Merv, by contrast, remained peaceful down to the latter days of Seljuk rule. It, too, experienced unprecedented prosperity, so much so that an entirely new city was created just west of the ancient center to house the burgeoning population, manufacturers, and trading houses. Under the surface Merv, too, experienced social and religious tensions, not to mention fears of invasion by yet another wave of nomads arriving in Central Asia from China. But overall a sense of security prevailed, thanks in part to the nine miles of walls with internal walkways that encircled the new quarters. Behind their high new brick ramparts, the residents of Seljuk Merv enjoyed the good life replete with iced drinks in summer, thanks to the many large, dome-shaped brick icehouses situated just outside the city walls. Seljuk Nishapur, by contrast, was unfortified.

No city in the Muslim East boasted so many libraries as Seljuk Merv, and these drew scholars from near and far. The Arab geographer Yaqut (1179–1229) spent three years studying in Merv's twelve libraries, just one of which, he reported, contained twelve thousand volumes.[48] Other Merv-based scholars worked in fields as diverse as mineralogy and philology.[49] Merv was also home to several historians, in spite of the fact that historical writing went into decline under the Seljuks.[50]

Another attraction at Merv was the centuries-old observatory. Among the astronomers and physicists employed there were many local talents, such as Abu Fazli Sarakhsi, who hailed from nearby Sarakhs.[51] Abu al-Rahman al-Khazini was a Byzantine Greek who was captured and enslaved by the Seljuks but went on to study under Omar Khayyam before pursuing a distinguished scientific career on his own.[52] Besides astronomy, Khazini's great passion was the science of weighing things, a branch of quantified mechanics. His *Book of the Balance of Wisdom*, written in Merv, has been called "the most comprehensive work on [weighing] in the Middle Ages, from any cultural area."[53] With admirable dispassion Khazini faulted the ancient Greeks for failing to differentiate between weight, mass, and force. He then elaborated how Archimedes's principle of flotation could be used to determine specific gravities. He even showed how the specific gravities of liquids could be determined, and he demonstrated for the first time that air, too, has weight, which diminishes with altitude. Khazini appears not to have known of Biruni's earlier research on these same issues.

Khazini was also a master mechanic. He and his team at Merv devised the world's most precise instrument for weighing ordinary objects, determining specific gravities, and even examining the composition of alloys. On the basis of his detailed description, it has been possible to reconstruct his complex mechanism, which he dubbed "The Comprehensive Balance." Modern study affirms his claim to its extraordinary accuracy of 1:60,000.

That Khazini's interests included metallurgy is not surprising when one considers that archaeologists have discovered at Merv one of the first known furnaces for the production of crucible steel. The Indians either discovered independently or borrowed this same technique a few centuries later, but their process was less sophisticated and efficient than the Central Asians'; European steel makers, by contrast, did not master what was called "co-fusion" until six centuries after the Merv furnace began production.[54]

The unrivaled genius of Central Asian science and mathematics during the Seljuk era was Omar Khayyam (1048–1131). Born in Nishapur to a family of tent-makers (he wrote that he had "stitched the tents of

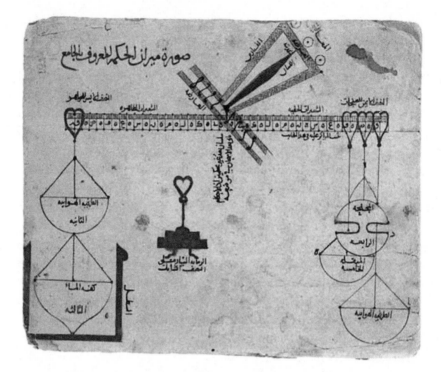

Figure 12.3. Working at Merv, Abu al-Rahman al-Khazini designed a "comprehensive balance" capable of weighing objects to an extraordinary accuracy of 1:60,000. ■ From *Kitab Mizan Al-hikmah* (Book of the Balance of Wisdom), c. 1270. In Arabic, ms. on paper, LJS 386, Rare Book & Manuscript Library, University of Pennsylvania.

science"), Khayyam is far more widely known in the West as a poet and author of the *Rubaiyat* than as a mathematician or astronomer. But like so many of the region's greatest minds, he was a polymath who wrote treatises on mechanics, philosophy, geography, and music, in addition to his pathbreaking works on algebra and the heavens.

By education as well as birth Khayyam was a true son of Central Asia. His earliest training was under a renowned scholar in Balkh, after which he studied with one of the most sought-after teachers in Nishapur. Thanks to the patronage of a prominent and wealthy jurist named Abu Tahir, the twenty-two-year-old scholar launched his career in Samarkand, where he wrote his landmark *Treatise on Demonstration of Problems of Algebra*, the greatest work in that field before modern times. After a sojourn in Bukhara, he returned to his native Nishapur, where he soon attracted the attention and patronage of Nizam al-Mulk, his elder by thirty years. With the support of his fellow Khurasani, Khayyam was to spend eighteen productive years in the Seljuk capital as the leader of the community of scientists assembled there. Khayyam was of a somewhat morose temperament, which he expressed in occasional poems:

Today is the time of my youth,
I drink wine because it is my solace:
Do not blame me, although bitter it is pleasant.
It is bitter because it is my life.[55]

But this by no means prevented his pursuit of pathbreaking research along several lines and, as we shall see, exploration of philosophical questions as well.

The death of the Sultan Malikshah I 1092 sparked a crisis in Khayyam's life, as it did for Ghazali. Having lost his patron, Khayyam eventually moved to the final Seljuk capital at Merv and placed himself under the new ruler there, Sultan Sanjar; then, at the end of his life, he moved back to Nishapur, where he died in his ninth decade.

At Isfahan Khayyam busied himself with astronomy. He charted out a map of the stars and is also said to have built a model that demonstrated how Earth might revolve on its axis. Thanks to Nizam al-Mulk, he played a key role in the astronomical research that led to the new solar calendar, mentioned above, which lasted in Iran to 1925. The eighteenth-century English historian Gibbon wrote admiringly that Khayyam's system "surpasses the Julian [calendar], and approaches the accuracy of the Gregorian [calendar]."[56] Typical of Khayyam as astronomer (and directly recalling Biruni), he carried out his computation of the solar year to the eleventh decimal point: 365.24219858156 days. The minute difference between this figure and the modern calculation is due entirely to the very slight shortening of the solar year that occurs over time, not to any error on Khayyam's part.

Omar Khayyam's most enduring achievements were in mathematics and geometry and derived directly from the clarity and precision with which he framed each problem he addressed.[57] Thus, when barely out of his teens he challenged himself to "find a point on a quadrant of a circle in such manner that when a normal is dropped from the point to one of the bounding radii, the ratio of the normal's length to that of the radius equals the ratio of the segments determined by the foot of the normal." He then realized that if he could do this, he would also solve a second problem that had vexed him, namely, to "find a right triangle having the property that the hypotenuse equals the sum of one leg plus the altitude on the hypotenuse." Advancing step by step, Khayyam proceeded to solve the cubic equation $x^3 + 200x = 20x^2 + 2000$ and then to find a numerical solution by interpolation in trigonometric tables. The Scottish mathematicians O'Connor and Robertson marvel at the fact "that Khayyam states that the solution of this cubic requires the use of conic sections and that it could not be solved by ruler and compass methods, a result which would not be proved for another 750 years."[58]

In his *Treatise on Demonstration of Problems of Algebra*, written in Samarkand, Khayyam offered a complete classification of cubic equations with geometric solutions that he found by intersecting a hyperbola with a circle. Sweeping beyond Greek, Arab, and other Central Asian mathematicians like Khazini, he became the first to conceive a general theory of cubic equations and an exact process for solving them. This included the realization that a cubic equation can have two different solutions.

Such insights have led some historians of science to suggest that Khayyam might also have come up with a theorem that would permit the algebraic expansion of a binomial up to any power. Whether or not Khayyam hit on a binomial theorem, as it came to be called, only much later did such Italian mathematicians as Scipione del Ferro (1465–1526) and Giordano Vitale (1633–1711) advance beyond Khayyam to propose that cubic equations can have three solutions. Summarizing Khayyam's contribution, the distinguished historian of mathematics E. S. Kennedy concluded, "In his major work in algebra, Khayyam exploited [the technique he had put forward] to work out solutions for all possible types of cubic equations."[59]

Khayyam's was the kind of mind that approached the problem at hand with such breadth and boldness that he would reframe the problem itself. Thus, even though he was a committed follower of Euclid, the fourth century BC father of geometry, his attempt to prove Euclid's postulate that parallel lines cannot meet led him to reject as invalid the proofs advanced by Greek and Arab scholars. In his treatise *On the Difficulties of Euclid's Definitions*, he did not propose an alternative postulate to Euclid's, but he was the first to sense that a non-Euclidian postulate on parallel lines might be possible. In an effort to devise an alternate geometry to Euclid's, the seventeenth-century Jesuit priest Giovanni Girolamo Saccheri (1667–1733) addressed the same set of postulates, but without achieving a breakthrough. Only much later, in the 1820s, did a Russian, Nikolai Lobachevskii, succeed in producing the systematic non-Euclidian geometry to which Khayyam first opened a path.

THREATENING IDEAS, FROM WITHOUT AND WITHIN

The Seljuk era should have been a period of calm on the religious front. A majority of Central Asians had by now converted to Islam,[60] and the Sunni Seljuks had dethroned and conquered all the Buyid Shiite rulers across the north of Persia. Also, since both the sultans and caliphs were Sunnis, outward accord reined between them. But the Ismailis still maintained in Cairo their rival Fatimid caliphate. True, many of the Fatimids' lands had begun to fall away, and the dynasty itself entered a period of intolerance toward both Sunni Muslims and Christians. The Berber and Turkic mercenaries who ran their army threatened the Fatimid state from within, even as crusaders from Europe threatened them from without. Worse, the Sunni Seljuks were subjecting Ismailis to massive pressure throughout their realm.

Rather than fade quietly away, radical elements among the Fatimids proposed to mount a bold counteroffensive against the orthodox Sunnis and the Seljuks who protected them. Withdrawing into heavily fortified castles, they created a decentralized state within the state. As Baghdad intensified the pressure on them, one branch of the Ismailis began launching suicide attacks against leading figures of both the Baghdad caliphate and the Seljuk state. Their Muslim enemies insisted that the Ismaili hitmen were pumped with hashish to embolden them. When Crusaders heard of this, they called them "assassins,"[61] after the drug they purportedly used.

Whatever physical danger the Ismailis posed, and historians continue to argue over this,[62] they presented a far more formidable ideological challenge to the Seljuk rulers and to their allies among the orthodox Sunni establishment. They discredited the scriptural justification for the granting of appanages (*iqta*) on which the entire Seljuk tax system depended. They unapologetically applied rationalist logic and Greek Fal-safa to theological questions, sought out hidden meanings in the Quran that were unacceptable to the orthodox religious scholars, and even de-nied the Quranic arguments on which Sunni scholars and imams based their claim to guide the Muslim faithful and the state.

The fact that many leading thinkers in Central Asia and Persia had been attracted to the Ismaili cause only made matters worse. Some who did not actually join the Ismailis admired the intellectuality and open-ness of their thought, and even their Neoplatonic mysticism. Others, like Ibn Sina, were attracted by their use of both Greek philosophy and faith to resolve matters of religion.

The figure of Nasir Khusraw (1004–1088) epitomized the challenge that the Ismailis presented to Sunni orthodoxy in the Seljuk period. Born in the village of Qabodiyon in northern Afghanistan, Khusraw bene-fited from a thorough education in science, medicine, and the rationalist philosophy of Farabi and Ibn Sina. But down to his forty-second year he was no dissident. A Shiite Muslim, he chose instead the solid and quiet life of a civil servant in the Seljuks' revenue office in Merv. With no small pride he boasted of the comfortable home and garden he had been able to purchase in the ancient capital.

In 1046 Khusraw's solid existence came unglued. We do not know the nature of his personal crisis, but it led him abruptly to quit his job, sell his house, and set out on what turned out to be a seven-year journey that covered 12,500 miles and extended from Baghdad and Syria to the Holy Lands, Mecca, and Egypt. His detailed description of the places he visited is not only one of the most informative travel accounts from the Middle Ages but also one of the most gracefully written.[63] Fatimid Cairo was already in the last stages of decline when Khusraw arrived there, but its impact on him was profound. Encounters with the Ismaili scholars at their center of learning, al-Azhar, left him an ardent convert to their teachings. Henceforth his sole object in life was to disseminate their faith among the people of his native Central Asia.

This put Khusraw on a collision course with the Seljuk authorities. During his extended trip back from Cairo, he missed no opportunity to preach against what he considered the dogmatism of Seljuk rule and the stifling of free thought by their officials. Seljuk authorities treated him as a troublemaker and punished him accordingly. But Khusraw was by temperament less a stump speaker than a poet, and a very thoughtful one at that. His new missionary role inspired him to write Persian poetry that weighs the moral issues of daily life with touching simplicity and at times stunning directness. At one point he even versified against God for not accepting responsibility for the bad behavior of his creatures.[64]

In keeping with Ismaili doctrine, Khusraw saw no contradiction between rational philosophy and revelation and viewed them as being equally valid avenues to truth. By propagating such doctrines, Khusraw opened himself to charges of heresy, if not apostasy. From the moment he returned to Central Asia, he was everywhere persona non grata, hounded from city to city by orthodox Sunni officials. Eventually he found refuge in the high Pamir Mountains of northeastern Afghanistan. There, in the village of Yamgan, which is still almost unreachable, he spent the final three decades of his life instructing disciples and writing poetry that veers between religious ecstasy and raw bitterness. One such lyric begins with a paean to spring and then explodes in the following:

> Bah! Enough of such futile nonsense! Such blather
> merely embarrasses me! Spring has returned
> as my guest now sixty times—it will be the same
> if I live to be six hundred. Those whom Fate
> has stripped of all adornment can take no joy
> in the garden's decorations; to me its loveliness,
> this Spring of yours, is but a daydream
> concealing pain beneath its charming robes,
> poison in its sugar, thorns in its roses.[65]

And yet this same Khusraw, down to the end, affirmed knowledge:

> Make a shield from knowledge, for there is
> No stronger shield against calamities.
> Whosoever owns the shield of knowledge
> will not suffer from the blows of Time.[66]

And, too, down to the end Nasir Khusraw affirmed his abiding faith:

> O Lord, O my creator high:
> I am so grateful for your grace,
> For in old age's days I have

No confidant, but "Thanks for You."
With piety and obedience,
I sing a hundred thanks to You![67]

With alarm verging on panic, Nizam al-Mulk followed the activities of Ismailis like Nasir Khusraw and especially their more radical coreligionists. The vizier had no doubt that they were corrupting Islam and undermining the caliphate that the Seljuks had vowed to defend. Their strong stand against social abuses found support among many common people, especially in Khurasan. Did the appearance of the Ismailis not mark the End of Time? In this grim state of mind, the vizier appended ten further chapters to his *Book of Government*, tracing the Ismailis' rise, exposing their nefarious activities in city after city, and cursing them in God's name.[68] He was convinced that Islam itself and the Seljuk state were under assault, and by a Muslim enemy that was as clever as it was ruthless.

Nizam al-Mulk Counterattacks: The Nizamiyya Schools

Scarcely less alarming in Nizam al-Mulk's eyes were the fierce conflicts being waged in Nishapur and other cities of Central Asia between the mutually hostile Muslim legal schools and among the contending schools of theology. In Nishapur in particular, these had long since ceased to be purely theological issues and had descended into armed confrontations between discontented members of the lower classes and the merchant elite, which in turn became a turf battle between neighborhoods.[69]

At the risk of oversimplifying a landscape that was rich with subtle distinctions, the basic juxtaposition was between the rising conservative legal experts and everyone else. The former, mainly followers of Ibn Hanbal, rejected reason, philosophy, and logic out of hand as sources of ultimate truths. Some of them allowed a modest role for reason but still emphasized the body of traditions tracing to the Prophet's time. Only slightly less restrictive were the Shafi'i jurists, whose influence rose throughout the eleventh century. Opposed to both of these schools of law were the Hanafi jurists, who, from their base in Central Asia, allowed a slightly greater role for reason in addressing questions of faith. True, they, too, fully accepted the authority of the Hadiths but were open to somewhat broader interpretations of them. Several prominent Hanafi experts, one from the Ferghana Valley in what is now Uzbekistan and the other from Sarakhs, had recently sharpened their arguments, which only inflamed the yet more deep-dyed traditionalists.[70]

Deeply traditionalist in his views and hence unwilling to broaden the sphere of "interpretation," Nizam al-Mulk saw in the entire debate a threat to the stability of the Seljuk realms. Pessimistic to the core, he feared that the legal and theological conflicts welling up on every side presaged the end of Islamic civilization. To counter these threats, Nizam al-Mulk lent his full support to the Shafi'i legal experts and the Asharites in theology. Knowing that fine words alone would not suffice, Nizam al-Mulk resolved to establish a series of madrasas where Asharite theology and Shafi'i legal doctrines would be inculcated in the leaders of the rising generation. Such schools were to be called Nizamiyyas, after their founder.

The madrasa as such was nothing new. Schools for studying Muslim holy writ had been common for centuries, particularly in Central Asia. The two large, domed chambers of the tenth-century Khwaja Mashhad madrasa in southern Tajikistan housed a tomb and mosque, respectively. But the adjacent large courtyard and perimeter buildings that housed students in the Samanid era both anticipate the standard madrasa design of later centuries.[71] During the years the Shiites controlled Baghdad, they, too, had set up madrasas for their coreligionists; and in the years immediately preceding Nizam al-Mulk's initiative, rival Hanafi officials had founded their own school there as well, and also a hostel for their young supporters.[72] Within a few decades there were no fewer than thirty such "colleges" in a confined area in East Baghdad.[73] Nizam al-Mulk boldly entered this battle zone.

Three features defined the new Nizamiyya: First, they were devoted not to education as an open-ended process of questioning but to the inculcation of doctrinal purity and its maintenance against the claims of Shiites, Ismailis, Mutazilites, and other groups that challenged the orthodox Sunni hegemony. The fact that each of these groups included effective polemicists, like the hard-hitting Shiite Abu Jafar from Tus,

Figure 12.4. Fragment of inscription from the façade of Nizam al-Mulk's Nizamiyya madrasa at Khargird, Iran. ▪ The Ernst Herzfeld papers, Freer Gallery of Art and Arthur M. Sackler Gallery Archives, Smithsonian Institution, Washington, DC. Gift of Ernst Herzfeld, 1946. Glass plate negative, neg. 2736-70.

only made the madrasas' task the more challenging. Second, since the Asharites had placed strict limits on the use of reason and defined many leading philosophers and scientific thinkers as heretics, it is fair to assume that a skeptical stance on all forms of scientific inquiry was part of the curriculum, even without Ghazali's diatribes against the excesses of science and philosophy.[74] And third, reflecting the high importance the Seljuk state assigned to dogma, the Nizamiyya madrasas were state institutions, funded by, and subordinate to, the Seljuk sultanate.

Nizam al-Mulk called on the popular thirty-three-year-old teacher-scholar Ghazali to set up the first Nizamiyya madrasas at Baghdad, Nishapur, and Khargird (on the Iran-Afghanistan border) in 1065, to be followed soon thereafter by another in Shiraz. With three thousand students each by the 1090s, the Nizamiyyas at Nishapur and Baghdad ranked as the largest educational institutions in the world at the time.[75]

These new institutions fundamentally redefined the purpose and form of Muslim education and provided a pattern that was to be emulated throughout the Muslim world. They had arisen amid doctrinal conflicts that the Seljuk government feared could destroy the state. Nizam al-Mulk's purpose was to create intellectually militant institutions that would train young men to oppose and suppress nonstandard ideas and to advance the correct ones. In other words, they represented a shift from analysis to indoctrination, from defense to offense on the intellectual front.

Just three years before the Baghdad Nizamiyya opened its doors, the first European university had been established at Bologna, with many more to follow over the next century in France, Italy, and England. But the Nizamiyya schools pursued a far narrower mission than these institutions, their goal being to defend Muslim orthodoxy and prepare young men to promote it. As such, they bear comparison not with universities but with the institutions Ignatius Loyola established in the sixteenth century to advance the cause of the Catholic Counter-Reformation. But unlike Loyola's schools, which broadened over the years, the new-type madrasas remained true to their founding mission.

It is tempting to view the establishment of the Nizamiyyas as the moment at which Central Asian learning, and perhaps Islamic learning as a whole, recoiled from the outward-looking and bold intellectuality of Khwarazmi, Farabi, Biruni, and Ibn Sina and began turning against itself. To the extent this is true—and thoughtful people differ sharply on this point—it represents the culmination of a process that began with the reaction to Caliph Mamun's ill-conceived effort to purge learning of all the enemies of reason.[76]

However one views the Nizamiyya, it is worth noting again that, for better or worse, it was a Central Asian—Nizam al-Mulk—who devised this new approach to education, and that he did so in response to the spread of heterodoxy in Central Asia and especially in his native Khurasan. Moreover, the man who successfully developed it was another native of Central Asia, Ghazali, who set up one of his prototype madrasas in Nishapur, the capital of Khurasan and headed the one in Baghdad. In other words, as surely as the great Age of Enlightenment had arisen in Central Asia and spread outward from there to other parts of the Muslim world, so did the most powerful movement against it.

Khayyam the Philosopher

It is hard to imagine a less favorable moment than this to take up problems of philosophy and metaphysics, but this is precisely what Omar Khayyam did. Nizam al-Mulk had banned philosophy from his madrasas,[77] and Ghazali was railing daily against the abuses of the philosophers from his academic lectern. Amid this barrage, Khayyam quietly toiled away on a number of philosophical and theological questions that vexed him. His writings in these areas are unknown to all but a few Muslim theologians and scholars. Yet they provide an interesting perspective on the era and present Khayyam himself in a challenging new light.

The starting point for Khayyam's reflections was his core field, mathematics. Immersed in the subject, he was acutely aware of the existence of an orderly world of number and measure. Similarly, he was keenly conscious of the great significance of abstract postulates in geometry. He could not help asking himself where these orderly phenomena came from, and how it was that they corresponded so precisely to the observed natural world. His probes into non-Euclidean geometry both reflected and reinforced his interest in such questions.

Khayyam's answer was to trace both to God. In this spirit he turned to the very thinker who was the target of ceaseless attacks at Nizam al-Mulk's Nizamiyya—Ibn Sina—and made a translation and commentary of his *Splendid Sermon* in praise of God's unity. Sometime thereafter Khayyam, in response to some questions he had received from a judge in the Iranian province of Fars, penned an earnest *Treatise on the Realm of Existence and Human Responsibility*. In this work he again traced the sense of responsibility toward others to human nature and to the Creation, which were both God's work. This prompted him to acknowledge both the role of prophesy in our quest for certainty and the possibility of the kind of direct knowledge of God's world that Sufis seek to attain. In an astonishing summation, he exhorted his reader to "tell the reasoners that for lovers of God [Gnostics], *intuition* is the guide, not discursive thought."[78]

In these and other passages, Khayyam revealed himself as a practitioner of philosophy, a faylasuf of the very sort Ghazali attacked daily. Worse, in his critically important view of *intuition* as a kind of bridge between man and God, he directly championed Ibn Sina'a solution to the problem of faith. At the same time, Khayyam made clear that he accepted revelation and sympathized with the basic quest of the Sufis.

How did all this fit with the views of his patron, Nizam al-Mulk, or his friend Ghazali? We don't know. These aspects of Khayyam were barely known in his own day and have been largely forgotten since. How, also, did it fit with the skeptical and hedonistic Khayyam of the well-known quatrains? We will turn to this intriguing question shortly. At the very least, these three Khayyams—mathematician-astronomer, philosopher theologian, and hedonist skeptic—reveal the kind of multifaceted and complex personalities that proliferated during Central Asia's Age of Enlightenment but became all too scarce in the region thereafter.

THE INWARD ALTERNATIVE: SUFISM

For the philosopher Hegel, the key to history was the constant dialectic of change, with each new phase of human experience arising in the bosom of what preceded it and then overwhelming it. Something like this occurred in Central Asia and much of the Muslim world at the time Nizam al-Mulk was planning his new institutions for the inculcation of right thinking and dogma. If his ideal was outward respect for Muslim law and dogmas and communal solidarity as defined and regulated by religious scholars and imams, the countermovement stood for inwardness rather than outwardness, asceticism rather than worldliness, and direct communion between each believer and God, with no role for the ulama: in short, ecstasy rather than doctrine.[79]

"Sufism," as it was called, took its name from the cloaks of white wool worn by early Muslim mystics. When Sufism first appeared in Arabia, Iraq, and Central Asia in the ninth century, it was not clear whether it was a form of deviance from true Islam or its fulfillment. After many struggles and several martyrs, Sufism emerged at the heart of Islam during the Seljuk period. It was incubated above all in Central Asia, drew many of its most revered masters from there, and gained theological legitimacy thanks to Nizam al-Mulk's protégé from Nishapur, Abu Hamid Muhammad al-Ghazali.

We have already encountered the diverse sources of Sufism. Suffice it to recall that Christian teachers and converts brought rich traditions of Byzantine mysticism and the spiritual poetry expressing love for the deity to which it gave rise. Among the earliest Muslim Sufis had been Tirmidhi, whose doctrine of "friendship with God" drew on both the Christian and Buddhist mystical traditions that thrived in his native city of Tirmidh (Termez).[80] The ninth-century Sufi Bistami from Khurasan (848–874), whose "intoxication" with God made him one of the greatest Sufi teachers, had as his early master a Hindu from Sind in what is now Pakistan.[81] Such emotions found their natural expression in poetry. Early Arab Muslims had little sympathy for versifying, but Sufism would eventually produce two magnificent and world-renowned poet-mystics, Rumi and Jami—both from Central Asia—not to mention other equally brilliant but lesser-known poets.[82]

The urge for direct communion with God is to be found among many religions, especially in times when social cohesion is fraying. For all their differences, the *hesychast* movement in Byzantine Christianity and its powerful offshoot in Russia, the Hasidic movement among seventeenth-century Jews in Poland, the early Quaker movement in seventeenth-century England, the Protestant Moravian Brethren in central Europe during the eighteenth century, and early Methodists in England and America all summoned the pious to abandon wooden dogma in favor of inner purification and spiritual riches.[83]

In the end, formal education and doctrine matter little to such movements, nor did they concern the Sufis, who instead embraced spirituality and the mysteries of intuition as the sole valid source of faith. Their path to truth was to emancipate themselves from all worldliness, in which condition they would then be open to beauty, love, and the understanding of God.[84] The charismatic Sufi master Kharaqani (d. 1033), a Turk from Herat in Afghanistan, attracted some of the most brilliant students of the age to this spiritual journey, in spite of his lack of formal education. Rejecting all concern for Heaven and Hell, this ardent mystic called instead for total concentration on the Creator. Contemporaries viewed him as the greatest mystic of all times.[85] Teachers like Kharaqani rendered Sufism attractive not only to the masses in areas of Central Asia where Iranian languages were spoken but also to newly converted Turkic peoples, who were quick to detect commonalities between their native shamanism and the capacious Sufi worship.

As early as AD 1000, Central Asia emerged as the heartland of the Sufi tide that was eventually to sweep the Islamic world. A handful of the region's leading Sufis were descendants of Arab immigrants, but most were locals. Soon Central Asia's reputation as the center of Sufi thought attracted immigrants from elsewhere. Many leading mystics at Merv, Bukhara, Balkh, and Nishapur had fled there in search of a more hospitable environment for their new faith. This led inevitably to different forms of worship, rival Sufi masters or *pirs*, and the emergence of the first competing schools of Sufism in the Muslim world.[86] A typical Sufi sect called itself grandly "the Nishapur path."[87] Of the Sufi orders or *tariqas* that came eventually to prevail throughout the Islamic world, three of the largest and most important trace their roots to the soil of Central Asia.[88]

Down to the Seljuk era, Sufism's rise was slow and evolutionary, but beginning in the eleventh century it spread with revolutionary speed. Popular preachers attracted large crowds in the bazaars. Common folk thrilled at the fact that these new types of Muslims did not demand that they read learned treatises or memorize dogma, were quick to acknowledge human frailty, and laid out a path to salvation that was accessible to all honest seekers.

Against the prevailing background of contention and urban strife, the Sufi preachers, with their call to individual faith, presented a simple alternative that was accessible to all. In response to chaos in civil society ,they invited people to withdraw into a private realm, seeking respite through direct communion with God. Even when Sufism, too, became an element in the political strife of the day, as happened by the next century, its appeal did not diminish.[89]

True to their ancient traditions, Central Asians were also the first to group the emerging strands of Sufism and to categorize them according to their affirmations and practices. Just as they had earlier submitted Zoroastrianism, Greek religion, Buddhism, and Islam to such analysis, they now studied Sufism. The two leaders of this effort were Abu Abd al-Rahman al-Sulami (d. 1024), a descendant of Arab immigrants to Nishapur, and Abdullah Ansari (d. 1089) from Herat.[90] The benign and prolific Sulami penned admiring biographies of scores of leading Sufis and analyzed their teachings. For forty years his house and library in

Nishapur was a magnet for aspiring Sufis from throughout the Muslim world. Ansari, by contrast, stood directly in the great tradition of Khurasan polemicists, railing in the name of Sufism against the intellectual pretensions of both the radical Mutazilites and the conservative Asharites.

THE CRISIS OF 1092–1095

The last five years of Sultan Malikshah's rule were prosperous; trade boomed and new buildings arose throughout the region. But grave problems festered just beneath the surface of daily life. Nomadic armies posed new threats from the east and north. Insurgents seized Basra on the Red Sea, and a radical branch of Ismailis fortified themselves in the mountains of northern Persia. The Seljuk army, which should have addressed these challenges, increasingly served the political ambitions of its generals as much as the sultan and his vizier.[91] To make matters worse, the sultan and caliph were at odds over Nizam al-Mulk's assertion of religious authority through his Nizamiyya schools and other measures. The vizier had carefully built his relationship with religious scholars and imams, and now they sided with him as he openly supported the conservative and militant Hanbali and Shafi'i schools of Islamic law, as opposed to the more moderate Hanafis with whom the caliph was aligned.

During the 1090s these tensions erupted in chaos on the streets of Baghdad and other cities. Hanbali mobs attacked cafes and prostitutes, smashed chessboards, and drove women back into their homes. Anyone suspected of Shiite sympathies, including any scholar, was fair game, while Ismailis could be killed on the spot. Students from the Nizamiyya near the Friday mosque poured into the streets to do battle with young Hanafi supporters. The civil authorities did nothing to curb them. Similar clashes occurred in Nishapur and Merv. And at Bukhara and Samarkand, the now hereditary religious authorities found themselves under attack for excessive lenience in the implementation of the Sharia, the law of Islam.

So deep was the mood of intolerance to which Nizam al-Mulk's policies gave rise that many of the remaining Zoroastrians left the region, emigrating to Gujarat and the northwest coast of India, where they became known as Parsis. Christians and Jews retreated into their houses, praying for better days.

For twenty years the young sultan Malikshah had docilely signed each new decree presented to him by the hyperactive Nizam al-Mulk. But Nizam al-Mulk was haughty in manner, arrogant, and an easy butt of satire. By the 1090s the sultan, now thirty-eight years old, had had enough. The dislike was mutual, for he had infuriated his vizier by sacking one of Nizam al-Mulk's protégés and then by criticizing his vizier for placing his many sons in prime jobs. But just as Malikshah was poised to sack Nizam al-Mulk, an Ismaili assassin preempted him, murdering the great vizier on October 14, 1092.[92] Contemporaries whispered that this was the work of the sultan himself. Then, barely a month later, Sultan Malikshah died (see plate 18).

The ensuing succession struggle plunged the entire region into civil war and fractured the Seljuk realm. Sanjar, an uncle of the new sultan, laid claim to Khurasan and made clear his intention to rule all Central Asia as an independent state from his capital at Merv. Even though he was careful to acknowledge his nephew's title as sultan, Sanjar's assertion of authority led to a further reign of terror by the new sultan, Berk-Yaruk, in the course of which the population of Balkh rose up en masse against the invading army from Baghdad.[93]

Revisiting Ghazali's Nervous Breakdown

The philosopher and theologian Ghazali was at the epicenter of these momentous struggles. As the rector of the flagship Nizamiyya school in Baghdad, he was the reigning authority on religious matters through sheer force of intellect. With political cover from Nizam al-Mulk, he could stand up to anyone. How curious, then, that the spiritual crisis and nervous breakdown described at the start of this chapter occurred just three years after the deaths of both Nizam al-Mulk and Sultan Malikshah.

We noted at the outset that Ghazali had described his crisis as an "illness" and searched in vain for medical help. But in his impressively candid autobiography, *Deliverance from Error*, he touched on more immediate causes: "I examined my motives in my work and teaching and realized that it was not a pure desire for the things of God but that the impulse moving me . . . was the desire for an influential position and public recognition." Now he found himself "tossed between the attractions of worldly desires and the impulses toward eternal life."[94] Which would he choose?

In reality, Ghazali had no choice, and he knew it. Down to his patron's death, he was on top of the world. At thirty-eight he headed the most powerful educational institution on earth and was the most popular lecturer in Baghdad. Wealthy, with full access to the highest reaches of power, and with what was ostensibly a happy family, he could do no wrong. His lectures at the Nizamiyya had struck just the right balance between swashbuckling irreverence toward authority and idealism.

His specialty had been to attack the philosophers and thinkers who had heretofore been the glory of Muslim civilization. He had pored through their works, and it is said that as a young man he even studied with a well-known Nishapur freethinker. But he was no sympathizer and was said by friends to show up early in the morning for lessons with this skeptic in order to escape the notice of his friends. This deceitfulness disgusted Khayyam, who suggested that a bell should have been rung whenever Ghazali left the freethinker's house.[95] Such behavior on Ghazali's part indicates that he adopted his pose of antirationalism early. Thereafter he missed no opportunity to savage what many considered the reigning truths of reason. Yes, the thinkers had achieved much with their clever mathematical formulas and their use of spherical geometry and trigonometry to measure the heavenly bodies. And, yes, their studies in geology, mineralogy, medicine, and pharmacology were useful to businessmen and ordinary citizens. But these thinkers had overstepped themselves by applying the same kind of reasoning to the eternal questions of God and religion.

Whatever his intention, Ghazali gave his students an excuse for ignoring these and other difficult studies on the grounds that they were of only secondary importance and did not, and could not, touch the great questions of existence. Seekers of truth should instead focus squarely on religion. But what is the true religious life? Down to his crisis, Ghazali reveled in the easy role of acerbic critic. Yet he offered no clear path out of the blind alley into which he drove his auditors, and he knew it. His spiritual crisis changed this.

Certain details of Ghazali's breakdown have gone largely unnoticed.[96] Immediately before he lost his speech, he had penned a vitriolic polemic against the Shiite Ismailis and their faith, one of four works he devoted to dissecting and attacking that group.[97] It was a nasty piece, in which the author simply gathered between two covers every scabrous accusation that had ever been directed against the Ismailis, without bothering to consult a single Ismaili source.[98]

It is sad that Ghazali would have stooped so low, and very much to the credit of the Ismailis that they calmly responded with a point-by-point refutation of his many charges. Ghazali doubtless believed what he wrote, yet it is clear that he was also doing the bidding of his patron and the founder of the madrasa he headed, Nizam al-Mulk, whose blind rage against the Ismailis knew no bounds. Ghazali's screed stood as clear evidence that, as he himself later confessed, "the impulse moving me . . . was the desire for an influential position and public recognition."

The murder of Nizam al-Mulk and the death of the sultan robbed Ghazali of patronage, destroyed his prospects, and took away his political cover. Ghazali tried to save himself by working out an arrangement with the emerging leaders that would preserve his favored status. Indeed, he spent more than two years laboring at this. Nothing is known of his maneuvers during that time, but it is easy to imagine him conducting himself with the new rulers as if his power, money, and privileges were his by right; it is equally easy to imagine them brushing him off as an unwelcome holdover from the previous administration. It soon became clear that the new sultan wanted to sweep out every remnant of the old regime, including Ghazali. Nor did Nizam al-Mulk's many enemies and his fractious and greedy sons have the slightest interest in saving this cheeky young academic. Only after three years of painful struggle did Ghazali admit to himself that he was out, and his fortunes gone.

Where could he turn now? He knew he would be no more welcome in his hometown of Tus or in Nishapur than in Baghdad. Once he acknowledged to himself that he could not go home again, Ghazali had no choice but to consider a completely new life. He did so, choosing for himself the role of mendicant. In his autobiography he claimed to have launched his new life by giving away all his wealth,[99] but this had already vanished with the murder of his patron. His decision to abandon his family, don the white coat of a Sufi, and head for the Holy Lands was arguably the only way out for Ghazali, short of suicide.

To account for Ghazali's spiritual, professional, and psychological crisis is not to explain the profound body of writings he produced after those painful months in 1095. Still less does it explain why Ghazali's formidable treatises on religion came to exert so profound an influence on the entire Islamic world, and even on Christianity and Judaism. For this we must dig deeper into his works themselves.

THE INCOHERENCE OF THE PHILOSOPHERS

Ghazali claimed that from his student days in Nishapur he had been obsessed with a desire to find truths that could stand up to attack from any quarter. There is no reason to doubt his words. This naturally led him into the study of science and philosophy, especially the works of Aristotle and Plato, which he claimed to have completely mastered. He also studied the writings of their recent disciples, especially Farabi's cosmology and Ibn Sina's discourses on reason and faith. When eventually he wrote on these subjects, Ghazali did so as a thoroughly informed insider. The fact that he was not a disciple of any master freed him from pressures to conform. He offered his revolutionary conclusions in lectures at the Nizamiyya and then in a passionate book entitled *The Incoherence of the Philosophers*.

In the first lines of his preface, Ghazali makes clear his purpose: to attack unbelief.

> I have seen a group who, believing themselves in possession of a distinctiveness from companion and peer by virtue of a superior quick wit and intelligence, have rejected the Islamic duties regarding acts of worship, disdained religious writs pertaining to the office of prayer and the avoidance of prohibited things, belittled the devotion and ordinances prescribed by the divine law. . . . On the contrary, they have entirely cast off the reins of religion.
>
> The source of their unbelief is their hearing high-sounding names like "Socrates," "Hippocrates," "Plato" and "Aristotle." [Influenced by such writers, they deny] revealed laws and religious confessions and

[reject] the details of religious and sectarian [teachings,] believing them to be man-made laws and embellished tricks.[100]

Ghazali claimed to have nothing against mathematics and science. Indeed, in his later autobiography he gladly admitted that they were capable of producing useful results. But mathematics and Aristotle's logic are totally irrelevant to revealed religion and matters of faith because in the end they refer only to themselves. If the philosophers had actually arrived at truth, why did the philosophers and scientists, both ancient and modern, so often disagree with one another?

Regarding God, the best these sages could come up with was the notion of the First Cause or the Creator. But this hung God's existence on nothing more substantial than the proposition that an infinite regression is impossible. And having fulfilled his function as First Cause, what was left for God to do thereafter? Such a philosophy has no place for God to know, see, or hear, or to have any knowledge of particulars. Yet the Quran explicitly stated (34.3) that nothing escapes God's knowledge, "not even the smallest particle in heaven or on earth."[101]

Worse, some of these same experts then proceeded to argue that the world is eternal, with neither beginning nor end, in which case there was no need for God to have intervened in the first place. Of course, Farabi and Ibn Sina were not blind to this objection, which is why Ghazali focused on what he considered their incomprehensible doctrine of reality as an *emanation* of God rather than his direct creation, and particularly on the tortured doctrine by which Ibn Sina had asserted in one breath that the world is eternal as *concept* but that God had intervened to bring it into *material* reality. All such arguments were vacuous and exposed their advocates as enemies of Islam's central doctrines of the earth's createdness and of God as the source of every action, no matter how small.

No notion was dearer to medieval scientists and philosophers than the direct connection between cause and effect. Clearly, they claimed, fire *causes* cotton to burn. Ghazali denounced this as quackery and fraud. To say that events *a* and *b* occur sequentially is not to say that *a causes b*. Fire, Ghazali argued, is inanimate and cannot be the cause of anything. The best that can be said is that *b follows a*. God may constantly intervene to join them, but we know from miracles that God can also intervene to produce quite different outcomes. He can, for example, enable a person to live after decapitation. God, Ghazali concluded, is not bound by any worldly order, causal or otherwise. "The connection between what is habitually believed to be a cause and what is habitually believed to be an effect is not necessary."[102] In the end, God alone has the power to create things in a certain order but has the power also to disconnect them at will.[103]

By such arguments Ghazali dismantled what he saw as the fraudulence, immorality, and "hideous absurdities" of the scientists and philosophers and laid bare their heretical essence.[104] By denying or qualifying God's creation of the world, and by questioning miracles and the possibility of physical resurrection, they had crossed the Rubicon. Students must have delighted in what a recent philosopher called Ghazali's "skeptical games," the more so since other teachers would have required them to plow wearily through the difficult writings of Ibn Sina, Farabi, and the other great thinkers of the age whom he was attacking. Ghazali's lectures left the terrain on which science and philosophy had blossomed a desolate wasteland: "The world is two worlds, spiritual and corporeal or, if you will, sensible and intelligible or, if you wish, higher and lower. This all comes to much the same thing. The differences are merely differences of terminology."[105] *Everything*, he concludes, is spiritual, and the intelligible world is as rich with angels as it is with peoples and minerals.

Ghazali was anything but a closeted theologian. Raised in the vigorous tradition of Nishapur polemics, he always directed his arguments against some person or some group. Besides Aristotle, Plato, and the ancient Neoplatonists, the special targets of his wrath were Farabi, Ibn Sina, and the Ismailis because they embraced the ancients. In their case he assigned himself the role of both judge and jury, and his findings were harsh to the point of ruthlessness. Those who follow the philosophers were not merely heretics but, he averred, *apostates*. The Quran quite clear states that apostates must be punished with death.

But what if the apostates should repent? Ghazali, with the bit between his teeth, did not relent:

> The meaning of "repentance" of an apostate is his abandoning of his inner religion. [But] the *secret apostate* does not give up his inner confessions when he professes the words of the *shahada* [Confession

of Faith]. He may [still] be killed for his unbelief because we are convinced that he stays an unbeliever who sticks to his unbelief.[106]

He left no doubt that he was speaking explicitly of the great thinkers of the present age. Pious Muslims should not be fooled by their professed "conversion." In an ominous passage he singled out the scientists who engaged in philosophy, the philosophers who based their arguments on Aristotelian logic, and the Ismaili thinkers who embraced them—in short, all the faylasufs: "Whoever claims [what they claim] annihilates the achievement of the religious laws. He sets limits to the possibility of receiving guidance from the light of the Quran, and he hinders the forming of one's own moral conduct according to the Sayings of the Prophet."[107]

Ghazali was not engaged in a mere academic exercise. When a very senior cleric branded someone an apostate or heretic, he in effect issued a death warrant, which is precisely what Ghazali's words amount to. It is pointless to ask by what right he did this, or by what right he could condemn others for doing what he himself had done over the course of several decades. Ghazali, convinced that he was in possession of divine truth, proceeded to pass judgment on all those who, in his view, were not.

In so doing, Ghazali administered the coup de grâce to those philosophers and scientists who pretended to deal with issues of faith that are the proper domain of the imams. But what positive alternative to these false prophets did he offer? Down to his crisis, none, besides the stern regimen of the Shafi'i school of Muslim jurisprudence.[108] And even though one of his own teachers had been a renowned Sufi, he did not evince any sympathy for the Sufi approach, at least until after his crisis in 1095. From then on, however, he zealously embraced Sufism and devoted himself singlemindedly to incorporating the Sufis' insights into mainstream Muslim theology. The very title of his autobiography, *Deliverance from Error*, attests to the mood of newfound certainty in which he pursued this mission.

It was at this time that Ghazali wrote his greatest work, *The Revival of Religious Learning*. Consisting of a modern printed edition of four thick volumes, this contained the positive vision to which Ghazali's entire experience had led him. With his customary boldness and clarity,

Ghazali set forth the most authoritative account of mainstream Sunni thought ever written, a status it still enjoys today. In contrast to his earlier practice, he now called for toleration and mutual understanding. In addition to theology, it provided advice on daily life, much of it refreshingly moderate. He welcomed sports and singing for relaxation and supported with Quranic verse his call for the "golden mean in belief and conduct." He even defended the return of science, logic, and medicine to the curriculum—stripped, of course, of the heretical excesses in which the philosophers had indulged. Finally, in his *Alchemy of Happiness* he returned once more to his main theme, namely, the loving relationship between the believer and God.[109]

Ghazali's advice or, more accurately, his emphatic injunction, was for people everywhere to reject the restless seeking of the philosopher and the relentless curiosity of the scientist and instead embrace the quietistic life of a Sufi mystic. In his autobiography Ghazali had written:

> It had become clear to me that I had no hope of the bliss of the world to come save through a God-fearing life and the withdrawal of myself from vain desires. It was clear to me, too, that the key to all this was to sever the attachment of the heart to worldly things by leaving the mansion of deception and returning to that of eternity and to advance towards God Most High with all earnestness.[110]

These fine words came from a man who from earliest childhood had shown himself to be omnivorous in his pursuit of worldly knowledge of all sorts, who had spent four decades of his life in these very "mansions of deception," and who had adopted the Sufi way only when the elaborate structures of social success and public acclaim had collapsed around him. Ghazali, in other words, exhorted others to begin their lives where he was ending his, in mystical contemplation of the Divine.

Ghazali's "new life" arose from the collapse of his old one. Like everyone who has passed through a professional or personal crisis, once back on his feet he was determined to demonstrate that he had emerged from it stronger and better than ever, and that the new life into which he was thrust by the collapse of his old life was in fact the best path for everyone on earth. With a convert's zeal and deep conviction in the righteousness of his cause, he spent his final years in what was transparently a campaign of self-justification. Ghazali, who spoke so much about serenity, was incapable of practicing it.

There were three important reasons that Ghazali's message found a powerful welcome among Muslims, and especially among the religious scholars and mullahs. First, he provided devastating weapons for dealing with the pesky philosophers and scientists who dared challenge the authority of the Muslim divines. Second, he provided a neat justification for moving Sufism and its adherents, many of them Turkic, from the periphery to the center of Muslim life. Third, he vindicated the orthodox conception of the faith and provided a solid rationale for shifting more and more of life's decisions from the realm of reason to the realm of law, the ever more prescriptive Sharia. On all three points he resolved concerns that had been mounting throughout the Age of Enlightenment, and on all three points he came down firmly on the sides of the opponents of reason, science, and logic.

Ghazali's message found as warm a reception among many Christians and Jews as it had among Muslims. Thomas Aquinas (1225–1274), the "Doctor Angelicus," encountered the works of Ghazali while studying at the University of Naples. While he differed sharply with Ghazali on the ancient philosophers—Ghazali rejected them while Thomas embraced them—he found that Ghazali's stress on revelation rather than reason and his call for direct and loving communion with God through mystical faith were in full accord with Christianity and the new piety he himself espoused.[111] Even in economics Thomas Aquinas drew on Ghazali as he expounded his views on usury, fraud, pricing, and private property.[112] Many later European writers also turned to Ghazali, including the French mathematician and Catholic thinker Blaise Pascal (1623–1662), who warmed to Ghazali's views on intuition as opposed to formal reason, and the Scottish economist David Hume (1711–1776), who found Ghazali's rejection of causality akin to his own thoughts on the subject.

Ghazali had even more Jewish readers, thanks to several translations of his major works into Hebrew. Among those directly inspired by Ghazali was the great Ibero-Egyptian thinker Maimonides (1135–1204). Like St. Thomas Aquinas, Maimonides, who read Ghazali in Arabic, brushed over his invectives against Aristotle, Plato, and the sciences but was in full accord with Ghazali's notions of the unity of God, prophesy, resurrection, and the absolute centrality of faith.

Only from the Arabic world did a thoroughgoing refutation of Ghazali arise, and that focused solely on his *Incoherence of the Philosophers*. Writing a bare century after Ghazali, the great Iberian philosopher Ibn Rushd, known in Latin as Averroes (1126–1198), mischievously entitled his rebuttal *The Incoherence of the Incoherence*.[113] Ibn Rushd was an ardent champion of Aristotle and felt called upon not only to refute Ghazali's attack on his idol but to defend Ibn Sina as well. He did so with a point-by-point rebuttal that is as exhausting as it is devastating. He left Ghazali's assault in tatters.

But Ibn Rushd was too late. Ghazali's youthful denunciation of science and philosophy had long since become a best-seller with the Arabic-reading public. Never again would open-ended scientific enquiry and unconstrained philosophizing take place in the Muslim world without the suspicion of heresy and apostasy lurking in the air. Happily, more than a few pious Muslim thinkers proceeded anyway and made valuable contributions to world civilization by so doing. But the damage had been done. Against the background of modern development, it is hard not to conclude that Ghazali's great insights on faith and the moral life were qualified, some might say negated, by his untimely early diatribes.

The Seljuks' End: Sultan Sanjar

The end of Seljuk rule coincided with the nearly sixty-year reign of Ahmad Sanjar (1085–1157). Sanjar assumed power in 1118 after the deaths of Sultan Malikshah, Nizam al-Mulk, and the next heir, Sanjar's elder brother.[114] He promptly moved the capital of the entire Seljuk state back to Merv. Because Sanjar could neither read nor write, it is tempting to dismiss him as irrelevant to intellectual life and the arts. Yet his reign was in many respects a glorious one, with significant achievements in culture before its tragic end (see plate 19).

Under Sanjar, Merv and the other great cities continued to blossom, thanks to a thriving economy fed by manufacturing and long-distance trade. China dropped out of the picture in these years, but the Indian trade burgeoned, as did trade with Europe, including the Baltic region, and the Middle East. Clever merchants and freight forwarders refined their bookkeeping and figured out ways to charge interest in spite of Quranic prohibitions against usury.[115] Prosperity opened ever new possibilities. That it utterly failed to reignite Central Asia's intellectual and cultural flame is therefore of great significance. Something important had been lost. Sanjar's reign provides useful insights into what that was.

Sanjar, while illiterate, was no fool. The letters he dictated reveal a shrewd and on the whole benign ruler. He was adroit at rewarding and punishing subordinate rulers and, while fabulously rich, was not greedy. Formed in a world of strife, he focused single-mindedly on building and maintaining his own authority. This caused him to be dismissive of ordinary men and women but not to disrespect them. After all, he said,

> They do not know the language of kings, and any idea of agreeing with their rulers or revolting against them is beyond them; all their efforts are devoted to one aim: to acquire the means of existence and to maintain their wife and children; obviously, they are not to be blamed for this, nor for [their desire to] enjoy constant peace.[116]

Nizam al-Mulk had loathed panegyrics, but Sanjar welcomed poets who sang his praises in verse. He did not object that the large gold coins flowing from his mint proclaimed him the "Great Sultan, Glorying Peace and Religion."[117]

The prosperity of Sanjar's era produced some remarkable instances of upward mobility. A local seller of fruits and aromatic herbs named Abu Bakr Atiq al-Zanjani prospered to the point that he came to Sanjar's attention and ended up a key adviser to the throne. Abu Bakr Atiq used his wealth to amass a library of some twelve thousand volumes, one of two great collections attached to the central mosque adjacent to the tomb being built for Sanjar. Collections like Abu Bakr Atiq's were welcoming to scholars and offered generous terms of access. The Greco-Arab geographer Yaqut, as mentioned earlier, spent several years after Sanjar's death working in Merv's libraries and considered them the best in the world. From only one of them, the collection at the Dumayriyya madrasa, he was able to borrow two hundred books without putting down any security. He blamed Merv's libraries for causing him to neglect his wife and children but declared that he would gladly have spent his whole life working in them.[118]

Sanjar's misfortune was to rule at a time when diverse regional and cultural forces were tearing at the Seljuk and Islamic worlds. Islam was splitting into a multitude of mutually hostile creeds and factions, with the most contentious of many fault lines between Sunni and Shiite believers growing ever deeper. Social discord coalesced around these and other differences and erupted in bloody conflict among classes and neighborhoods in the main cities. The tribal character of Seljuk rule was painfully manifest as dynastic power centers in Iraq, Turkey (Rum), and western Persia all aspired to rule the whole.

Threats from abroad arose on every point of the compass. To the east the Karakhanids, though weakened, were still able to mount serious assaults on Seljuk territory from their base at Balasagun.[119] To the west crusading armies from Europe arrived for the first time on the eastern Mediterranean to claim, or reclaim, territories that had recently fallen to the Seljuks and other Muslim powers. Other dangers arose from the revived shahs of Khwarazm to the north and from Mamluk Egypt in the southwest. Together these internal and external threats caused the weakened Seljuk and Muslim worlds to turn inward, in spite of their many trade links abroad.

By contrast, western Europe in these years was opening its windows to the world after a long period of looking inward. Christian Europe was expanding eastward into the Baltic and across the plains of Poland. On the Iberian Peninsula, Castile was pushing forward the reconquest (*Reconquista*) that traced its roots back hundreds of years to Charlemagne, while the Normans reestablished Christian rule in Sicily in 1072. Paralleling this geopolitical expansion was the reopening of European intellectual life to influences from abroad, including from the Muslim lands, and the revitalization of European civilization from within. With good reason, these developments came later to be known as the "Renaissance of the Twelfth Century."[120]

SANJAR THE BUILDER

Sultan Sanjar forged ahead, using all the tools available to him. He strengthened the administrative system put in place by his Seljuk forebears[121] and revived long-neglected taxes to the point that Samarkand rose up in rebellion against what its merchants viewed as confiscatory duties.[122] His expanded army was able to rush to Iraq, Afghanistan, or the northern steppes whenever danger threatened. A highly visible part of this surge of activity were the many construction projects that Sanjar initiated, all on the grandest scale.

For the merchants he erected immense caravanserais. Typical was the vast Ribat-i-Sharif, on the road between Nishapur and Merv in Iran, which featured an impressive portal or *pishtaq*, an immense courtyard defined by high perimeter walls, and a maze of interior chambers, each with its own underground cistern, all of which indicate that it was designed to accommodate hundreds of travelers.[123] The ruins of another great caravanserai, the Ribat-i Malik, stands in the Khurasan Desert. Several architectural historians have judged the massive row of cylindrical piers that formed the walls to be one of the most imposing structures of the entire Islamic period.[124] Long bridges with up to twenty graceful arches of baked brick or stone sped traffic along the major routes.[125]

Sanjar's capital at Merv was not the ancient center around the Erk-Kala but a new city that his predecessors had begun to build directly alongside the old. The late Tertius Chandler, in his study *Four Thousand Years of Urban Growth*, concluded that by 1150 Merv was the largest city in the world, with a population of 200,000.[126] The heart of Seljuk Merv was Sanjar's new fifty-room palace, which included quarters where imperial craftspeople plied their arts. New congregational mosques arose at several points throughout the city, as did large bazaars, governmental offices, and baths. Attached to one of the principal mosques and adjoining a public bath was the sultan's massive tomb, which began to rise long before Sanjar's death at age seventy-two.[127] Since this structure, which still stands, is arguably the most imposing building of Central Asia's Age of Enlightenment, and since it incorporates innovations that still resonate in architecture today, it warrants a closer look (see plate 20).

Even though nomadic Seljuks had lived in pointed tents, their Persian architects favored traditional rounded domes over tent roofs.[128] But if architects of the Seljuk era were glad to emulate such now antique prototypes as the Tomb of Ismail in Bukhara, commercial wealth inspired them to construct ever-larger buildings with the broadest and highest domes possible. In their quest for a solution to this challenge, they eventually hit on an epochal innovation: the double dome. Instead of a single heavy dome, which would be particularly vulnerable in seismic zones, this simple arrangement called for two lighter domes, one nesting inside the other. Not only did this make it possible to cover a larger area, but it greatly simplified the process of construction.

We have noted that the Romans had devised a double dome for Hadrian's Pantheon in Rome. Whether, or how, this feat of virtuoso engineering was transmitted to Central Asia is unknown. In the absence of evidence to the contrary, it is possible that this was an instance of independent discovery. Whatever the case, the oldest known double dome in the region—or anywhere outside Rome, for that matter— is at the Abu 'l-Fadl mausoleum in southern Tajikistan, dating from the late eleventh century.[129] The technique reached its full development and was applied on a grand scale first at Sanjar's tomb. In many respects Merv, rather than the earlier capitals at Isfahan, Hamadan, and Konya in Anatolia, was the spiritual capital of the Seljuk state. Two of the four previous sultans had already been buried there. The architect of Sanjar's monument to himself was Muhammad ibn Atsiz al-Sarakhsi, from nearby Sarakhs on the border between modern Turkmenistan and Iran. Sarakhsi conceived the structure on the grandiose, even ponderous scale typical of Seljuk architecture in Central Asia: the cube-shaped mass stretches 88 feet on each side and is 131 feet tall.[130] The double dome of baked brick, at 62.5 feet across, broke all previous records. An ornamental arcade surmounts the cube and provides an eye-catching transition between the cube and the drum of the dome.

Emulating the model of Sanjar's tomb, lesser Seljuk rulers erected a number of great domed tombs in the west of Iran, all of them based on the same engineering concepts that were so monumentally applied at Sanjar's tomb in Merv.[131] From these the concept spread to the Caucasus, then to Anatolia and eventually to the Mediterranean. In 1367 the city fathers of Florence selected a plan for completing their cathedral that included a double dome of the type pioneered by the Seljuks. Completed by Filippo Brunelleschi in 1436, the dome marked a turning away from the Gothic style of architecture and hence a decisive step leading to the Renaissance. In the early nineteenth century the French-born architect Auguste de Montferrand (1786–1858), planning the Cathedral of St. Isaacs in St. Petersburg, Russia, drew inspiration from Brunelleschi's dome but changed the construction material from brick to cast iron.

In 1866 the American architect Thomas U. Walter completed the new dome of the United States' Capitol in Washington, DC, which he based on Montferrand's dome in Russia. Thus it can be said that three of the most renowned buildings in the West are all lineal descendants of Sanjar's tomb in Central Asia.

SANJAR AS PATRON

Aside from the usual formulaic verses praising his piety, there is nothing to suggest that Sanjar was particularly religious. He was deeply superstitious, however, and frequently consulted dream interpreters, of which there were no fewer than forty in Merv.[132] Without doubt he valued social peace more highly than dogmatic purity. Even before he came to rule the entire Seljuk realm, he reached out to Ghazali, who had returned from the Holy Lands and was now back in Khurasan. Sanjar's invitation recognized Ghazali's service under Sanjar's father but reflected no particular piety on the ruler's part. In spite of all his resolutions, Ghazali accepted and set out for Merv. But by the time he reached Mashad he had had second thoughts, which led him to cancel his move. Explaining his turnabout in a letter to Sanjar, the still vain Ghazali did not fail to mention the seventy books he had authored, but he concluded that he must remain loyal to his pledge, made at the tomb of Abraham in Jerusalem, never again to attend the court of a king or do anything that might provoke religious controversy.[133] Ghazali therefore returned home.

There is little evidence that Sanjar embraced the role of patron of science and learning. To be sure, Merv's great libraries were still attracting scholars, and its ancient observatory was still functioning, presumably with support from the throne. The mere existence of these distinguished institutions assured for the Seljuk capital a rich intellectual life, whether or not the sultan took much interest in it. The mathematician and Sufi Ayn al-Qudat al-Hamadani (1098–1131) is known to have lived and worked in Sanjar's Merv, as did an astronomer named Hakimi, a Jewish convert. More noteworthy is the fact that Omar Khayyam settled for several years in Merv before returning to his native Nishapur.

Scarcely noticing the sciences, Sanjar lavished generous support on poetry. The great poet Anvari (Awhad al-Din Anvari, ca. 1126–1189) was a highly visible adornment to Sanjar's court, the most luminous of a large group of poets living in Merv.[134] Beautifully educated in the sciences and philosophy, Anvari was erudite to a fault and not hindered by modesty:

> In every science known to any of my contemporaries,
> Whether pure or applied, I am competent.
> Of Logic, Music and Astronomy I know something . . .
> Neither am I a stranger to the effects and influences of the stars.
> If you don't believe me, come and test me: I am ready.[135]

Anvari chose instead the life of a public poet and courtier. This perfectly fit his expansive personality. He had squandered a handsome inheritance before arriving at Merv from his native Sarakhs, on the border between modern Turkmenistan and Iran. Broke and dissolute, he turned to the role of court poet out of sheer necessity. But his taste for riotous living did not abate in the capital, and he is remembered for having once fallen off a rooftop while drunk.[136]

Anvari became a bosom comrade of Sanjar yet turned out surprisingly few of the usual panegyrics to his employer. Instead, after promising that "You need have no fear of being bored," he wrote lengthy and lyrical *qasidas*, short verses on events of the day, exquisite depictions of nature, and short "fragments," in which he vented his spleen against enemies, real and imagined, often with splendid and earthy obscenity.

Anvari's own bragging about his skills at astronomy and astrology led to his demise. In 1185, on the basis of astronomical data, he confidently predicted a major earthquake and conflagration. Frightened residents of Merv dug basements in which to hide. When the day passed without the mosque candles even flickering, Sanjar sacked his overconfident poet, who moved to Balkh. Just when he arrived, booksellers began offering a collection of unsigned verses entitled *A Book of Asses*, abusing Balkh for being "filled with rogues and libertines." Anvari was not the author, but everyone assumed he was. Paraded through the streets in a woman's headdress, he barely escaped with his life. Disgusted, Anvari retired, vowing no longer to "fling the filth of poetry to the winds."[137]

Specialists argue over the depth of cultural and intellectual life in Sanjar's Merv. On the one side, E. G. Browne, the pioneer English student of

Persian literature, claimed that the age of Sanjar was as brilliant as any preceding or following it in the region.[138] On the other, one might suggest that with a reign of more than half a century and enough money to build endless monuments, Sanjar had ample time and resources to support a large stable of the greatest scientists and thinkers as well. That he fell short in this task suggests that by the early twelfth century one could rule an empire stretching from the Mediterranean to the borders of China, and from the Persian Gulf to the steppes of what is now Ukraine, without feeling any particular need to serve the world of learning. Or is it possible that Sanjar did his best, but the supply of rising geniuses was not what it had been in earlier times?

SANJAR'S CAPTIVITY AND DEATH

Sanjar's impressive but flawed rule ended by the same process by which the Seljuks themselves had come to power: by the intrusion of nomads into the intense and sophisticated world of urban Central Asia. The finale occurred in two stages. First, a large and heretofore unknown group of nomads from northern China, pushed westward by the Chin ruler,[139] appeared on the Seljuks' eastern border. Unlike the Seljuks, the Karakhitais, who spoke a Mongolian language, remained shamanists and Buddhists and showed no interest in converting to Islam.[140] That an unknown non-Muslim people was attacking the Muslim Seljuks from the east convinced more than a few otherwise sane Europeans that a mysterious Christian kingdom under the leadership of a man named "Prester John" had arisen in Asia and was coming to their rescue. It was a preposterous legend, but kings and popes bought into it.[141]

Meanwhile, the Karakhitais' conduct during early incursions convinced some Central Asians that they were less rapacious than many earlier invaders, and that in time coexistence with them might be possible That time never came. Sanjar rejected their proposal to negotiate, and on September 9, 1141, on the wasteland northwest of Samarkand, the Karakhitai utterly routed the entire Seljuk army. Corpses of up to 100,000 Seljuk solders were strewn over a six-mile area near the village of Katavan.[142]

Worse was yet to come. Over the previous generations a large number of Oghuz (Turkmen) nomads had settled in the area around Balkh in Afghanistan. Sanjar had managed to strike a deal with them and, at a price of twenty-three thousand sheep annually for the royal kitchen, allowed them to come under the direct and personal rule of the sultan. The deal stuck and the Oghuz loyally supported Sanjar in several conflicts. But then overeager local officials in Balkh decided that this arrangement made no administrative sense and imposed themselves between the Oghuz and the throne. This breach infuriated the nomads, who attacked Seljuk forces in Balkh. This brought Sanjar and his remaining troops to the scene. The Oghuz captured the hapless sultan and held him for three years, during which they repeatedly sacked Merv and pillaged Nishapur and other major centers. By the time they finally released Sanjar, he was a broken man. He died in 1157, but his empire had already fallen to pieces.

During Sanjar's captivity, Anvari watched with cold horror as the Oghuz warriors ravaged what had been among the most sophisticated centers of culture in all Central Asia. Samarkand had been spared, and its prince was the last hope for any retribution against the nomads. The grieving Anvari sent him a deeply moving poem, *The Tears of Khurasan*. Here are its opening lines:

> Oh, morning breeze, if you pass by Samarkand,
> Bear to the Prince this letter of the people of Khurasan;
> A letter whose opening is grief of body and affliction of soul,
> A letter whose close is sorrow of spirit and a burning heart,
> A letter in whose lines the sighs of the miserable are manifest,
> A letter in whose folds the blood of martyrs is concealed,
> The characters of its script are dry as the bosoms of the oppressed,
> The lines of its address moist from the eyes of the sorrowful.

Anvari concludes:

> Over the great ones of the age the small are lords,
> Over the nobles of the world the mean are chiefs;
> At the doors of the ignoble the well-born stand sad and bewildered,
> In the hands of libertines the virtuous are captive and bound.
> You see no man glad save at the door of Death,
> No girl a maiden save in her mother's womb.[143]

OMAR KHAYYAM'S LAST JUDGMENT

About the same time Sanjar invited Ghazali to Merv, he also invited his father's most famous scientist, Omar Khayyam, to his capital. Things had gone badly for Khayyam even before the murder of his patron, Nizam al-Mulk, and the death of Sultan Malikshah. His astronomical observatory in Isfahan was a distant memory, and he himself was broke. So he welcomed Sanjar's invitiation and used his brief sojourn in Merv to set down several further works, including an algebraic method of using the known weight of silver and gold to determine their ratio in alloys.[144] But Sanjar had a long memory and recalled Khayyam's prediction of his rapid demise from a boyhood disease years before. They soon had a falling out, and Khayyam departed.

Omar Khayyam spent his last years back in Nishapur, his soul torn between despondency and wine. Abandoning science, he turned to poetry:

> Now, when only the name of happiness is left,
> No ripe comrade remaining but the rough wine,
> Keep the happy hand clenched to the wine-jug,
> Today when the jug is all the hand has got.[145]

A strain of melancholia had been evident even in the four-line stanzas or *rubaiyat* Khayyam had occasionally tossed off as a youth. This depression persisted throughout his middle years:

> Alas, the book of youth is finished,
> The fresh spring of life has become a winter;
> That state which they call youth,
> It is not perceivable when it began and when it closed.[146]

Over the years the many personal attacks to which he had been subjected took their toll, and his depression turned into angry defiance:

> If I'm drunk on forbidden wine, so I am!
> If I'm an unbeliever, a pagan or an idolator, so I am!
> Every sect has its own suspicions of me,
> I myself am just what I am.[147]

He had watched with horror as Ghazali, his near-contemporary from Khurasan, had become the darling of right-thinking Baghdad by savaging his idol, Ibn Sina, and all he stood for. Khayyam particularly loathed Ghazali's dogmatic and self-serving defense of Creation:

> The cycle which includes our coming and going
> Has no discernible beginning nor end:
> Nobody has got this matter straight—
> Where we come from and where we go.[148]

It was fine for Ghazali and his triumphantly orthodox backers to proclaim the misery of earthly life and the delights of heaven but, he asked:

> How long shall I lay bricks on the face of the seas?
> I am sick of idolators and the temple.
> Khayyam, who said that there will be a hell?
> Who's been to hell, and who's been to heaven?[149]

In another verse, Khayyam shifted to outright ridicule:

> They say lovers and drunkards go to hell,
> A controversial dictum not easy to accept:
> If the lover or the drunkard are for hell,
> Tomorrow Paradise will be empty.[150]

In several further quatrains he attacked the scholars of Sharia and imams who so confidently passed judgment on their fellow men:

> Oh Canon Jurists, we work better than you,
> With all this drunkenness, we're more sober:
> You drink men's blood; we, the vine's.
> Be honest—which of us is the more bloodthirsty?[151]

> A religious man said to a whore, 'you're drunk,
> Caught every moment in a different snare.'
> She replied, 'Oh Shaikh, I am what you say,
> Are you what *you* seem?'[152]

And in a particularly angry stanza he defended a local agnostic:

> I saw a waster sitting on a patch of ground,
> Heedless of belief and unbelief, the world and the faith—
> No God, no Truth, no Divine Law, no Certitude:
> Who in either of the worlds has the courage of this man?[153]

By these steps Omar Khayyam, the most brilliant mathematician and astronomer of the Seljuk century and the author of important philosophical works that embraced revelation and the Sufi path to wisdom, proclaimed a skepticism so thoroughgoing that it extended even to the great intellectual quest to which the best minds of recent centuries had devoted their lives:

The captives of intellect and of nice distinction,
Worrying about Being and Non-Being themselves become nothing;
You with the news, go and seek out the juice of the vine,
Those without it wither before they're ripe.[154]

As if to remove all doubt that he included himself in this declaration of bankruptcy, he added:

My mind has never lacked learning,
Few mysteries remain unconned;
I have meditated for seventy-two years night and day,
To learn that nothing has been learned at all.[155]

During the late twentieth century a number of respected experts rushed to Khayyam's defense (as if he needed one). Through careful textual analyses they cast doubt on the authenticity of some, but not all, of the dark quatrains on which Khayyam's star status as a freethinker and skeptic was based. They also reminded us of Khayyam's early and unjustly neglected philosophical works, in which he emerged as a believer and sympathizer with the Sufis' quest. They then proceeded to denounce anyone who might suggest that Khayyam's views on many matters remained unsettled. "To accuse Khayyam of blatant hypocrisy while seeking to make of him a cultural hero for modern skeptics is itself the worst kind of hypocrisy."[156]

But this goes too far. People, even men and women of genius, are complex and sometimes infuriatingly inconsistent. Khayyam, who lived into his ninth decade, had ample time to embrace different positions at different times and in different moods. What he might write as a young man on a sunny afternoon to a judge in Fars may well have differed from what he would set down for himself in old age, sitting over a candle late at night. Finally, those who defend Khayyam's Muslim faith neglect the fact that his quatrains are almost as corrosive of his scientific work as of the expressions of piety contained in his philosophical writings. Is it fair to cry out when Khayyam expressed disgust at the fossilized official Islam of his day but remain silent when he implied that his own career in science had been for naught?[157]

A number of *rubaiyat* strongly support the view of those who seek to affirm Khayyam's consistency:

> It is a flash from the state of non-belief to faith,
> There is no more than a syllable between doubt and certainty:
> Prize the precious moment dearly,
> It is our life's only fruit.[158]

One probably early quatrain is also cited in support of the continuity of Khayyam's underlying faith:

> Thou hast said that Thou wilt torment me,
> But I shall fear not such a warning.
> For where Thou art, there can be no torment,
> And where Thou art not, how can such a place exist?[159]

Assuming that both these quatrains are genuine (the authenticity of the first has been questioned), where did Khayyam really stand? Like his hero Ibn Sina, Khayyam would have turned to intuition to affirm what reason suggests; like Ibn Sina, too, his faith would have had no place for ponderous religious scholars, for one or another school of law, or for dogma of any kind. Yet his version of Sufism, if such it was, reflected precisely the kind of belief that led Ghazali to brand Ibn Sina an apostate.[160] And then there were Khayyam's many quatrains expressing relentless skepticism and doubt, and proclaiming the utter emptiness of human existence and the abyss that yawns beyond. To dismiss these is to diminish and demean the man.

In spite of all this, Khayyam clung to life's fleeting moments of light. In his most widely quoted quatrain, gloriously but quite inaccurately translated into English in 1859 by the gifted eccentric Edward FitzGerald, he describes the small island of happiness that is left after all the deceptions of rationality and of fossilized religion have evaporated. Here is FitzGerald's classic verse:

> A Book of Verses underneath the Bough,
> A Jug of Wine, a Loaf of Bread—and Thou,

Beside me singing in the Wilderness,
Oh, Wilderness were Paradise enow.[161]

Here is a more literal rendering, which may serve as a postscript to the entire intellectual adventure of the waning days of Seljuk rule:

Anybody who in this world has half a loaf
And a home in which to live
Is no man's master and no man's slave;
Say to him "Be happy always," for he possesses a world of
    happiness.[162]

CHAPTER 13

❀

The Mongol Century

The Arrogance of Power: The Shahs of Khwarazm

In the autumn of 1219 Chinggis Khan and a force of 150,000 Mongols and Uyghur Turks appeared suddenly beneath the high walls of Otrar on the Syr Darya near the southern border of present-day Kazakhstan.[1] After enduring a five-month siege, the defenders launched a desperate sally to break the Mongol ring. It failed and the Mongols killed all the warriors. Then the invaders drove all the civilians onto the plain, pillaged and burned the city, and put to the sword the entire population, estimated at 100,000.[2] Over the following three years, this grim process was repeated in most of the major cities of Central Asia, in Persia and the Middle East, and across modern Ukraine and Russia, clear to the borders of Poland (see plate 21). In some cases craftspeople were spared and sent off to work for the Mongols, and a few women and children were saved by being enslaved. The rest were dispatched by the Mongol soldiers, each of whom was assigned a certain number of people to butcher.

Both the cause and the effects of the Mongol invasion of 1219–1222 and of Mongol rule in Central Asia, which endured for twelve decades, are much debated. Many at the time agreed with the Bukhara scholar who, told of the destruction of the city's Friday mosque and of horses trampling the pages of the Quran, counseled submission: "Be silent. The wind of God's night blows."[3] Modern observers in many countries tend to view the Mongol invasion as an inevitable and unstoppable social whirlwind, the last great nomadic outpouring from Asia. In Central Asia itself, there are many who see it as the decisive end of the region's vaunted golden age. But others, both abroad and within the region itself, are less apocalyptic. They point out that, after the initial wave of destruction, the Mongols fostered free trade (of which Marco Polo was a beneficiary) and patronized learning in various fields. The Mongols, according to this view, were among the builders of the modern world.

Such sweeping and mutually incompatible claims demand specific responses. They call for a close look at Central Asia on the eve of the Mongol assault, an assessment of the causes and effects of the invasion itself, and an evaluation of the region's economic and cultural conditions between the death of Chinggis Khan and the final collapse of the Mongol empire barely a century later. Such an examination gives rise to some surprising conclusions.

The story begins not at Otrar but with the death of Sanjar, the last Seljuk sultan, in 1157. This left Greater Central Asia divided among three dynasties. In Afghanistan and Khurasan the descendants of Mahmud of Ghazni hung on until 1187 but had in fact had ceded nearly all power to another dynasty based in Ghor in Afghanistan. In the East the Karakhitai nomads had settled down at the old Karakhanid capital at Balasagun after conquering most of what is now Kyrgyzstan, eastern Kazakhstan, and Xinjiang. Many local residents lauded these nomads, who included shamanists, Buddhists, and Christians, for not interfering with the practice of religion. Finally, the entire northern and central zone of the region was under the control of the most recent dynasty of Turkic shahs of Khwarazm, who ruled from their revived capital at Gurganj.

Over the years immediately preceding the Mongol invasion, the rulers of Khwarazm managed to destroy all their rivals and extend their rule to the western border of Iran clear to India. In 1215 they even demanded that the weakened Abbasid caliph recognize their hegemony, thus bringing nearly all Islam east of the Mediterranean under a single Central Asian ruler. By that year the shah of Khwarazm ruled by far the largest realm in the Islamic world.[4]

The Life of the Mind under Khwarazm's Rule

The visiting Arab geographer Yaqut judged Gurganj, the capital of Khwarazm, to be one of the richest cities he had ever seen.[5] After the assault by Mahmud of Ghazni in 1017, the shahs of Khwarazm had restored and greatly expanded the city, but the rich cultural and intellectual of the court of the last Mamun rulers never revived. Theirs was above all a military state based on an army of mercenaries and a series of complex arrangements with the surrounding tribes of nomads. The old Samanid and Seljuk administration decayed under their watch,[6] as did cultural life.

A writer at the Khwarazm court, Nizami Arudi from Samarkand, left us a small book, *Four Discourses*, in which he argued that a good ruler's intellectual stable should include four categories of wise men: secretaries, poets, astrologers, and physicians.[7] The poets and secretaries could praise the ruler and record his words; the astrologers could find the best time for him to act; and the physicians could cure him of illness. Conspicuously absent from the list of requisite wise men were writers, philosophers, historians, and scientists.

The most notable scientist to garner official support in Khwarazm during the two centuries before the Mongol conquest was the court physician, Zayn al-Din Jurjani (1040–1136), who assembled a massive compendium of medical knowledge, the *Khwarazm Shah's Treasure*.[8] Building on the works of Muhammad ibn Zakariya al-Razi and Ibn Sina, Jurjani focused particularly on the needs of the practicing doctor. Overall his work is solid but unoriginal. The same can be said of Fakhr al-Din al-Razi (1149–1209), who used his time in Gurganj to pen a widely read *Compendium of the Sciences* covering some fifty-seven different fields, including both the natural sciences and humanities.[9]

The record of official poetry is similarly unimpressive. The twelfth-century versifier from Balkh, Rashid al-Din Umari (1114–1177), specialized in dull panegyrics extolling the greatness of the shah at Gurganj, for which he was amply compensated. Dubbed by his enemies as "Rashid al-Din the Bat" (*watwat*) on account of his bald head, he railed against philosophers and defended religious fanatics. Rashid favored a learned style that was unruffled by emotional or poetic subtlety.[10] It is not clear whether he owed his great wealth to the breadth of his knowledge or to the fact that he had been rewarded for missing no opportunity to attack Mutazilites, Ismailis, and other opponents of the prevailing mainstream Sunnis.

The real innovators in this final era before the Mongol conquest were craftspeople and architects who worked for the many local courts that continued to thrive across Central Asia. During Seljuk times artisans in several centers in Khwarazm developed their own version of blue-and-white Chinese ceramics. When the Karakhitai invaders blocked trade with China, these Central Asian knockoffs dominated ceramics markets throughout the Muslim world. Architects, too, found a decentralized market for their skills. Unfortunately the Mongols were soon to destroy most of the impressive monuments erected in the era of the Khwarazm shahs. One that survived is an enormous madrasa at the village of Zuzan near Nishapur. But the masterly carvings on the exterior of this gem owed more to the local patrons than to any inspiration from Gurganj.[11]

A SUFI RENAISSANCE

The price that Central Asians paid to host the Khwarazmians' continental empire was to live in a society that was almost constantly at war. Even when the battle zone shifted elsewhere, Central Asians still paid dearly in the form of onerous taxes, forced recruitment into the army, and economic disruption. When this state of affairs persisted into the third generation, thousands lost hope in the government and focused instead on building "inner kingdoms of the mind."[12] This was the Sufis' moment.

The personal piety that Sufis had long championed contrasted starkly with the social regulation and Sharia-based conformity that the more hard-line Sunnis tried to impose on society. Nowhere was Sufism more popular than among the newly converted Turkic peoples who lived along the northern fringes of Central Asia. Not only did it conform more closely to the individualism of the nomads and former nomads, but it dovetailed nicely with many practices of their traditional shamanist religion and their worship of Tengri, the god of the Blue Sky.[13] Over the century preceding the Mongol conquest the lands of northern Central Asia witnessed the appearance of many Sufi divines whose teachings shaped the practice of Islam down to the present.

First among this group was Ahmad Yasawi (1093–1166) from Isfijab (Sayram) on the southern border of modern Kazakhstan east of Shymkent. When Yasawi was a mere eight years old a local holy man, Arslan Baba, encountered the lad on a road near Isfijab and immediately concluded that the young orphan had been miraculously anointed by the Prophet Muhammad himself in order to purify and revive the Muslim faith. Studies with Arslan Baba and then with noted Sufis in Bukhara prepared Yasawi for his divine mission. By feeding ninety-nine thousand people from a single piece of bread and removing an entire mountain, Yasawi validated the authenticity of his prophetic mission.[14]

Soon he had attracted a group of followers, whom he formed into a brotherhood, or *tariqa*. Carving wooden spoons for a living, Yasawi continued to preach down to his sixty-third year. Then, on the grounds that Muhammad had died at that age, Yasawi retired to an underground cell he had dug for himself in the town of Yassi (now Turkestan, in Kazakhstan) and remained there until his death ten years later.

Yasawi's core message was simple and direct: to love God and be linked to him through private prayer and contemplation. He promoted this program in a collection of poetic quatrains written in his native Turkic language. In this *Compendium of Wisdom* he soulfully bemoaned the prevalence of oppression and misery in the world but counseled a life of humility and acceptance as the only response that was compatible with the divine will. These verses, still widely known throughout the Turkic world, and Yasawi's life of renunciation and self-purification did much to spread popular mysticism among the nomads and other Turkic peoples across Central Asia and on the Eurasian steppes.[15]

More radical in his inward turn was another Khwarazm native, Najmuddin Kubra (Najm al-Din Kubra), (1145–1220), whose model for the saintly life attracted a large following of disciples over the next century. More than any Muslim before him, Kubra stressed that ordinary mortals had direct access to God through dreams and visions.[16] He did not come to this easily. Beginning as a student of theology in his hometown of Khiva, Kubra traveled to the Middle East for further study, only to turn his back on the traditional practice of Islam. Along the way he became a relentless polemicist before setting down his bold new path to faith in a series of closely argued works. In *The Path of God's Bondsman* and other treatises, he focused not on correcting the ills of the social order but on the inner enlightenment that comes through self-isolation, ritual, and contemplation. The goal was to become like a newborn infant—which caused one of his disciples to be nicknamed "The Wetnurse."[17] Even though he went out of his way to defend the Quranic

Figure 13.1. The Gurganj mausoleums of Sufi divine Najmeddin Kubra and Sultan Ali at Gurganch are still pilgrimage sites that thousands visit. ▪ Shahina Biznes Travel Ltd.

sources for his approach,[18] Kubra's method, which stressed the physical dimensions of contemplation and prayer, had much in common with Tibetan Buddhism and even with yoga.

Yasawi had shown that poetry far surpassed prose as a tool for evoking the deep emotions that lay at the heart of the Sufi message. The great Sufi poet Farid al-Din Attar from Nishapur (1145–1221) combined Sufi mysticism with the magic of the storyteller's art to create extended works of great subtlety and beauty.[19] By far the best known of Attar's poems is his *Conference of the Birds*, an allegory in which the birds of the world take wing together in search of their king, the mythical Simurgh, or Truth.[20] The journey was difficult, for they had to fly through seven valleys representing Searching, Love, Knowledge, Independence, Unity, Bewilderment, and Annihilation before reaching their goal. In the end, only thirty birds managed to survive all seven temptations, but when they entered the Simurgh's palace he was not there. Only then did they realize that they themselves, the "thirty birds" (*si murgh* in Persian), were the Simurgh, and that their goal was not outside of themselves but within, in the form of self-awareness. Only by searching the entire world could they find themselves.[21]

Even this dry enumeration of the poem's main sections suggests the drama and ecstasy that suffuse the work. Attar is said to have fallen into a trance as he composed it. The message of his *Conversation of the Birds* anticipated Lev Tostoy's *The Kingdom of God Is within You*; it is a Sufi's *Pilgrim's Progress*. Many have found in Attar's work a pervasive pantheism. While this is certainly true, it is much more. In his quest for God, Attar brushed aside the Muslim preoccupation with God's unity, which turns out to be the same as multiplicity; with submission—the very essence of Islam; and even with eternity, for God is beyond eternity as well. In one passage Attar described a moth seeking "to learn the truth about the candle's light":

> He dipped and soared, and in his frenzied trance
> Both Self and fire were mingled by his dance—
> The flame engulfed his wing-tips, body head;
> His being glowed a fierce transluscent red;
> And when the mentor saw the sudden blaze,
> The moth's form lost within the glowing rays,
> He said: "He knows, he knows the truth we seek,
> That hidden truth of which we cannot speak."
> To go beyond all knowledge is to find
> That comprehension which eludes the mind,
> And you can never gain the longed-for goal
> Until you first outsoar both flesh and soul.[22]

Attar was not modest about his artistic achievement. The Quran had declared Muhammad to be the "Seal (or last) of the Prophets"; Attar presented himself to the world as the "Seal of the Poets." He came both to his Sufism and to his vocation as a poet later in life. A wealthy pharmacist, he inherited and managed a Nishapur drugstore with thirty employees; his name, Attar, derives from Attar of Roses, a perfume and medicine. For unknown reasons he abandoned all interest in medicine and science and explicitly turned his back on the entire Aristotelian tradition of inquiry that Central Asian intellectuals had nurtured for three hundred years. There is a direct link between Ghazali and Attar, based on their common rejection of learning acquired through study and their belief in the possibility of direct knowledge from God, unmediated by logic, science, theology, or any known form of organized religion.[23] Attar represented a profound reaction against what he felt to be the excesses of reason in all its forms, an urgent call to rescue the self from a world in political and intellectual turmoil.

Yasawi, Kubra, and Attar established Central Asia as a gold mine of inspired intuition. Standing behind them was a reservoir of millions of people in search of something more than was offered by science, the humanities, the state, and organized religion. As people became aware of this, the region became the target of Sufi proselytizers from elsewhere.

In some cases the local converts outshone their masters in the popular mind, as can be seen in the dramatic career (and enormous shrine) of the Central Asian Sufi teacher Sheikh Zaynuddin Baba of Tashkent, the locally born disciple of a Syrian Sufi.

By the time of the Mongol invasion, Sufism had become the most powerful movement of thought in Central Asia. Promising personal emancipation from suffering and despair through mystical reunion with God, it offered a solution to the challenges of life based on the individual, not society as a whole. As such it was attractive to all classes. When the armies of Chinggis Khan arrived in Central Asia, Yasawi, Kubra, and other Sufi leaders had already experienced their visions of salvation, and their followers had begun the process of organizing and systematizing them. The number of their adepts and followers was growing rapidly.[24] Whether they would eventually have been confronted by a counter-movement of rationalists or of legalistic and conservative Muslim clerics and scholars will never be known. For just at the moment when such a reaction might have occurred, the Mongol invasion threw the entire region into turmoil.

INVITING DISASTER

By 1210 the shah of Khwarazm, Qutb al-Din Muhammad, felt supremely self-confident. With little difficulty and much good luck, he had brought all Central Asia, Afghanistan, Iran, and parts of India under his rule. Were it not for the fact that the Karakhitai still blocked him on the east, he could even think of reopening trade from his capital at Gurganj along the great northern route to China.[25] But in that year a puzzling event occurred. After first reconnoitering the area, a heretofore unknown nomadic force swept across the Tian Shan from Xinjiang and overwhelmed the Karakhitais at their capital of Balasagun, in modern Kyrgyzstan.[26] Some overtaxed local Muslims initially received the pagan Mongols as liberators.[27] Then, for unknown reasons, the invaders had swiftly decamped again for China. Suddenly the path to the East lay open to Qutb al-Din Muhammad. But who were these new invaders?

Over the previous two decades, an enormously gifted organizer and strategist named Chinggis Khan had managed to unite the diverse

Mongol tribes of the North Asian steppes into a single unit under his control. His task was facilitated by a temporary vacuum of power in the region, which Chinggis rushed in to fill.[28] Crowned leader by the Mongol nobles in 1206 and faced with declining average temperatures in Mongolia, which diminished the supply of grass for foraging,[29] he immediately cast his glance southward toward China. An initial attack revealed China's weakness, after which Chinggis Khan withdrew in order to make a move on the Karakhitais. Restiveness in China forced him to abandon his toehold across the Tian Shan, however, leaving a mood of hope mixed with fear across Central Asia. In 1215 he conquered Beijing (see plate 22).

Chinggis Khan followed the nomads' age-old formula of hitting settled communities with a devastating blow and then withdrawing once they agreed to pay tribute thereafter. He also followed the well-tested steppe custom of organizing his forces into units of tens, hundreds, and thousands.[30] In other respects, though, he was no ordinary steppe nomad. He imposed the strictest discipline on his forces and allowed no deviation. Realizing that written communications were essential for effective administration, he had adopted the Turkic Uyghurs' script based on the ancient Aramaic language that Syrian Christian traders had brought from the West. Unlike prior nomad fighters, he was also preparing to establish a fixed capital for himself at Karakorum, deep in the Orkhon River valley in Mongolia.[31] Very early he also grasped the importance of minting currency, which he standardized across the conquered territories, but with locally based production.[32]

Sensing both opportunity and danger, Qutb al-Din Muhammad sent a delegation of merchants to meet with this fast-rising Mongol ruler. Chinggis Khan received the group amicably and went so far as to state that he recognized the shah of Khwarazm as the ruler of the West, just as he, Chinggis, was ruler in the East. In the ensuing mood of euphoria, the Muslim merchants on the delegation showed Chinggis their goods, unwisely jacking up the prices well beyond their actual values. Chinggis calmly asked if they really thought that Mongols had never before seen luxury products and were ignorant of prices. The Muslims then offered the goods as gifts, but Chinggis insisted on paying the asking price for them.[33]

In the summer of 1218 Chinggis Khan reciprocated by sending a Mongol delegation to Central Asia. The participating diplomats and traders were drawn from conquered Muslim communities in Xinjiang. The caravan of five hundred camels paid a peaceful visit to Bukhara, selling and buying goods, but when they stopped in Otrar on their return the local head of the Khwarazmian government accused them of spying. He forthwith beheaded the diplomats and murdered all the merchants.

The official who committed this momentous deed was named Inalchug, and he was drawn from the nearby nomadic tribe of Kipchaks, from whom the rulers in Gurganj drew many talented officials. Qutb al-Din Muhammad had given orders for Inalchug to seize the caravan's goods but not to kill the traders. It was a disastrous move. A single camel driver survived to report to Mongol headquarters what had happened. Hearing the news, Chinggis Khan at first showed astonishing restraint, merely demanding that Qutb al-Din Muhammad turn over to him his official from Otrar.[34] When the headstrong shah of Khwarazm refused, Chinggis Khan took it as a casus belli, which it surely was.[35] By September 1218, he had crossed the Tian Shan with an army of 150,000 Mongols and Uyghur Turks. After capturing Otrar and killing most of its inhabitants, he executed the greedy Inalchug by pouring molten silver down his throat.

This strange combination of forbearance and raw brutality was to be the hallmark of Chinggis Khan's entire campaign in Central Asia. The eastern Uyghurs submitted and he left their cities untouched; the same happened at a number of Central Asian cities and towns. But resistance or, worse, reneging on capitulation doomed the population to extermination by Chinggis's forces.

Mongol Vengeance

Having reduced Otrar to ruin, the Mongol forces beset Bukhara. The local garrison retreated to the citadel and fought fiercely, which assured that the Mongols would take revenge by killing most of the population and devastating the city itself. Thirty thousand refugees struggled to reach a neighboring town but were captured and killed on the way.[36] The

Figure 13.2. It is hard to imagine Chinggis Khan actually lecturing at Bukhara, as shown here in an early miniature, but his conquests taught Central Asians bitter lessons. ■ From Serik Primbetov, *Atlas Turan* (Almaty, 2008). Courtesy of Serik Primbetov.

same followed in Samarkand, where, as usual, the Mongols designated artisans for Mongolia and certain women and children as slaves. Here, though, a group of Muslim clergy offered their surrender, which saved some of the mosques from destruction.

The routes the various Mongol armies followed across Central Asia and Afghanistan were defined not by prospects of booty but by the paths of Qutb al-Din Muhammad and his son, Jalal al-Din. The main force drove Qutb al-Din Muhammad to an island in the Caspian, where he died so impoverished that he was buried without a shroud. Chinggis himself chased Jalal al-Din clear across Afghanistan to Multan in the Indus Valley, where the heir to the throne of Khwarazm dealt a punishing blow against the Mongol forces before escaping once more.[37] As opposed to those who see the Mongol conquest as an inevitable hurricane from the East, this phase of the campaign suggests instead that it began as a calculated punitive expedition directed against the shah of Khwarazm, his son and heir, and those loyal to them.[38]

No sooner had the Mongols wreaked their vengeance against the two leaders than they confronted the Central Asians' readiness to fight rather than surrender, and to pretend submission in order to strike back later. The response of both the Iranian and Turkic-speaking natives of the region recalls the behavior of their forebears when facing the Arabs five centuries early. The Mongols responded with a war of extermination that reached its culmination in the attacks on Khojent, Tirmidh, Nisa, Ghor, Balkh, Bamiyan, Nishapur, Tus, Herat, Merv, and Gurganj.[39] They employed storms of arrows, flanking tactics, and feigned retreats in open country and used siege engines and pots of burning naphtha against cities. A contemporary observed that they fought "like trained wild beasts after game."[40]

Tirmidh (Termez) and Khojent both resisted and were destroyed outright. At Tirmidh a woman who had pleaded for mercy with the promise that she would produce a jewel she had swallowed instead had her belly cut open. The city fathers of both Balkh and Merv faked capitulation but then reversed ground, which led to the annihilation of their cities.[41] At Merv a particularly degrading spectacle was played out before the final devastation. It included an abject letter of capitulation penned by a local "sheikh of Islam" that was intercepted by city residents; efforts by local leaders to trade surrender for high office under the new rulers; and drunken celebrations during the respite prior to the final destruction.[42] Meanwhile, in Nishapur the desperate local leaders veered between opposite strategies, now refortifying the city, then staging a tactical capitulation, then fighting (killing a Mongol general in the process), and then capitulating again. At nearby Tus the Mongols sent a request for surrender, but then a local rabble-rouser stirred up the populace and they sent an abusive reply, which led to the extermination of the entire population.[43]

Each attack left devastation in its wake. When a Daoist monk from China passed by Balkh a year later, he reported that not a soul remained, but that "we could still hear dogs barking in the streets."[44] At Nishapur, Mongol forces did not retire until they had built two mountains of severed heads, one of men and the other of women. The victorious general commanded for the city to be destroyed "in such a manner that the site could be ploughed upon." This time even the dogs and cats were put to the knife. At Gurganj only two buildings remained standing. What only weeks earlier had been a large and wealthy metropolis was reduced to being "the abode of the jackal and the haunt of the owl and the kite." At Balkh, Merv, and elsewhere, a few thousand residents survived by hiding in the ruins, only to be killed by Mongol bands that were dispatched to the site to wipe out survivors. At Herat sixteen people (all of whose names were carefully recorded by a local historian) survived by fleeing to a nearby mountaintop. But when they returned they could neither feed nor clothe themselves in the devastated land and turned to cannibalism. Otherwise, all those residents except the few who were enslaved or deported were butchered outright.[45]

Should the many instances of resistance be ascribed to confidence in their own military prowess or to the fact that the locals believed that even capitulation would not save them from destruction and death? Many towns and cities, including Nur north of Bukhara, Qarshi in southern Uzbekistan, and Sarakhs, chose to surrender and as a result survived. The taxes the Mongols subsequently imposed on these places were generally considered fair. Many surrendered because they had been so exploited and humiliated by the Khwarazmians that they assumed the Mongols could not be any worse.[46] The Khurasan historian Juvayni, admittedly in the pay of the Mongols, stated categorically that Chinggis Khan left in peace those cities that capitulated, saying that "whenever towns along the way submitted he in no way molested them."[47] But reports of the terrible fate of other cities convinced doubters that not even unconditional surrender would save their lives. While there were many motives for the Central Asians' decision to fight, sheer desperation was surely among them.

Just as the Arabs had done, the Mongols skillfully played on internal divisions among the Central Asians. More than the Arabs, however, the Mongols remained united, at least to the death of their leader, Chinggis Khan. Tens of thousands of surrendering Central Asians were dragooned into the Mongol army, which deployed them to lead the attacks on their kinspeople. Thus residents of Sarakhs had long competed with Merv for trade; after capitulating, they joined forces with the Mongols and outdid them in their savagery toward the people of the nearby capital.[48] At Merv itself, a force of some ten thousand Oghuz Turks had only recently forced its way into town and inflicted heavy damage before being repulsed. Then the Turks camped outside the walls, hoping to make common cause with the Mongols. But in this case the Mongols refused to play their assigned role and slaughtered them all.[49]

Against this, one can cite countless instances of striking boldness and heroism on the part of the defenders. Inalchug, the hapless governor of Otrar, whose foolish actions triggered the entire conflict, fought on until he lost his sword, at which point he hurled bricks handed to him by women of the town, until the Mongols finally captured him.[50] At Khojent on the Syr Darya in present-day Tajikistan, members of the local court managed to escape by building armored boats that had also been fireproofed against the Mongols' catapulted pots of burning naphtha. They eventually made their way to Syria disguised as Sufi pilgrims.[51] And Chinggis Khan himself, after watching Jalal al-Din, heir to the Khwarazmian throne, launch himself on horseback off a cliff into a river below, admitted that he would have been proud to have such a young man as his son. Above all, the numerous cases of whole cities deciding to fight rather than capitulate attest to the resoluteness of the Central Asian defenders.

Consolidation in Beijing

Chinggis Khan's onslaught of 1219–1221 left all Central Asia, Afghanistan, and Iran under Mongol rule. His death in 1227 might have triggered the kind of competition over succession that was typical of nomadic societies, but Chinggis had taken care to name his son Ogodei as his heir. Nonetheless, it took a decade for the Mongols to resume their western conquests, this time focusing on what is now Ukraine, Russia, and eastern Poland. But succession problems now intervened.

Finally, in 1255 the new ruler charged his brother, Hulegu, to destroy all the Muslim powers between Iran and the Mediterranean. Hulegu gathered a massive army and managed to sack Baghdad and capture and execute the caliph. He then conquered all Syria before rushing off to Mongolia to participate in the struggle for the throne. The Uyghur general he left behind suffered a terrible defeat in Galilee in 1260 at the hands of the Egyptian Mamluks.[52] Meanwhile, the contested selection process in Mongolia resulted in the naming of Khubilai, a grandson of Chinggis, as khan.[53]

Khubilai Khan's epochal conquest of South China changed forever the eastern Mongol state. Khubilai focused his attention on consolidating his hold on China, which he did by establishing a new rectangular-shaped capital of Ta-tu ("The Great City") near Beijing and organizing a new administrative system.[54] Further reflecting Khubilai's solid eastern orientation were his attack on Korea and his two large-scale attempts in 1274 and 1281 to conquer Japan from the sea, which failed when a "Divine Wind" or *kamikaze* arose and destroyed the Mongol fleet.[55] While Khubilai Khan labored at these projects, the Mongol rulers in Iran and Russia pursued their own very different interests. Hulegu and his successors in Iran established capitals at Tabriz and Sultaniyya and set about building the Ilkhanid state. As part of the same process of indigenization, Hulegu's successors converted to Islam. The Mongols of the Golden Horde in southern Russia tried to extend their rule into Central Europe but were stopped in Hungary, after which they focused not on building a new regional state but on extracting tribute from their far-flung domains.

Amid this general decentralization of Mongol power, Central Asia, too, went its own way, but with consequences that were far less favorable to the region than what occurred in China or Iran. In the end the region remained in a pitiable state for a century and a half, during which time important advances were occurring in both the Chinese and Iranian parts of the Mongol empire. Even the herculean military and civic undertakings of Tamerlane (Timur), which followed this era of torpor, failed to put Central Asia back on a path of sustained development.

To understand this unfortunate turn of events, it is necessary, first, to appreciate the two success stories of Mongol rule. The advances that took place in both China and Iran under the Mongols were due significantly to the fact that in those countries the new rulers chose to assimilate with the local culture, which enabled them to draw on local skills and expertise in many fields. By contrast, the regressive nature of Mongol rule in Central Asia arose from their failure or refusal to effect a transition to settled life and to embrace urban civilization.

Sheer necessity forced the Mongols in China to engage directly and deeply with Chinese life. As nomads, China's new Mongol rulers knew they had to rely on the administrative skills of others. Since they could not count on the loyalty of the Chinese, from the outset they turned for help to Turkic Uyghurs, many of whom were Nestorian Christians, and also to Central Asian Muslims.

Typical of this latter group was Sayyid Ajjay Shams al-Din, an aristocrat from Bukhara who had surrendered to Chinggis and signed on as a provincial administrator. His rise was stunning and was capped by his being named governor of the important province of Yunnan.[56] To serve as his vizier, or prime minister, Khubilai appointed another Central Asian, remembered simply as "Ahmad," who hailed from the outskirts of Tashkent. Soon this adroit manager and manipulator was in control of the entire tax and financial apparatus of the Mongol state in China and was wrangling to place his son in a high position as well. For twenty years Ahmad was de facto ruler of the empire, an updated version of the Seljuks' Nizam al-Mulk, but without the latter's religious program. So much control over the Chinese state did Ahmad wield that Marco Polo thought this Tashkent merchant had bewitched the khan. Angry Chinese rivals, looking askance at Ahmad's peculation and womanizing, branded him one of the "three villainous ministers" and finally assassinated him.[57]

Whether because or in spite of such senior advisers, Khubilai Khan managed to put in place effective administrative and tax systems, which continued for years to use the Uyghur language as well as Chinese. No less important, he pushed to reopen the great east-west caravan routes that had been closed since the Karakhitai had severed them. Trade flourished and enriched many Chinese cities. Marco Polo, who showed up in Beijing in the fifteenth year of Khubilai's reign, was just one of thousands of merchants from abroad who rushed to take advantage of this Mongol-sponsored boom. Many Muslim architects and craftspeople captured during the conquest of Central Asia were also deployed across China, building and ornamenting the Mongols' administrative centers.

CENTRAL ASIAN CULTURE IN MONGOL CHINA

Once he had accepted a settled and urban life, Khubilai Khan's dependence on Central Asian talent was even more in evidence. The intellectual horizons of the new Mongol conquerors were not broad, being focused mainly in the four utilitarian areas that the writer Nizami Arudi had defined as essential for rulers: astronomy/astrology, medicine, panegyric poetry, and historical writing. Of these, the Mongols asigned particular importance to astrology and medicine.

Khubilai sent to India for doctors, some of whom were by now practitioners of the new "Greek" techniques introduced into India by Ibn Sina's *Canon*. Nestorian Christian doctors from the Seven Rivers region of present-day Kyrgyzstan and Kazakhstan were no less consequential, to the extent that they influenced the highly developed field of Chinese pharmacology. These same doctors from Central Asia introduced their Chinese colleagues to Ibn Sina's *Canon*, which Khubilai ordered to be translated into Chinese.[58]

As to astrologers and astronomers, the Mongols had their own practitioners in both fields. But while Mongol astrologers were able to predict eclipses, their knowledge of astronomy was so limited, and their confidence in their own predictions so weak, that they greeted every eclipse with the anxious beating of drums, followed by drunken celebrations once it was over.[59] Again, Khubilai had to look abroad.

It is curious that at the time of the Mongol conquest, Samarkand's observatory was headed by a Chinese astronomer named Li.[60] But with this exception, local talent dominated the field across Central Asia and provided a large pool from which Khubilai Khan could choose his astronomical team. For several generations thereafter, Central Asians dominated the field of astronomy in Beijing. And if the first were brought there as captives, they were succeeded by luminaries whom Khubilai himself invited to his court.

Of prime importance among the invitees to Beijing was Jamal al-Din of Bukhara, who was charged with setting up an "Islamic Astronomical Institution" to operate alongside the traditional Chinese bureau and, inevitably, in competition with it. Jamal al-Din, who had already worked in Mongolia, arrived at Khubilai's new capital with detailed plans for seven astronomical instruments, which he commissioned a Chinese engineer to construct and then presented to Khubilai Khan. These included an advanced astrolabe, sundials that indicated both equal and unequal hours, an armillary sphere or spherical astrolabe, a celestial globe, and a terrestrial globe.

Chinese astronomers had been building armillary spheres since at least 1092, and they were also familiar with celestial globes, so neither of these models presented anything new. But this was their first contact with the advanced astrolabe, which applied discoveries by Biruni,

Khayyam, and a host of other Central Asian and Arab astronomers over the previous three centuries. Also completely new to them was the concept of unequal hours embodied in one of the two sundials. Similarly surprising to them was the fact that Jamal al-Din's spherical globe showed rivers, oceans, and even the distances along roads. This globe, of course, built directly on the project of young Biruni at Gurganj two centuries earlier. For all the sophistication of Chinese astronomy, this device, too, was entirely new to Chinese stargazers.[61]

Jamal al-Din also prepared for Khubilai a new calendar system extending ten thousand years into the future. This involved the use of Ptolemaic tables that had been corrected by such Central Asian astronomers as Biruni and Omar Khayyam and then adjusted to the latitude of Beijing. Jamal al-Din himself had begun these twenty years earlier while engaged as an astronomer-astrologer at the Mongol capital of Karakorum (now Kharkhorin). The task proved so daunting that he brought in Nasir al-Din al-Tusi, the dean of Central Asian astronomers, who was then working for the Mongols in Persia. Now, with the help of his Islamic Astronomical Institution and the observatory associated with it, he was able to bring the project to completion. In addition to this, Jamal al-Din, a typical Central Asian polymath, carried out a prodigious survey of the entire Yuan empire, said to have consisted of 755 volumes, of which only the introduction survives.[62]

For nearly a century thereafter, the Astronomical Institution and its observatory continued to attract Central Asian luminaries to Beijing. The observatory still stands today. In the 1360s several of Shams al-Din's translators from Kunduz in Afghanistan were employed to make Chinese translations of Greek and Arabic astronomical texts, including Euclid and Ptolemy, as well as Biruni's *Chronology of Ancient Nations* and more recent works by Nasir al-Din from Tus. Also at that late date yet another Samarkand astronomer, Abu Muhammad Khwaja Ghazi, was preparing astronomical tables and predicting eclipses for the Mongol viceroy in Tibet.[63]

CENTRAL ASIAN CULTURE IN MONGOL IRAN

While the main branch of the Mongols was consolidating its rule in China, centrifugal forces were growing throughout the newly captured territories to the west. It took less than a generation for contending descendants of Chinggis Khan to be at sword's point, each asserting his claim over a major region and, in some cases, over the empire itself. Meanwhile, Mongol conquests in what is now Ukraine and Russia opened the way for one branch, under Chinggis Khan's grandsons Batu and Berke, to dominate both the steppe country and forest belt that extended clear to the arctic permafrost. This "Golden Horde" ruled Muscovy for 270 years, more than enough time to exert a profound influence on the subsequent society.

Meanwhile, the Mongols had destroyed the caliphate and sacked Baghdad in 1258, an epochal event in the history of the entire region. This conquest, followed by successful campaigns in western Iraq and Syria, created the geographical base for another dynamic Mongol regime that was to be based in the city of Tabriz in the Azerbaijan region of what is now western Iran.

Unlike the Mongols of the Golden Horde, who retained their traditional shamanist religion, the Mongols of Persia converted to Islam by the 1290s and opened themselves fully to the riches of Persian urban culture. Even before that date, armies of this so-called Subordinate Khanate (Ilkhan) were fighting both the Golden Horde to the north and a third branch of the dysfunctional dynasty, the descendants of Chinggis Khan's second eldest son, Chaghadai, in Central Asia.

The fractious state of intra-Mongol relations and the many clashes of arms to which it gave rise force us to discard the popular notion of a Pax Mongolica that is said to have prevailed across Eurasia once the initial conquests had run their course. To be sure, trade with China reopened and briefly enriched those cities that participated in it, especially Tabriz and the other commercial centers of the Ilkhanate in Persia. But the sheer distance between the various Mongol legacy states, widening economic and religious differences among them,[64] and the centrifugal force that afflicted all nomadic societies caused the Mongol empire to begin splitting apart even while it was forming.

In spite of this, the Mongols of Persia experienced a brief period of great cultural effervescence.[65] Their support of industry and agriculture produced new wealth, which, under Persian influence, they lavished on new cities and construction projects, most of them heavily influenced by Central Asian prototypes. Thus the splendid domed tomb of Oljeitu (Uljaytu Khudabanda) at Sultaniyya, with its octagonal base supporting a 161-foot-high double cupola, is a lineal descendant of the Tomb of Sanjar at Merv.[66] Artists and poets were generously rewarded for glorifying the new leaders in the traditional manner.

The one negative was that under its Mongol rulers, Ilkhan Persia looked more to the East than to the West, which minimized its contact with the latest developments in west European civilization.[67] This reorientation affected Central Asia as well, since its ties with Europe had been mainly through Iran. Thus, at the very moment when fresh breezes were blowing in European intellectual life and culture, Central Asia found itself increasingly out of touch with the West.

In many respects Mongol Persia paralleled developments in far-off Beijing. For the same combination of practical reasons and superstition, the Ilkhan court patronized astrology and astronomy, while medical doctors toiled to preserve the rulers' health. Nomad traditions of tolerance both before and after the Ilkhans' conversion to Islam attracted both Sunni and Shiite Muslims, as well as many Christians and Jews, to the Mongols' western capital. Indeed, the only group that was not welcome there were Ismaili Muslims, whom the Mongols considered a political threat and had driven from the country. As at Beijing, many scientists and artists from Central Asia found this atmosphere attractive and thronged there in such numbers as to give this autonomous western bastion of Mongol power a decidedly Central Asian cast.

Thus the man who designed and implemented the Mongols' successful economic strategy was Minister of Finances Shams al-Din from Khurasan, whose father had preceded him in that post.[68] Present at the pillaging of Baghdad was Shams al-Din's brother, Ala al-Din, known as Juvayni, who then became governor of the former Abbasid capital. This is the same Juvayni who penned the massive *History of the World-Conqueror*, a detailed and highly readable source on the Mongol conquest. Once condemned to death by the Mongols and saved only by the intervention of another native of Khurasan, Juvayni had to tread lightly as he wrote his history.[69] Successfully maneuvering between horror at the events he described and his duty to his Mongol employers, Juvayni brilliantly engaged his reader, offering dramatic crescendos, explosive and often bone-chilling action, and thoughtful asides designed to challenge common assumptions. His uncertainties were reflected in his production of two often contradictory versions of the same events.[70]

Among writers paid to flatter the Mongol rulers with panegyrics were poets from Isfara, Marginan, and Faryab, in modern Tajikistan, Uzbekistan, and Kazakhstan, respectively. A more colorful and independent poet came from the Ferghana Valley. Writing under the pseudonym "Athir," he criticized all who disagreed with him and attacked one alleged plagiarist for being a "robber of the caravan of poetry."[71]

None of these held a candle to the great Saadi from Shiraz and Nizami from present-day Azerbaijan, two stars of the firmament of poetry. But there was one Central Asian poet who outshone even these two giants in terms of sheer breadth and depth of expression, and of his immense attraction to readers even today. Jalaluddin (Jalal al-Din) Muhammad Balkhi (ca. 1195–1273) is known everywhere today as "Rumi," that is, "One from Rome." He acquired that name because for more than four decades he was a member of the colony of Central Asian exiles at Konya in what had earlier been Byzantine Anatolia—hence "Rome" or "Rumi." But he was born in Central Asia and lived there down to the time he and his family fled west to escape the Mongol onslaught. Although said to have been from Balkh (thanks to which he is known in Afghanistan not as "Rumi" but as "Balkhi"), Rumi was more probably born in the Tajik town of Vakhsh, where his father, a Hanafi jurist, mystic, and struggling preacher, had vainly sought employment. Failing in Vakhsh, the family shifted to Samarkand just in time to experience the devastating assault by the shah of Khwarazm, Qutb al-Din Muhammad. Seven years later the family hastily fled Samarkand before Chinggis Khan's forces reached the city in March 1220. This time they headed first to Balkh and then, as the Mongols approached that city, to Nishapur, where legend holds that the young man met the Sufi poet Attar. When citizens of Nishapur killed Chinggis Khan's son-in-law, Rumi's family, accompanied by a band of his father's disciples, once more took flight, this time to avoid the large Mongol force that was hell-bent on punishing the city for murdering their leader's kinsman. After a visit to Mecca, the refugees arrived at the Anatolian town of Konya about the year 1225.

How old was Rumi when he left Central Asia? The traditional dating, supported by Emory University's Franklin D. Lewis, would make him fourteen years old at the time. But the German scholar Annemarie Schimmel argued that he was at least nineteen–twenty-one years old. Either way, it is clear that Rumi grew up in, and was formed by, an atmosphere permeated by the passions and concerns of Central Asia, not Konya. His father took inspiration from Ghazali's younger brother, Ahmad, an even more uncompromising foe of rationalism than his better-known sibling. Rumi received formal schooling in the intellectually charged city of Samarkand, where he was exposed to the surging currents of Sufism. And his outlook on life took shape in a society that was collapsing around him, a world suffused with fear from which even outward flight offered no respite.[72]

Rumi's world was one of complete darkness that contrasted with what a modern scholar has called "a hitherto unknown brightness on the spiritual plane."[73] To the brutality and soullessness of daily life as he had experienced it, Rumi responded with a capacity for ecstasy that exceeded that of any poet before him. In his verse he lost himself in recitation, a profane love, a beautiful face, God's unfathomable essence, in the whirling dance that became the hallmark of his dervish heirs in Konya, and in music:

> Oh, music is the meat of all who love,
> Music uplifts the soul to realms above.
> The ashes glow, the latent fires increase:
> We listen and are fed with joy and peace.[74]

In the large body of poems inspired by an illiterate goldsmith with whom he lived, Rumi oscillated back and forth between earthly and divine love, writing with a precision and delicacy that contrasts wonderfully with the roiling tumult of feelings that impelled him:

> By love what is bitter becomes sweet,
> Bits of copper turn to gold.
> By love dregs are made clear,
> And pain begins to heal.
> By love the dead come alive,
> And a king becomes a slave.
> This love, moreover, is the fruit of knowledge;
> No fool will ever sit on the throne of love.
> No, ignorance only falls in love
> With what is lifeless.
> It thinks it sees in something lifeless the appearance of the one it
>   desires,

As if it heard the beloved whistle.
A lack of knowledge cannot discern;
it mistakes a flash of lightning for the sun.
Lightning is transient and faithless;
without clearness you will not know
The transient from the permanent.[75]

In the end the object of all love is God. But Rumi's was not the God of the mainstream Sunni or Shiite, or of sectarians of any stripe:

The blind religious are in a dilemma, for the champions on
Either side stand firm: each party is delighted wih is own path.
Love alone can end their quarrel, Love alone comes to the
Rescue when you cry for help against their arguments.

Lo, for I to myself am unknown, now in God's name what
Must I do?
I adore not the Cross nor the Crescent, I am not a Giaour ["Kafir" or
unbeliever] nor a Jew.
East nor West, land nor sea, is my home; I have kin [neither] with
angel nor with gnome.[76]

For a people set adrift by upheavals that were destroying and creating whole civilizations before their eyes, Rumi offered the consolation of faith, in the form not of passive submission but of the active love that is the heart of the Central Asian Sufi message:

This world is a trap and desire is its bait:
Escape the traps, and quickly
Turn your face towards God.[77]

THE GREAT TRADITION BRIEFLY REVIVED: NASIR AL-DIN AL-TUSI

Late in November 1256, a highly consequential conversation took place in the ruins of Alamut, a mountaintop castle in northwestern Iran. For months the Mongol general and khan Hulegu (1217–1265) had besieged this stronghold of the Shiite Ismailis, having been told that it was the home of the dread "Assassins" and hence an impediment to Mongol control of Iran. This siege followed by only thirteen years a similar siege by papal forces of the fortress of Montsegur in southern France, where the heretical Christian sect of Cathars or Albigensians had fled. The capture of Montsegur was followed by a mass killing of the schismatics. The same fate awaited Ismaili Shiites once Hulegu finally broke their last resistance and forced them to surrender all their fortresses. One of the few to survive the siege at Alamut was the astronomer, philosopher, and polymath Nasir al-Din al-Tusi, who had been hiding there from the Mongols.[78] Alamut's well-educated Ismaili ruler had constructed an observatory on the mountaintop for Tusi's use. During the 1990s Iranian archaeologists excavating at Alamut discovered the charred remains of this structure.[79] Hulegu, a grandson of Chinggis Khan, had by the age of thirty-eight scored a string of military victories, which he attributed to wise advice from astronomers and astrologers. Now, having captured one of the most highly regarded astronomers of the age, he placed him in charge of religious affairs and astronomical predictions.

After the death of Omar Khayyam, Nasir al-Din al-Tusi (1201–1274) was the dean of Central Asian and Muslim astronomers and the greatest living exponent of Ibn Sina's philosophy and of Biruni's natural science. He was yet another brilliant intellectual to arise from Tus, the city in Khurasan that had given the world the poet Ferdowsi, the vizier and political scientist Nizam al-Mulk, the theologian Ghazali, the poet Asadi, and the astronomer, mathematician, and teacher of Nasir al-Din's own teacher, Sharaf al-Din al-Tusi.[80] Nasir al-Din came from an Ismaili family and received a broad religious education. But then he concluded that religion and science constitute two entirely different realms and that each should be left to its respective experts.[81] As a young scholar in Nishapur he turned out a formidable mass of work in scientific fields as diverse as chemistry, logic, mathematics, and even biology (see plate 23).[82]

Evidence for the breadth of Tusi's interests is his early treatise on how humans adapt to their surroundings. "As a result [of adaptation]," he wrote, humans "gain advantages over other creatures."[83] Joining the ranks of Central Asian thinkers who anticipated Darwin, he observed that:

> [ape-like] humans live in the Western Sudan and other distant corners of the world. They are close to animals by their habits, deeds

and behavior. . . . Before [the appearance of humans], all differences between organisms were of natural origin. The next step will be associated with spiritual perfection, will, observation and knowledge. . . .

All these facts prove that the human being is on the middle step of the evolutionary stairway. According to his inherent nature, the human is related to the lower beings, and only with the help of his will can he reach a higher level of development.[84]

To enable people to reach this happy state, Tusi, following Plato, Farabi, and Ibn Sina, proposed that society be guided by a single, wise, and all-powerful leader. Whether he saw this as Plato's lawgiver or a Shiite imam is less important than the fact that he rejected any form of self-government on philosophical grounds. Unfortunately Tusi had no time to pursue this theory further.

As Chinggis Khan's army approached Nishapur, Tusi fled to a nearby Ismaili castle, one of a string of fortresses the Ismailis had established to protect themselves from Nizam al-Mulk's crusade against them. From there he moved on to Alamut, where he survived for more than two decades before the arrival of Hulegu's army. In an early meeting with Hulegu, Tusi explained that he could not carry out the astronomical tasks Hulegu had assigned him until all available astronomical tables had been adjusted to the latitude and longitude of Hulegu's new capital at Tabriz. Hulegu responded by agreeing to fund the equipment needed to do this. This resulted in a four-story stone observatory, ninety-two feet in diameter, which was erected on a high promontory at a site called Maragha, forty-five miles west of Tabriz. The facility also included a library of forty thousand books plundered by the Mongols from Alamut, and rooms for staff and a hundred students.

The researchers at Maragha were drawn from across the Islamic world, and also from Byzantium and China. From the outset Central Asians were especially prominent among them. Thanks to such luminaries, Maragha became the world's most advanced institution for astronomy at the time. After Tusi himself, the leading astronomer to work at Maragha was Sadr al-Sharia Bukhari (d. 1347) from Bukhara, who did not arrive at Maragha until after Tusi's death. It was this Bukhari who drew together in lucid form the entire centuries-long effort to refine the Ptolemaic system of the heavens and, in the process, transcend it.[85]

The actual science that flowed from Tusi and his colleagues far surpassed the efforts of the Bukharan and Chinese astronomers whom Khubilai had assembled in Beijing. Tusi alone produced over 150 treatises, including the first treatment of spherical trigonometry independent of astronomy. Beyond his work in astronomy and mathematics, he made contributions to mineralogy and medicine and penned theological works and poetry.[86] Like his predecessors, Tusi took the Ptolemaic system as his point of departure and focused not on overthrowing it but on ironing out its inconsistencies and wrinkles. In this spirit he faulted Ptolemy's claim that Earth is at rest. What particularly bothered him was the assertion that this could somehow be established through observation. To the contrary, Tusi argued that there was no observation that could establish *either* that Earth is stationary *or* that it is rotating. As to those in the Arab philosophical tradition who continued to champion a stationary Earth on the basis of religious or secular authority, he rejected their position as pure dogma. Tusi already suspected that Earth is rotating but did not pursue this opening. Two centuries later Copernicus was to follow Tusi's lead to its conclusion, by proposing that Earth in fact turns. Tusi's role was less to establish a new paradigm than to undermine the old, which he did most effectively, and with important consequences for the future.

The effort to bring mathematical hypotheses and empirical observation into harmony with each other lay behind Tusi's questioning of Aristotle's claim that all motion in the universe is either linear or circular.[87] He and his colleagues had confronted this issue as they sought to describe planetary orbits. Using an elegant geometric model, Tusi showed that if one circle rolls within the periphery of another, and if the larger one is twice the size of the smaller one, then a point on the smaller circle can be made to describe a straight line that is a diameter of the larger circle. In other words, he showed that the sum of two circular motions can generate linear motion.

This model, called the "Tusi Couple," enabled Tusi to predict the longitude of planets without the use of Ptolemy's equant. Two centuries later Copernicus called on the Tusi Couple to address the problem of Mercury's eccentric orbit and used it also to construct his heliocentric model of the solar system.[88] It is fair to say that this epochal revolution began when Tusi, like other astronomers in the Central Asian tradition, struggled to square the mathematically based analyses of Ptolemy with

Figure 13.3. By describing the motion of two circles rolling within a third circle, Tusi corrected an assertion by Aristotle and simplified the prediction of the longitude of planets. Copernicus later drew on Tusi as he developed his heliocentric model of the solar system. This early manuscript describes Tusi's research. ▪ Nasir al-Din al-Tusi, *Tadhkira*. From Anthony Grafton, ed., *Rome Reborn: The Vatican Library and Renaissance Culture* (Library of Congress: Washington, DC; Yale University Press: New Haven, in association with Biblioteca Apostolica Vaticana: Vatican City, 1993), plate 117. In Arabic, paper, 14th c. Vat. ar. 319, fol. 28.

the purely philosophical assertions on which Aristotle based his physics. Ptolemy himself had rejected Aristotle's philosophizing, but Tusi and his colleagues tried to embrace both—such was the authority of Aristotle among medieval Central Asians. Tusi and his colleagues set out to improve Ptolemy's mathematical approach to astronomy by adding a dose of Aristotle, but in the end they were unsuccessful.[89] Nonetheless, their "Maragha revolution" was a significant step in the emancipation of astronomy from natural philosophy, and toward its full embrace of mathematics and observation, which occurred in Europe during the scientific revolution.

Tusi's interest in the relationship among mathematical hypotheses, empirical observation, and natural philosophy brought him into territory that Ibn Sina had explored in his later works. Tusi systematically defended Ibn Sina against his critics, in the process revalidating the classical Greek thinkers whose works had so inspired Ibn Sina.[90] As part of this effort, he revised key works of Greek science, including Euclid's *Elements* and Ptolemy's *Almagest*, and wrote works on classical Greek and Arabic-language mathematics that for centuries were standard everywhere. He also found time to pen a massive study of logic and treatises on ethics, causality, and the soul.

THE DEVASTATION OF CENTRAL ASIA

What was happening in Central Asia while the Mongols in China and Persia were building up their powerful states and resplendent cultural life? Did Central Asia also experience a postinvasion boom, thanks to its location at the heart of the Mongols' continent-wide trading empire? Several bronze Mongol *yarliks* or visas preserved in museums have encouraged this optimistic conclusion. These badges, which gave their bearer safe passage across the length of Eurasia, tempt one to conclude that great numbers of merchants were criss-crossing Central Asia throughout the Mongol era. Also encouraging such a conclusion is the story of Marco Polo, who twice traveled unimpeded from Venice to Beijing in the 1270s. But it should be noted that Polo avoided most of Central Asia, taking instead a more southern route through Balkh and then heading northwest up the Wakhan Corridor into Xinjiang. Another Western traveler, William of Rubruck, heading to the Mongol capital in 1253–1255, chose a northern route that also avoided Central Asia. He reported that "there is no counting the times we were famished, thirsty, frozen, and exhausted," but he clearly preferred this to the dangers of taking a more direct route through the heart of the region.[91] That Central Asia might have long remained a dangerous zone between the opposite ends of the Mongol empire implies that the numbers of merchants traveling across the region bearing bronze visas may have been quite limited.

There were good reasons for the negative assessments of Rubruck and Polo. The same powerful centrifugal forces of distance, clan differences, chaotic succession, and personality that pulled Hulegu's western empire away from Beijing assured that Central Asia, too, would go its own way.[92] Unlike the heirs of Chinggis Khan who ruled in Persia and China, the grandson to whom Chinggis entrusted Central Asia never abandoned the nomadic life style and therefore never established a capital city. As a result, no Khubilai Khan or Hulegu arose in Central Asia to embrace urban life and spark economic and cultural revival.

No less important as a cause of the prolonged economic and cultural slump into which Central Asia fell after the Mongol invasion was the sheer destructiveness of Chinggis Khan's military operations there. This can be measured by the numbers of major thinkers who were killed or driven into exile. A few could doubtless have saved themselves by making a deal with the conqueror. The Sufi divine Najmuddin Kubra refused and died on the ramparts, while the poet Attar arrogantly declined to be ransomed because the price set by the Mongols was too low.[93] Others perished anonymously.

Still others chose flight. We have noted the careers of Jamal al-Din, "Ahmad," the Nestorian doctors, and Abu Muhammad Khwaja Ghazi in Beijing, and of Juvayni, Rumi, and Nasir al-Din al-Tusi in the western Mongol realms and beyond. Some headed in other directions, with Minhaj al-Siraj Juzjani from Balkh ending up in India, where he wrote a major history before finally signing on with the conquerors.[94] The list of Central Asian refugees in India alone includes luminaries from practically every field of knowledge.[95] Tens of other intellectual exiles shared the plight of these notables. Once abroad, they tended to huddle together in colonies for mutual support. Significantly, none of those who fled the Mongol terror are known to have returned to Central Asia. This emigration of talent left the region bereft of cultural and intellectual leadership.

But to what would they have returned? Balkh, Nishapur, Tus, Gurganj, and Herat had all been reduced to uninhabited desert. A returnee would have had to pick among the pillaged or burnt-out ruins to find a dwelling. Even had someone tried to return, as occurred at Herat, the death or deportation of the entire class of tradespeople, provisioners, and craftspeople would have rendered survival difficult, if not impossible.

Reconstruction proceeded at a snail's pace. The first Mongol governor of Bukhara ordered the city rebuilt, but internal feuding and the lack of skilled workers delayed the work interminably. The Arab traveler Ibn Battuta, visiting the city a century after the invasion, found the markets, mosques, and madrasas still in ruins.[96] In Nishapur, half a century passed before a governor ordered the start of reconstruction. Work at Samarkand commenced earlier but made little progress, while a century passed before rebuilding began at Balkh.[97] At Herat a clever local ruler managed finally to bring about renewal in spite of the Mongol governor who appointed him, but this was a rare exception.[98]

The demographic impact of the Mongol invasion of Central Asia was enormous. Unlike the invasions of Persia and Russia, the Mongol assault on Central Asia was highly personal, a war of revenge calculated to punish the shah of Khwarazm and the entire population for their defiance and perfidy. The result was a war of extermination. Estimates of the death rate at Samarkand, which was treated relatively mildly, approach three quarters of the population.[99] Though thirty thousand craftspeople were taken captive and thousands of women enslaved, the death toll still stood at seventy thousand.[100] Juvayni, the Mongol's Nishapur-born court historian, reported that a group of surviving noblemen at Merv counted corpses for thirteen days and nights and arrived at a total of 1.3 million dead.[101] A more cautious contemporary estimate still places the losses in this one city, heretofore the largest on earth, at 700,000.[102] The same sources report that the number of those massacred at Nishapur and Herat was also above one million.[103] A partial list of other devastated Central Asian cities for which we do not have even exaggerated estimates includes Faryab (Otrar), Gurganj, Chach (Tashkent), Balasagun, Uzgend, Khojent, Bamiyan, Kabul, and Tirmidh (Termez).

Even allowing for gross exaggeration by contemporary writers, the number of those who perished was clearly enormous. David Morgan, a cautious and prudent scholar, did not hesitate to call it "an attempted genocide."[104] Demographers estimate that the Black Death killed off a third to two-thirds of Europe's population in the 1350s, setting back development for generations.[105] By comparison, the best evidence on the demographic loss from the Mongol invasion in Khurasan alone (including Merv, Balkh, Nishapur, etc.) was a horrific 90 percent.[106] If one ignores the evidence on which this estimate is based and halves it, the demographic consequences of the Mongol invasion of Central Asia still surpassed what the Black Death wrought in Europe. And of course this does not factor in the Black Death itself, which further devastated Central Asia before reaching Europe. A modern Tajik historian was therefore not far off the mark when he declared that the Mongol invasion and Black Death "emptied" Central Asia.[107]

Depopulation alone does not account for Central Asia's slow and halting recovery after the Mongol invasion. All efforts at renewal would have been stymied by the fact that the Mongols had systematically attacked the hydraulic systems that were essential to the very existence of the great cities of Central Asia.[108] At Bukhara and Gurganj they destroyed the dams that channeled water to the city, which at Gurganj led to the flooding of the entire metropolitan area.[109] Historians shortly afterward recorded that the Mongols broke down all the great dams and dikes on the Murgab River near Merv, and also the main irrigation dam at the fortress of Markha.[110] Such focused attacks were repeated many times at other urban centers. The replacement cost for the destroyed irrigation infrastructure would have been immense, certainly far beyond the greatly reduced manpower and material resources of the region in the post-Mongol era.

Beyond all these measures of the cost of the Mongol conquest was a civilizational loss of incalculable dimensions. The destruction of libraries, bookshops, observatories, endowed institutions, archives, schools, and *scriptoria* where copyists published the latest works in all fields was devastating. That the Mongols preserved at least one library—the collection the Ismailis had built up at Alamut—did not change this. Yet more crippling to intellectual life was the destruction of the class of people who maintained these institutions. After such devastation, how many parents were left who, like Ibn Sina's father, would sacrifice all to educate their offspring and who enjoyed the web of relations with other highly educated people that would enable them to do so? Two of the earliest works by the young Ibn Sina had been written for neighbors in Bukhara. How many such neighborhoods still existed after the Mongol conquest, and how many residents would have had the education, resources, and inclination to encourage a bright young neighbor to pursue open-ended studies? This sharp tear in the civilizational fabric of Central Asia is the most tragic legacy of the Mongol incursion.

Mongol Central Asia

Strictly speaking, Central Asia did not exist under the Mongols, for they divided it into three parts. The region stretching from the west side of the Tian Shan across the Uyghur lands of Xinjiang was ruled directly from Beijing,[111] while the western parts of Khurasan came under the control of Hulegu's successors in Iran. This left a broad strip from the steppes of Khwarazm in the north to Afghanistan's Helmand Valley in the south to the Chaghadai clan, which Chinggis Khan had named to rule the area. From the outset the Mongol rulers of this rump of Central Asia sought to assert their independence both from Beijing and from Tabriz.

The Mongols of Central Asia were not builders. Chinggis Khan is said to have constructed a road through the Tian Shan passes with forty wooden bridges and with inns at a day's distance from each other.[112] But even if this really existed, it did not foster commerce on a permanent basis. It is also true that a later Mongol prince built a modest palace a few miles from Qarshi in what is now southern Uzbekistan[113] and sometimes used a rebuilt Uzgend in what is now Kyrgyzstan as a headquarters. But these second-tier cities fulfilled mainly administrative functions and did not become major centers of commerce or culture.

During the early postconquest years the Mongol overlords turned the administration of the region over to Mahmud and Masud Yalavach, traders from Khurasan, who took up residence in Khojent. As de facto viceroys, these brothers introduced steep new taxes that were proportionate to income and used the ample yield to pay for the garrison and the administration. Typical tax farmers, they did not even bother to do a census, instead grasping for whatever they could get.[114] Even though this directly contradicted the usual Mongol practice of imposing only moderate taxes, the Mongol overlords were content with this arrangement and left the Yalavachs free to oppress their fellow Muslims.[115] The rulers of Khurasan quickly copied the system.

The hard-fought rise of Khaidu, a cousin of Khubilai Khan, as the Mongols' chief for the central part of the region immediately reignited tensions with Beijing. Khaidu remained a steppe nomad to the core. He clung to traditional tent life and assumed his purpose on earth was to expand his powers at the expense of the local Chaghadais and of Khubilai in China.[116] One of his beloved daughters refused to marry until she found a man who could beat her in a fight. The popularity of Chinese culture at Khubilai's court appalled Khaidu, and he took advantage of Khubilai's focus on the East to fill the vacuum left in the Eurasian heartland. In short order he had raised an army of 100,000 horsemen, attacked Samarkand and Bukhara, and then moved on to conquer Khurasan. Meanwhile he was plotting to invade Mongolia and Khubilai's realm in China.[117]

Khaidu's stubborn adherence to old Mongol values was in stark contrast to the processes of consolidation and stabilization taking place in both China and western Iran. Under Khaidu's backward-looking regime, trade declined and the value of the currency he minted fell. His response was to issue gold-plated bronze coins and to punish tradespeople who refused to accept them.[118] Only his silver coins were trusted. Faced with this fiscal mess, and with the continuing threats of marauders and popular rebellions in the region,[119] caravan traffic shifted to the south through Afghanistan and to the old northern route across Khwarazm.[120] The rest of Central Asia was left an island of relative poverty amid the reviving economies to the east and west. Surviving local dynasties continued to rule out of inertia.

Central Asia's intellectual life reflected this prevailing stagnation. Printing, both with inscribed blocks and with movable type, had penetrated from China into East Turkestan by 1300. By that date Uyghurs at Turfan and elsewhere were issuing full books in their native Turkic language. Not long after that, printing appeared also in the Mongols' western realms, where the Ilkhan dynasty held sway. It was to die out again in Persia, yet at least it gained an early toehold there. By contrast, in most of Central Asia, including Khurasan, printing remained unknown.

CULTURE UNDER THE MONGOL YOKE

Khaidu is known to have employed a poet, an astrologer, and a doctor, but his cultural horizons seem not to have extended beyond this. Unlike Khubilai and Hulegu, he built no capital, patronized no great historians or men of letters, and established no astronomical observatory. Like most medieval rulers, East and West, his idea of a good time was to hunt and drink. To be sure, the Christian wife of one of his Mongol colleagues sponsored a madrasa at Bukhara, but it employed no significant thinkers. Muslim imams and judges may have debated fine points of doctrine, but they posed no questions that had not been asked a hundred times before.[121] Even historical writing, the last refuge of vainglorious rulers, died out, with the exception of a single independent writer in the city of Herat.[122] Only two poets left a trace: one from the town of Qarshi in present-day Uzbekistan, who wrote seditious quatrains; and the other from Bukhara, who used verse to rail against the civil and religious authorities.[123] Neither paired his dissident spirit with talent. Since Islam was no longer the religion of state, Christianity briefly revived under Mongol rule, but without producing significant thinkers or writers. One may argue with Russian historian Bartold when he declared that Chinggis Khan "remained a stranger to all culture," but it is hard not to agree with his judgment that in the end it proved impossible to reconcile nomadism with urban-based traditions of intellectual culture.[124]

The one exception, and a darkly engaging one, was the poet Nizari Quhistani (1247–1285), an aristocrat from south of Nishapur whose family lost everything during the Mongol onslaught.[125] Utterly impoverished, Nizari took work with the Mongols as a traveling fiscal official in the Caucasus, which left him plenty of time to reflect on his existence, and then as a court poet in his home region. Raised in a devout Ismaili family, Nizari sank into a mood of profound bitterness after enemies conspired to have him fired and his savings confiscated. After pouring out his troubles in a grim *Book of Day and Night*, he turned to savage irony. A collection of verses supposedly written for his young son parodied traditional advice books, praising not abstemiousness but "gluttons and swillers of wine." In other works Nizari presented himself as an unapologetic freethinker and blasphemer in the spirit of Razi and Omar Khayyam, whom he deeply admired and to whom many of his own poems were later incorrectly ascribed. At the end of his life Nizari turned to farming and reclaimed some of his earlier faith.[126]

Nizari's life and works are a grim monument to the cultural and intellectual devastation wrought by the Mongols. No less, they mark the closing chapter in the story of free thought in Khurasan and Central Asia as a whole that began four centuries earlier with the likes of Hiwi, Ibn al-Rawandi, and Razi.

CULTURE BEYOND THE STATE: FOLKLORE AND RELIGION

Neither Khaidu nor his successors established a permanent capital in Central Asia, instead using as their main places of assembly a site in the Talas Valley on the border of present-day Kazakhstan and Kyrgyzstan. Here, amid rolling plains broken by low mountains, the Mongol rulers convened their *kuriltai* or council, fought over who should succeed whom, and struggled in vain to preserve a semblance of unity in their decaying empire. But theirs was not the only activity in the valley. In 1334 Kyrgyz settlers in the region erected a mausoleum on the banks of a tributary of the Talas River. Known locally as the "Manastin Kumbuzi" or "Tomb of Manas," the building is small—barely over by thirteen feet square.[127] Yet this modest structure of burnt clay with a high pyramidal dome has an importance all out of proportion to its size, for it is said to have been built to celebrate the legendary national hero of the Kyrgyz tribes, Manas. Never mind that an inscription on the wall states that it was the mausoleum of the daughter of a local khan; this, it is claimed, was just a ruse to divert looters.

When, where, and even whether Manas actually existed have been the subjects of passionate debate since 1858, when a Kazakh officer in the tsarist Russian army chanced upon a traditional "Manas singer" or *Manaschi* and listened to him sing a long passage of what appeared to be part of a much larger narrative in verse. He promptly published a Russian translation of what he had heard.[128] Shoqan Walikkhanuli (1835–1865), or Chokan Chingisovich Valikhanov, as he was known to Russian friends like the novelist Dostoevskii, had cast his lot with the Russian colonial project in Kazakhstan. He believed in Western education and had turned his back on Islam. But he was a sensitive folklorist and a knowledgeable historian of the Turkic peoples. Now, as he jotted down the verses sung by the *Manaschi*, he concluded that what he had stumbled on was a fragment of an ancient Kyrgyz national epic covering the life of the hero Manas and his progeny. Here, declared Valikhanov, was nothing less than the "*Iliad* of the steppes."

Soon folklorists with notebooks and, later, recording machines were sticking their noses into every Kyrgyz yurt in which a *Manaschi* was to be found. By the late twentieth century Valikhanov's brief passage had

Figure 13.4. Mausoleum (1334) in Kyrgyzstan's Talas Valley, hailed by tradition as the tomb of the Kyrgyz epic hero Manas, but more likely the tomb of a local khan's daughter. ▪ Photo by Michael Padraic Murphy.

mushroomed into half a million lines of verse, drawn from three score separate versions of the Manas epic. But skeptics abounded. Scholars discovered that many of the events described in *Manas* had occurred not in medieval times but in the sixteenth and seventeenth centuries. Others pointed out that Manas, if he existed at all, spent his life not in what became Kyrgyzstan but far to the east, near the sources of the Yenisei River that rises in northern Mongolia and flows to the Arctic Sea. Still others argued that most of the epic dated only to the century and a half before Valikhanov, while at least one skeptic ascribed the whole thing to an eighteenth-century forgery.[129] Meanwhile twentieth-century archaeologists even questioned whether the Manasim Gumbaz had anything at all to do with Manas.[130]

Confronted with all this, the Russian historian Bartold concluded that the entire Manas legend was just that and chose to ignore it completely when he wrote his history of the Kyrgyz people.[131]

This was surely an overreaction. There is no doubt that the various Kyrgyz clans had an ancient and highly developed tradition of heroic poetry, and that this epic tradition traces at least to the fourteenth century and possibly earlier. The first written reference to the Manas epic appeared in a *Compendium of Histories* that dates from the late fourteenth or fifteenth century.[132] Far from fading into oblivion, this tradition of untutored bards singing versified tales gained strength with the passage of time. Valikhanov, in other words, encountered a living tradition that had been enriched by all the elaborations, inventions, and improvisations that are inherent in organic evolution.

What concerns us here is that the gradual breakup of the Mongol empire opened new territories to the Kyrgyz and other nomads. The struggle to prevail against their many intractable enemies was the work of centuries. This created ample opportunities for the kind of heroism of which epics are made. Sung in a yurt to a circle of intent auditors, these ballads affirmed the values of the extended family and tribe and at the same time posed challenging questions of morality and ethics. Inevitably they provided models of behavior for the young. Arising from the populace, the message of the epic ballads reached the khans and tribal leaders as well, assuring the continuity of shared values through all levels of society. Much remains to be learned about the folkloric heritage of the Mongol age in Central Asia. But it is clear that the kind of popular creativity that later found expression in *Manas* and other sung poems of the region is one of the most enduring legacies of an otherwise bleak period in the history of culture.

The other outpouring of cultural energy during the Mongol era occurred in the sphere of religion, namely, the second flowering of Sufism. Like the singers of ballads and epics, the leaders of this movement arose directly from the people, with little or no help from any government. And like the Kyrgyz epic, Sufism came eventually to affect directly leaders at every level.

The spread of Islam among the Turkic peoples along the northern and eastern borderlands of Central Asia had not been rapid. True, it had made important advances during the Seljuk era. But only during the period of Mongol rule in the fourteenth century did many of the Turkic folk on Central Asia's northern border accept it, and then very much on their own terms. Mongols who had settled in Central Asia converted at the same time, and in the same manner.[133] Neither Turks nor Mongols wanted any part of the Islam of strict rules, rote memorization, and conformity. The fact that Islam under early Mongol rule was no longer the established faith made it all the easier to avoid the orthodoxy of the religious scholars and their madrasas. The learning that mattered to these new Muslims was inwardly focused and personal, like their traditional Tengrianism, and spoke as much to the emotions as to the mind. The Sufism of Ahmad Yasawi and Najmuddin Kubra responded perfectly to such values.

For its rapid spread and sheer vitality, no cultural or intellectual movement in Mongol Central Asia surpassed Sufism. Once Ghazali had opened a door for good Muslims to embrace the new "Islam of the heart," millions rushed in. All the groundwork had been laid prior to the Mongol conquest. Now this current of inward piety blossomed into a mass movement of both nomads and urban dwellers. By the end of the Mongol era it had merged completely with orthodox Islam and with the state.

All the various strains of Sufism participated in this important development. For example, a follower of Kubra from Gurganj, Sheikh Sayf al-Din Bakharzi, moved to Bukhara, ingratiated himself with the Mongols, and set up a large madrasa founded on Sufi principles.[134] The dominant figure of this reunion among Sufism, orthodoxy, and the state was a native of a village near Bukhara, Bahaudin al-Din Naqshband Bukhari (1318–1389). Having shown an early inclination toward the spiritual life, Bahaudin Naqshband was taken by his grandmother to a Sufi master in a nearby town. He later recalled the miraculous event that followed:

In the beginning of my travel on the Way, I used to wander at night from one place to another in the suburbs of Bukhara. I visited cemeteries by myself in the darkness of the night, especially in the wintertime, to learn a lesson from the dead. One night I was led to visit the grave of Shaykh Ahmad al-Kashghari and to recite *Surat-al-Fatiha* for him. When I arrived, I found two men, whom I had never met before, waiting for me with a horse. They put me on the horse and they tied two swords on my belt. They directed the horse to the grave of Shaykh Mazdakhin. When we arrived, we all dismounted and entered the tomb and mosque of the shaykh. I sat facing the *qiblah*, meditating and connecting my heart to the heart of that shaykh. During this meditation, a vision was opened to me and I saw the wall facing the *qiblah* come tumbling down. A huge throne appeared. A gigantic man, whom no words can describe, was sitting on that throne. I felt that I knew him. Wherever I turned my face in this universe, I saw that man. Around him was a large crowd in which were my shaykhs, Shaykh Muhammad Baba as-Samasi and Sayyid Amir Kulal. I felt afraid of the gigantic man while at the same time I felt love for him. I had fear of his exalted presence and love for his beauty and attraction. I said to myself, "Who is that great man?" I heard a voice among the people in the crowd saying, "This great man who nurtured you on your spiritual path is your shaykh. He looked at your soul when it was still an atom in the Divine Presence. You have been under his training. He is Shaykh Abdul Khaliq al-Ghujdawani, and the crowd you see are the caliphs who carry his great secret, the secret of the Golden Chain."[135]

Bahaudin Naqshband vowed total submission to his Sufi master. Over time the master instructed him on the path by which he could merge his being with God. At the same time he enjoined the young seeker to adhere meticulously to all the legal strictures of the Sharia. Again, Naqshband's commitment to absolute obedience assured that this order would be followed. By this step, the form of Sufism that Naqshband came to espouse circled back to the strict law of the faith, embracing it down to the smallest detail but placing it within a world of inner enlightenment and direct communion with God.

In other words, Naqshband evolved a synthesis between the inner and outer worlds, between faith and law, the individual and society. No less than other forms of Sufism, the Naqshbandi vision relied on the heart, and on direct communion between the believer and God. But to a far greater degree than that of earlier Sufis, Naqshband's vision was built on submission, first to one Sufi master or *pir* and the entire succession before him, then to the civil authorities and the Sharia, and finally to God. In time, Naqshband's approach came to dominate the region and to define its political culture as well as its religious life. Sufism had made its peace with power and with the status quo.[136]

This was easier because of the great pains Naqshband's new Sufi order—the Naqshbandiyya—took to show that its message flowed by a direct "Golden Chain" of occult succession from the founders of the faith. Just as Yasawi was said to have been named by Muhammad himself, so the Naqshbandiyya claimed that their founder had miraculously received his vocation directly from the ninth-century imam Hasan al-Askari, a figure acceptable to both Sunni and Shiite Muslims. Leaders of the new Sufi order gladly used this miraculous laying on of hands to trump any and all objections raised on the basis of reason.[137]

It also trumped innovation. Earlier, Sufi masters arose from all orders of society. Their charisma and authority were rooted in the piety of their lives and the impact of their teachings, and not in their lineage or official connections. But during and after the Mongol years the transmission of sacred authority from one generation to the next became a family affair, with sanctity deriving from bloodlines rather than from charisma. A *pir* or holy man was one whose father had been a *pir*, just as secular rulers passed their scepter of rule to their sons. Innovators were not welcomed.

Thanks to Bahaudin Naqshband, Sufism, which had begun as an ascetic and passivist oppositional movement, embraced the worldly life and became reconciled with the prevailing political order. It is revealing of this new turn of events that when Naqshband died in 1389, the rulers of Bukhara took over the care of his school and mosque and provided them with permanent endowments. Naqshband's vision quickly spread throughout the Islamic world. Its numerous followers from India to the Magreb attest to the enduring appeal of his vision down to the present day.[138] Both as a spiritual and an intellectual tradition, this current from Bukhara came to exercise a profoundly important influence on Islamic thought, not least in the twentieth and twenty-first centuries.

BALANCE SHEET FOR A DARK AGE

Inward-looking spiritual movements enable people to survive the most difficult times. That these arose and flourished during the century of Mongol rule in Central Asia, a time when other intellectual and cultural activity in the region all but ceased, attests to the creativity and resource-fulness of ordinary men and women. Sufism provided consolation and spiritual balm to a society that had been pulled up by its roots and nearly extinguished. That it gave rise to beautiful poetry and song, and that it nourished inner worlds in a way that modern men and women still find attractive, is much to its credit.

And yet this legacy of popular culture during the Mongol era has a less positive side. As the Czech scholar Jan Rypka observed, it was borne of pessimism over the prospects of life and nurtured passivity and quiet-ism, not only during the era of Mongol hegemony but for many centu-ries thereafter.[139] Notwithstanding their many thoughtful treatises, the Sufis disdained the intellect, which they rejected in favor of intuition. Like Ghazali, they considered rationality in all its forms to be a lower form of knowledge and were dismissive of its practitioners. Their laud-able search for spiritual guides and moral exemplars ended in a cult of saints across Central Asia that quickly degenerated into superstition.

In the end, the Sufi quest for inner fulfillment did much to suppress and marginalize intellectual life. For all the beauty of its music and po-etry, Sufism was the enemy of logic, mathematics, science, philosophy, and rational inquiry, including theology. Equally, it was often the enemy of civic impulses, including those of classical writers like Plato and Aris-totle, or of Central Asians like Yusuf of Balasagun, Farabi, and Ibn Sina. Rather than adjudicate among the various philosophies, the Sufis re-jected them all in favor of an anticivic self-centeredness. When Bahudin Naqshband effected a reconciliation between Sufism and the state, he did so not on the basis of any broad and open civic idea but by embrac-ing the detailed prescriptions of the Sharia.

During the Mongol era in Central Asia, Europe's universities and other centers of learning were burgeoning and their autonomy from political and spiritual authorities was being affirmed.[140] True, Aristotle's doctrines were condemned in thirteenth-century Paris, but this was not a state decision and was not repeated. St. Francis of Assisi, who a cen-tury earlier had espoused his own form of poverty-based spirituality, ended by reaching out to the world, even sending his friars to teach at the rising universities. It is hard not to conclude that the Mongol con-quest greatly diminished intellectual life across all Central Asia, even as it crippled a great tradition of urban economic and cultural life. Fragments of the great tradition revived thereafter, but never for long, and with little of the former intellectual creativity and brilliance.

Acknowledging this, clear signs of decay were evident in Central Asia's intellectual life long before the Mongol invasion. Rulers and their courts had greatly reduced their support for thinkers and writers other than those who flattered them. Ghazali and others had long since undermined the deeper rationale for open-ended intellectual inquiry and thrown aspiring scientists and thinkers on the defensive. And the law-bound keepers of the prevailing faith were solidly in control long before the Mongol armies showed up. In light of this, the Mongols may be blamed for deepening and completing a process of cultural destruction that had begun earlier, but certainly not for initiating it.

CHAPTER 14

❀

Tamerlane and His Successors

The last great outburst of cultural and intellectual activity from Central Asia was released by Timur the Lame, known in the West as Tamerlane (1336–1405). Like the cultural effervescence triggered by Mahmud of Ghazni, the early Seljuks, and Chinggis Khan, Timur launched his renaissance—if it can be called that—with a ferocious round of conquest that continued with only brief interruptions throughout his life. After this came a respite between the founder's destruction of all who opposed his rule and the final breakup of his empire into minor states ruled by warring descendants. It was during this century-long interval that the cultural flowering occurred (see plate 24).

Emerging from the rolling hills of what is now southern Uzbekistan, Timur was the humble product of a century and a half of intermarriage between a minor branch of Chinggis Khan's Mongols and local Turks.[1] Like his Central Asian forefathers on both sides—and in sharp contrast to the Mongols' descendants in China and Iran—Timur remained at heart a nomad and a warrior. His lameness was the result of an early fall from a horse. When the Spanish ambassador Ruy Gonzales Clavijo visited Timur's palace, he was told that the great amir or captain had begun his career stealing sheep and horses and dividing the spoils among members of his band.[2] But it proved easier to rustle sheep than to conquer the region's cities, which were in no mood to return to Mongol-type rule. To subdue Samarkand alone, it took Timur nine battles over eighteen years.[3] But by a string of adroit moves and sheer tenacity, he and his hearties gained control of all Central Asia. Timur was officially installed as ruler at Balkh in 1370.

The timing of Timur's imperial project could not have been better. The Black Death had thinned populations across Persia and the Caucasus in the early 1350s; the Persian–Mongol Ilkhan state collapsed about the same time, leaving a power vacuum across the Middle East and Persia; and the Russians defeated the Mongols of the Golden Horde in 1380. With his army swollen with recruits attracted by booty, Timur marched into these vacuums and then moved on the Middle East, successfully laying siege to Baghdad, Antioch, and Damascus. Outside Damascus he met with the great North African historian and scholar Ibn Khaldun, who pleaded for him not to pillage the old Umayyad capital. Timur plied the elderly thinker with probing questions, and Ibn Khaldun responded with flattering remarks, tracing Timur's ancestry back to Nebuchdnezzar.[4] But Timur sacked Damascus anyway.

Following the age-old routes pioneered by the Kushans, Mahmud of Ghazni, and Chinggis Khan, Timur then swept deep into India, devastating Delhi and other cities. Reversing course once more, he attacked the Ottoman Turks and captured their hapless sultan, Bayazit, in 1402. By unintentionally delaying the fall of Constantinople for half a century, Timur gained Europe's gratitude, which found expression in diplomatic contacts between him and England and France and in visits like that of the diplomat Clavijo from Spain.[5]

After spending only a scant few years in his capital, Samarkand, Timur hatched a plan to conquer China as well. Even Indians in distant Bengal were now paying tribute to Timur, but not China. Now he wanted to correct this deplorable situation. Since the new Ming dynasty had defeated the Mongols only in 1368 and was still consolidating its power, he might well have succeeded. But just as he was launching his China campaign, Timur came down with a fever at Otrar and died in 1405.

Timur's ceaseless conquests were accompanied by a level of brutality matched only by Chinggis Khan himself. At Isfahan his troops dispatched some 70,000 defenders, while at Delhi his soldiers are reported to have systematically killed 100,000 Indians. At Damascus Timur herded thousands of residents into the Friday mosque and set it ablaze. At Izmir on the eastern Mediterranean he beheaded all the captured soldiers of the defending Ottoman army and then lobbed their heads by catapult onto the ships on which others were fleeing the port. At Aleppo, Baghdad, Tikrit, Isfahan, Delhi, and other conquered cities, Timur ordered the construction of what he called "minarets" of the skulls of the defeated populace.[6]

For all his ferocity as a conqueror, Timur was notably lax at establishing effective and loyal governments in the conquered lands. Some would say this was the result of his conviction that he, a mere captain, was not truly the khan and that only Chinggis Khan's anointed heir should assume the duties of ruling. Others would defend Timur by noting that conquered lands had their own governing bodies, and that he was content to leave them be. But none of this compensates for the fact that Timur often had to reconquer his own cities or territories, which doubled the devastation.

No less striking than Timur's thirst for blood was the fact that nearly all his enemies and victims were Muslims. Even the Indian states he destroyed had Muslim governments, including the Delhi Sultanate and other local potentates that traced their rule to Mahmud of Ghazni. Timur attacked orthodox Sunni and Shiia alike. That he killed or enslaved the Christian populations of several Genoese trading colonies scarcely qualified him as a warrior for the faith. And even though he constructed a magnificent mausoleum at the grave of Ahmad Yasawi, his attitude toward Sufism seems not to have gone beyond a pro forma bow to saintliness.

Strange to say, Muslim preachers and scholars of the day raised no outcry against Timur's mass murder of Muslim believers. In fact, the ulama remained quietly in the background throughout Timur's reign, a situation that was to change only during the more conspicuously pious and narrowly orthodox reigns of his son and grandsons.

After destroying the great Syrian city of Aleppo, Timur is said by Edward Gibbon to have claimed that "I am not a man of blood."[7] Was he serious? History is rich with examples of leaders who were blind to their own savagery because it was carried out in the name of constructing a new and presumably better order. Also, it must be acknowledged that the fourteenth through sixteen centuries in particular were a time when people across Eurasia became accustomed to mass deaths. Some of these resulted from natural causes like the Black Death, which swept through Central Asia when Timur was ten year old. Others were the deliberate acts of leaders like Ivan the Terrible in Russia, Vlad III (Dracula or "the Impaler") of Wallachia, and Cortez and Pisarro in the New World, all of whom gained their ends through mass murder. Still others were caused by massive social and political upheavals like the one in Yuan China during Timur's lifetime that led to the deaths of thirty million people.[8] No one winced when Kyrgyz bards of the era sang of their hero, Manas, that, "You say 'Take off your hat' and he takes off a head."[9]

Europeans, grateful that Timur defeated the Ottomans, underplayed or ignored his depredations and cast him instead in the more familiar role of a prince driven by powerful inner passions. Both the English Renaissance playwright Christopher Marlowe, in his *Tambulame the Great*, and the operas of Handel and Vivaldi devoted to Timur focus not on his murderous career but on the relationship between Timur and his Ottoman captive, Sultan Bayazit, as mediated by Timur's supposed love for Bayazit's daughter. More recently many Central Asians point out that his conquests were followed by a century of highly sophisticated life in his main capitals. For his patronage of culture he is treated as a national hero in Uzbekistan.

THE ARCHITECTURE OF POWER

Friends and foes of Timur both acknowledge the diligence with which he rounded up craftspeople in all fields and sent them off to his capital at Samarkand.[10] And well he might have, since Chinggis Khan and his successors had utterly depleted Central Asia's cadres of skilled artisans by killing them or by deporting them by the thousands to Beijing or Tabriz. This destruction of Central Asian craft traditions that had been built up over millenniums now forced Timur to commandeer talent wherever he could find it. He assembled the most highly skilled manpower from many countries and traditions, an astonishingly rich assemblage of masters in virtually every field of the arts and crafts. At Timur's court these artists and artisans had no choice but to interact with one another, which over time led to the creation of brilliant new syntheses in fields as diverse as ceramics, sculpture, painting, woodwork, metalwork, and glass blowing. The new styles, and the technologies that underlay them, were inextricably connected with the circumstances of conquest and cultural appropriation that were Timur's hallmark.

Given Timur's assiduous culling of craftspeople from all the conquered societies, his utter disinterest in drawing intellectual talent to his court is all the more noteworthy. In striking contrast to Mahmud of

Ghazni, local rulers under the Seljuks, and the Mongols of China and Iran, Timur made no effort to identify and recruit scientists, scholars, men of letters, or even sycophantic poets and bards. In light of this, it is fair to conclude that this last "world conqueror" was more interested in things than ideas. In the one field in which he took a real interest and on which he showered money—architecture—his enthusiasm stemmed precisely from its ability to dramatize a very specific idea: that of his own power and greatness.

Calling on the combined talents of architects, plasterers, woodcarvers, and joiners, as well as manufacturers of baked bricks faced with brilliantly colored ceramics, Timur ordered the construction of structures that were to become his most lasting legacy.[11] We know the names of specific architects who were drawn from Iran, Georgia, Azerbaijan, Anatolia, India, and the Middle East to work for the amir. But to a greater extent than any Central Asian ruler before him, Timur became his own architect-in-chief.

Proof of this can be seen in the soaring entrance arches of the cavernous palace he built for himself in his hometown, formerly called Kesh but after Timur's reconstruction known as Shahrisabz, or "The Green City."[12] These immense arches spanned 73 feet and are still visible from twenty-five miles away.[13] The pylons that supported them originally soared to nearly 164 feet, the equivalent of a fourteen-story building. Even though the nave of the cathedral at Amiens in France, which was being built during Timur's lifetime, was 220 feet above the floor, the height of Timur's palace is nonetheless impressive.

The White Palace, or Ak Serai, as the ensemble was called, was only one of the many edifices, mainly in the form of domed cubes or polygons, with which Timur adorned the heretofore modest hill town he called home. The arches themselves collapsed soon after they were constructed, as did most of the largest arches and domes that Timur commissioned in Shahrisabz, Samarkand, and elsewhere. The stock explanation for this is that exceptionally strong earthquakes shook them down. But before ascribing the collapse of these structures to unexpected seismic events, it should be remembered that architects throughout these seismically active zones had long since developed methods for preserving domes, arches, and minarets against earthquakes.[14] These technologies appear not to have been rigorously applied at Timur's major buildings.

Figure 14.1. The monumental and forbidding ruins of the main gate to Timur's palace, Aksarai, at Shakhrsabz, Uzbekistan. For scale, note the tourists atop the structure.
■ Photo by Brian J. McMorrow.

The fact that so few of his great constructions outlasted him must be traced above all to the manic haste with which they were built.

Then there is the factor of sheer size. Specialists take pains to point out the several areas in which Timur's architects innovated. For instance, at the spacious mausoleum of Ahmad Yasawi at Turkestan, they combined transverse arches with traditional Central Asian arches in new ways, creating an awe-inspiring effect.[15] His architects also went beyond their predecessors in opening up interior spaces, in the construction of ribbed domes with cornices of complex plaster stalactites, and in ornamenting the exteriors with dazzling multicolored tile work (see plate 25). But taken together, Timur's buildings were conservative, dazzling more with their size and with the brilliance of their surface ornamentation than with any structural innovations.

Following in the tradition of Mahmud of Ghazni, Timur confronted visitors to his buildings with large terra-cotta or enameled brick billboards embedded into the walls at conspicuous places. At the Ak Serai

Figure 14.2. The looming mass of the never-completed Yasawi tomb in Turkistan, Kazakhstan. It was typical of Timur to create this gigantic memorial to a Sufi master who was so retiring that he chose to live out his last years in a subterranean cell. ▪ Photo by Marcia Newton.

in Shahrisabz, the messages were particularly blunt. One inscription announced that "The Sultan is the Shadow of God," while another, atop the portal in large letters, advised guests that "If you have doubts about our grandeur, look at our edifice."[16] Architecture was power, and Timur's addiction to it was an extension of his addiction to power.

By Timur's calculus, larger buildings signified greater power. Timur's gigantomania took full flight when he designed the Kok Gumbaz in his hometown as a tomb for his father. Here the main arch reached a breathtaking 151 feet. He went even further with the construction of the so-called Bibi Khanym mosque at the heart of his new city of Samarkand (see plate 26). The total area of the complex, at 358 by 548 feet, is two and a half times the size of an international football field. The walled area was flanked by a forest of 480 tall stone columns, which had been hauled to the site by ninety elephants captured in India.[17] Even this was not big enough for Timur who, ailing, supervised the army of workmen from a stretcher (see plate 27). The Castillian ambassador, Clavijo, reported that several times Timur ordered the demolition of completed arches and had them replaced with yet higher or broader ones. While thus engaged, Timur was also planning what was to have been his ultimate exercise of power, the conquest of China.

Alisher Navai, the greatest poet to serve Timur's dynasty, perfectly understood the purpose of these great undertakings:

> Whoever builds a structure . . . [with his] name inscribed thereon,
> For as long as the structure lasts, that name will be on the lips of
> people.[18]

Unfortunately for Timur, both the Bibi Khanym arches and his planned conquest of China soon collapsed.

POST-TIMUR: PEACE AND TRADE

Timur's death in 1405 gave rise to the kind of bloody internecine struggle over succession that by now seemed to accompany every change of rule in Central Asia. In the end Shahrukh, his fourth and youngeast son and the offspring of a concubine, won out. Having served as governor in Herat, Shahrukh felt himself an outsider in Samarkand and Shahrisabz and therefore moved his capital to the Afghan city. With Shahrukh's rise began what has often been called a golden age, a last great flowering of Central Asia, that extended down to the end of the fifteenth century.

But was the post-Timur century truly a creative age? Or was it, rather, a final and very incomplete reprise of certain older achievements that had already been waning for more than a century? Alternatively, could it instead mark the start of a fundamentally new cultural ideal, one that looked forward in time more than backward, but which lacked some of the most important features of Central Asian intellectual life and culture as they had existed over the preceding millennium?

In the political and economic spheres, it is easy to discern the fundamental continuity between the earlier nomadic empires of the Seljuks and Mongols and Timur's successors. Like his predecessors, Timur understood the value of trade and made haste to reopen the great continental transport corridors that had long been a mainspring of Central Asia's wealth. Samarkand once again dominated the China trade, but the capital at Herat quickly established itself as the main entrepôt to India, replacing Balkh, which still lay in ruins. The impact of these revived links with both India and China was immediately felt across Central Asia and the Middle East.

Like their nomadic predecessors, too, members of Timur's clan were quick to grasp the importance of a regularized tax system.[19] They knew that squeezing the cities would backfire. To some extent the problem was mitigated by the reemergence of slavery, which had diminished during the Seljuk era but revived thereafter.[20] But if hundreds of thousands of slave laborers reduced the state's expenditures, they did not solve the income problem. Timur's bureaucrats therefore resorted to the old trick of handing out vast tracts of land to relatives and favorites on the sole condition that the recipients make regular payments to the treasury. The effect of this appanage system was as it had always been: to create a new class of rich and autonomous grandees who were largely beyond the control of the central government. Certain of these personages were to become major patrons of culture, but they also stimulated the centrifugal forces that eventually led to the breakup of Timur's empire.

Within a century of Timur's death, these centrifugal forces were to gain the upper hand, fracturing all Central Asia into a maze of contending emirates, which proceeded further to fragment the region by imposing high tariffs on trade while failing to provide the security that might have justified such tariffs. Many times over the previous fifteen hundred years, the great transport corridors had been closed down, whether by domestic or international events. Yet the barriers to trade that the successors to the House of Timur imposed from within were the most destructive of Central Asia's economic viability and the most enduring. Indeed, they created the essential precondition for the opening of sea links between Europe and Asia and the eventual colonization of the region by Russia and Britain.

Culture under Timur's Grandsons

Before its early fall, Timur's dynasty gave rise to a stunning, if one-sided, cultural effervescence. Empire builders rarely succeed in passing to their sons the great passions that impelled them as founders. More common is for the successor to focus on building stability and prosperity. Timur's son and grandsons, as well as society at large, were so exhausted by his ceaseless warfare that they placed a high value on stability. His son, Shahrukh, ruled for forty-two years. After a chaotic interlude, several more decades of relative quiet followed. This created ideal conditions for culture to flourish.

Two of the three greatest patrons of culture in the post-Timur century were themselves members of the royal house, while the third was a vizier and close relative of the ruler. The first two were Shahrukh's sons, Baisunghur Mirza and Mirza Muhammad Taraghay, known as Ulughbeg, both of whom were fated to sit on the sidelines for decades as Shahrukh lived out over his long reign. Baisunghur whiled away these years in the capital of Herat, actively patronizing and even practicing the arts but in the end drinking himself to death by age thirty-seven. His older brother, Ulughbeg, passed the time in Samarkand, where he rose quickly from the post of governor to become the ruler of all Central Asia except for his father's power base in Khurasan. Neither brother had much interest in politics, let alone military affairs. Both found ample time to engage in the cultural activities that were more to their taste.

For centuries before the Mongols destroyed it, Herat had been overshadowed by Balkh to its northeast and Nishapur to its northwest. But under Shahrukh's benevolent rule, Herat came fully back to life and throughout the fifteenth century was the unrivaled political and cultural capital of Central Asia, Iran, and large parts of the Middle East and India. Shahrukh's powerful wife, Goharshad, and his son, Baisunghur, were his full partners in his effort to make Herat a city worthy of such a role.

On a large tract of land just north of the historic center they constructed several madrasas and mosques. These were to embody the solid but intellectually narrow Sunni orthodoxy that Shahrukh considered essential for the maintenance of social order.[21] The gem of this complex was the Musalla, an enormous madrasa for girls funded by Goharshad. Little of this majestic ensemble remains today, but six surviving minarets stand like unblinking watchguards over the ruins of fifteenth-century Herat. One is decorated with brilliant turquoise stars. Nearby is the stately mausoleum of Goharshad herself, which includes also the tomb of Baisanghur Mirza.[22]

Surrounding Herat, Samarkand, and the other capitals of Timur's empire were rural gardens. These invariably were defined by walled zones
with water flowing peacefully through them in stone-lined channels;
there, too, were pavilions, usually of two stories and with an open upper
terrace commanding a view of the rich plantings. The terraces were frequently the scene of royal receptions, entertainments, and drinking parties. Over time they were imitated throughout the Muslim world and
became the prototypes for what we know today as an "Islamic" garden.

No less important, Baisunghur set up a large and opulently outfitted
royal library, the first new one to be built in Central Asia since Mongol
times. In connection with the library, he established a center for all the
arts involved in the copying, ornamenting, and binding of manuscript
books. This *Kitabkhana* or "Book Workshop" employed forty copyists to
turn out fresh editions of ancient masterpieces. Artists from the royal
workshop developed elegant new scripts for Arabic and Persian and invented the art of decoupage. Others squinted over miniature paintings
to illustrate or adorn the pages, while still others found new ways to em-
boss leather covers with gold, and to apply block-printed ornaments to
the endpapers. Baisanghur himself was a dedicated and talented callig-
rapher. The deluxe editions from his workshop featured stunning miniature paintings, some of which portrayed events described in the texts
but others of which were fully independent works of art with no other
purpose than to delight the viewer.[23]

Baisunghur assembled his team of artists at almost the same moment
the Duc Du Berry was commissioning his gorgeous volume *Très Riches
Heures* in France. Each in his separate world produced the most important illuminated manuscripts of the fifteenth century. Baisunghur's
painters, like Timur's architects, were drawn from all the centers that
had fallen under Timur's rule. Indeed, the revolution in painting that
began under Sultan Shahrukh traces its artistic roots to the Ilkhan Mongol era in western Iran, whence several of the most prominent Herat
artists had been drawn.[24] These models, along with fresh contact with
Chinese painting, stimulated Herat artists to a degree not seen since the
centuries preceding the advent Islam in the region.[25]

The contrast between Timur's gigantomania and his successors' passion for depicting small worlds and miniaturized action is striking.
Was the latter part of the general reaction against Timur? Did all the

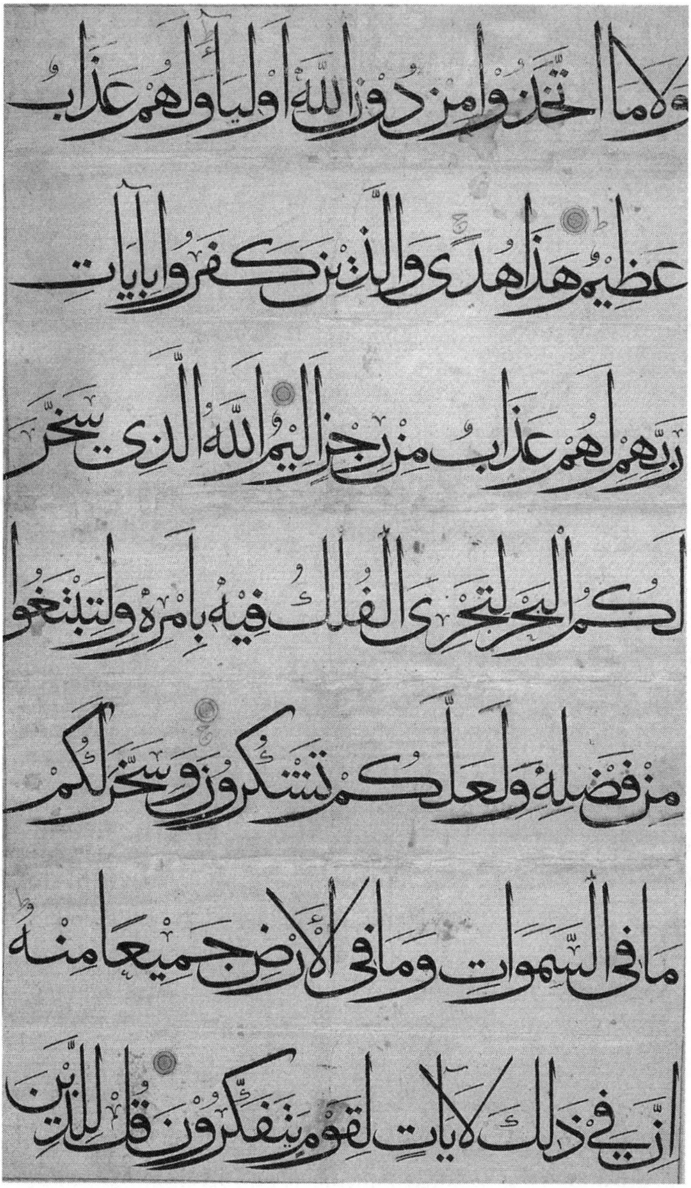

Figure 14.3. Prince Baisunghur, grandson of Timur, founded a library and book workshop at Herat and practiced the art of ornamental calligraphy, as in this page of the Quran from his hand. ▪ The Art and History Collection, courtesy of the Arthur M. Sackler Gallery, Smithsonian Institution, Washington, DC. LTS 1995.2.16.2.

refinement of the courts at Herat, Samarkand, and Tabriz, in which even the smallest painted space or architectural surface was elaborately ornamented, arise from an effete revulsion against the surging energy by which Timur had formed the world in which the courtiers lived so comfortably?

The rise of a cadre of virtuoso artists at Herat led also to the production of large-format paintings and murals. Timur himself had launched this new movement by covering the walls of several of his garden palaces with large frescoes depicting heroic and awe-inspiring moments of his career.[26] Shahrukh expanded and broadened the themes of such paintings. Only a few fragments of these works have survived, but descriptions make it clear that they were on a scale and of a prominence not seen since the great pre-Muslim murals at Panjikent, Samarkand, Balalyk-Tepe, and Topraq Qala. In a remarkable instance of cultural continuity, the ancient heroes of the *Shahnameh* once more graced the walls of palaces and public buildings, along with depictions of rulers, visiting diplomats, and other dignitaries. Both the official patrons and the painters blithely ignored all Muslim prohibitions against the depiction of the human form. The preachers and religious scholars, if they objected at all, were cowed into silence.

Merely to mention the establishment of educational institutions and libraries, and the assembling, editing, and reissuance of rare books in magnificent editions, is to conjure up a heady intellectual milieu. But Baisunghur was no more energetic than his grandfather Timur in locating and corralling great minds. For all the gorgeous books he commissioned, Baisunghur's intellectual interests seemed not to have gone far beyond producing an authoritative edition of Ferdowsi's *Shahnameh*—a worthy task, to be sure, but hardly pathbreaking work. Even if he located and made copies of a few books by great scientists of the region, Baisunghur's interest was mainly antiquarian. Khwarazmi, after all, had been dead for more than half a millennium, and Abu Sina for four centuries.

The problem here is not so much the age of the books Baisunghur assembled but the fact that he and his circle found in them no genuine intellectual challenge, let alone an urgent one. Doubtless, bookmen of the post-Timur era respected the great scientists and thinkers of the past, but less as bold thinkers who addressed inconvenient questions than as the providers of definitive answers. They were sure that Khwarazmi,

Farabi, Biruni, and Ghazali had ferreted out the secrets of nature and human life to the extent that it was humankind's lot to do so. The ageing volumes of these masters merited respect, not as living examples of human struggle but as the last words on their respective subjects. This attitude had been implanted and reinforced by the literalist and intellectually timid version of Islam that had taken root across Central Asia and Muslim lands to the west since the time of Ghazali. This set itself not against the claims of Christians, Hindus, or Confucians but against those fellow Muslims who were suspected of theological deviance, if not apostasy. Because it was so militantly on guard against schisms within the faith, the reigning worldview at Herat sharply narrowed what was permissible in the realm of thought.

Unstated constraints were felt even in the nonthreatening realm of aesthetics. The case of the Arab scientist al Haytham (965–1040) is instructive. Known in the West as Alhazen, Haytham was a contemporary of Ibn Sina and Biruni who did pioneering work in physics, mathematics, and optics. Western readers were latecomers to his studies on optics, but when they finally read them during the early Renaissance they immediately realized that this four-hundred-year-old research spoke directly to their own quest for a more accurate way of presenting perspective. Thanks to this, Haytham played a role in the development of "Renaissance perspective." But at Herat, which boasted the most talented painters in the Islamic world, some of whom professed an interest in the geometric and algebraic structure of physical space,[27] the entire Eastern tradition of optics, geometry, and trigonometry was ignored. The same can be said of much of the learning that had been amassed in Central Asia and the Arabic-reading world over the preceding half millennium. By the time the Timur's dynasty came to power, the reigning paradigms across most fields of science and philosophy were so deeply entrenched, their unresolved loose ends so few and seemingly insignificant, and the discordant voices so weak, that stasis set in.

Most creative people of Timur's era acted as if they had inherited a finished system of knowledge, whole and complete. Because it no longer demanded their attention, they allowed themselves instead to concentrate on ornamenting and beautifying their world. This is not to diminish the aesthetic achievement of Baisunghur's artists, craftspeople, and architects. But their achievement was aesthetic, not scientific or philosophical. With only one outstanding exception, to be discussed shortly, the civilization of Timur's successors was concerned more with beauty than with exploring the world of nature, or of humans' relation to God and the universe. In this respect they differed radically from the mainstream of Central Asian civilization over the preceding millennium.

ULUGHBEG: THE ROYAL SCIENTIST

The fact that various treatises on medicine continued to appear during the reigns of Timur's successor does not qualify this judgment. After all, the rulers and their close associates needed doctors to look after their health. The doctors, for their part, could not point to the ruler's good health as proof of their skills, for they would then open themselves to blame when the ruler's health flagged. And so they validated their expertise by pulling together dry compendiums of received medical knowledge. During Shahrukh's reign an extended discourses on medicine entitled *The Requirements of Surgery* appeared in Herat, while another volume on *The Curing of Diseases* appeared a few years later. Meanwhile, in Samarkand, Shahrukh's son, Ulughbeg, employed the erudite physician Burhan al-Din Kirmani, who rehashed the works of Hippocrates and Galen in ways that would have been old-fashioned in Ibn Sina's time, and added commentaries on minor works by recent Central Asian medical men. Even courts in such remote provincial areas as Afghan Badakhshan employed physicians, who paraded their expertise in grandiosely titled works like *A Book of Knowledge of the World*.[28]

The one exception to the general decline of scientific knowledge under Timur and his successors—but a very worthy one indeed—was Timur's own grandson, Mirza Muhammad Taraghay, dubbed Ulughbeg, or "The Great Leader." Here, long after the waning of Central Asian intellectual life, we find a sitting ruler who not only commissioned and received scientific treatises from the learned men he had brought to his court but wrote them himself. Unlike the "doctors" who rehashed the work of others in their compendiums, Ulughbeg based his astronomical writings on fundamental research that he himself conceived, organized, and carried out. If in no other area, in astronomy, at least, the age of Timur was indeed one of culmination.

The fact that Shahrukh entrusted control over all of Central Asia except Khurasan to his sixteen-year-old son suggests that young Mirza Muhammad had displayed signs of leadership talent. But early in his new life at Samarkand, Ulughbeg dealt so ineffectively with a rebellion in Otrar that his father had to rescue him.[29] In his early thirties he scored a major victory against Uzbek tribes near Lake Issyk-Kul in present-day Kyrgyzstan but then went on to suffer a stinging defeat at the hands of the same Uzbeks in Khwarazm two years later. This marked the Uzbeks' debut as a rising power and the beginning of the end for Ulughbeg and the direct descendants of Timur. Ulughbeg soldiered on doggedly, but his heart was not in being a field commander or even a sultan. He must have envied his younger brother, Baisunghur, whom his father never saddled with major responsibilities, which left him free to build his elegant madrasas in Herat, gather his ancient books, assemble his artists, and drink.

Ulughbeg grew up in the shadow of his grandfather, Timur, shifting from place to place with his father, Shahrukh, as the military campaign changed directions.[30] Born in the Ilkhan Mongols' second capital, Sulimaniye, he traveled throughout what is now Iranian Azerbaijan. In the course of these peregrinations, young Ulughbeg was taken to visit the site of Nasir al-Din Tusi's old observatory at Maragheh. Destroyed and abandoned for some two centuries, it was a forlorn yet, for the young boy, inspiring ruin. After this it was a foregone conclusion that Ulughbeg would study mathematics and astronomy, which he pursued in Herat under leading Khurasan scientists and also under an Anatolian immigrant, Kazi Zade Rumi, who himself had been drawn there by the fame of the Khurasan astronomers.

Ulughbeg was arguably the most serious educator to appear in Central Asia since the tenth century, when the ruler Abdallah attempted to promote popular education across the breadth of Khurasan from his capital at Nishapur. At one point Ulughbeg was providing financial aid to ten thousand students at twelve institutions, with fully five hundred of them specialized in mathematics.[31] His greatest efforts were reserved for the capital. At the noble Registan Square at the center of modern

Samarkand stands the great madrasa that Ulughbeg founded in 1417. With an uncompromising bluntness that must have startled even his most ardent supporters, Ulughbeg inscribed into the door the words of Muhammad, as recorded in the collection of Hadiths by the ninth-century scholar Tirmidhi,[32] that "The Search for Knowledge is the Duty of Every Muslim."[33] It is difficult, if not impossible, to square this exhortation with the stifling religious orthodoxy that his father, Sultan Shahrukh, was seeking to instill at his own new madrasas at Herat and which prevailed wherever Timur's dynasty ruled.[34]

The curriculum at Ulughbeg's madrasa focused on the sciences, especially mathematics and astronomy. Since we know Ulughbeg took a keen interest also in history and literature, as well as music, it may be assumed that these subjects, too, were in the curriculum, further diluting—but not replacing—the usual diet of religious studies. All these areas of knowledge became once more accessible to the educated elite of Samarkand, thanks to the library that Ulughbeg established at the same time, the first such collection in the city since the Mongol depredations.[35]

To head the teaching staff, Ulughbeg brought his old mentor, Kazi Zade Rumi. Also on the faculty was a remarkable immigrant to Samarkand from an Iranian town near Isfahan, Jamshid al-Kashi (1380–1429). And among the earliest students was fifteen-year-old Ali Qushji (1402–1474), the son of Ulughbeg's falconer and already renowned locally for his mathematical virtuosity. Both of the latter two were to make significant contributions to astronomy and mathematics.

Kashi left us solid proof that the spirit of inquiry was alive and well at Ulughbeg's madrasa. A few months after arriving there, he wrote to his father a pair of lengthy letters, which, remarkably, have been preserved.[36] He reported on the open and fiercely competitive intellectual environment, saying that one arrogant new staff scientist who boasted of the knowledge he had brought with him from Egypt was so thoroughly humiliated in these battles of wit that he quit and fled into the safer world of pharmacy.[37]

As to Ulughbeg, Kashi made clear that it was the ruler himself who was the intellectual driver behind the entire enterprise. Ulughbeg showed up almost daily in the classroom, teaching courses, posing

Figure 14.4. Ulughbeg's madrasa at Samarkand welcomed visitors with the Hadith "To strive for knowledge is the duty of every Muslim." ▪ Photo by Jennifer Hattam.

questions, challenging students and faculty alike, and taking delight in finding answers, even when they differed from his own conclusions.

Kashi's published work, most of it dating from his twelve years at the madrasa, attests to the intellectual level of the place. In mathematics he continued the long process initiated by Khwarazmi to implant the decimal system, providing a systematic method of calculating with decimal fractions. In *A Treatise on the Chord and Sine* Kashi advanced a new method for solving cubic equations and computed the sine of one degree to an accuracy that was not surpassed for two centuries. In his *Key to Arithmetic* he worked out the value of pi to a degree of precision twice greater than the Greeks or Chinese had done and greater than any European was to do for another 150 years.

Kashi was also to devise an utterly unique mechanical instrument—a kind of analog calculator—for doing linear interpolation, a purely mathematical operation of great use in all branches of mathematical astronomy. He also invented a "planetary equatorium" for identifying the position within the celestial sphere of any planet at any time.[38] This task had challenged ancient tinkerers and various medieval inventors as well. A half century before Jamshid al-Kashi, even the English poet Chaucer had taken an interest in building one. But of all these, Kashi's was far the most advanced, for it was the only one that included a method for calculating planetary latitudes and longitudes.[39]

Given Ulughbeg's personal interest and the financial and human resources at his command, it was inevitable that he would construct a copy of the old Maragheh observatory at Samarkand. To this end he began construction on a low, round hill a few miles northeast of the old city of Afrasiab. To keep a close watch on the process, he built himself a villa near the construction site. The result was a three-story structure of almost the same dimensions as Tusi's observatory a century and a half earlier. The facility was well equipped, which is scarcely surprising considering Kashi's pathbreaking inventions and the fact that he himself had written a treatise on astronomical observational instruments. Kashi ably supervised the design and construction of the observatory's scientific equipment.

At the heart of Ulughbeg's observatory was a large and carefully calibrated sextant of brass, with a radius of more than a hundred feet. The section of this immense instrument that was embedded in a semicircular channel in the rock of the hill is still preserved, complete with its precise calibration marks, and can be seen at the site. As finally constructed, this huge instrument differed significantly from what the ruler had originally planned. When Kashi pointed out to Ulughbeg that he must have misremembered important details of the system used at Maragheh, Ulughbeg, acting as scientist rather than imperious sultan, accepted the criticism and had the faulty elements reconstructed.

Like Tusi's observatory, Ulughbeg's was a monument to the proposition that greater accuracy can be gained through the use of ever larger instruments.[40] The alternate view, increasingly championed by Europeans as they miniaturized clocks and other instruments to use them on ships, was that accurate results arise from precisely constructed instruments, irrespective of their size. Nonetheless, Ulughbeg's great sextant was to yield remarkably precise results.

Ulughbeg's greatest passion as an astronomer was to determine the exact locations of all the major stars in the sky. This challenge had engaged Tusi at Maragegh, not to mention every Central Asian astronomer

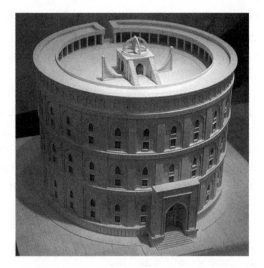

Figure 14.5. Reconstructed model of Ulughbeg's observatory at Samarkand, where a staff of gifted professionals achieved remarkable precision in their star charts. ▪ Courtesy of OrexCA, Tashkent, Uzbekistan, www.OrexCA.com. Photo by Inna Tsay.

back to Abu Mahmud Khujandi at Rayy in the tenth century. Thanks to Kashi and the other brilliant colleagues he assembled, Ulughbeg succeeded wonderfully at this task. His compendium of data, called the *Zij* or collection of astronomical tables, was clearly a collaborative work involving especially Kashi. In nearly three hundred pages of charts and quantitative data, it fixed with precise figures the locations of 992 stars. The star catalog included in the *Zij* was more comprehensive than any previous catalog, and far more precise. Indeed, it is considered the most authoritative guide to the heavens between Ptolemy's in the second century and Tycho Brahe's (1546–1601) at the end of the sixteenth century. An admiring Oxford scholar, the Arabist Thomas Hyde, translated and issued Ulughbeg's star catalog work in 1665.[41]

Ulughbeg's intentions in this project had been thoroughly traditional. Far from challenging Ptolemy's geocentric view of the universe, he thought he was confirming the Ptolemaic paradigm and extending it into new territory. But Ulughbeg, by the very nature of his investigation, changed the course of astronomy. Prior to him, even the best astronomers combined observation with data drawn from what were assumed to be authoritative sources, mainly Ptolemy of Alexandria. Biruni, for example, had followed this path in his *Masudi Codex*.[42] This led to important corrections and refinements of the classical data but not to their general overhaul. Under Ulughbeg, however, the main focus shifted radically toward producing new observations.

Of course, the development of precise data on nearly a thousand stars demanded tremendous resources, not only to build the astronomical equipment but also to pay for staff time. After Tusi at Maragheh, who was supported by the patronage of the Seljuk Sultan, only Ulughbeg commanded such funds, thanks to his exalted official rank. This is how his fixation with numbers came fully to supplant the philosophizing that had accompanied so much astronomy before his time. In this respect Ulughbeg moved in the same direction as Biruni and Tusi, advancing a process that freed astronomy from the last vestiges of philosophy and focusing it fully on observation and hard data.

That those investigating the universe were breaking their last ties with philosophy upset many members of the ulama of Samarkand, the mullahs and religious scholars who were the arbiters of truth. And they erupted with fury when Ulughbeg celebrated his son's circumcision with a public wine party, charging that he had "destroyed the faith of Muhummad and introduced the customs of infidels."[43] Members of the ulama bided their time. Only after Ulughbeg's death in 1449 did they move to purge the madrasa of its heretical faculty and curriculum and attack the observatory. When the site was excavated between 1909 and 1967, archaeologists found clear evidence that it had been torn down to its very foundations and most of the building material hauled away.

The story did not end there. Shahrukh died in 1447. By the time Ulughbeg finally left Samarkand to serve briefly as sultan, Ali Qushji was forty-three years old, with much of life still before him. Two years later Ulughbeg was beheaded by his own son. Seeing the rising tide of opposition to all that Ulughbeg had stood for, Qushji headed first to Herat. Finding the capital no more receptive to scientific learning than Samarkand after Ulughbeg's demise, he traveled to Kerman and Tabriz before settling finally in Constantinople, which the Ottoman Turks had only recently conquered from the Byzantines. Over the centuries prior to Qushji's appearance, the Ottomans had largely ignored mathematics and astronomy.[44] But during the year between his arrival in the Ottoman capital and his death in 1474, Qushji set up a madrasa next to the Hagia

Figure 14.6. Tomb of Ulughbeg at Ghazni in Afghanistan. This and other Timurid structures deeply influenced architects in Mughal India and Saffavid Iran, even serving as models for the Taj Mahal. ▪ Josephine Powell Photograph, courtesy of Special Collections, Fine Arts Library, Harvard University.

Sofia patterned after Ulughbeg's in Samarkand, with its intensive focus on mathematics. He also penned a tract on astronomy that combined conservatism and bold innovation in surprising ways.

At one level, *Concerning the Supposed Dependence of Astronomy on Philosophy* was a concession to those orthodox Muslims who continued to battle the influence of Aristotle and his Muslim disciples like Kindi, Farabi and the other faylasufs. But Qushji's declaration of independence cut two ways, for it also moved astronomy beyond the interfering reach of Muslim philosophy.[45] Of course, Central Asian astronomy had been moving in this direction for centuries, culminating in Ulughbeg's decision to ignore the natural philosophers and focus solely on the data before him. But now the divorce from natural philosophy was systematized and finalized. Thus emancipated from all constraints of speculative philosophy and theology, Qushji proceeded to repeat Biruni's old argument that it was as plausible to assume that Earth moves as it is to assume that it stands still. Indeed, he noted, empirical evidence supported such a view, even though it may conflict with Ptolemy, Aristotle, and their followers, whether Muslim or Christian. Following in Omar Khayyam's footsteps, he also argued that there is no empirical reason to assume that all motion in the universe occurs in geometrically tidy circles, as Aristotle had assumed. Both elements of this argument trace directly to what Qushji had learned during his years with Ulughbeg in Samarkand.

NAVAI: PATRON AND POET

All transitions of leadership are difficult, but those of nomadic or formerly nomadic clans particularly so. Even before the death of the earnest Shahrukh, a bevy of sons, grandsons, and nephews began maneuvering for the top position.[46] After Ulughbeg was murdered by his son, an ambitious offspring of Baisunghur killed Ulughbeg's son and seized power. All this turmoil invited the so-called Black Sheep (Kara Koyunlu) Turkmen tribes of western Iran to assert themselves, which they did by freeing all Iran from the rule of Timur's dynasty. In the end Baisanghur's son controlled only Khurasan, having been forced to hand over the rest of Central Asia to Abu Said, a nephew of Ulughbeg.

Exhausted by this bout of internecine warfare, the entire region once more subsided into peace. Abu Said (1424–1469) proved serious and competent, winning popular support by attending to the restoration of irrigation and agriculture. But in the cultural sphere he spent all his energies on building up the Naqshbandiyya order of Sufis and drawing it ever closer to the state. While this doubtless contributed to social stability, it did nothing to support learning and culture. The leading Samarkand Naqshbandi, one Khoja Ahrar, so opposed scientific learning that he refused all medical treatment until he lay on his deathbed.[47]

Following another messy transition in Herat, the top position fell to a fourth generation descendant of Timur, Husayn Bayqara (1438–1506), whom inertia preserved in power from 1470 down to 1506. Turmoil continued, but its worst effects were kept at a distance from the capital. The years of Husayn's rule marked the greatest period of cultural effervescence under Timur's dynasty, and also its swan song.[48]

The foundations for this last burst of creative life in Herat had been laid earlier in the century, when Ulughbeg's brother had founded his library and assembled illuminators of books and other painters. General prosperity assured that craft and industrial production in many areas would continue to flourish. The arrival in Herat and Samarkand of the best artisans and craftspeople from the Middle East, Central Asia, and India assured that their products met the highest world standards.[49] Most important, the generous distribution of appanage lands to grandees and favorites created a number of patrons with seemingly unlimited wealth.[50] True to the ancient traditions of Khurasan, much of this money was spent on culture and intellectual life.

No one more fully epitomized this final burst of creativity in Central Asia than Nizam al-Din Alisher Harawi (1441–1501), known as "Navai," who singlehandedly elevated his native Turkic language, Chaghatay, to the same high level as Persian, the traditional language of poets, and thereby opened the way for a burst of creative writing in Turkish that rolled outward from Herat as far as the Bosphorus and India. Navai, while not a descendant of Timur, was an insider's insider. Over several generations his forebears had accumulated influence and money, to the point that Alisher would have been someone to reckon with even without his particular good fortune.[51] Because Alisher's father had served as a provincial governor under the Timur dynasty, the sultan himself adopted the young man when he was orphaned and promoted his education, alongside young Husayn Bayqara, the future ruler. Alisher then continued his studies in Mashad, Herat, and Samarkand while the future sultan spent a decade living first as a free spirit and then as governor of Merv.[52] But the friendship continued, and when Husayn came to the throne he named Navai keeper of the seal and then amir and governor of Herat in the sultan's absence. By such steps Husayn Bayqara elevated his friend to the status of the second or third most powerful person in the realm.

Navai used his personal resources to build and endow traditional madrasas in Herat and elsewhere[53] and to patronize music, art, and especially painting. Sustained by Navai's knowledgeable support, the quality and popularity of figural painting soared. While many of Navai's artists achieved unprecedented levels of refinement in their work, pride of place goes to the legendary Kamoliddin Bihzad (1450–1537), whose book illustrations, separate scenes, and portraits of high officials redefined the artistic ideal throughout the Muslim world (see plate 28). He doubtless did murals and other larger works as well, but they have not survived. Bihzad and his contemporaries in Samarkand and also in the Persian city of Shiraz achieved greater naturalism than any had previous artists in the Muslim world, and they were much more attuned to the human personalities they depicted than was anyone before them. At the same time, they organized their busy pictures in highly rhythmic ways, lavishing rich colors on them. They did not, however, address the challenge of perspective, perhaps because of their preoccupation with ornament.[54]

Navai was liberal in his use of the public treasury, as well as his own funds, becoming the greatest patron of architecture Herat had ever known. In the capital alone he built a congregational mosque, a madrasa, a meeting hall for a Sufi brotherhood, a hospital, and public baths.[55] The future conqueror of India and founder of the Mughal dynasty, Babur (1483–1530), spent forty happy days as a tourist there and was amazed by the splendid facilities and wealth of its civic and religious institutions, most of which traced to Navai's patronage.[56] And beyond the capital, Navai built many caravanserais and mosques, as well as a monumental tomb for the poet Attar at Nishapur, which still stands. These and scores of other projects nearly bankrupted the sultan, who exiled Navai for a year to a provincial governorship on the Caspian but then brought him back into his inner circle as vizier.

By the 1480s the clash of rival factions at court and the obstreperous behavior of minor descendants of Timur in the provinces were steadily weakening the state. Sultan Husayn Bayqara had become a drunkard and by the end of the century was senile besides. But by then so was his vizier, and when the two met for the last time in 1501, both were on stretchers, barely able to communicate.

Throughout his life, Navai wrote poetry that gained him as much renown as his civic works. This was a period in which scores of people in every walk of life were amusing themselves by turning out quatrains or qasidas. A pedant from Samarkand put together a memoir touching on no fewer than 350 contemporary rhymers.[57] "Extremely affected" is but one of the kinder epithets that modern critics use to describe most of their work.[58] But Navai stood apart, and for a very good reason: he had before him the example of a very great poet from the immediately preceding generation, Nuradin Jami (1414–1492).

Like Navai, Jami was a public figure, thanks to his leadership role in the Naqshbandiyya Sufi order in the capital. He had been born in a town near Herat but graduated from Ulughbeg's madrasa in Samarkand and stayed in that city to be initiated into the Naqshbandiyya order (*tariqat*) there. When he returned to Herat, he worked to expand the order and met Navai when the latter appeared among those aspiring to join. He initiated Navai into the Naqshbandiyya, but Navai continued to engage in high politics, involve himself in all the arts, especially music and calligraphy, and enter into homosexual relationships, usually with young male slaves.[59] Jami meanwhile wrote on subjects as diverse as history and irrigation technology, in addition to voluminous works on spiritual topics and Sufism. He returned again and again to the power of love and divine mercy, the role of the miraculous in human affairs, sainthood, and God's ubiquitous presence in the world.[60] He also penned a vast compendium entitled *The Breaths of Fellowship*, a collection of biographies of more than nine hundred great Sufi teachers, most of them from Central Asia, written in a clear and readable prose that was a model in its era.

Jami was a master poet who worked successfully in all genres, from epics to mystical allegories. Dense with symbolism, his works are utterly accessible at one level yet, like a Sufi exercise, challenge and stretch the reader's understanding at another. Particularly appealing are his sets of verses depicting the stages of life. Navai knew well Jami's views on this subject, and he understood the path to spiritual enlightenment they embodied. He applied these ideas to his own poetry when he organized his life's work into four sequential *divans*, or collections, and named each after a season of life, from childhood to old age.

Neither Jami nor Navai hesitated to draw inspiration from other poets and retell their stories in his own language. When he was barely seven, Navai encountered Attar's long narrative poem, *The Conference of Birds*. He loved the work and would later produce his own imitation of it. Navai's great distinction, however, was to render it not in Persian, which he spoke and wrote fluently, but in his native Chaghatay Turkish.

No aspect of Navai's artistic career met with more favorable comment in his own day and had a more lasting impact on the future than his championing of the Chaghatay Turkish language. Descended from Uyghur Turks, Navai grew up speaking the one Turkic language that had successfully adapted to western Eurasia's literary and administrative culture. But no major writer had used Chaghatay for poetry or literature in a Western mode. Navai did precisely this, effectively and consistently throughout his career. The fact that he wrote in a variety of poetic modes and genres, and that the language of nearly all his works was accessible to audiences beyond the palace grounds, assured his central place in the history of Turkic letters and his status as the national poet of the Uzbeks.

During the last year of his life, Navai penned a passionate defense of his native Chaghatay language, which he juxtaposed to the Persian in which much court life was conducted.[61] Continual references to his own experience throughout the text gave it the character of an apologia. At some points Navai resorted to blatantly dubious arguments. For example, the fact that Turkic languages have many words for saddles and ducks while Persian has only one for each does not diminish Persian's value as a medium for poetry. Nor does the large number of loan words in Persian, which may in fact strengthen its standing as a flexible and expressive tongue.

But the sheer passion and tenacity with which Navai defended the Turkic languages added force to his argument. Never since Mahmud of Kashgar had anyone made this case so directly, and no Turkic poet since Yusuf of Balasagun had taken up cudgels in behalf of his native tongue. That Navai, unlike Kashgari or Balasaguni, was an important member of the courtly circles he sought to convince added weight to his argument. It is no wonder that Navai was the founder of a long line of Turkish poets extending to our own day, and that both his name and his poetry are still revered wherever Turkic languages are spoken.

End of the Aesthetic Century

Timur's dynasty in Central Asia ended in 1506, overwhelmed by the same kind of nomadic cavalry as Timur himself had once led, only this one headed by horsemen from the Turkic Uzbek tribes. Samarkand had fallen to the Uzbek leader Muhammad Shibani Khan Sultan six years before. It was no consolation that this Uzbek conqueror was himself a descendant of Chinggis Khan, via the great ruler's eldest son, Jochi. Now

Herat, where the aged sultan, Husayn Bayqara, had sunk into a life of heavy drinking, cockfights, and ram fights, offered no resistance. Husayn's two sons both fled before the conquering Uzbeks.

Timur's dynasty, founded amid bloodshed and savagery, ended in a glow of beauty, rich in all that delights the eye and ear. The refined taste of its leading courts, manifest in its music, painting, and poetry, and also in the work of its weavers, goldsmiths, ceramicists, and architects, set a high standard throughout the Muslim East. A synthesis of Persian and Turkic cultures, it achieved the merging of city and steppe, agrarianism and nomadism, trade and conquest that had been in the making in Central Asia for half a millennium. It was a radiant achievement, glorious in its way, but one-sided.

Except for the projects of the great Ulughbeg, science and learning subsided into dull routine or vanished altogether. The education offered within the walls of the well-funded court madrasas was narrow and grounded not on reason but on authority and publicly enforced orthodoxy. Outside the madrasas this was a culture of the senses, not the intellect; of the subjective, not the objective; of feeling more than of thinking. Its greatest achievements lay in the realm of beauty, not truth.

Though they were in contact with India, China, and, to a much lesser extent, Europe, Timur's heirs found in them no challenge either to the mind or to the security of the state. The late fifteenth century was a time of epochal change elsewhere, with Europeans mastering the sea routes to the East and to new worlds in the West. No less dramatic was their rediscovery of the achievements of the pagan Roman and Greek civilizations. This process, which frontally challenged their settled ideas, in many respects repeated what the Arabs and Central Asians had undergone from the ninth to the twelfth centuries.

Could Central Asians have generated their own Renaissance during these years? Perhaps so, by rediscovering and evaluating with fresh eyes the achievements of their Zoroastrian, Hellenistic, Buddhist, and Christian forebears, or by reexamining some of the many lost threads of their Islamic heritage. But this did not happen and could not have happened under the prevailing orthodoxy and traditionalism. How else can we explain the fact that the Uyghur Turkic cities of East Turkestan (Xinjiang) were already printing religious and other texts with both woodblocks and movable type by the early fourteenth century but their immediate neighbors to the west, with whom they had been in direct and constant contact for nearly two millenniums through the caravan trade, did not take up printing until long after the last heir of Timur had died?

This is all the more striking when one recalls that printing reached Europe during Timur's lifetime and that Johanes Gutenberg, working in Strasbourg and Mainz, was using movable type to print by 1439, exactly when Ulughbeg's research was at its peak. Gutenberg was a weaver and goldsmith, two fields in which the artisans of Herat and Samarkand stood at the highest world level. And yet it was Gutenberg, not a craftsman in Timur's realm, who perfected the most efficient method the world had ever seen for reproducing the written word. Had the conviction that the great questions of life could be addressed only through faith so dampened the quest for new knowledge as to smother all interest in, or demand for, this clever innovation? For all the serene beauty and spiritual depth of the civilization of Herat and Samarkand, it failed to produce any new Khwarazmis, Farabis, Ibn Sinas, Birunis, Khayyams, or Tusis.

TIMUR'S STEPCHILDREN: MUGHALS, SAFAVIDS, AND OTTOMANS

In spite of this, the legacy of Timur's dynasty proved extraordinarily rich. Far from ending in 1506, many aspects of its culture continued on, finding new homes in the three great Muslim empires that reached their zenith over the next two centuries: Mughal India, Safavid Iran, and Ottoman Turkey. All three became great regional powers, ruling diverse peoples over a vast area. All three were phenomenally successful down to at least the 1630s for the Ottomans and Safavids and 1700 for the Mughals. But all three entered thereafter into a prolonged decline.

Most important, all three were ruled by Turkic dynasties that felt the powerful influence of the culture of Herat, Samarkand, and Tabriz, the third of Timur's capitals. The founder of the Mughal state, Babur, was himself a direct descendant of Timur and would have ruled Central Asia had he not been driven out of the region by the ascendant Uzbeks. A century and a half later his heirs were still dreaming of reuniting their empire with Central Asia and returning to their ancestral homeland.[62]

The Safavids were Turkicized Iranians, probably of Kurdish origin, whose power base was among the Turkic tribesmen in northwestern Iran, Anatolia, and parts of Syria. This Turkish-speaking dynasty built its capital at Tabriz in Iran, where Timur's culture had held sway, and then at Isfahan. Aside from imposing Shiite Islam on the entire population of Iran, the Safavids were quick to absorb the culture of the great Timurid centers of western Iran. And the Ottomans, descended from the Osmanli Oghuz tribes, were defeated by Timur in 1402 but came under the strong influence of his heirs thereafter. Together these three ruled most of the cradle of civilization from the Atlantic coast of Africa to Madras in southern India.

All these great empires were essentially military enterprises; when one historian called the Mughal empire "a war state," he could have been speaking of all three.[63] Each viewed kingship in military terms, and each built its power on the solid rock of Turkic cavalry. But all three were quick to master the technologies needed to cast cannons and muskets, thanks to which they have sometimes been called "gunpowder states."[64] All derived great riches from taxing agriculture, but all encouraged their populations to engage in trade, from which the states drew tax revenues. And all three took over cities that were already great centers of culture—Delhi, Isfahan, and Constantinople—and transformed them in their own image.

The Mughals, Safavids, and Ottomans came increasingly to reflect the sharply different cultures of the people they ruled. Yet important commonalities should also be noted, not only because these trace to the legacy left by Timur and his heirs but, more important, because they have today come to define the conception of Islamic civilization held by the general public, not only in the world at large but by most Muslims themselves. Countless museum exhibitions and books unwittingly leave the inaccurate impression that the main cultural features of these three empires define Islamic civilization as a whole.

In no area are the commonalities among these three empires more striking than in architecture. The great New Delhi tomb of Humayum, son of Babur, the founder of the Mughal dynasty, was the launching point for all Indian monumental architecture of the next centuries. Built in 1562–1572, it was the work of two Central Asian architects from Herat. The Taj Mahal in Agra, the epitome of Mughal architecture, also derived directly from such Central Asian prototypes as Timur's tomb (the Gur-e-Amir) in Samarkand and the tomb of Ulughbeg and Abd al-Razzaq in Ghazni, Afghanistan.[65] Turning to the Ottomans, the Cinili Kosk or "Chinese Kiosk" and other structures built shortly after the Ottoman conquest of Constantinople reveal what one German expert on the era called a "startling resemblance" to Central Asian prototypes, even though the first hints of Western influences also appeared then.[66]

Beyond this, the serene gardens of Isfahan, Kabul, Agra, Delhi, and Lahore were all based on the gardens of Herat and Samarkand.[67] In Safavid Persia the architects of Tabriz and other centers had long since mastered the idiom of Central Asian architecture beginning with the era of the Seljuk sultan Sanjar, and also their own variant that arose under the Ilkhan Mongols. Thus it is fair to see the sublime architecture of Isfahan of the era of Shah Abbas (1571–1629) as a fulfillment of Central Asian models, some of which Iranians had helped create in the first place.[68] And the starting point for the great Ottoman architect and innovator Mimar Sinan was the Ottoman mosques and mausoleums that had been patterned after Timurid propotypes in Herat and Tabriz.

The aesthetic impulse of Timur's dynasty found its highest expression in book illustrations and other forms of miniature painting, which spread quickly to all three of the gunpowder empires. The Ottoman court avidly collected the work of Bihzad and other Herat artists and encouraged local painters to emulate them, as did the leaders in Isfahan. The Mughal sultan Akbar went so far as to establish a *Kitabkhana* in imitation of Baisunghur's in Herat.[69] Persian painters of the Safavid period developed a new sensitivity to personality and began to individualize their subjects to an extent without precedent in Muslim art.

The literary cultures of all three empires also were greatly indebted to Timur's heirs in Central Asia.[70] Babur was directly inspired by Navai to use Chaghatay for his memoirs, the first memoir written in any Turkic tongue. The Ottoman sultans collected and reissued many of the poems of Navai and Jami, while Jami also found readers and imitators throughout India. The works of earlier poets from Central Asia, notably Anvari, were also widely disseminated in the three empires in lavishly produced copies.[71] This was easier because all three dynasties, following Timur's model, conducted their court business in Persian, which greatly extended the geographic range of the Persianate culture of Khurasan and the former Ilkhanate.

In none of the three, however, did a strong and contemporary school of either philosophy or science develop. True, Persian philosophers developed what has recently been dubbed the "School of Isfahan." This produced many highly competent texts, which only recently have come to the attention of scholars. But both the questions they asked and the answers they reached derived less from the contemporary world than from Central Asia's long-passed Age of Enlightenment.[72] Absent are a systematic presentation of the new ideas that had emerged in the West during the sixteenth and seventeenth centuries and a thoroughgoing exploration of their implications for received wisdom. In other words, the old style of philosophizing did not end, but it stagnated.

We have noted how all three of these empires eventually mastered the technologies needed to cast bronze and build modern weaponry. Their openness to new technology extended to many other areas but did not go deep. Typically, the Ottoman sultan Bayazit II solicited from Leonardo da Vinci a technically bold design for a bridge over the Golden Horn in Constantinople but in the end backed away from building it. Both Ottomans and Mughals proved adept sailors, and their sea captains provided detailed information to native-born cartographers. Yet the resulting maps were technically far behind those of the West. A partial exception was the Ottoman admiral and cartographer Hacı Ahmed Muhiddin Piri (1470–1554), who drew on his own experience and his large collection of European maps to prepare his representations of the coasts of Central and South America (1513), North America (1528), and a world map.[73] Safavid maps also fell short, not only of those of Western mapmakers but of maps by Arab and Central Asian cartographers from a half millennium earlier.[74] In India the emperor Akbar toyed with gears, air conditioning, and handguns, but nothing came of this curiosity about technology.[75] And neither lens grinding nor watchmaking advanced in any of the three kingdoms, which came to rely instead on foreigners resident in their capitals for these miniaturization technologies.

The parallels between these three empires and the culture of Timur's Central Asia are particularly striking in mathematics and the natural sciences. Ottomans were quick to claim Ulughbeg's teacher Kazi Zade Rumi as their first scientist, and then to hail Bukhara-born Qushji as their first astronomer, even though he spent barely a year in Constantinople. Astronomy did not really get off the ground in Constantinople until 1576, when Taqi al-Din, a Damascus-born Turk and a gifted mathematician and astronomer, persuaded the sultan to fund an observatory patterned after Ulughbeg's (see plate 29). In the tradition of Bukhari and Central Asian Sufism, with their emphasis on the lines of descent and transmission rather than scientifically verifiable facts, Taqi al-Din sought out and studied with Qushji's grandson. At his new observatory Taqi al-Din did useful research for four years, his goal being to carry on Ulughbeg's mapping project.[76] But the sultan, egged on by a suspicious vizier, had the entire structure torn down. Astronomy remained a dead letter in Constantinople for centuries thereafter, proving that this had been not simply a clash of personalities. No wonder that the first Turkish translation of any European book that spoke of Nikolaus Copernicus's heliocentric view of the solar system was not done until 1660, 210 years after the Polish astronomer's death.

Mughal rulers who wanted to update their calendar also followed Ulughbeg in constructing an observatory—five of them, in fact. But this took place three centuries after Ulughbeg and left no permanent legacy except for some very exotic structures, which can still be seen in Delhi and Jaipur. In Iran, as at the Ottoman and Mughal courts, medicine continued to thrive for the obvious reason that it was needed to maintain the rulers' health. And while there were laudable advances in surgery and ophthalmology, the Persian medical profession was greatly depleted by wholesale emigration.[77]

The muted progress of mathematics and the natural sciences under the three Turkic empires demands an explanation. After all, even though they long relied on their cavalry, they had mastered the gunpowder revolution by the sixteenth century. Only the Safavids lagged in this, the price of their tardiness being defeat by the Ottomans in 1514. Thereafter the Persian court at Isfahan rushed to catch up. Why, then, did the related fields of ballistics and trigonometry not advance more quickly than they did?, Success on the battlefield may have played a major role. All three maintained military superiority over their major rivals down to the mid-seventeenth century. Self-satisfaction, which was amply manifest in the grandiose court ceremonies in Delhi, Isfahan, and Constantinople, may have diminished their readiness to master new fields of knowledge, and even their interest in learning foreign languages, especially those of seemingly insignificant peoples in far-off Europe.

Beyond this, the world of knowledge was powerfully constrained in all three empires by religious orthodoxy. Both the Mughals and Ottomans were staunch Sunnis, forging a close alliance between the Sharia law and the military, while the Safavids were equally staunch Shiites. The Shiite state in Isfahan was a theocracy in all but name,[78] while the Ottoman seizure of what remained of the caliphate and Sultan Selim's 1522 decision to move it to Constantinople gave a theocratic cast to that empire as well.

All three empires ruled over large numbers of people of other religions. The Ottomans practiced a high degree of tolerance, but the Shiite Safavids faced the reality of a majority Sunni population, even in Persia.[79] In the Mughal lands of India, the existence of a Hindu majority led one extraordinary ruler, Akbar (1542–1605), to devise his own synthesis of Islam, Hinduism, and Christianity. Idiosyncratic and independent, Akbar hoped that his new faith might supplant existing religions and end cultural conflict forever.[80] In planning it, he returned to the great questions of the ideal society that Farabi and Ibn Sina had addressed half a millennium earlier. He saw himself as the head of a new kind of Sufi order, and hence a new kind of theocrat, albeit a benign one.[81] A successor, the stern and intolerant Aurangzeb (1618–1707), promptly returned the Mughal state to the narrow Sharia-based military system that had prevailed after Babur, underscoring this point by opening in Lahore the largest imperial mosque ever constructed. Thus religious diversity in Persia and India prompted leaders to assert a common orthodoxy from above, while the relative tolerance that the Ottomans permitted did not spark significant intellectual interchange.

In all three empires, madrasas enshrined and perpetuated religious orthodoxy. All three had acquired their first madrasas during the era of Nizam al-Mulk, who had used his power as the caliph's vizier to open educational institutions that would stamp out heterodoxy and impose a strict religious orthodoxy. In the absence of universities, these institutions in all three empires retained their original character as protectors and perpetuators of the reigning orthodoxy in the face of perceived deviations, whether from the Muslim opposition or from modern science and philosophy. True, the Mughal emperor Akbar made mathematics a required subject at madrasas in his realm, but this innovation, like most of Akbar's initiatives in the sphere of culture, did not outlive him.[82]

Stated differently, none of the three post-Timur empires challenged the limits on reason that Ghazali had propounded in the twelfth century and which had become pillars of Muslim orthodoxy ever since. Having accepted the possibility of direct communion between the individual believer and God, Ghazali did not so much reject reason as marginalize it, accepting it as a tool for solving practical problems but rejecting it as means of addressing existential issues. The religious scholars and states in all three of the great Turkic empires were united in their acceptance of Ghazali's relegation of mathematics, science, and logic to a secondary role that had no bearing on the more urgent issue of humankind's relationship to God.

THE UNPRINTED WORD

All of these characteristics were preserved over time by one more factor that significantly retarded intellectual life in the three great Turkic empires of the sixteenth and seventeenth centuries, namely, their extreme reluctance to embrace movable-type printing. In fact, the fate of printing in all three is remarkably similar.

Thanks to a printing press with Arabic letters that the Vatican sent to Isfahan in the 1620s, the Safavids knew of printing well before the middle of the seventeenth century, when Armenian Christians issued the first printed book in Persia, the *Psalms*. But the Persians themselves did not take up the "new" invention, which by then was already two hundred years old. It took another two generations before the first printed book in Persian was issued within the Safavid state.[83]

The Ottomans meanwhile allowed Sephardic Jews to print a volume of Jewish laws in 1493, but only in Hebrew. The first movable type book published by a Muslim in Constantinople—or anywhere else, for that matter—appeared in 1729. The publisher was a Transylvanian polymath and sometime Ottoman diplomat named Ibrahim Muteferrika, who used his position in the bureaucracy to gain the sultan's permission to issue several volumes.[84] But thereafter printing in the Ottoman lands lapsed for several more generations.

It was the Jesuits who introduced movable type printing to India, with a volume on the Christian life issued in a Latin edition at Goa in 1556. Meanwhile, within the Mughal empire the normally inquisitive emperor Akbar, around 1575, was shown type fonts that could print books in Persian, but he evinced absolutely no interest in them. Armenians, Danes, and English were all printing books in India long before Indians themselves began doing so. and so Indians had to wait until the end of the eighteenth century before books in Persian were issued there, and even longer for books in local languages.[85]

The refusal of the Muslim leaders of these three empires to allow printing, let alone to encourage it, had enormous consequences for their societies. Their self-satisfied disinterest in foreign innovations, which turned into stubborn opposition, severely limited the circles of men and women who could engage in the quest for, and acquisition of, new information and knowledge. This greatly diminished public discourse on new topics as they arose, which in turn hampered civic development as a whole.

HEIRS, AND A LOST INHERITANCE

These, then, are some of the common features of the three great Turkic-led empires that flourished after the fall of Timur's dynasty. In some ways all three recall the Timurid state that preceded them and from which all drew inspiration. The Ottomans, Mughals, and Safavids were all founded as traditional Turko-Persian monarchies. All were built on their armies, and prone to centrifugal forces, and all were fundamentally urban, valuing internal trade and international commerce. As pious Muslims, their leaders all saw it as government's task to defend the faith. And all valued human creativity as expressed in music, art, poetry, and the crafts and were generous in their support of it.

On one further issue the three gunpowder empires directly recall the Timurids, and on that same issue all four stand in sharp contrast to the earlier societies of the Age of Enlightenment: very few of the gifted people in the Timurid elite or the three successor empires gained distinction through achievements in mathematics, the natural sciences, social science, or philosophy. And even if they were open to new technologies, they were lax in generating them on their own. In all these respects thinkers in the three great early modern empires, and the Timurids before them, lagged far behind their predecessors in Bukhara, Nishapur, Merv, Gurganj, Tus, Ghazni, Samarkand, or Balasagun. By comparison with Central Asia's Age of Enlightenment, these three vast early modern empires appear intellectually impoverished and backward.

During the eighth through the twelfth centuries, Central Asian thinkers had shown that civilized people could pursue truth and beauty simultaneously, without having to choose between them, and that both were compatible with faith. Beginning with Ghazali, though, more and more people in Central Asia became convinced that a choice was necessary: either the unfettered intellect or faith as filtered through a specific body of tradition. They chose the latter or had it chosen for them. This decision had been taken long before the appearance of Timur and the opulent civilization that followed him. The contribution of Herat and Samarkand was to combine faith and a rich aesthetic culture, but at the price of an increasingly marginalized and stunted intellectual culture. This is what the Timurids passed on to the Mughals, Safavids, and Ottomans. It is also what most distinguishes all these problematic late bloomers from Central Asia's Age of Enlightenment five centuries earlier.

CHAPTER 15

❀

Retrospective: The Sand and the Oyster

Events themselves defined the starting point for this investigation. The Arab conquest of Central Asia between 680 and 740, the Central Asians' central role in the Abbasid revolution in 750, and Caliph Mamun's capture of Baghdad in 819 opened a new phase in the life of the already ancient civilization of Central Asia. For several hundred years thereafter the area between what is now eastern Iran and western China and extending from Kazakhstan south through Afghanistan—in other words, Greater Central Asia—was the center of the world. Here were concentrated the world's greatest entrepôt of trade and its most prosperous cities. A rich heritage from the past, worldly cultural contacts in the present, and affluence sufficient to support a bevy of thinkers and artists—all this created a nearly ideal environment for study and reflection. A passion for inquiry and an openness to innovation resulted in an astonishing number of firsts, extending over a score of fields. By any measure, this was one of humanity's great ages of thought and creativity, a true Age of Enlightenment.

Since we have at least touched on some of the specific achievements of Central Asia's scientists and thinkers in these centuries, there is no need to enumerate them all again. Let us instead take note of some of their common features. Among these, none is more striking than the fact that the same thinker often turned out seminal work in three, four, or more fields. This was the norm, not the exception, and its impact was to focus the attention of the literate public not on the peculiarities of one field or another but on knowledge as such, and on the paths to its attainment—epistemology.

Here the Central Asians recognized two great tools: first, geometry and mathematics; and, second, formal logic based on the model that Aristotle had set forth. These two instruments for reaching truth were closely explored and meticulously elaborated by a bevy of thinkers from the region, from Khwarazmi to Khayyam and beyond, and from Farabi through Ibn Sina and Ghazali. Quite independent of their many important insights on natural phenomena or philosophy, these thinkers greatly clarified and expanded our understanding of how human beings know what they know.

Nearly all Central Asian philosophers and scientists during the Age of Enlightenment were comfortable with some form of religion. Islam, with its uncompromising monotheism and notions of a First Cause, seemed to many to affirm the truths of science and philosophy. True, there were skeptics, as well as agnostics and atheists, but they were not numerous and met with a fairly high degree of toleration, as witnessed by the great medical doctor and scientist Razi, whose skepticism would have been considered scandalous even in eighteenth-century France.

But if some form of faith was fairly general, there was no agreement on how faith related to reason. Indeed, growing polarization was the order of the day. There were maximalists on both sides, with some, like Ghazali, who granted reason some modest role in practical affairs but considered it irrelevant to the great questions of existence; and others, like Biruni, who rarely paused to ask what role, if any, God played in the issue at hand. Finally, there were those who, following Sijistani, accorded due respect to both science and religion, reason and faith, but insisted that the two constituted separate realms and counseled that anyone pursuing the one must steer clear of the other.

In many centers across Central Asia there emerged leaders who embraced the creation of knowledge as one of the goals of their rule. Among them were some who took the exercise of reason to what were considered dangerous extremes. Thoroughgoing rationalism gained a zealous patron in the caliph Mamun, whose power base and first capital were both in Central Asia. His effort to impose the Mutazilites' rationalist doctrines through intimidation and force elicited the fierce opposition of many theologians, first among them the Central Asian Ibn Hanbal, who called for the rejection of all teachings not based on the authority of tradition, as recorded in the Quran and the Hadiths of Muhammad. Partly because of this warning to both sides, the scientist/philosophers and theologians improvised a modus vivendi that separated the two realms and allowed them to coexist for several centuries.

Only when this tacit understanding broke down around 1200 were the forces of Enlightenment thrown into retreat.

For all the differences among them, the scientists and philosophers of Central Asia's Age of Enlightenment agreed that human beings were endowed with all the capacities needed to arrive at truth. Some believed that divine revelation and faith were sufficient. But many others, Farabi chief among them, went so far as to claim that logic and reason could affirm the truths of revealed religion. Some, again including Farabi and others, believed emphatically that society itself could and should be ruled by Reason and not by the dogmas espoused by the powerful or by the superstitions of the multitude. No task elicited more intricate arguments than the effort, pursued by Ibn Sina and others, to chart out some sort of understanding or method that would enable reason and revelation to coexist in a mutually beneficial manner. All those who participated in these quests, and who insisted that reason be part of the equation, were emphatically humanists, three to five hundred years before the term gained wide use in Florence and then throughout Europe.

WHY DID IT HAPPEN?

The sheer scale of Central Asia's Age of Enlightenment cannot escape notice.Here were scores of talented writers from a fairly confined geographical area producing thousands of treatises and books on that broad range of issues pertaining to science and the theory of knowledge that Bertrand Russell dubbed "The Problems of Philosophy."[1] Since nothing quite like this occurred in the region either before or since, it is worth asking again, as we did in the beginning, why this intellectual effervescence took place. Even now, after following the story through several centuries, clear answers are elusive.

Thus, while wealth was clearly a necessary condition for the growth of a class of intellectuals who were free of practical obligations, it by no means assured that such a class would actually appear, let alone that its members would produce anything worthwhile. The sources of Central Asia's wealth may be as important as the wealth itself, for continental trade generated intercultural contacts, while irrigated agriculture fostered a high level of technical skill and a general predilection for intensive rather than extensive development that was readily translated into other activities. When later states, beginning with Mahmud's Ghazni, came to reap their wealth mainly from conquest and booty, they denied themselves the broadly international perspective and practical competence that had long been their predecessors' birthright.

Specific cultural factors affected how that wealth was used. Throughout the Persianate cultural sphere, of which Central Asia was in these centuries the heart, it was considered normal for rulers to use their wealth to foster the life of the mind and the arts. This conviction, born and thoroughly developed in the Zoroastrian and Buddhist past, continued under Islamic rule. Even when the motives of a particular ruler were banal and self-serving, such patronage was considered normal. It did much to create the necessary conditions of a cultural flowering.

Religion was also important, although sometimes less for what it did than for what it did not do. Islam's monotheism encouraged comprehensive explanations of natural and intellectual phenomena. Moreover, its clear and emphatic positions on such important questions as creation, nature, and human freedom posed a challenge equally to those who accepted and sought to confirm them, and those who questioned them. Much the same could be said of the other religions of the book—Judaism and Christianity—or, for that matter, Zoroastrianism. It must be admitted that few rulers who patronized learning did so in the hope of proving the truths of Islam, or any other faith. Yet to the extent that Muslim leaders in the early centuries found no reason to object to the work of writers and thinkers and many reasons to support it, they did an immense service to philosophy and science.

Another important causal factor was the prevalence of religious pluralism down to the end of Central Asia's Age of Enlightenment. This pluralism remained on display in every city in the region where churches, fire temples, or stupas continued to function. Pious Muslims had to acknowledge that those with other beliefs were often as bright and well informed as they themselves, and that the views of others could not be merely brushed aside. It may not be a coincidence that the sharpest conflicts between Sunni and Shiite Muslims arose after the year 1000, just as the old pluralism was everywhere beginning to fade.

The fact that people of the region had absorbed several of the world's major religions prior to the advent of Islam shaped how they received the Islamic message and, later, how they dealt with the appearance of forgotten books by classical Greek authors. In every case the Central Asians began by compiling the holy writings of each new religion to which they were exposed, or the various treatises of the classical authors, editing them, and, in some cases, translating them into local languages. This practice gave them a deep knowledge of each body of thought and an independent perspective from which they could analyze the next incoming set of ideas. Through this process Central Asians became masters at *adapting* new religions and ideas to their lives, and not passively *adopting* them. Thanks to this, Central Asia shaped Buddhism and Islam almost as much as those world religions shaped Central Asians.

Each new religion brought its own language, whether the Old Persian of Zoroastrianism, Greek for the Hellenic pantheon, Sanskrit for Buddhism, or Syriac (Aramaic) for Christianity. Islam brought Arabic, which, more than any of the tongues that preceded it, became the chief language of intellectual discourse regionally and a vehicle for international communication. Persian could have fulfilled the same function, and was moving in that direction at the time of Islam's rise, thanks to the Sassanian empire and to the Nestorian Christians working in Persia and Central Asia, who translated many classical Greek texts into Persians. But Persian would not have had as long a reach geographically as Arabic, thanks to the Arab conquests. The decision of Arab, Persian, and Central Asian officials to lend their support to the great work of translation into Arabic stands as an extremely enlightened act of leadership.[2] Overwhelmingly because of these translations, initially by Syrian Christians and later by many others, the Arabic language became a vehicle for the introduction of new ideas in Central Asia. When it ceased to fulfill this function, the intellectual effervescence dissipated.

It is tempting to dismiss the process of studying, codifying, and editing of new texts, whether religious or secular, as a worthy enterprise but scarcely one that could lead to intellectual breakthroughs. Indeed, on this basis more than one student of science and philosophy in the Islamic Middle Ages has criticized the process for being a brake on change rather than a stimulus to it. Such an interpretation, while understandable, misses the fundamental character of the study, codification, and editing of texts as it was carried out by Central Asian scientists, philosophers, and theologians. This in turn derives from a misunderstanding of the origins of innovative thinking and scientific breakthroughs in general.

Steven Johnson, in *Where Good Ideas Come From*, suggested that an innovative idea often arises when someone simply pursues one step further what is already before him or her.[3] This slow shift of learned attention to the "adjacent possible" is indeed a cautious process and is almost never intended as a challenge to the status quo, let alone as a call for a revolution in thinking. In the case of Central Asian scientists, the goal was not to challenge Aristotle, Galen, or another classical authority but better to understand their writings and therefore to appreciate their wisdom.

Thomas Kuhn made the same point in his *Structure of Scientific Revolutions*, to which we referred in the opening chapter. Someone who discovers an "anomaly" in any reigning scientific paradigm may well begin by assuming that he himself made the mistake and that the anomaly was more apparent than real. Only through the pursuit over many years of what Kuhn called "normal science" does the weight of the anomaly reach the point at which people begin groping for a new paradigm.

At the start of this book a number of innovations by Central Asian scientists were enumerated. Many others could well have been added to the list. The purpose, of course, was to convince the skeptical reader that a region that today includes one of the most backward countries on the planet—Afghanistan—could once have produced leading scientists and thinkers. Without diminishing the value of these and other breakthroughs achieved by Central Asian thinkers, we must also appreciate the immense contribution they made to world civilization by intelligently practicing "normal science" and "normal philosophy" over half a millennium. By this intensive and interactive process, which involved scores of researchers reading each other's work, many key issues were identified, analyzed, and handed on to successor generations for further analysis and resolution. If the world is largely unaware of innovations by Central Asian scientists and thinkers, or if it attributes their achievements to others, it is fair to say that even the experts have grossly undervalued the enormous importance of the "normal science" and "normal philosophy" practiced in Central Asia for half a millennium. During the scientific revolution in the sixteenth and seventeenth centuries, many so-called breakthroughs were achieved by Europeans who simply followed to its conclusion the logic of earlier research carried out by Central Asians and others. Not until the seventeenth century did Europe attain the kind of "normal science" that existed in Central Asia in the eleventh century.

The Sand in the Oyster

It is difficult to appreciate the importance of an anomaly when the acknowledgment of its existence does not lead immediately to a new paradigm. But what about the much more common situation in which anomalies are seen as either an inconvenience or a threat and are therefore swept under the rug or repressed? Unfortunately Kuhn did not delve into this question. He did not try to define the cultural and intellectual conditions that lead some people to resist what does not conform to their understanding of things, and others to acknowledge inconvenient evidence and try to deal with it.

Viewing the sweep of Central Asian history, it is clear that over the millennium before the Arab invasion and for at least four hundred years thereafter, the culture of the region was able to deal thoughtfully with even the most inconvenient new information or insights in either the secular or religious sphere. In this respect it was a self-confident civilization, accustomed to change and expert at dealing with it. One wonders, of course, whether or how it would have dealt with the classical masterpieces that were *not* translated into Arabic or Persian, works like Thucydides' *The Peloponnesian War*, Herodotus's *The Histories*, Plutarch's *Lives*, or the Greek tragedies. But this counterhistorical exercise cannot change the fact that the best thinkers of the region showed themselves remarkably open to the unfamiliar and the inconvenient and ready to deal with it.

Pearls, which Central Asians long imported from India and the Middle East, do not grow easily. Most oysters never encounter the grain of sand that will become the core of a pearl. Many oysters receive the sand, but with no effect at all, while still others try to reject it, sometimes dying in the process. It is a marvel of Central Asian culture that for several centuries it was both open to receiving the alien grain of sand and had the capacity to use it to produce pearls. This implies both an openness to the outside world and a healthy social organism that can rework and adapt the new rather than simply absorb it. So long as this prevailed, Central Asia was at the forefront of intellectual life and culture globally. When it lost this ability to face constructively what is alien and to turn it into pearls, its civilization went into decline.

Abdus Salam, the noted Pakistani physicist and first Muslim scientist to receive a Nobel Prize, stated the matter quite succinctly. Asked how he would strengthen scientific research in the Muslim world, he repeated the response of the director of Israel's renowned Weitzmann Institute to a query from a United Nations committee: "We have a very simple policy for science growth. An active scientist is always right and the younger he is, the more right he is."[4]

This book began with the scientific exchange between an eighteen-year-old Ibn Sina and the twenty-eight-year-old Biruni. Both were convinced that they were dealing with some of life's most important questions and, significantly, that they had at hand all the tools needed to answer them. Within less than three generations Ghazali was to lash out against this claim. In the process, he explicitly denounced Ibn Sina as a heretic. The most important truths are already fully accessible in the Quran and Hadiths and are embodied in the Sharia. Faith, not science and rigorous thought, is the key to unlocking them. Faith, unlike the exercise of reason, comes to a person only after a prolonged process of self-purification. It is most likely to appear among the elders of the community, the "white beards" (*aksakals*) or Sufi masters whose lifelong pursuit of wisdom gives them the right and duty to teach the young. Therefore, he concluded, let them impart wisdom and let the young listen. The kind of spiritual and intellectual jewels that the imams and Sufi elders offered required submission, not sand in the oyster.

WHAT HAPPENED TO IT?

Why did the Age of Enlightenment wane, and when did this happen? This, the third of the three questions posed at the beginning of this book, is no easier to answer after fourteen chapters than it was at the beginning. But there exists one hypothesis that, if valid, would enable us to declare the task accomplished: some students of the problem simply deny that the decline occurred. Typical is the scholar Mohamad Abdalla, who hauled out the fourteenth-century Moroccan historian Ibn Khaldun to make the case that there were still solid scientists and thinkers in the Arab world after their supposed disappearance from Central Asia. But on closer reading, it turns out that Ibn Khaldun himself argued exactly the opposite, namely, that even though science and thought went into decline in the Arab world, they continued to thrive "in the Eastern part of the Empire," namely, in Central Asia[5]—which in fact was not the case.

Sonja Brentjes, an outspoken scholar in Berlin, has denounced the very idea of decline in Islamic civilization as a "value-laden" and "inappropriate" category that was both a "nineteenth-century invention" and an offspring of the post-9/11 war on terror.[6] Amid all this vitriol, Brentjes, a talented Arabist, offers an interesting defense of Ghazali, whose attacks, she claims, had little impact, and of the madrasas, which she argues were less hostile to rationalism and science than most writers suggest.

More common is the argument that the speculative and scientific thought that had flourished in Central Asia lived on in the three Muslim empires that drew most heavily on the Central Asian tradition, namely, the Mughals in India, the Safavids in Iran, and the Ottomans in Turkey. Beyond doubt, these empires amassed formidable wealth in their heydays and built great capitals in Agra, Delhi, Lahore, Isfahan, Shiraz, Constantinople, and Cairo. All three attained a degree of sophistication that invariably impressed foreign visitors. But we have also noted that the science that existed in these empires was limited in both quantity and quality, and that the philosophy consisted more of commentaries than original work. In the end, the greatest cultural achievements of these great Islamic empires lay not in ideas but in the aesthetic realm.

In all three cases, art was less the sister of thought than a surrogate for it. Beauty reigned, while truth was treated as a complete and perfect whole that needed no further modification or elaboration. This conviction gave the intellectual life of these empires a backward-looking cast and denied it the steady prodding that new knowledge alone provides. As a result, they dissipated their great intellectual heritage, even to the point of allowing most of the brilliant manuscript books from the Age of Enlightenment to crumble physically and disappear from their libraries and collections. It is important to stress that the appalling loss of books and treatises from the Age of Enlightenment was mainly the result of the attitudes and priorities of the heirs to that civilization, not of the actions of outsiders, and that it was in fact Europeans who launched the effort to locate and preserve the few manuscripts that survived.

All three of the late successor empires suffered from the absence of modern universities and other centers of learning; from the tragically slow adoption of printing; and from a lack of interaction with the most vital centers of thought elsewhere.

Let us admit, then, that Central Asia experienced a profound intellectual decline at some time after AD 1100. Let us also accept, with the Beirut historian Ahmad Dallal, that this decline, while real, was not inevitable and culturally determined, but the result of specific historical circumstances.[7] But there is little consensus on the dating of this decline, for the simple reason that different analysts employ different metrics for gauging it. Thus the author of a volume on *Freethinkers of Medieval Islam* adopted a strict standard based on the ability of outspoken skeptics and agnostics to express themselves publicly. On this basis, she wrote, "It appears that after the tenth century, blunt prose expression of freethinking was no longer possible. The preoccupation of intellectuals with prophecy then found very different expressions. Philosophical parables like Ibn Sina's . . . offered ways of discussing this preoccupation that were deemed safer for the writers, and perhaps also intellectually more rewarding."[8] Others focus instead on the quantitative decline of fundamental research in the sciences and philosophy. This drop in production became clearly evident by late Seljuk times and was a recognized fact a century prior to the Mongol invasion. The diminished intellectual productivity of the powerful and wealthy Khwarazmian empire of the twelfth century epitomizes the striking change that had occurred. The decline was qualitative as well as quantitative, as the issuance of compendiums, encyclopedias, and collected works in many fields replaced treatises announcing the results of original research or analysis. The appearance of a Nasir al-Din al-Tusi or the brief burst of astronomical research by Ulughbeg and his colleagues did not affect this larger negative trend.

SOME ENVIRONMENTAL HYPOTHESES

After a century of Marxist thought, there is an automatic tendency to explain intellectual changes in terms of "objective" shifts in the physical or economic environment in which thinkers function. Since it is clear that Central Asia's intellectual flowing coincided with a great era of prosperity fed by high agricultural productivity, successful manufactures, and long-distance trade, arguments based on the weakening of any of these are ipso facto attractive. Jared Diamond, in his popular study of *Collapse: How Societies Chose to Fail or Succeed*, goes a step further by alerting us to the ways in which societies can unwittingly or even knowingly destroy the ecological systems that sustained their economic and social existence.

In the second chapter we saw how the overall climate of Central Asia has remained nearly constant since around 500 BC. The so-called Little Ice Age that began around 1550 came too late to affect Central Asia's economic and social conditions in the period under consideration. Further investigations may change this picture, but for now it appears that changes caused by humans were more significant than any climatological shifts that may have occurred. The large-scale destruction of forests and hillside vegetation led to increased seasonal changes in river flows, longer dry periods, and desertification. In earlier centuries oasis communities rose to the challenge of encroaching deserts, building walls of a hundred miles or more to thwart the spread of sand. They organized armies of workers and armed them with the latest technologies to enable them to water the oases and transform them into verdant gardens. But the Mongol destruction of irrigation systems and the inability of the post-Mongol societies to organize and deploy new armies of workers to rebuild these systems tipped the balance in favor of the deserts. This led to reduced agricultural productivity, declines in the oasis economies, and a less favorable environment for the pursuit of any intellectual concerns beyond the most immediately practical matters. But it bears repeating that the intellectual downturn took place prior to the era in which these environmental changes came strongly into play.

Playing the Mongol Card

By far the most popular explanation for the decline of Central Asia, especially within the region itself, is to blame it on the Mongols. We have seen how the region felt the brunt of the Mongol assault and paid an awful price in terms of manpower, treasure, and intellectual resources. Worse, Chinggis Khan's successors in the region clung to their nomadic ways, refusing to embrace the local urban culture as did their Mongol counterparts in China and Iran. They created a low-trade zone at the very center of what many recent authors have inaccurately lauded as a miraculous zone of free trade stretching from the Mediterranean to China. To be sure, trade with China boomed after the Mongol conquest, but it largely bypassed Central Asia. This contributed to a "lost century" in Central Asian civilization that continued down to the rise of Timur.

More enduring was the impact of the Mongols' systematic destruction of Central Asia's complex and sophisticated irrigation systems. We have noted how, following the Mongols' departure, demographic losses and the greatly reduced technical and organizational capacities of the local populations prevented Central Asians from rebuilding the hydraulic systems on anything like their former scale. This multiplied the economic and cultural cost of the Mongol invasion.

Yet the intellectual decline that is our main concern had already occurred by the time the first Mongol horsemen rode up to the walls of Otrar. Destruction by Mongol forces may have slowed or even prevented a return to prosperity based on agriculture, manufacturing, and trade, and it enormously complicated the task of rebuilding the region's diminished intellectual resources, had anyone wished to do so. But it did not *cause* the falling-off of intellectual life, which antedated it by at least a century. The "Mongol card," with the sense of victimization that goes with it, must therefore be rejected.

Vasco da Gama as Villain

A more sophisticated form of economic determinism is to blame the region's intellectual decline on the eclipse of caravan trade and the rise of sea routes connecting Europe and Asia. The implication is that ingenious Europeans brought about the collapse of trade, which impoverished the region and drastically reduced the resources that might have been available to support intellectual and cultural life.[9]

There are two problems with this line of argument. First, it confuses cause and effect. Central Asians were not the innocent victims of Western scheming, as the hypothesis implies. True, for centuries they masterfully managed the economic and security conditions essential for trade. But then, at the end of our period, Central Asians failed to provide security along the old trade routes; worse, they greedily raised tariffs to the point that transporters began searching for alternative routes. Timur, rather than addressing the cause of the decline of the land routes, attempted vainly to cut back its effects by actively working to close down the emerging maritime routes to the Far East.[10] The process of decline gained momentum with the breakup of Timur's realm into competing emirates and the rise of a Shiite Persia. Thus it is fair to say that unwise policies on the part of the Central Asians killed the already ailing golden goose of free trade,[11] and that Vasco da Gama merely dealt with the consequences.

Even if this helps explain the long-term economic decline of the region, it is irrelevant to the intellectual decline, which predated by a century or two the onset of economic decay. Indeed, the Seljuks were building some of Central Asia's largest and most impressive caravanserais, and trade was pouring gold into the coffers of the shahs of Khwarazm, precisely while the process of intellectual decay was intensifying. The eventual decline of the old continental transit trade isolated the region from intellectual developments elsewhere on the Eurasian land mass. It impoverished the region and diminished the funds that were available to support purely cultural and intellectual pursuits. Finally, it negatively affected the region's ability to rebound economically and helps explain its long-term economic backwardness. But it cannot be invoked to explain the waning of the culture of inquiry and enlightenment that had existed earlier.

Did Wealthy Patrons Close Their Pocketbooks?

This line of argument is advanced by the American Richard Frye, who proposed that the decline of intellectual innovation was caused by the waning power and resources of the landed aristocratic class (*dikhans*), who had patronized much cultural activity locally and from whose numbers were drawn many of the thinkers themselves.[12] It is undeniable that down to Seljuk times there existed a landed class throughout the central and western reaches of Central Asia, and that its members cultivated a sophisticated style of life that included support for music, the arts, and intellectual life. It is true, also, that this support diminished over time. Frye's argument gains support from recent studies, which see Muslim inheritance law as breaking up estates and inherited wealth.[13] These same laws have also been blamed for the fragmentation of fortunes in all the Muslim lands and the shrinking supply of capital for investment. This process surely hindered the perpetuation of individual patronage.

Yet it should be remembered that these same laws had been in place during the centuries of cultural vitality. Even in later periods, a person who wished to support culture could still transfer his fortune to a foundation (*waqf*) and designate it for the support of research in science or philosophy. But few chose to do so, preferring instead to set up madrasas with closely prescribed pedagogical agendas. Overall Frye surely overestimated the patronage provided by wealthy *dikhans* and undervalued the changes that reoriented private giving away from science and philosophy.

THE LARGESSE OF RULERS

Those who would trace the decline of intellectual culture in Central Asia to the rise and fall of patronage usually base their case not on the aristocracy but on the royal courts that existed on the local and regional levels, and especially on the major courts at Balkh, Merv, Samarkand, Baghdad, Bukhara, Gurganj, Ghazni, and Samarkand. Their ample resources attracted the cream of the intellectual and artistic intelligentsia from far and wide and placed them in highly competitive environments that were only partly softened by formal rules of protocol and traditions of civility.

In a more democratic age it is tempting to minimize the role of individual rulers, let alone of tyrants like Mahmud of Ghazni or Timur, in the creation of culture. Yet it is impossible to deny the importance of that deeply rooted pre-Islamic tradition that withheld honor from even the most powerful ruler until he had gathered around himself a coterie of poets, artists, and thinkers and demonstrated an interest in their work. Fully developed under the Greeks of Bactria, the Kushans, and the Sassanians, this tradition of patronage transformed the most savage warrior into a culture-maker. It mattered little that the resulting encomiums came from the pens of paid poets, historians, and propagandists.

In spite of this tradition, the decline of intellectual life in Central Asia cannot be blamed on the waning of empires and imperial patrons there and their replacement by small-scale amirs and emirates. It is true that the last great empire based in the region—that founded by Timur—spent far more lavishly on culture and the arts than did the emirates of Bukhara, Khiva, and Qokand that succeeded it. But long before Timur, most imperial patrons had withdrawn support from mathematics, the sciences, systematic philosophy, and other core fields of learning. The later rulers of Khwarazm, for example, commanded almost unlimited resources but did not see fit to spend them in support of learning. Medicine and astronomy/astrology limped on in some places, but for the sole reason that they protected the ruler's health and foretold his future. Other fields of learning withered and in some cases died out entirely—all of this during an era when great empires still thrived.

Writing in the eleventh century, Biruni took note of the start of this development. "It is quite impossible," he rued, "that a new science or that any new kind of research should arise in our days. What we have of sciences is nothing but the scanty remains of bygone better times."[14] Royal patrons shifted their support from science and philosophy to architecture, painting, music, and the crafts. The aesthetic cultures that resulted were splendid but lacked the inquiring spirit and rigorous habits of thought that had flourished at many of the region's courts in earlier centuries.

Those Nonspeculative Turks

There is no more sensitive hypothesis regarding the decline of science and philosophy in Central Asia than the one that lays the blame at the doorstep of Turkic culture. According to this view, Central Asia's intellectual effervescence arose overwhelmingly from the urban civilization of the Persianate East, while the incoming Turks over the centuries oversaw its decline and disappearance. We have seen that the main champion of this view was the Russian historian Bartold, who expounded it several times in the many works on the region that he published before and after the Russian Revolution. Bartold considered the nomadic culture of the Turkic peoples to be alien to the rigorous logic and mathematical sophistication of the settled oasis dwellers. But the fact that he was prepared to acknowledge the achievements of such Turkic savants as Mahmud of Kashgar, Ulughbeg, Ali Kushji, or Navai suggests that Bartold felt the problem lay not with the Turkic *ethnos* per se but with nomadism, of which the Turkic peoples were long the most prominent exemplars.

The removal of the ethnic issue does not clinch the argument that the incursion of nomadic cultures caused the intellectual decline. After all, there were as many nomads during the Age of Enlightenment as before or after, and most of the settled oasis dwellers had themselves been nomads earlier. What changed over time was the ability of the oasis dwellers to preserve the intellectual culture that had long thrived in their cities while selectively assimilating newcomers to it. The gradual erosion of the economic and political life of cities doubtless weakened their cultural role as well. This diluted the cultural preeminence of the cities and made their relations with the nomadic populations more an exchange among equals. However, this was due as much to the erosion of urban Central Asia as to the rise of Turkic nomads.

MUSLIM AGAINST MUSLIM

Even this brief review casts serious doubt on all the most widely disseminated explanations for the waning of Central Asia's intellectual vitality. Neither environmental change, the Mongol devastation, the rise of sea routes to Asia, the shrinking of noble patronage, the waning of empires, or the purportedly less speculative culture of the Turkic nomads explains the quantitative and quantitative decline of scientific and philosophic life in Central Asia. We must therefore seek other causes.

Central Asia's Age of Enlightenment did not occur at a time of religious and intellectual tranquility. Quite the contrary. Its starting point was the epochal arrival in the region of Islam, with the armies of the Arab nomads. This unleashed a complex process of interaction between Muslims and other confessional groups that continued over several centuries. Simultaneously, throughout the ninth, tenth, and eleventh centuries tensions mounted between groups that eventually crystallized into contending Sunni and Shiite branches of Islam. Especially during the late tenth and early eleventh centuries, Shiite thinkers exercised a powerful influence across Central Asia and even affected the Sunni empire of the Samanids in Bukhara. Yet the Sunnis were quick to respond to the Shiites using many of the same intellectual tools the Shiites used. Above all, they increasingly stressed what they believed should be the decisive voice of tradition. Within both parties a broad range of views existed.

The rise of Mahmud of Ghazni and then of the Seljuks gave a tremendous boost to mainstream Sunni traditionalists and gave rise to frontal confrontation between them and the Shiites. By the eleventh century the religious and intellectual climate of all Central Asia, and of Iran as well, crackled with polemics. Notwithstanding their agreement on many points, the Shiites continued to plea for an authoritative imam who was blessed by lineal descent from the Prophet to interpret the Quran and other holy texts. And the Sunnis ever more insistently demanded close adherence to the strictures spelled out in the Quran, Muhammad's Hadiths, and the laws or Sharia. Over time this division split asunder whole communities, states, and empires. So complete was the polarization that the mere suspicion that someone belonged to the other camp was pretext for the most severe punishment and exclusion. The proliferation of uncompromising currents on both sides, the ultrarationalist Ismailis among the Shiites and among the Sunnis the Asharites, whose antirationalism extended even to denying normal cause and effect, contributed to, and also reflected, the state of extreme polarization that existed.

It was once thought that the eclipse of scientific and speculative thought that occurred in the Islamic world during these years was in response to the confrontation with the European crusaders, who had arrived on the eastern shores of the Mediterranean to wrest control of the holy sites from the Muslims. But that conflict was far away from Central Asia and even from Iran, and quite irrelevant to the intellectual climate in either region. By contrast, the Sunni-Shiite split became so charged with conflict that anyone dealing with ideas—whether a scientist, mathematician, astronomer, philosopher, historian, astrologer, metaphysician, or poet—was forced to run the gauntlet. During the late ninth century the Shiites had the upper hand. But once the Sunnis gained the backing of the newly ascendant Seljuk Turks, the balance of power between the two sides shifted in their favor. By the late eleventh century a full-blown cultural war was under way, with Sunni watchdogs of the faith making sure that no thinker strayed beyond the strict bounds of tradition, and Shiite watchdogs of the faith responding in kind. Free inquiry was caught in the crossfire.

It is no exaggeration to say that strife within the community of Islam, the *umma*, the struggle of Sunni versus Shiite, was more than anything else responsible for the closing of the Muslim mind in Central Asia. This lay behind Nizam al-Mulk's fulminations, and also the diatribes of the young Ghazali. Increasingly, ideological tests were brought to bear not only against individual thinkers and specific treatises and books they had written, but against whole fields of inquiry and, finally, against specific modes of thought. As mainstream orthodoxy hardened, the price of ideological deviance increased. At first the thinkers responded with various forms of self-censorship, but over time they simply abandoned those fields in which they were most likely to come athwart the guardians of correct thinking. This is not to deny that the traditionalist mainstream produced thoughtful and highly moral thinkers of their own, or that the writings of these people were anything but sincere. But because they were no longer subjected to rigorous and open criticism, even the writings of members of the tradition-bound mainstream tended over time to become lax and ritualistic.

Ghazali, the Dark Genius

It is customary to recount this evolution in such broad and general terms that it comes to resemble an inexorable geological process. Like a glacier, slowly crushing and grinding up everything in its way, the rise of a narrow ansd tradition-bound orthodoxy in Islam is seen as an inevitable process. Maybe this was indeed the case. But maybe not. For each stage in the rise and fall of the Age of Enlightenment was epitomized by the activities of powerful and unique individuals, whose specific life experiences, rather than some general forces or trends, defined their actions and thoughts.

We followed the remarkable "purge in the name of progress" instituted by the brilliant and imprudent Caliph Mamun and examined how it led to a showdown with the quiet but resolute Ibn Hanbal. In this confrontation we saw the most direct juxtaposition of the Mutazilite doctrine of free will with the new forms of determinism. We watched as Ibn al-Rawandi, Hiwi, and Razi engaged in astonishing exercises of free thought, and also the lifelong work of Bukhari as he painstakingly gathered what he believed were the authentic words of Muhammad and established them, with the Quran, as the only legitimate source of authority and law. We observed how Ibn Sina endeavored to square the circle with his grand but ultimately failed system of epistemology and theology, and how Biruni chose to walk away from these endless disputes and focus instead on his scientific work. Finally, we saw Omar Khayyam penning his ringing but enigmatic poetic declarations of freedom from dogma but then holding them back from wider circulation.

Without these and other real people, without their specific writings and without their lives and decisions, the general trends and forces that we invoke to explain the rise and fall of end of the Age of Enlightenment would have no existence. This being the case, it is imperative to ask if the end of this age of innovation can also be traced to the work of a specific individual or individuals. There are many candidates, among whom the most prominent was Nizam al-Mulk, the Seljuk vizier from Tus, who took it as his holy mission to roll back and destroy the forces of Shiism in general and of the Ismailis in particular. It was he, after all, who rallied the full power of the Seljuk state to promote the cause of conservative and traditionalist Sunni Islam, and it was he who founded a new type of educational institution to inculcate right thinking and correct thoughts, and promoted the spread of such institutions throughout the Islamic Muslim world.

But it is doubtful we would remember Nizam al-Mulk were it not for his resident philosopher and theologian, the hyperactive, combative, yet profound thinker Muhammad al-Ghazali. Both the negative and positive aspects of Ghazali's very diverse achievement are directly relevant to the decline of the Age of Enlightenment in Central Asia. On the one hand, his brilliant and devastating attack in *The Incoherence of the*

Philosophers put the entire enterprise of independent speculation on the defensive, not only in his own day but for centuries thereafter. True, he delineated a certain realm in which mathematics could go forward and carved out a limited sphere for science that had mainly to do with practical applications. But otherwise Ghazali succeeded in marginalizing the philosophers, cosmologists, epistemologists, mathematicians, and theoretical scientists. Henceforth they lived as if in a building with very low ceilings. In Central Asia, it was not until the so-called *Jadid* reformers of the late nineteenth century, with their enthusiastic embrace of modern science and education, that the old spirit of inquiry began to revive. It is the more tragic, then, that the new Soviet rulers suppressed or killed these true Islamic reformers.

Ahmad Dallal in Beirut quite bluntly links the decline in scientific thinking with the rise of Sufi mysticism: "let me note that the decline in scientific activity . . . is often coupled with a return of interest in cosmology; only this time the traditional philosophical cosmology was replaced with a religious/Sufi one."[15] Ghazali did not cause the rise of Sufi mysticism, which, as we have seen, had deep roots in the culture of Central Asia and in the difficult circumstances of the twelfth and thirteenth centuries. But he did more than anyone else to legitimize it within Islam and to give it the central place among ways of knowing. In so doing he helped push aside reason and logic. His writings on faith are justly regarded as seminal explorations of this most human realm and as relevant to Christianity and Judaism as to Islam. But Ghazali proposed a clear hierarchy, in which the rational intellect was reduced to a subordinate status from which it was neither able nor allowed to challenge knowledge gained through mystical intuition and tradition.

Moreover, Ghazali strongly supported the suppression of all he opposed:

> The majority of men, I maintain, are dominated by a high opinion
> of their own skill and accomplishments, especially the perfection of
> their intellects for distinguishing true from false and sure guidance
> from misleading suggestions. It is therefore necessary, I maintain, to
> shut the gate, so as to keep the general public from reading the books
> of the misguided as far as possible . . . on account of the danger and
> deception in them. Just as the poor swimmer must be kept from the

slippery banks, so must mankind be kept from reading these books; just as the boy must be kept from touching the snake, so must the ears be kept from receiving such utterances.[16]

It took a full generation before anyone dared to rebut Ghazali. We have noted how in far-off Muslim Spain the judge and philosopher Ibn Rushd (1126–1198), known in the West as Averroes, eventually penned a response, mischievously entitled *The Incoherence of the Incoherence*. In this tract Averroes culled the body of Central Asian scholarship from Farabi onward to defend reason and logic against Ghazali's assault. In the end he convinced few Muslims, who largely ignored his book. Only in Europe did Averroes's rejoinder, translated into Latin from a Hebrew translation, find a sympathetic audience.

The seventeenth-century English thinker John Locke faced a similar challenge to accept revealed religion and at the same time preserve a realm where reason could be freely exercised. He deftly accomplished this by acknowledging a realm of cognition that was "above reason" and then plunging into the vast realm that was left, including science, philosophy, and all social and economic life. The severity of Ghazali's attack on reason and the vehemence of his many followers seemed to close this avenue to speculative philosophers and scientists in Central Asia and the Islamic world. On the other side, the protectors of faith and tradition set out to create purely Islamic forms of the sciences, free from the errors of Greek and alien thought.[17] Meanwhile, as more and more aspects of life were brought within the embrace of Islamic law, a rigid legalism came to dominate the intellectual sphere, and innovation became a term of opprobrium.[18] The fading of Central Asia's Age of Enlightenment can be neatly calibrated in terms of the number of open questions that remained and the willingness of the intellectuals to explore them.

Worse, as soon as states in the region embraced Ghazali's concept of mystical truth and intuition, as happened over the next centuries, and combined it with the hegemony of an ever-expanding but highly restrictive version of Islamic law, there was in place a complete mechanism for suppressing the free exercise of the intellect. The result was not a theocracy in the strict sense, because the religious establishment remained formally independent of the state. Also, unlike the Catholic Church, Islam lacked a focused authority and a bureaucracy to enforce that authority's will. In spite of this, the extension of the Sharia body of law into every area of civil and religious life turned regional states into theocracies in spite of themselves; they used their full power to enforce conformity with Islamic law, which became the sole arbiter of the claims of science, philosophy, and logic. This persuasive thesis was first systematically advanced in 1862 by the French Arabist Ernest Renan (1823–1892) in his inaugural lecture at the Collège de France, developed further by the Russian historian Bartold in 1902, and championed more recently by the New Zealand–based scholar John Joseph Saunders.[19]

From the founding of the caliphate, Muslims had assumed that the temporal and spiritual realms were inseparable. Gradually this eroded, especially under the early Abbasids. But then, in a bitter paradox, Caliph Mamun forced religion and the state back together when he used his civil authority to carry out what amounted to a purge of those conservatives and traditionalists who failed to embrace rationalism. Mamun triumphed in the short term, but the confrontation energized the opposition, which in due course came to power across Central Asia. A straight line connects Ibn Hanbal's heroic stand against Caliph Mamun with Ghazali's less-than-heroic ranting against Ibn Sina and the rationalists.[20]

As the range of open questions narrowed, the Arabic language, which had once been an instrument of connectivity and a powerful force for the dissemination of the latest inquiries, was constricted to a largely religious role. We have noted the very slow penetration of printing into the three great successor states to the Timur dynasty: the Mughals in India, the Safavids in Iran, and the Ottomans in Turkey. Knowledge of Arabic came to be restricted mainly to the mullahs and religious scholars, but no new lingua franca took its place as a link between the region and the world's most dynamic societies. Nor did any new Birunis appear, eager to plunge into unfamiliar cultures, learn their languages, and extract their wisdom.

Fatal Timing

The story of Central Asia's Age of Enlightenment began around the fifth century BC and reached its apogee in the ninth through the twelfth centuries AD. Ghazali, it will be recalled, died in 1111 and Omar Khayyam in 1131. From that period forward, the region as a whole was in intellectual but not economic decline under the Khwarazmi rulers, in crisis because of the Mongols, and finally in a state of brief revival under Timur's dynasty.

Western Europe during the same period presented a very different picture. Beginning from a very low starting point, major population growth was accompanied by agricultural expansion and improvements, and also the development of towns. A final boom period in the economy of Byzantium enriched the Italian cities and sparked a commercial revolution across much of Europe. In Italy this led to the development of new popular assemblies and legal institutions to regulate civic life. Beginning in Italy and then extending into France, Roman law revived, as scholars delved into long-forgotten legal texts with the same enthusiasm that Central Asians had earlier plunged into Aristotle, Plato, Euclid, and Ptolemy. Soon schools of law were assembling talented students in Bologna and elsewhere, transforming European law in the process. This was also the period in which the English common law came to be systematized.

Even if we acknowledge all this, as late as the twelfth century Europe had nothing to match the large and sophisticated cities of Central Asia, with their rich intellectual resources. The enormous impact of Ghazali's published works on the thought of both Arabs and Europeans is clear proof that as of his death in 1111, Central Asia retained its long intellectual lead over Europe and, equally, over much of the Arab world. But western Europe and Central Asia were following opposite trajectories. Even without the Mongol invasion of Central Asia, the lines charting their development would have crossed by 1300 and would probably have proceeded in opposite directions thereafter. To be sure, there was a period of renewal under Timur's successors. But this turned out to be politically unsustainable and intellectually one-sided and gave rise to the usual dynastic quarreling and decline. In other words, the beginning of the end of Central Asia's leadership in both the intellectual and economic spheres dates to the twelfth century. In the longer perspective, timing is important.

Is There a Need to Explain?

Having reviewed diverse theories and hypotheses on the waning of the Age of Enlightenment in Central Asia, it is now time to step back and raise a larger question: does it really require an explanation? The assumption behind our search for causes is that if one or another factor had not come into play, the movement of thought would have continued. But that great period of intense cerebration, that age of inquiry and innovation, had lasted for more than four centuries. If more information on the centuries preceding the Arab invasion had survived, we might confidently extend that period of flowering even further back in time. Even without this addition, the Age of Enlightenment was five times longer than the lifetime of Periclean Athens; a century longer than the entire history of the intellectual center of Alexandria from its foundation to the destruction of its library; only slightly shorter than the entire life span of the Roman Republic; longer than the Ming or Qing dynasties in China and the same length as the Han; about the same length as the history of Japan from the founding of the Tokugawa dynasty to the present; and of England from the age of Shakespeare to our own day.

As they say in the theater world, it had a long run. It is well and good to speak of causes of the decline of the passion for inquiry and innovation, or of some supposed exhaustion of creative energies. But just as we feel little need to discover the cause of a nonagenarian's death, we need not inquire too urgently into the cause of the waning of this remarkable age. Of course, the question of why the region as a whole remained in a state of backwardness from the end of the Age of Enlightenment down to recent times is vitally important, but it involves many factors besides those that came into play in the intellectual decline. It should form the subject of another book.

A related and far more urgent question involves the relevance of the Age of Enlightenment for the region today. The new states of Central Asia have proven that their sovereignty is viable. Prosperity is spreading, albeit at different rates. A remarkable group of younger men and women from across the region have received impressive educations, which have exposed them both to the thoughts and habits of mind of the larger world and to their own heretofore neglected intellectual and cultural heritage. Central to the latter is the Age of Enlightenment, when their forefathers led the world in many fields of inquiry. This heroic age may have been lost, but there are signs today that it is being rediscovered by the young. As that happens, old habits and expectations revive, and eyes are raised to farther horizons.

Meanwhile, it is well for the rest of the world, both East and West, to reflect on the fact that a region that some persist in viewing as marginal and backward was, over a number of centuries, the pivot of the political and economic world and the center of science, philosophy, and intellectual life on the Eurasian land mass. Is it not far wiser, then, to ask how this great movement of culture and ideas arose and endured as long as it did than to focus narrowly on its demise?

Notes

❀

PREFACE

1. Edward Gibbon, *The Decline and Fall of the Roman Empire*, 8 vols. (London, 1854), 1:xix.

CHAPTER 1

1. Excerpted in Arthur Hyman and James J. Walsh, eds., *Philosophy in the Middle Ages* (Indianapolis, 1973), 283–92.

2. A notable but, in the end, incomplete effort to break out of this framework is Elton L. Daniel, "The Islamic East," in *The Formation of the Islamic World, Sixth to Eleventh Centuries*, ed. Chase F. Robinson, *The New Cambridge History of Islam* (Cambridge, 2010), 1:459–79. See also S. Frederick Starr, "In Defense of Greater Central Asia," Policy Paper, Central Asia-Caucasus Institute / Silk Road Studies Program, 2008, http://www.silkroadstudies.org/new/docs/Silkroadpapers/0809GCA.pdf.

3. Daniel, "The Islamic East," 449.

4. B. A. Litvinsky and Zhang Guang-da, "Historical Introduction," in *History of Civilizations of Central Asia*, ed. B. A. Litvinsky (Paris, 1992–2005), 3:25–26.

5. Starr, "In Defense of Greater Central Asia."

6. John. J. O'Connor and Edmund. F. Robertson, "Abu Ali al-Husain ibn Abdallah ibn Sina (Avicenna)," http://www-history.mcs.st-andrews.ac.uk/Biographies/Avicenna.html; E. S. Kennedy, "Al-Biruni (or Beruni), Abu Rayhan," in *Dictionary of Scientific Biography*, ed. C. G. Gillispie (New York, 1981), 2:152.

7. D. King has recently questioned Khwarazmi's participation in this project, but the issue remains open; under any circumstances, Khwarazmi was the key figure in Islamic astronomy in this period. D. King, "Too Many Cooks . . . A New Account of the Earliest Muslim Geodetic Measurements," *Suhayl* 1 (2000): 207–41.

8. Samuel Eliot Morrison, *Admiral of the Ocean Sea*, 2 vols. (Boston, 1942), 1:87.

9. G. B. Nicolosi and J. L. Berggren, "The Mathematical Sciences," in *History of Civilizations of Central Asia*, ed. M. S. Asimov and C. E. Bosworth (Paris, 1992–2005), 4:192.

10. A. Youschkevitch and B. A. Rosenfeld, "Al-Khayyami (or Khayyam)," in *Dictionary of Scientific Biography*, 7:330.

11. A. Dzhalilov, *Iz istorii kulturnoi zhizni predkov tadzhiksgogo naroda i tadzhikov v rannem srednevekove* (Dushanbe, 1973), 50–56.

12. Edward Sachau, *Alberruni's India*, 2 vols. (London, 1887), 1:400.

13. Lenn E. Goodman, *Islamic Humanism* (Oxford, 2003), 202–3.

14. See chapter 11; also see Akbar S. Ahmed, "Al-Beruni, The First Anthropologist," *RAIN*, Royal Anthropological Institute of Great Britain and Ireland, 60 (February 1984): 9–10.

15. Yusuf Khass Hajib, *Wisdom of Royal Glory (Kutadgu Bilig): A Turko-Islamic Mirror for Princes*, trans. Robert Dankoff (Chicago, 1983); Nizam al-Mulk, *The Book of Government or Rules for Kings*, trans. Huburt Darke (London, 1960).

16. Goodman, *Islamic Humanism*, 8.

17. M.J.L. Young, J. D. Latham, and R. B. Serjeant, eds., *Religion, Learning and Science in the Abbasid Period* (Cambridge, 1990), 395.

18. Adam Mez, *Die Renaissance des Islams* (Heidelberg, 1922), quoted by Joel Kraemer, *Humanism in the Renaissance of Islam* (Leiden, 1986), 3.

19. Rosamond McKitterick, "Eighth Century Foundations," in *The New Cambridge Medieval History*, ed. McKitterick (Cambridge, 1995), 2:709.

20. Seyyed Hossein Nasr, "The Achievements of Ibn Sina in the Field of Science and His Contributions to Its Philosophy," *Islam and Science* 1, 2 (December 2002), http://find.galegroup.com/itx/printdoc.do?contentSet=IAC-Documents&docType-IAC&isll.

21. A good exposition of this idea is by Peter Dolmnick, *The Clockwork Universe: Isaac Newton, the Royal Society, and the Birth of the Modern World* (New York, 2011).

22. C. E. Bosworth, "Legal and Political Sciences in the Eastern Iranian World," in *History of Civilizations of Central Asia*, 4:133.

23. Rosamond Mack points out this and other striking influences in her volume *Bazaar to Piazza: Islamic Trade and Italian Art, 1300–1500* (Berkeley, 2002).

24. K. B. Nasim, *Hakim Aiuhad-ud-Din Anwari* (Lahore, n.d), 167n.

25. J. Burckhardt, *The Civilization of the Renaissance in Italy* (New York, 1958), 143; cf. Joel L. Kraemer's interesting discussion of this point in *Humanism in the Renaissance of Islam*, 11–12.

26. See, for example, Peter Adamson and Richard C. Taylor, eds., *The Cambridge Companion to Arabic Philosophy* (Cambridge, 2005); Roshdi Rashed, ed., *Encyclopedia of the History of Arabic Science*, 3 vols. (London, 1996); Jim al Khalili, *The House of Reason: How Arabic Science Saved Ancient Knowledge and Gave us the Renaissance* (London, 2011).

27. Jacques Boussard, *The Civilization of Charlemagne* (New York, 1968), 18.

28. Richard N. Frye, *The Golden Age of Persia* (London, 1975), 150.

29. Abu Mansur al-Thalibi (961–1039) , cited in A. Afsahzod, "Persian Literature," in *History of Civilizations of Central Asia*, 4:370.

30. See chapter 6. See also Dzhalilov, *Iz istorii kulturnoi zhizni predkov tadzhikskogo naroda i tadzhikov v rannem srednevekove*, 42.

31. On the prehistoric origins of the distinction between Central Asia and Iran and on Khurasan's identity as part of the former, see Fredrik T. Hiebert and Robert H. Dyson, Jr., "Prehistoric Nishapur and the Frontier Between Central Asia and Iran," *Iranica Antiqua* 37 (2002): 113–29.

32. Herodotus, *The Histories*, trans. Robin Waterfield (Oxford, 1998), 210.

33. Charles K. Wilkinson, *Nishapur: Some Early Islamic Buildings and Their Decoration*, Metropolitan Museum of Art (New York, 1986), 43.

34. Biruni, *Alberuni's India*, ed. Edward C. Sachau, 2 vols. (London, 1910), 1:xv.

35. Souren Melikian, "Islamic Culture: Groundless Myth," *New York Times*, Special Report, November 5–6, 2011.

36. Judah Rosenthal, "Hiwi al-Balkhi, A Comparative Study," *Jewish Quarterly Review*, New Series, 38, 3 (January 1948): 317–42; Sarah Stroumsa, "Ibn al Rawandi and His Baffling 'Book of the Emerald,'" *Freethinkers of Medieval Islam* (Leiden, 1999), chap. 2.

37. Dzhalilov, *Iz istorii kulturnoi zhizni predkov tadzhiksgogo naroda i tadzhikov v rannem srednevekove*, 128ff.; Jean-Claude Chabrier, "Musical Science," in *Encyclopedia of the History of Arabic Science*, 2:594ff.

38. Tibor Bachmann advanced this claim in *Reading and Writing Music* (Ann Arbor, 1969), 1:137–54; cited by Harvey Turnbull, "A Sogdian Friction Chordophone," in *Essays on Asian and Other Musics Presented to Laurence Picken*, ed. D. R. Widdess (Cambridge, 1981), 197.

39. Henry George Farmer, *Al-Farabi's Arabic-Latin Writings on Music* (New York, 1965).

40. Al-Nadim, *The Fihrist of al-Nadim*, ed. and trans. Bayard Dodge, 2 vols. (New York, 1970), 2:735–36.

41. Boaz Shoshan's study on "High Culture and Popular Culture in Medieval Islam," *Studia Islamica* 73 (1991): 67–107, does not focus on Central Asia but suggests interesting avenues for further investigation.

42. Edward H. Schafer, *The Golden Peaches of Samarkand: A Study of T'ang Exotics* (Berkeley, 1985).

43. Richard Ettinghausen, Oleg Grabar, and Marilyn Jenkins-Madina, *Islamic Art and Architecture 650–1250* (New Haven, 2001), 135.

44. Frances Gies and Joseph Gies, *Women in the Middle Ages* (New York, 1978), chap. 5.

45. See chapter 8.

46. Camille Adams Helminski, ed., *Women of Sufism: A Hidden Treasure* (Boston, 2003), 46ff.

47. Manuela Marín, "Women, Gender and Sexuality," in *Islamic Cultures and Societies to the End of the Eighteenth Century*, ed. Robert Irwin, *New Cambridge History of Islam* (Cambridge, 2010), 4:372.

48. Richard N. Frye, "Women in Pre-Islamic Central Asia: The Khatun of Bukhara," in *Women in the Medieval Islamic World: Power, Patronage, and Piety*, ed. Gavin R. G. Hambly (New York, 1999), 63–64.

49. Richard N. Frye, *Narshaki, The History of Bukhara*, (Cambridge, 1954), 37–39; W. Barthold (V. V. Bartold), *Turkestan Down to the Mongol Invasion*, trans. T. Minorsky and C. E. Bosworth (London 1928), 252; Frye, "Women in Pre-Islamic Central Asia," 65–67.

50. Lenn E. Goodman, *Avicenna* (London, 1992), 27.

51. Priscilla Soucek, "Timurid Women: A Cultural Perspective," in *Women in the Medieval Islamic World*, 199–226.

52. Michal Biran, *Qaidu and the Rise of the Independent Mongol State in Central Asia* (London, 1997), 2.

CHAPTER 2

1. V. M. Masson, *Strana tysiachikh gorodov* (Moscow, 1966); Pierre Leriche, "Bactria, Land of a Thousand Cities," in *After Alexander: Central Asia before Islam*, ed. Joe Cribb and Georgina Herrmann, *Proceedings of the British Academy*, no. 133 (London, 2007), 124–25.

2. Estimated by Philippe Marquis of the Délégation archéologique Française en Afghanistan (DAFA), personal communication, June 4, 2012.

3. The most intensive research to date has focused on the walls. See Rodney S. Young, "The South Wall of Balkh-Bactra," *American Journal of Archaeology* 59, 4 (October 1955): 267–76; and M. Le Berre and D. Schlumberger, "Observations sur les remparts de Bactres," in *Monuments preislamiques d'Afghanistan, Memoires de la Délégation archéologique Française en Afghanistan*, ed. Bruno Dagens, Marc le Berre, and Daniel Schlumberger (Paris, 1964), 9:67ff.

4. A Roman writer claims that the small stream that flowed past Balkh to the Oxus River (Amu Darya) was still navigable, enabling merchants to ship goods from India clear to the Caspian Sea. See B. V. Lunin, *Zhizn i deiatelnost akademika V.V. Bartolda* (Tashkent, 1981), 43ff.; Bartold's exposition is in "Khafiz-i Abru i ego sochineniia," *Muzafariia, sbornik statei uchennikov barona Victora Romanovicha Rozena ko dniu dvatsatiletiia ego pervoi lektsii* (St. Petersburg, 1897), 1–28, reprinted in V. V. Bartold, *Sochineniia* (Moscow, 1983), 8:74–97.

5. Young, "The South Wall of Balkh-Bactra," 275.

6. For Arab accounts of Balkh, see Schwartz, "Bemerkungen zu den arabischen Nachrichten uber Balkh," in *Oriental Studies in Honor of Cursetji Erachji Pavry* (London, 1933), 434–43.

7. W. Barthold, *An Historical Geography of Iran*, ed. C. E. Bosworth (Princeton, 1984), 17.

8. G. V. Shishkina, "Ancient Samarkand: Capital of Soghd," *Bulletin of the Asia Institute*, New Series, 8 (1994): 81–100. For an authoritative review of these data, see A. M. Belenitskii, I. B. Bentovich, and O. G. Bolshakov, *Srednevekovyi gorod Srednei Azii* (Leningrad, 1973).

9. Y. F. Buryakov et al., *The Cities and Routes of the Great Silk Road* (Tashkent, 1999), 54ff.

10. Besides the studies of Georgina Herrmann, cited below, a useful starting point is N. G. Bulgakov, "Iz arabskikh istochnikov o Merve," *Trudy iuzhno-turkmenistanskoi arkheologicheskoi kompleksnoi ekspeditsii*, vol. XII, Ashgabad, 1963, 213–24.

11. A. B. Yazbderdiev, "The Ancient Merv and Its Libraries," *Journal of Asian Civilizations* 23, 2 (December 2000), 138.

12. For a comprehensive overview of the monuments today, see Georgina Herrmann, *Monuments of Merv: Traditional Buildings of the Karakum* (London, 1999).

13. Two studies of Central Asian urbanism are V. A. Litvinskii, "Drevnii sredneaziatsii gorod," in *Drevnii vostok, Goroda i torgovlia (I–III tysiachiletiia do n.e.)* (Yerevan, 1973), 99–125; and V. M. Masson, "Protses urbanizatsii vi drevnei istiorii Srednei Azii," in *Drevnii gorod Srednei Azii: Tezisi i doklady* (Leningrad, 1973). G. A. Fedorov-Davydov's seemingly low population estimates (150,000 for Merv, 80,000 for Bukhara, etc.) are summarized in "Archaeological Research on Central Asia of the Muslim Period," *World Archaeology* 14, 3 (February 1983): 394; the special case of Khwarazm is considered by S. Tolstov, *Drevnii Khwarizm* (Moscow, 1948) and his *Po sledam drevnekhorezmiiskoi tsivilizatsii* (Moscow, 1948).

14. D. A. Alimova and M. I. Filanovich, *Toshkent tarikhi/Istoriia Tashkenta* (Tashkent, 2009), 89, 122ff.

15. K. M. Baipakov, *Srednevekovye goroda Kazakhstana na Velikom Shelkovom puti* (Almaty, 1998), 47–60, 145–48.

16. Buryakov et al., *The Cities and Routes of the Great Silk Road*, 106.

17. In addition to the well-known examples at Merv, see G. A. Pugachenkova, *Puti razvitiia arkhitektury iuzhnogo Turkmenistana pory rabovladeniia i feodalizma* (Moscow, 1958), 149–67.

18. See the description for the sixth- to eighth-century Binjakat castle at Ura Tyube, Tajikistan, in N. N. Negmatov, "Utrushana, Ferghana, Chach, and Ilak," in *History of Civilizations of Central Asia*, 3:259, 264.

19. G. L. Semenov, "Excavations at Paikend," in *The Art and Archaeology of Ancient Persia*, ed. Vesta Sarkhosh Curtis, Robert Hillenbrand, and J. M. Rogers (London, 1998).

20. K. M. Baybakov, *Srednevekovaia gorodskaia kultura iuzhnogo Kazakhstana i Semirechia* (Moscow, 1986), 88; see also O. G. Bolshakov's estimates of population densities in Merv, Bukhara, Termez, etc., *Goroda iuzhnogo Kazakhstana i Semirechiia (vi–xiii v.)* (Alma Ata, 1973), 256–68.

21. Baybakov, *Srednevekovaia gorodskaia kultura iuzhnogo Kazakhstana i Semirechiia.*

22. V. A. Litvinsky, "Cities and Urban Life in the Kushan Kingdom," in *History of Civilizations of Central Asia*, ed. J. Harmatta (Paris, 1992–2005), 2:291–312; see also V. G. Gafurov, ed., *Istoriia tadzhikskogo naroda* (Moscow, 1964), 1:471ff., on Sogdian slavery.

23. Discussed in detail by Eric Gustav Carlson, "The Abbey Church of Saint-Etienne at Caen in the Eleventh and Early Twelfth Centuries," PhD dissertation, Yale, 1967.

24. Peter Draper, "Islam and the West: The Early Use of the Pointed Arch Revisited," *Architectural History* 48 (2005): 1–9.

25. Viktor Sarianidi, *Gonurdepe, Turkmenistan* (Ashgabat, 2006), 36; also his *Margush, Turkmenistan* (Ashgabat, 2009).

26. Jim G. Shaffer, "The Later Prehistoric Periods," in *The Archaeology of Afghanistan from Earliest Times to the Timurid Period*, ed. F. R. Allchin and Norman Hammond (London, 1978), 114.

27. Fredrik T. Hiebert with Kakamurad Kurbansakhatov, *A Central Asian Village at the Dawn of Civiliztion: Excavations at Anau, Turkmenistan*, University Museum Monograph, no. 116 (Philadelphia, 2003); Peggy Champlin, *Raphael Pumpelly, Gentleman Geologist of the Gilded Age* (Tuscaloosa, 1994); S. Frederick Starr, "Raphael Pumpelly: Founder of U.S.–Turkmen Cultural Ties," *Miras* 2 (2010): n.p.

28. Daron Acemoglu and James A. Robinson, *Why Nations Fail: The Origins of Power, Prosperity, and Poverty* (New York, 2012), 148.

29. Sarianidi, *Gonurdepe, Turkmenistan*, 35.

30. Sophia R. Bowlby, "The Geographical Background," in *The Archaeology of Afghanistan*, 17; Fredrik T. Hiebert, *Origins of the Bronze Age Oasis Civilization in Central Asia* (Cambridge, 1994), 10–11.

31. Published in *Bibliotheca geographorum Arabicorum* (Leiden, 1870–1894). Bartold accepted de Goeje's conclusion: V. V. Barthold, *Four Studies on the History of Central Asia*, trans. V. and T. Minorsky (Leiden, 1956), 1:13.

32. Richard W. Bulliet, *Cotton, Climate, and Camels in Early Islamic Iran* (New York, 2009).

33. Sarianidi, *Margush, Turkmenistan*, 264–67.

34. See Muhammed Mamedov, *Drevnaia arkhitektura Baktrii i Margiani* (Ashgabat, 2003).

35. Naomi F. Miller, "Paleoethnobotanical Evidence for Deforestation in Ancient Iran: A Case Study of Urban Malyan," *Journal of Ethnobiology* 5 (1985): 1–21.

36. Jared Diamond, *Collapse: How Societies Choose to Fail or Succeed* (New York, 2005).

37. On the concept of hydraulic civilizations, see Karl A. Wittfogel, "The Hydraulic Civilizations," in *Man's Role in Changing the Face of the Earth*, ed. William L. Thomas, Jr. (Chicago, 1956), 152–64.

38. On Central American achievements before Columbus in this and other areas, see Charles C. Mann, *1491: New Revelations of the Americas before Columbus* (New York, 2005).

39. For a summary focusing on Otrar oasis, see Renata Sala, "Historical Survey of Irrigation Practices in West Central Asia," Laboratory of Geo-Archaeology, Centre of Geologo-Geographical Research, Ministry of Education and Science, Kazakhstan, http://lgakz.org/Texts/LiveTexts/7-CAsiaIrrigTextEn.pdf.

40. On breaching Balkh's dam, and on Balkh's irrigation system as whole, see Akhror Mukhtarov, *Balkh in the Late Middle Ages* (Bloomington, 1993), 70–79.

41. A. M. Mukhamedzhanov, "Economy and Social System in Central Asia in the Kushan Age," in *History of Civilizations of Central Asia*, 2:272.

42. The urban water system of Aktobe in Kazakhstan is thoroughly examined by U. Kh. Shalekenov and A. M. Orazbaev, "Nekotorye dannye o vodoprovodnoi sisteme srednevekovogo goroda Aktobe," in *Istoriia materialnoi kultury Kazakhstana* (Alma-Ata, 1980), 24–48.

43. For this information, C. Edmund Bosworth cites the tenth-century Arab geographers Ibn Hawqal and al-Maqdisi: "Merv," in *Historic Cities of the Islamic World*, ed. C. Edmund Bosworth (Leiden, 2007), 401ff.

44. See chapter 12.

45. The most thoroughgoing exposition of this view is Louis Gardet, *La cité musulmane, vie sociale et politique* (Paris, 1954); also Eckart Ehlers, "The City of the Islamic Middle East: A German Geographer's Perspective," in *Papers in Honor of Professor Ehsan Yarshater, Iranica Varia, Textes et mémoirs* (Leiden, 1990), 16:167–76; Gustave E. von Grunebaum, *Islam and Medieval Hellenism: Social and Cultural Perspectives* (London, 1976) v, 25–37; and the papers in R. B. Sergeant, ed., *The Islamic City* (Paris, 1980).

46. E. E. Kuzmina, *Prehistory of the Silk Road* (Philadelphia, 2008), 66ff.; Richard W. Bulliet, *The Camel and the Wheel* (New York, 1990), 27. This discussion is based on Bulliet's research.

47. A. Foucher, *La vieille rue de l'Inde de Bactres à Taxila, Mémoires de la Délégation archéologique Française en Afghanistan* (Paris, 1942), vol. 1.

48. Barthold, *Turkestan Down to the Mongol Invasion*, 137.

49. For a relatively successful attempt at such mapping, see Buryakov et al., *The Cities and Routes of the Great Silk Road*.

50. Denis Sinor, *Inner Asia and Its Contacts with Medieval Europe* (London, 1977), chap. 4; Edvarde Rtveladze, *Civilizations, States, and Cultures of Central Asia* (Tashkent, 2008), 258ff.

51. Bosworth, "Merv," 402.

52. Valerie Hansen's interesting *The Silk Road: A New History* (New York, 2012) concentrates solely on the China trade. A history of the trade routes from Central Asia to India has yet to be written.

53. See chapter 6.

54. Needham, *Science and Civilization in China*, 1:195.

55. B. Y. Bichurin, *Sobranie svedenii o narodax obitavshikh v Sredneii Azii v drevnie vremena* (Moscow, 1950), 2:272.

56. A useful bibliographic introduction is by Frantz Grenet, "The Pre-Islamic Civilization of the Sogdians," *The Silk Road Newsletter* 1, 2:1–13, http://www.slk-road.com/newsletter/december/pre-islamic.htm; also B. I. Marshak and N. N. Negmatov, "Sogdiana," in *History of Civilizations of Central Asia*, 3:233–80. Étienne de la Vaissière, *Sogdian Traders: A History*, trans. James Ward (Leiden, 2005), is the best point of departure for anyone interested in this topic.

57. Edwin G. Pulleyblank, "A Sogdian Colony in Inner Mongolia," *T'oung Pao*, 2nd series, vol 41, bk. 4/5 (1952): 318–56.

58. Jonathan Karam Skaff, "The Sogdian Trade Diaspora in East Turkestan during the Seventh and Eighth Centuries," *Journal of the Economic and Social History of the Orient* 46, 4 (2003): 475–524; Étienne de la Vaissière, "Sogdians in China: A Short History," in *Science and Civilization in China*, 1:187.

59. B. I. Marshak and V. I. Raspopova, "Sogdiitsy v Semirechie," in *Drevnii i srednevekovyi Kyrgyzstan*, ed. K. I. Tashkaeva et al. (Bishkek, 1996), 124–43.

60. Nichlas Sims-Williams, "The Sogdian Merchants in China and India," in *Cina e Iran da Allesandro Magno alla Dinastia Tang* (Florence, 1996), 45–67; Rtveladze, *Civilizations, States, and Cultures of Central Asia*, 258–65.

61. Needham, *Science and Civilization in China*, 1:179.

62. DeLacy O'Leary, *How Greek Science Passed to the Arabs* (London, 1979), 70–71, makes the point about Roman seamen, but it applied equally in this era.

63. Needham, *Science and Civilization in China*, 1:187. On recent Sogdian finds in China, see Judith A. Lerner, "Les Sogdiens en Chine—Nouvelles découvertes historiques, archéologiques et linguistiques," *Bulletin of the Asia Institute*, New Series, 15 (2001): 151–62.

64. Abdukakhor Saidov et al., "The Ferghana Valley: The Pre-Colonial Legacy," in *Ferghana Valley: The Heart of Central Asia*, ed. S. Frederick Starr (Armonk, 2011), 18.

65. On Merv, see Dafydd Griffiths and Ann Feuerbach, "Early Islamic Manufacture of Crucible Steel at Merv, Turkmenistan," *Archaeology International* (1999–2000): 36–38. Feuerbach is completing a book on the early history of crucible steel.

66. All aspects of the earliest metallurgy in the region are considered by Benoît Mille and David Bourgarit, "The Development of Copper Metallurgy before and during the Indus Civilization," Centre de Recherche de Restauration des musées de France (Paris, 1993).

67. Needham, *Science and Civilization in China*, 4:108–9; see Samuel Kurinsky's interesting *The Glssmakers: An Odyssey of the Jews*, New York, 1991, 267 ff.

68. Needham, *Science and Civilization in China*, 1:243.

69. Peter B. Golden, *Central Asia in World History* (Oxford, 2011), 41–42.

70. Needham, *Science and Civilization in China*, vol. 4, pt. 3, 136–37. Frances Wood dates the first chairs in China to the second century but traces them to Rome and Byzantium without considering the process of transmission: *The Silk Road: Two-Thousand Years in the Heart of Asia* (Berkeley, 2002), 86–87.

71. Needham, *Science and Civilization in China*, 6:71, 163, 235, 271, 425, 516. The seed drill originated in Mesopotamia but became a Central Asian export to China.

72. Schafer, *The Golden Peaches of Samarkand*.

73. Rtveladze, *Civilizations, States and Cultures of Central Asia*, 220–21.

74. Svetlana V. Lyovushkina, "On the History of Sericulture in Central Asia," in *Silk Road Art and Archaeology* (1995/1996): 4:143–49. Etsuko Kageyama confirms Sogdian silk from the sixth century, i.e., prior to the Battle of Taraz (Talas), in "Use and Production of Silks in Sogdiana," *Webfestschrift Marshak* (2003), http://www.transoxiana.org/Eran/Articles/kageyama.html.

75. Needham adheres to the traditional eighth-century date in *Science and Civilization in China*, 5:297ff.

76. A. A. Hakimov, "Arts and Crafts in Transoxonia and Khurasan," in *History of Civilizations of Central Asia*, vol. 4, pt. 2, 440.

77. Jonathan M. Bloom, "Lost in Translation: Gridded Plans and Maps along the Silk Road," in *The Journey of Maps and Images on the Silk Road*, ed. Philippe Forêt and Andreas Kaplony (Leiden, 2008), 84–85.

78. Herodotus, *The Histories*, 420, 554, 587.

79. Richard N. Frye, "Achaemenid Centralization," in *The Heritage of Central Asia from Antiquity to the Turkish Expansion* (Princeton, 1996), chap. 5.

80. Buryakov et al., eds., *The Cities and Routes of the Great Silk Road*, 9.

81. Herodotus, *The Histories*, 210.

82. Ibid., 587.

83. The cities of East Turkestan did not issue their own currency. Frye rightly stresses the role of the later Kushans in the spread of currency. Richard N. Frye, *Notes on the Early Coinage of Transoxonia* (New York, 1949); see also Joe Cribb, "Money as a Marker of Cultural Continuity and Change in Central Asia," in *After Alexander*, 333–45; E. V. Zeimau, "The Circulation of Coins in Central Asia during the Early Medieval Period (Fifth–Eighth Centuries AD)," *Bulletin of the Asia Institute*, New Series, 8 (1994): 248–65; Rtveladze, *Civilizations, States, and Cultures of Central Asia*, 112–23.

84. On the formation of the Bactrian state, see H. Sidky, *The Greek Kingdom of Bactria: From Alexander to Eurcratides the Great* (Lanham, 2000); and Frank L. Holt, *Alexander the Great and Bactria* (Leiden, 1988), esp. 3. W. W. Tarn's *The Greeks in Bactria and India* (Cambridge, 1938) is still of value.

85. Needham, *Science and Civilization in China*, 1:233.

86. Ibid., 1:195.

87. Sidky, *The Greek Kingdom of Bactria*, 222–26.

88. This relationship, not covered here, is detailed by Christopher I. Beckwith, *The Tibetan Empire: A History of the Struggle for Great Power among Tibetans, Turks, Arabs, and Chinese during the Early Middle Ages* (Princeton, 1987).

89. One of several excellent overviews on China in Central Asia in the classical period is Rtveladze, *Civilizations, States, and Cultures of Central Asia*, 130–39, 272–82.

90. In Xinjiang Chinese currency played the same role as Persian currency did in more western parts of Central Asia. Helen Wang, "How Much for a Camel? A New Understanding of Money on the Silk Road before AD 800," in *The Silk Road: Trade, Travel, War and Faith*, by Susan Whitfield et al. (London, 2004), 24–33.

91. Masson, *Strana tysiachikh gorodov*, 75.

92. Rtveladze, *Civilizations, States, and Cultures of Central Asia*, 236.

93. Ibid., 244–45.

94. Ibid., 79ff.; see also the annual reports in the journal *Parthica* written by Antonio Invernizzi, Carlo Lippolis, and others, as well as Invernizzi's "The Culture of Parthian Nisa between Steppe and Empire," in *After Alexander*, 163–73; and his "New Archaeological Research in Old Nisa," in *The Art and Archeology of Ancient Persia*, ed. Vesta Sarkosh Curtis, Robert Hillenbrand, and J. M. Rogers (London, 1998), 8–13.

95. For a competent but critical review of research on Begram, see Pierre Cambon, "Begram: Alexandria of the Caucasus, Capital of the Kushan Empire," in *Afghanistan: Hidden Treasures of the National Museum, Kabul*, ed. Fredrik Hiebert and Pierre Cambon (Washington, 2008), 145–208; also R. Ghirshman, *Bégram: Recherches archéologiques et historiques sur les Kouchans, Mémoires de la Délégation archéologique Française en Afghanistan* (Paris, 1946), vol. 12; and J. Hackin, *Nouvelle recherché archeologique a Bégram* (Paris, 1954).

96. L. A. Borovkova, *Tsarstva "Zapadnogo kraia"* (Moscow, 2001); and her *Kushanskoe tsarstvo po kitaiskim istochnikam* (Moscow, 2005).

97. B. A. Litvinsky, "Cities and Urban Life in the Kushan Kingdom," in *History of the Civilizations of Central Asia*, ed. Janos Harmatta (Paris, 1994), 2:291–312.

98. Mukhamedzhanov, "Economy and Social System in Central Asia in the Kushan Age," 265ff.

99. Frye, *The Heritage of Central Asia from Antiquity to the Turkish Expansion*, 154.

100. Rtveladze, *Civilizations, States, and Cultures of Central Asia*, 118.

101. Victor Sarianidi, *The Golden Hoard of Bactria from to Tillya-Tepe: Excavations in Northern Afghanistan* (New York, 1985).

102. B. N. Puri, "The Kushans," in *History of Civilizations of Central Asia*, 2:261; Rtveladze, *Civilizations, States, and Cultures of Central Asia*, 77.

103. Authoritative accounts of these groups are provided by Christopher I. Beckwith, *Empires of the Silk Road* (Princeton, 2009); by Denis Sinor's syllabus, *Inner Asia: History, Civilization, Languages* (The Hague, 1971); and by Rene Grousset, *Empire of the Steppes* (New Brunswick, 1970). Many specialized studies exist on the Hephthalites, Kidarites, etc.

104. Sandra L. Olsen, "The Exploitation of Horses at Botai, Kazakhstan," in *Prehistoric Steppe Adaptation of the Horse*, ed. Marsha Levine, Colin Renfrow, and Katie Boyle (Cambridge, 2003), 83–103.

105. An excellent recent study is David Anthony's *The Horse, the Wheel, and Language: How Bronze-Age Riders from the Euirasian Steppes Shaped the World* (Princeton, 2007).

106. Ute Luise Dretz, "Horseback Riding: Man's Access to Speed," in *Prehistoric Steppe Adaptation of the Horse*, 197.

107. Robert Drews, *Early Riders: The Beginnings of Mounted Warfare in Asia and Europe* (London, 2004), chap. 5.

108. On the early phase of nomadism, see Claude Rapin, "Nomads and the Shaping of Central Asia: From the Early Iron Age to the Kushan Period," in *After Alexander*, 29–72. On nomadic migration and rule as a whole, see Peter B. Golden, *Nomads and Sedentary Societies in Medieval Eurasia* (Washington, DC, 1998); and Beckwith, *Empires of the Silk Road*.

109. Rtveladze, *Civilizations, States, and Cultures of Central Asia*, 62.

110. S. A. Viazigin, "Stena Antiokha Sotera vokrug drevnei Margiany," in *Trudy iuzhno-Turkmenistanskoi arkheologihcheskoi kompleksnoi ekspeditsii* (Ashgabat, 1949),

1:260–75; Andrei N. Bader, Vassif A. Gairov, and Gennadij A. Koselenko, "Walls of Margiana," in *In the Land of the Gryphons*, ed. Antonio Invernizzi (Florence, 1995), 39–50; Parvanch Pourshariati, "Iranian Traditions in Tus and the Arab Presence in Khurasan," PhD dissertation, Columbia University, 1995, 119–20.

111. Barthold, *Turkestan Down to the Mongol Invasion*, 76–77.

112. Beckwith, in his *Empires of the Silk Road*, thoroughly expounds this thesis.

113. Claude Cahen, "Nomades et sedentaires dans le monde muselman du milieu du moyen age," in *The Islamic City*, 1:93–104; David Christian, "Silk Roads or Steppe Roads? The Silk Roads in World History," *Journal of World History* 11, 1 (Spring 2000): 1–25; see also Philip D. Curtin, *Cross Cultural Trade in World History* (Cambridge, 1985); and Sh. S. Kamoliddin, "Etnokulturnoe vzaimodeistvie iranskikh i tiurkskikh narodov na velikom shelkovom puti," in *Identichnost i dialog kultur v epokhu globalizatsii* (Tashkent, 2003), 33–36.

114. Peter B. Golden, "War and Warfare in the Pre-Cinggisid Western Steppes of Eurasia," in *Warfare in Inner Asian History (500–1800)*, ed. Nicola DiCosmo (Leiden, 2002), 153–57.

115. See Gilbert Highet's delightful and profound *Poets in a Landscape* (New York, 1957).

116. Peter Brown, *The Making of Late Antiquity* (Cambridge, 1978), 3.

117. Von Grunebaum, "Observations on City Panegyrics in Arabic Prose," in *Islam and Medieval Hellenism*, 65ff., discusses slightly later panegyrics in Arabic, but many of these were to Central Asian or Persian cities.

118. Gafurov, ed., *Istoriia tadzhikskogo naroda*, 2:85.

119. Barthold, *Turkestan Down to the Mongol Invasion*, 180–81.

CHAPTER 3

1. Quoted by Dzhalilov, *Iz istorii kulturnoi zhizni tadzhikskogo naroda*, 38.

2. Bichurin, *Sobranie svedenii o narodakh, obitavshikh v Srednei Azii v drevnie vremena*, 2:310.

3. Simplified English by the author, based on authoritative translation by Nicholas Sims-Williams, *The Silk Road: Trade, Travel War, and Faith*, ed. S. Whitfield with U. Sims-Williams (London, 2004), 248–29. Both this document and the following are available, with introduction by Daniel C. Waugh, at http://depts.washington.edu/silkroad/texts/sogdlet.html; see also W. B. Henning, "The Date of the Sogdian Ancient Letters," in *W. B. Henning Selected Papers* (Leiden, 1977), 2:315–29.

4. Dzhalilov, *Iz istorii kulturnoi zhizni tadzhikskogo naroda*, 27.

5. Simplified English by the author, based on authoritative translation by Nicholas Sims-Williams "The Sogdian Ancient Letter II," in *Philologica et Linguistica. Historia, Pluralitas, Universitas. Festschrift für Helmut Humbach zum 80. Geburtstag, a, 4. Dezember 2001* (Trier, 2001), 267–80.

6. Acemoglu and Robinson, *Why Nations Fail*, 144–52.

7. Richard N. Frye, "Pre-Islamic and Early Islamic Cultures in Central Asia," in *Turko-Persia in Historical Perspective*, ed. R. L. Canfield (Cambridge, 1991), 37.

8. See, for example Nicholas Sims-Williams, "A Bactrian Deed of Manumission," in *Silk Road Art and Archaeology* 5 (1997/1998): 191–93.

9. Gafurov, ed., *Istoriia tadzhikskogo naroda*, 2:84.

10. Frye, *The Heritage of Central Asia from Antiquity to the Turkish Expansion*, 195.

11. A. A. Freiman, *Opisanie, publikatsiia i issledovanie dokumentov s gory Mug* (Moscow, 1962).

12. A. R. Mukhmejanov, "Natural Life and the Manmade Habitat in Central Asia," in *History of Civilizations of Central Asia*, vol. 4, pt. 2, 294–97.

13. B. V. Tsizerling, *Oroshenie na Amudare* (Moscow, 1927), 588ff.

14. Samuel Kurinsky, *The Glassmakers: An Oddesy of the Jews* (New York, 1991), 282–88.

15. National Museum, Afghanistan.

16. N. S. Asimov, "Nauka srednei Azii kushanskoi epokhi i puti ee izucheniia," *Mezhdunarodnaia konferentsiia UNESCO po istorii, arkheologii, i kultury Tsentralnoi Azii v kushanskuyu epokhu* (Dushanbe, 1968), 11.

17. Needham, *Science and Civilization in China*, 3:205.

18. O'Leary, *How Greek Science Passed to the Arabs*, 56.

19. Henning, "Zum soghdischen Kalendar," in *W. B. Henning Selected Papers*, 5:629–37; Dzhalilov, *Iz istorii kulturnoi zhizni tadzhikskogo naroda*, 40; Nicholas Sims-Williams and François de Blois, "The Bactrian Calendar," *Bulletin of the Asia Institute*, New Series, vol. 10, Studies in Honor of Vladimir A. Lifshits (1996): 149–53.

20. See chapters 8 and 9.

21. Needham, *Science and Civilization in China*, 2:204ff.

22. Semenov, "Excavations at Paikend," in *The Art and Archaeology of Ancient Persia*, 117.

23. Nicholas Sims-Williams, "The Sogdian Fragments of the British Library," *Indo-Iranian Journal* 18 (1976): 44.

24. B. N. Mukherjee, *India in Early Central Asia* (New Delhi, 1996), 27.

25. Biruni, *Alberuni's India*, 1:xxxii.

26. Lattimore applied the phrase only to Xinjiang. See his *Sinkiang and the Inner Asian Frontiers of China and Russia* (Boston, 1950).

27. On Bactrian and other regional languages, see J. Harmatta's thorough review, "Languages and Literature in the Kushan Empire," in *History of Civilizations of Central Asia*, 2:417–44.

28. On Khwarazmian, see Henning, "The Choresmian Documents," in *W. B. Henning Selected Papers*, 2:645–58; Dzhalilov, *Iz istorii kulturnoi zhizni tadzhikskogo naroda*, 19–32. On Bactrian, see Henning, "The Bactrian Inscription," in *W. B. Henning Selected Papers*, 2:545–53; and Ilya Gershevitch, "Bactrian Literature," in *The Cambridge History of Iran*, ed. Ehsan Yarshater (Cambridge, 1983), vol. 3, pt. 2, 1250–58. On "unknown scripts," see J. Harmatta, "Languages and Literature in the Kushan Empire," in *History of Civilizations of Central Asia*, 2:417–42; Rtveladze, *Civilizations, States, and Cultures of Central Asia*, 39.

29. Slightly simplified from translation by Vladimir Lifshits, "A Sogdian Precursor of Omar Khayyam in Transoxonia," *Iran and the Caucasus* 8, 1 (2004): 18.

30. Hansen, *The Silk Road*, 25ff.

31. Saidov et al., "The Ferghana Valley," 23.

32. Bloom, "Lost in Translation," 85. For a thorough discussion of these issues, see his excellent *Paper before Print: The History and Impact of Paper in the Islamic World* (New Haven, 2001).

33. Guitty Azarpay, *Sogdian Painting: The Pictorial Epic in Oriental Art* (Berkeley, 1981), 132–39; N. B. Diakonova and O. U. Smirnova, "K voprosu o kulte nany (Anakhity) v Sogde," *Sovetskaia arkheologiia* (1967): 74–83; M. M. Diakonov, "Obraz Siivasha," *Kratkie soobshcheniia instituta materialy kultury* 1 (1951): 34–44.

34. Sarianidi, *Margush, Turkmenistan*, 276ff.

35. A.D.H. Bivar, "Fire-Altars of the Sassanian Period at Balkh," *Journal of the Warburg and Courtauld Institutes* 17, 1–2 (1954): 182–83.

36. The dating of Zoroaster's life, as with so much connected with Zoroastrianism, is highly controversial. See C. F. Lehmann-Haupt, "Wann lebte Zarathustra?" in *Oriental Studies in Honour of Curseti Erachji Pavry*, ed. Jal Dastur Cursetji Pavry (London, 1933), 251–80.

37. For an excellent overview on Zoroastrianism, including its aesthetics, see K. Olimov and A. A. Shomolov, eds., *Istoriia tadzhikskoi filosofii (sdrevneishikh vremen do. xv v.)* (Dushanbe, 2012), 1:105–57, 383–427.

38. Mary Boyce, *Zoroastrians, Their Religious Beliefs and Practices* (London, 1979), 29.

39. R. C. Zaehner, *The Dawn and Twilight of Zoroastrianism* (New York, 1991), chap. 8.

40. Henning, "The Dates of Mani's Life," in *W. B. Henning Selected Papers*, 2:505–19; "Manichaeism and Its Iranian Background," in *The Cambridge History of Iran*, vol. 3, pt. 2, 963–90.

41. Samuel N. C. Lieu, *Manichaeism in the Later Roman Empire and Medieval China* (Manchester, 1985), 56.

42. O. M. Chunakova, ed., *Manikheiskie rukopisi iz Vostochnogo Turkestana: srednepersidskie i parfianskie fragmenty* (Moscow, 2011).

43. V. A. Lifshits, "Sogdian Sanak, a Manichaean Bishop of the 5th–Early 6th Centuries," *Bulletin of the Asia Institute*, New Series, 14 (2000): 47–54.

44. On Manicheanism, see Lieu, *Manichaeism in the Later Roman Empire and Medieval China*, 5–33.

45. Henning, "Sogdian Tales," in *W. B. Henning Selected Papers*, 2:169–71.

46. Henning, "A Fragment of the Manichean Hymn-Cycle in Old Turkish," in *W. B. Henning Selected Papers*, 2:537–39.

47. This theme is developed by Richard Foltz, *Religions of the Silk Road: Pre-Modern Patterns of Globalization* (New York, 2010).

48. Frantz Grenet, "Religious Diversity among Sogdian Merchants in Sixth Century China: Zoroastrianism, Buddhism, Manichaeism, and Hinduism," *Comparative Studies of South Asia, Africa, and the Middle East* 27, 2 (2007): 463–78.

49. Pierre Leriche points out that most Greek centers in Central Asia have yet to be identified. "Bactria, Land of a Thousand Cities," 122. On Greek Merv, see Masson, *Strana tysiachi gorodov*, 100–122.

50. Herodotus, *The Histories*, 210, 301.

51. On this colony of the "Branchidae" from Miletus, Rtveladze cites Strabo, Arrian, and Quintus Curtius Rufus, *Civilizations, States, and Cultures of Central Asia*, 43.

52. Figures from Micah Greenbaum, "An Ancient Coalition: The Composition of Alexander the Great's Army," 13, 17, http://ebookbrowse.com/greenbaum-doc-d32027256.

53. This temple at Takht-e-Sangin had four columns across the façade. See the annual reports by I. R. Pichikian in the Moscow series *Arkheologicheskie otkrytiia* for the

years 1976–1979; and R. A. Litvinskii and I. R. Pichikian, "The Hellenistic Architecture and Art of the Temple of the Oxus," *Bulletin of the Asia Institute*, New Series, 8 (1994): 47–66; also I. Pichikian, *Kultura Baktrii* (Moscow, 1991), 138ff.

54. J. H. Marshall, *An Illustrated Account of Archaeological Excavations Carried out at Taxila under the Orders of the Government of India between the Years 1913 and 1934* (Cambridge, 1951), 222–29.

55. J. Hackim, *Nouvelles recherches archéologiques à Begram (ancienne Kâpici), 1939–1940, rencontre de trois civilisations, Inde, Grèce, Chine. Mémoires de la Délégation archéologique française en Afghanistan* (Paris, 1954), 91ff.

56. B. A. Litvinskij and I. R. Picikian, "River-Deities of Greece Salute the God of the River Oxus-Vakhsh; Achelous and the Hoppocampess," in *In the Land of the Gyrphons*, 129–37.

57. National Museum of Tajikistan, Dushanbe.

58. B. A. Litvinskii, "Kushanskie eroty-odin iz aspektov antichnogo na tsentralnoaziatskuyu kulturu," *Vestnik drevnei istorii* 2, 148 (1979): 89–109.

59. Rtveladze, *Civilizations, States, and Cultures of Central Asia*, 233; English corrected on basis of the Russian original.

60. S. H. Nasr, "Life Sciences, Alchemy, and Medicine," in *The Cambridge History of Iran*, ed. Richard N. Frye (Cambridge, 1975), 4:403.

61. Figure 3.2, Ai Khanoum, is described by Guy Lecuyot (UMR 8546 CNRS-ENS) thus:

Computer graphic (CG) images are today part of our everyday life. A few years ago, when the documentary film "The illusive Alexandria" was produced by Masahiro Kikuchi for the Japanese television channel NHK, CG images were created by Osamu Ishizawa, a senior architect in the TAISEI Corporation, using 3D Studio Max software, under the archaeological supervision of G. Lecuyot. For this purpose we used the documentation gathered on the site of Ai Khanum in North-Eastern Afghanistan, which was excavated between 1964 and 1978 by the French Archaeological Delegation in Afghanistan (DAFA) under the direction of Paul Bernard. (Detailed reports on the excavations have been published by P. Bernard in the *Comptes rendus de l'Académie des Inscriptions et Belles-Lettres (CRAI)*, 1965-72, 74-76, 78 and 80 and in the *Bulletin de l'Ecole Française d'Éxtrême-Orient (BEFEO)*, LXIII (1976) and LXVIII (1980). The definitive publications are published in the *Mémoires de la Délégation archéologique française en Afghanistan (MDAFA), Fouilles d'Aï Khanoum.*)

We chose to give a general view of this Hellenistic city towards 145 BC, a city which was built at the confluence of the Koksha and Daria i-Panj rivers and whose architecture mixed Greek and Eastern features. Separate restorations of the monuments were then gathered together and combined in order to create several virtual views the city. However, when we are looking at these images we have to bear in mind that they are based on many hypotheses and only offer a *virtual reality* which has probably never existed (G.L., Paris, March 25th, 2013).

Bibliography

G. Lecuyot, « Essai de restitution 3D de la Ville d'Aï Khanoum », dans O. Bopearachchi et M.-Fr. Boussac (éd.), *Afghanistan ancien carrefour entre l'est et l'ouest, Actes du colloque international de Lattes 5-7 mai 2003, Indicopleustoi. Archaeologies of the Indian Ocean* 3, Turnhout, 2004, p. 187-196.

G. Lecuyot et O. Ishizawa, « Aï Khanoum, ville grecque d'Afghanistan en 3D », *Archéologia* 420, mars 2005, p. 60-71.

G. Lecuyot, « Ai Khanum reconstructed », dans J. Cribb et G. Herrmann (éd.), *After Alexander. Central Asia Before Islam, Proceedings of the British Academy* 133, Oxford, 2007, p. 155-162 et pl. 1-4.

G. Lecuyot, « La 3D appliquée à la citée gréco-bactrienne d'Aï Khanoum en Afghanistan », *L'art d'Afghanistan de la préhistoire à nos jours. Nouvelles données*, Actes d'une journée d'étude du 11 mars 2005, CEREDAF, Paris, 2005, p. 31-48.

G. Lecuyot et O. Ishizawa, « NHK,TAISEI, CNRS : une collaboration franco-japonaise à la restitution 3D de la ville d'Aï Khanoum en Afghanistan », dans R. Vergnieux et C. Delevoie (éd.), *Actes du Colloque Virtual Retrospect 2005, Archéovision* 2, Bordeaux, 2007 p. 121-124.

62. D. W. Mac Dowall and M. Taddei, "The Greek City of Ai-Kahnum," in *The Archaeology of Afghanistan*, 216–32; Paul Bernard, "The Greek Colony at Ai-Khanoum and Hellenism in Central Asia," in *Afghanistan: Hidden Treasures of the National Museum, Kabul*, 81–130; also Guy Lecuyot, "Ai Khanum Reconstructed," in *After Alexander*, 155–62.

63. Claude Rapin, *Indian Art from Afghanistan* (Manohar, 1996), 9ff.

64. Jeffrey D. Lerner, "The Ai Khanoum Philosophical Papyrus," *Zeitschrift für Papyrologie und Epigraphik* 142 (2003): 45–51.

65. Preserved in the National Museum, Kabul.

66. Translated text from J. Harmatta, with editorial revisions by the author, "Languages and Scripts in Graeco-Bactria and the Saka Kingdom," in *History of Civilizations of Central Asia*, 2:406.

67. The Kandahar inscriptions, housed at the National Museum in Kabul, disappeared or were destroyed during the Taliban occupation.

68. Needham, *Science and Civilization in China*, 2:204ff.

69. On the less widespread but nonetheless important presence of Hinduism, see Banerjee, "Hindu Deities in Central Asia," in *India's Contribution to World Thought and Culture*, ed. Lokesh Chandra (Madras, 1970), 281–90.

70. A solid overview of Soviet research is B. A. Litvinsky's *Outline History of Buddhism in Central Asia* (Moscow, 1968); also B. N. Puri, *Buddhism in Central Asia* (Delhi, 1987).

71. B. Stavisky, "Kara Tepe in Old Termez," in *From Hecataeus to Al-Huwarizmi*, ed. J. Harmatta (Budapest, 1984), 134.

72. A. N. Dani and Bernard, "Alexander and His Successors in Central Asia," in *History of Civilizations of Central Asia*, 2:96.

73. Deborah Klimburg-Salter, *Buddha in India: Die Frühindische Skulpture von König Asoka bis zur Guptazeit* (Milan, 1995), 111–24; Joe Cribb, "The Origins of the Buddha Image—the Numismatic Evidence," in *South Asian Archaeology, 1981*, ed. Bridget Allchin (Cambridge, 1984), 243.

74. A. Foucher, "Greek Origins of the Buddha Type," in *The Beginnings of Buddhist Art* (Paris, 1917), 111–37.

75. F. A. Pugachenkova and Z. Usmanova, "Buddhist Monuments in Merv," in *In the Land of Gryphons*, 51–84.

76. B. A. Turgunov, "Excavations of Buddhist Temple at Dal'verzin-tepe," *East and West* 42 (1992): 131–53.

77. Stobdan, The Traces of Buddhism in the Semireche," *Himalayan and Central Asian Studies* 7, 2 (April–June 2003): 3–24.

78. Deborah Klimburg-Salter, "Corridors of Communication across Afghanistan 7th to 10th Centuries," in *Paysages du Centres de l'Afghanistan: Paysages naturels, paysages cuturel* (Paris, 2010), 167–85.

79. More useful as an overview than the special publications on this site is Galina A. Pugachenkova's "The Buddhist Monuments of Airtam," *Silk Road Art and Archaeology* 2 (1991/1992): 23–41.

80. Rtveladze, *Civilizations, States and Cultures of Central Asia*, 169.

81. Daniel Schlumberger, Marc le Berre, and Gerard Fussman, *Surkh Kotal en Batriane; les temples, architecture, sculpture, inscriptions, Memoires de la Délégation archéologique française en Afghanistan* (Paris, 1983), vol. 25, pt. 1, 144–52; on Bactrian religion in the Kushan period, see B. Ia. Stavinskii, *Kushanskaia Baktriia: Problemy, istorii, i kultury* (Moscow, 1977), chap. 8.

82. Asadi Tusi, *Garshaspnameh*, quoted in Bijan Omrani and Matthew Leming, *Afghanistan* (London, 2007), 579.

83. B. Stavisky, "Bactria and Gandhara," in *Gandharan Art in Context: East-West Exchanges at the Crossroads of Asia*, ed. Raymond Allchin et al. (New Delhi, 1997), 51, 160.

84. B. J. Stavisky, "On the Formation of Two Types of Buddhist Temples in Central Asia," in *Orient und Okzident im Spiegel der Kunst; Festschrift Heinrich Gerhard Franz zum 70. Geburtstag* (Graz, 1986), 381–86.

85. Boris J. Stavisky, "The Fate of Buddhism in Central Asia," in *Silk Road Art and Archaeology* (1993/1994), 3:132–33. Sarvastivadins, also mentioned in connection with Central Asia, is a later term for a subgroup of the Vaibhasika.

86. Litvinsky, *Outline History of Buddhism in Central Asia*, 13.

87. Siroj Kumar Chaudhuri, *Lives of Early Buddhist Monks: The Oldest Extant Biographies of Indian and Central Asian Monks* (New Delhi, 2008).

88. Anykul Chandra Banerjee, "The Vaibhasika School of Buddhist Thought," http://himalaya.socanth.cam.ac.uk/collections/journals/bot/pdf/bot_1982_02_01.pdf.

89. Ibid., 69.

90. Ibid., 59.

91. M. A. Abuseitova, "Historical and Cultural Relations between Kazakhstan, Central Asia, and India from Ancient Times to the Beginning of the Twentieth Century," *Dialogue* 6, 2 (December 2004), http://www.asthabharati.org/Dia_Oct04/Abuseitova.htm.

92. Ali Asghar Mostafavi, "Iranians' Role in Expansion of Buddhism," Circle of Ancient Iranian Studies, http://www.cais-soas/.com/CAIS/Religions/non-iranian/budhiran.htm.

93. J. Duchesne-Guillemin, "Zoroastrian Religion," in *The Cambridge History of Iran*, vol. 3, pt. 2, 882ff.; also Stavisky, "The Fate of Buddhism in Middle Asia," 15–133.

94. Mariko Namba Walter, "Sogdians and Buddhism," *Sino-Platonic Papers* 74 (November 2006): 32.

95. Buryakov et al., *The Cities and Routes of the Great Silk Road*, 83.

96. Deborah Klimburg-Salter, "Buddhist Painting in the Hindu Kush c. VIIth to Xth Centuries: Reflections of the Co-existence of Pre-Islamic and Islamic Artistic Cultures during the Early Centuries of the Islamic Era," in *Islamisation de l'Asie Centrale: Processus locaux d'acculturation du VIIe au XIe siècle*, ed. Étienne de la Vaissière, *Cahiers de Studia Iranica* (Paris, 2008), 39:140–42.

97. Deborah Klimburg-Salter, *The Kingdom of Bāmiyān: The Buddhist Art and Culture of the Hindu Kush* (Naples-Rome, 1989), 87–92, figs. 21–22.

98. Klimburg-Salter is bringing the various strands of this argument together in *Zones of Transition: Reconsidering Early Islamic Art in Afghanistan*, forthcoming.

99. Scores of major Buddhist sites in Xinjiang continued to thrive down to the Karakhanids' rise to power in the eleventh century (see chapter 10). On the greatest known library and its destruction, see Rong Xinjiang, "The Nature of the Dunhuang Library Cave and the Reasons for Its Sealing," *Cahiers d'Extrême-Asie* 11 (1999–2000): 247–75.

100. Rtveladze, *Civilizations, States, and Cultures of Central Asia*, 188–90.

101. A concise and authoritative summary is Peter Brown, *The Rise of Western Christendom* (Oxford, 2003), chap. 10.

102. O'Leary, *How Greek Science Passed to the Arabs*, 63–64; Rtveladze, *Civilizations, States, and Cultures of Central Asia*, 184–86. Christian communities thrived down to Tamerlane's (Timur's) devastations in the fourteenth century: see J. Asmussen, "Christians in Iran," in *The Cambridge History of Iran*, vol. 3, pt. 2, 947–48. On the paucity to date of archaeological evidence, see Maria Adelaide Lala Comneno, "Nestorianism in Central Asia during the First Millennium: Archaeological Evidence," *Journal of the Assyrian Academic Society* 11, 1 (1997): 20–67.

103. For an example of writings by Christians in Turfan, see Nicholas Sims-Williams, "The Christian Sogdian Manuscript C2," in *Schriften zur Geschichte und Kultur des Alten Orients, Berliner Turfantexte* (Berlin, 1985), vol. 12.

104. I. Gillman and H. Klimkeit, *Christians in Asia before 1500* (Ann Arbor, 1999), 252–53.

105. B. A. Litvinsky, "Christianity, Indian, and Local Religions," in *History of Civilizations of Central Asia*, 3:424–25. The identity of this *kagan* is unknown.

106. Brown, *The Rise of Western Christendom*, 76–77.

107. See, for example, G. Ia. Derevianskaia, "'Obvalnyi dom khristianskoi obshchiny v starom Merve," *Trudy arkheologichesko-komplektnoi ekspeditsii* 15 (1974): 155–81.

108. Theodore was a monophysite, who approached Christ's dual nature from the side of his divinity, in contrast to the Nestorians, who approached it from the side of his human nature. O'Leary, *How Greek Science Passed to the Arabs*, 37.

109. Yazberdiev, "The Ancient Merv and Its Libraries," 142.

110. Amber Haque, "Psychology from an Islamic Perspective: Contributions of Early Muslim Scholars and Challenges to Contemporary Muslim Psychologists," *Journal of Religion and Health* 43, 4 (2004): 357–77.

111. Book of Esther, 3:6; 8; 8:5; 12; 9:20.

112. Rtveladze, *Civilizations, States, and Cultures of Central Asia*, 47–48.

113. Walter J. Fischel, "The Rediscovery of the Medieval Jewish Community at Firuzkuh in Central Afghanistan," *Journal of the American Oriental Society* 85, 2 (April–June 1965): 148–53.

114. *Daily Mail*, January 9, 2013.

115. Itzhak Ben-Zvi, *The Exiled and the Redeemed* (Philadelphia, 1957), 69–71.

116. Arthur Upham Pope, *Persian Architecture* (London, 1978), 78.

117. This and other influences are discussed by Boris A. Litvinskij and Tamara I. Zejmal, *The Buddhist Monastery of Ajina Tepa, Tajikistan* (Rome, 2004), 66ff.

118. Nargis Khidzhaeva, "Vliianie Zoroastrizma na formirovanie statusa zhenshchiny v islame (brak i razvod)," in *Samanidy, epokha, istoki, kultury, n.e.* (Dushanbe, 2007), 61–70.

119. Iu. N. Zavadovskii, *Abu Ali Ibn Sina: zhizn i tvorchestvo* (Dushanbe, 1980), 82.

120. Needham, *Science and Civilization in China*, 1:220–21.

121. On Jayhani's research methods, see Barthold, *Turkestan Down to the Mongol Invasion*, 12.

122. A. K. Tagi-Zade and S. A. Vakhalov, "Astroliabi srednevekovogo Vostoka," *Istoriko-astronomicheskie issledovaniia* 12 (1975): 169–225; A. K. Tagi-Zade, "Kvadranty srednevekovogo Vostoka," *Istoriko-astronomicheskie issledovaniia* 13 (1977): 183–200.

123. David Pingree, "Abu Mashar," in *Dictionary of Scientific Biography*, 1:31–39.

124. A comprehensive but much disputed study of Bozorghmer is A. Christensen, "Le sage Buzurjmihr," *Acta Orientalia* 8 (1930): 18–128; more balanced is the entry on Bozorghmer in the *Encyclopaedia Iranica*, "Bozorgmehr-e Boktagān," http://www.iranicaonline.org/articles/bozorgmehr-e-boktagan.

125. Abubekr Muhammed ibn Yahya, from Sul (hence "Assuli"), lived in the first half of the tenth century.

126. This account is based on Buzurgmehr Batakhon," *Istoriia tadzhikskoi filosofii s drevneishikh vremen do XV v.*, 3 vols. (Dushanbe, 2010), 1:465–70.

127. V. G. Shkoda, "The Sogdian Temple: Structure and Rituals," *Bulletin of the Asia Institute*, New Series, 10 (1996): 195–201.

128. For example, see the amplifications of Indian texts written in Xinjiang, in Mukherjee, *India in Early Central Asia*, 29.

129. On Abu Hanifa, founder of the Hanafi school of law, W. Madelung quotes an early source that noted that "all the people of Balkh were adherents of his [Abu Hanifa's] doctrine," *Religious Trends in Early Islamic Iran* (Albany, 1988), 18.

130. Litvinskij and Zejmal defend this connection and correctly trace the hypothesis to Bartold, in *The Buddhist Monastery of Ajina Tepa, Tajikistan*, 66–67.

131. Seyyed Hossein Nasr and Mehdi Aminrazavi, eds., *An Anthology of Philosophy in Persia*, 2 vols. (New York, 1999), 1:274–75.

132. H. Ritter, "Abū Yazīd (Bāyazīd) Tayfur b. Īsā b. Surūshān al-Bistāmī," in *Encyclopaedia of Islam*, 2nd ed. (Leiden, 2009), http://www.brillonline.nl/subscriber/entry?entry=islam_SIM-0275.

133. On the impact on Asia of Buddhist music from Central Asia, see Bo Lowergren, "The Spread of Harps between the Near and Far East during the First Millennium AD," *Silk Road Art and Archeology* 4 (1995–1996): 233–76.

134. Al-Hakim Al-Tirmidhi, *The Concept of Sainthood in Early Islamic Mysticism*, ed. Bernd Radtke and John O'Kane (London, 1996), introduction.

135. G. Gilliot, "Qu'ranic Exegesis," in *History of Civilizations of Central Asia*, vol. 4, pt. 2, 98–103.

136. Judah Rosenthal, *Hiwi al-Balkhi, a Comparative Study* (Philadelphia, 1949).

137. F. Gilliot, "Theology," in *History of Civilizations of Central Asia*, vol. 4, pt. 2, 121–22. See also Davlat Dovudi, "Zoroastrizm i Islam posle arabskogo zavoevaniia i v epokhu samanidov," in *Samanidy: epokha, istoki, i kultura*, 35ff.

138. Brown, *The Rise of Western Christendom*, 189.

139. Basing his conclusion on a study of six thousand biographies and five hundred genealogies from Nishapur, Richard W. Bulliet sees thoroughgoing Islamization as more or less complete by the beginning of the tenth century. However, this is almost certainly too early for the region as a whole. See his *Conversion to Islam in the Medieval Period: An Essay in Quantitative History* (Cambridge, 1979), 19–23.

140. Steven Johnson, *Where Good Ideas Come from: The Natural History of Innovation* (London, 2011).

CHAPTER 4

1. Hugh Kennedy, *The Great Arab Conquests* (London, 2007), 225. The following account draws on Kennedy's careful chronology, which is the most authoritative and, at the same time, engaging synthesis to date of these complex events. See also H.A.R. Gibb's account, *The Arab Conquest of Central Asia* (London, 1932).

2. See below on the Panjikent and Samarkand murals from this period.

3. G. A. Pugachenkova and E. V. Rtveladze, *Severnaia Baktriia-Tokharistan* (Tashkent, 1990), 131ff.

4. C. E. Bosworth, "Barbarian Incursions: The Coming of the Turks into the Islamic World," in *The Medieval History of Iran, Afghanistan and Central Asia* (London, 1977), chap. 23; Barthold, *Turkestan Down to the Mongol Invasion*, 183.

5. On local gods, see B. I. Marshak and F. Grenet, "Le mythe de Nana dans l'art de la Sogdiane," *Arts Asiatiques* 53 (1998): 5–18; N. V. Diakonova and O. U. Smirrnova, "K voprosu o kulte Nany (Anakhity) v Sogde," *Sovetskaia arkheologiia* 1 (1967): 74–83; Azarpay, *Sogdian Painting*, 132–39. On coinage, see Barthold, *Turkestan Down to the Mongol Invasion*, 180ff.

6. V. A. Zavyalov, "The Fortifications of the City of Graur Kala, Merv," in *After Alexander*.

7. A. I. Kolesnikov, "Social and Political Consequences of the Arab Conquest," in *History of Civilizations of Central Asia*, 3:483; Swetlana B. Lunina, "Die Stadt Merw, ein Zentrum des Kunsthandwerk im Mittelalterichen Orient," in *Orient und Okzident im Spiegel der Kunst*, ed. Guenter Brucher et al. (Graz, 1986), 221–27.

8. That the origins of Zoroastrianism are to be sought in the adjacent Karakum Desert is the thesis of Sarianidi's *Margush, Turkmenistan*.

9. Yazberdiev, "The Ancient Merv and Its Libraries," 141.

10. R. M. Bakhadirov, *Iz istorii klassifikatsii nauk na srednevekovom musulmanskom vostoke* (Tashkent, 2000), 6, 19ff., 144.

11. Azarpay, *Sogdian Painting*, plate 3.

12. On the Panjikent murals, see V. I. Marshak's contribution to ibid.; and A. N. Belenitskii et al., eds., *Drevnosti Tadzhikistana* (Dushanbe, 1985), 264ff.; also B. I. Marshak and V. I. Raspapova, "Wall Paintings from a House with a Granary, Panjikent," *Silk Road Art and Archaeology*, 1:123–76. On the Samarkand murals, see Azerpay, *Sogdian Painting*; also L. I. Albaum, *Zhivopis Afrasiaba* (Tashkent, 1975); and Sergei A. Yatsenko, "The Costumes of Foreign Embassies and Inhabitants of Samarkand on Wall Painting of the 7th c. in the 'Hall of Ambassadors' from Afrasiab as a Historical Source," *Transoxonia* (June 2004), http://transoxonia.com.ar/0108/yatsenko-afrasiab_costume.htm.

13. Frantz Grenet and Masud Samidaev, "'Hall of Ambassadors' in the Museum of Afrasiab" (Samarkand, 2002).

14. *The Cambridge History of Iran*, 4:18–26. For a convenient summary, see Hugh Kennedy, *The Prophet and the Age of the Caliphates: The Islamic Near East from the 6th to the 11th Century*, 2nd ed. (London, 2004), 112–22.

15. Needham *Science and Civilization in China*, 1:214.

16. Kennedy, *The Great Arab Conquests*, 171, 183, 187; also the chapter by Richard N. Frye, "The Islamic Conquests in Iran," in *The Great Age of Persia*, chap. 4.

17. Gibb dates the first raids to 652 in *The Arab Conquests of Central Asia*, 15; Barthold, *Turkestan Down to the Mongol Invasion*, 6, claims this occurred in the mid-650s, while A. H. Jalilov dates the start to the 680s: "The Arab Conquest of Transoxonia," in *History of Civilizations of Central Asia*, 2:456.

18. Barthold, *Turkestan Down to the Mongol Invasion*, 182–83.

19. Marshall G. S. Hodgson, *The Venture of Islam*, 2 vols. (Chicago, 1961), 1:208.

20. Kennedy, *The Great Arab Conquests*, 195.

21. Ibid., 239.

22. Ibid., 195–96.

23. Gibb, *The Arab Conquests in Central Asia*, chap. 4.

24. In addition to the accounts of Kennedy, *The Great Arab Conquests*, 255–76, and Barthold, *Turkestan Down to the Mongol Invasion*, 184ff., see Gibb, *The Arab Conquests in Central Asia*, chap. 3.

25. Kennedy, *The Great Arab Conquests*, 256.

26. Ibid.

27. V. G. Gafurov, *Tadzhiki: Drevneishaia, drevnaia, i srednevekovaia istoriia* (Moscow, 1972), 311.

28. Frye, *The Golden Age of Persia*, 95.

29. Barthold, *Turkestan Down to the Mongol Invasion*, 188.

30. Francis Henry Skrine and Edward Denison Ross, *The Heart of Asia* (London, 1899), 66.

31. G. Bulgakov, "Al Biruni on Khwarizm," in *History of Civilizations of Central Asia*, 3:229.

32. Dagmar Schreiber, *Kazakhstan: Nomadic Routes from Caspian to Altai* (Hong Kong, 2008). 293.

33. A. Biruni, *Pamiatniki minuvshikh pokolenii*, ed. M. A. Sale (Tashkent, 1957), 48, 63.

34. Rtveladze, *Civilizations, States and Cultures of Central Asia*, 167.

35. Kennedy, *The Great Arab Conquests*, 273–74.

36. K. Athamina, "Arab Settlement during the Umayyad Caliphate," *Jerusalem Studies of Arabic and Islam* 8 (1986): 187–89. Several authors equate the figure with fifty thousand families.

37. Venetia Porter, "Inscriptions of Companions of the Prophet in the Merv Oasis," in *Islamic Reflections. Arabic Musings: Studies in Honour of Alan Jones*, ed. R. Hoyland and Kennedy (Oxford, 2004), 290ff.

38. For a vivid account of the complexities of Islamization in Central Asia and Iran, see Jamsheed K. Choksy, *Conflict and Cooperation: Zoroastrian Subalterns and Muslim Elites in Medieval Iranian Society* (New York, 1997).

39. On Isfara, see Saidov et al., "The Ferghana Valley," 18.

40. Barthold, *Turkestan Down to the Mongol Invasion*, 203.

41. This charge is made by G. Le Strange, *Baghdad during the Abbasid Caliphate* (Oxford, 1924), 3.

42. Barthold, *Turkestan Down to the Mongol Invasion*, 191.

43. On the authority of the ninth-century Persian historian Tabari, see ibid., 188.

44. Le Strange, *Baghdad during the Abbasid Caliphate*, documents this with a quarter-by-quarter survey.

45. Kennedy, *The Great Arab Conquests*, 197–200.

46. Beckwith, *Empires of the Silk Road*, 132.

47. Barthold, *Turkestan Down to the Mongol Invasion*, 182.

48. Needham *Science and Civilization in China*, 1:214, 713, 726.

49. On Turkic relations with the region in these years, see Nargiz Akhundova, *Tiurki v sisteme gosudarstvennogo upravleniia arabskogo khalifata* (Baku, 2004), chap. 1.; Barthold, *Turkestan Down to the Mongol Invasion*, 188.

50. Barthold, *Turkestan Down to the Mongol Invasion*, 186.

51. For a solid account see Frantz Grenet and Étienne de La Vaissière, "The Last Days of Panjikent," *Silk Road Art and Archaeology*, 8:155–81.

52. Gafurov, ed., *Istoriia tadzhikskogo naroda*, 2:21.

53. Ibid., 2:107. On the Mug documents, see chapter 3.

54. I. M. Muminov, ed., *Istoriia Samarkanda*, 2 vols. (Tashkent, 1969), 1:83–115.

55. Gafurov, ed., *Istoriia tadzhikskogo naroda*, 2:111; Barthold, *Turkestan Down to the Mongol Invasion*, 192.

56. On taxes, see Frye, *The Great Age of Persia*, 89; on currency, see Barthold, *Turkestan Down to the Mongol Invasion*, 205ff.

57. Barthold, *Turkestan Down to the Mongol Invasion*, 191, 184–85.

58. A. K. Mirbabaev et al., "The Development of Education: Maktab, Madrasa, Science, and Pedagogy," in *History of Civilizations of Central Asia*, vol. 4, pt. 2, 33.

59. On the excavation of Varakhsha, see the seminal works of V. A. Shishkin, "Nekotorye itogi arkheologicheskoi raboty na gorodishche Varakhsha (1947–1953 gg.)," in *Trudy instituta istorii i arkheologii Akademii Nauk Uzbekskoi SSR* (Tashkent, 1956), 8:3–42; and especially his *Varakhsha* (Moscow, 1963); also A. N. Beleinitskii and B. I. Marshak, "Voprosy khronologii zhivopisi rannosrednevekovogo Sogda," *Uspekhi Sredneaziatskoi arkheologii* 4 (1979). For a convincing effort at chronology, see Aleksandr Naymark, "Returning to Varakhsha," http://www.silk-road.com/newsletter/december/varakhsha.htm.

60. Shishkin, *Varakhsha*, 150.

61. Naymark, "Returning to Varakhsha," 8–9.

62. Ibid., 160.

63. On Abu Muslim's origins, see Elton L. Daniel, "The 'Ahl Al-Taqadum' and the Problem of the Constituency of the Abbasid Revolution in the Merv Oasis," *Journal of Islamic Studies* 7, 2 (1996): 162–63.

64. Barthold, *Turkestan Down to the Mongol Invasion*, 195.

65. The Arab character of the Abu Muslim revolt was long asserted by most scholars but has been convincingly refuted by Daniel in his brilliant study, "The 'Ahl Al-Taqadum,'" 130–79.

66. Barthold, *Turkestan Down to the Mongol Invasion*, 192.

67. For an excellent account of the interplay of China, Tibet, and the Arabs, see Christopher I. Beckwith, *The Tibetan Empire in Central Asia*, 108–42; also Barthold, *Turkestan Down to the Mongol Invasion*, 195ff.

68. On An Lushan, see Edward G. Pulleyblank, *The Background of the Rebellion of An Lu-shang* (London, 1982), with source materials at R. des Retours, *Histoire de Lgan Lou-chan* (Paris, 1962); Steven Pinder makes this claim in his controversial *The Better Angels of Our Nature* (New York, 2011).

69. Kennedy, *The Prophet and the Age of the Caliphates*, 117.

70. The chief source for this is the thirteenth-century historian Ibn Khallikan. See Edward G. Browne, *A Literary History of Persia*, 4 vols. (London, 1908), 3:320; and Richard N. Frye, "The Role of Abu Muslim in the Abbasid Revolt," in *Iran and Central Asia (7th–12th Centuries)* (London, 1979), 29.

71. R A. Jairazbhoy, "The Taj Mahal in the Context of East and West: A Study in the Comparative Method," *Journal of the Warburg and Courtault Institutes* 24, 1–2 (January–June 1961): 98.

72. Skrine and Ross, *The Heart of Asia*, 88–89.

73. Hugh Kennedy, *When Baghdad Ruled the Muslim World* (London, 2004), 16.

74. Browne, *A Literary History of Persia*, 3:361; Frye, *The Great Age of Persia*, 128.

75. Barthold, *Turkestan Down to the Mongol Invasion*, 198ff.

76. For a comprehensive Marxist account, see A. Iu. Iakubovskii, "Vosstanie Mukanny—dvizhenie liudei v 'belykh odezhdakh,'" *Sovetskoe vostokovedenie* 5 (1948).

CHAPTER 5

1. Paul Kriwaczek, *Yiddish Civilization* (New York, 2006), 67–68.

2. Daniel, "The Islamic East," 469, 475.

3. Dimitri Gutas, *Greek Thought, Arabic Culture: The Graeco-Arabic Translation Movement in Baghdad and Early Abbasid Society (2nd–4th, 8th–10th centuries)* (New York, 1998), 33. For the writings of later descendants of this learned family, see K. van Bladel, "The Arabic History of Science of Abu Sahl ibn Nawbakht (fl. ca 770–809) and Its Middle Persian Sources," in *Islamic Philosophy, Science, Culture, and Religion: Studies in Honor of Dimitri Gutas*, ed. David Reisman and Felicitas Opwis (Leiden, 2012), 41–62. Another of the astrologers, Mashallah ibn Athari, was a Persian from Basra and the author of several books on both mathematics and astronomy, some of which showed the clear influence of Greek thought. O'Leary, *How Greek Science Passed to the Arabs*, 103.

4. J. Lassner, "The Caliph's Personal Domain: The City Plan of Baghdad Re-Examined," in *The Islamic City in Light of Recent Research*, ed. A. H. Hourani and S. M. Stern (Oxford, 1970), 104.

5. Saleh Ahmad El-Ali, "The Foundation of Baghdad," in *The Islamic City*, ed. A. H. Hourani and S. M. Stern, *Papers on Islamic History* (Oxford, 1973), 1:94.

6. Kennedy, *When Baghdad Ruled the Muslim World*, 38ff.

7. Le Strange, *Baghdad during the Abbasid Caliphate*, 422.

8. An authoritative account of the family is by Kevin van Bladel, "The Bactrian Background of the Barmakids," in *Islam and Tibet: Interactions along the Musk Routes*, ed. Anna Akasoy, Charles Burnett, and Ronit Yoeli-Tlalim (Farnham, 2010), 43–88;

also Lucien Bouvat, "Les Barmécides d'après les historens Arabes et persans," *Revue du Monde Musselman* 20 (1912): 1–131.

9. Barthold, *Turkestan Down to the Mongol Invasion*, 77.

10. O'Leary is one of several scholars who cite Arabic sources to argue that the Barmaks converted first to Zoroastrianism, which seems unlikely: *How Greek Science Passed to the Arabs*, 104.

11. Kennedy, *The Great Arab Conquests*, 183. On the round city of Gur (Firuzabad) in Fars, see Sheila Blair and Jonathan Bloom, "History," in *Islam Art and Architecture*, ed. Markus Hattstein and Peter Delius (London, 2001), 96. For a survey of round city plans in the ancient Middle East and Central Asia, see B. Brentjes, "The Central Asian Square in a Circle Plan," *Bulletin of the Asia Institute*, New Series, 5 (1991): 180–83.

12. O 'Leary, *How Greek Science Passed to the Arabs*, 103.

13. Among those claiming the larger size for Baghdad was E. Hertzfeld, *Archaeologische Reise im Euphrat und Tigris Gebiet* (Berlin, 1922), chap. 2. Cf. Richard Ettinghausen and Oleg Grabar, *The Art and Architecture of Islam 650–1250* (New Haven, 1987), 75–79.

14. Ross Burns, *Damascus: A History* (New York, 2005), 280n44.

15. Lassner, "The Caliph's Personal Domain," 108, 115.

16. Christopher T. Beckwith, "The Plan of the City of Peace: Central Asian Iranian Factors in Early Abbasid Design," *Acta Orientaliae Academiae Scientarum Hungaricae* 38 (1984): 126–47.

17. See, for example, Andre Clot, *Harun al-Rashid and the World of the Thousand and One Nights* (London, 1986), 33ff.

18. Ibid., 36; Kennedy, *When Baghdad Ruled the Muslim World*, 53.

19. C. E. Bosworth, *The Abbasid Caliphate in Equilibrium*, vol. 30 of *The History of al-Tabari* (Albany, 1989), xvii ff.

20. Clot, *Harun al-Rashid*, 218, 46.

21. This is not to deny the other, very different meaning of *jihad*, as a purely inward process.

22. Kennedy, *When Baghdad Ruled the Muslim World*, 211.

23. Kennedy, *The Prophet and the Age of the Caliphates*, 145.

24. Clot, *Harun al-Rashid*, 54–56.

25. Gutas, *Greek Thought, Arabic Culture*, 33.

26. D. I. Evarnitsky, cited by Yazberdiev, "The Ancient Merv and Its Libraries," 142.

27. Gutas, *Greek Thought, Arabic Culture*, 13; Cyril Elgood, *A Medical History of Persia* (Cambridge, 1951).

28. Van Bladel, "The Bactrian Background of the Barmakids," 43, 82–83.

29. Gutas, *Greek Thought, Arabic Culture*, 46–47; Kennedy, *When Baghdad Ruled the Muslim World*, 44; Ahmad Shalaby, *History of Muslim Education* (Beirut, 1954), 26ff.

30. John M. Cooper, *Pursuits of Wisdom; Six Ways of Life in Ancient Philosophy, from Socrates to Plotinus* (Princeton, 2012), 6–7.

31. Clot, *Harun al-Rashid*, 209.

32. Gutas, *Greek Thought, Arabic Culture*, 13.

33. Shalaby, *History of Muslim Education*, 26ff.

34. Peter Brown, *The World of Late Antiquity* (London, 1971), 201.

35. O 'Leary, *How Greek Science Passed to the Arabs*, 110.

36. Gutas, *Greek Thought, Arabic Culture*, chap. 5; also Bernard Lewis, *What Went Wrong? The Clash Between Islam and Modernity in the Middle East* (New York, 2002), 139.

37. Thomas Aquinas, *Commentary on Aristotle's Politics*, trans. Richard J. Regan (Indianapolis, 2007).

38. See Gutas, *Greek Thought, Arabic Culture*, 152ff.

39. On the fall of the Barmaks see Clot, *Harun al-Rashid*, 85ff.

40. Kennedy, *When Baghdad Ruled the Muslim World*, 79.

41. Clot, *Harun al-Rashid*, 82–84; Kennedy, *When Baghdad Ruled the Muslim World*, 70ff.

42. Freely adapted by the author from the translation by A. V. Williams Jackson, *Early Persian Poetry* (New York, 1920), 17.

43. Le Strange, *Baghdad during the Abbasid Caliphate*. The walls had actually been built in 250 BC by a successor to Alexander the Great.

44. Kennedy, *When Baghdad Ruled the Muslim World*, 88ff., 154.

45. D. R. Hill, "Physics and Mechanics, Civic and Hydraulic Engineering," in *Civilizations of Central Asia*, vol. 4, pt. 2, 252.

46. O'Leary, *How Greek Science Passed to the Arabs*, 83.

47. Kennedy, *When Baghdad Ruled the Muslim World*, 209.

48. A highly intelligent biography of Mamun is Michael Cooperson, *Al-Ma'mun* (Oxford, 2005).

49. Thus Jonathan Lyons uses the name as a general symbol of Arab achievement in his book *The House of Wisdom: How the Arabs Transformed Western Civilization* (New York, 2009).

50. For a more balanced perspective, see Gutas, *Greek Thought, Arabic Culture*, 5.

51. John J. O'Connor and Edmund F. Robertson, "Banu Musa Brothers," *MacTutor History of Mathematics Archive*, University of St. Andrews, http://www. history.mcs.standrews.ac.uk/Biographies/Banu_Musa.html.

52. Donald R. Hill, trans. and ed., *Book of Ingenious Devices* (Dordrecht, 1978), 19–25.

53. Teun Koetsier, "On the Prehistory of Programmable Machines: Musical Automata, Looms, Calculators," in *Mechanism and Machine Theory* (Amsterdam, 2001), 589–603.

54. One cannot exclude the possibility that the impressive drawings in the surviving manuscript were not copied from Jazari's originals but were developed by the medieval copyist or a colleague. Donald Routledge Hill ably describes the drawings accompanying Ahmad ibn Musa's and Jazari's works: *A History of Engineering in Classical and Medieval Times* (Düsseldorf, 1973).

55. On Mutazilism, see Richard C. Martin and Mark R. Woodward, *Defenders of Reason in Islam: Mutazilism from Medieval School to Modern Symbol* (London, 1997); and Henry Corbin, *History of Islamic Philosophy* (London, 1993), chap. 3. For Patricia Crone's thesis regarding Mutazilite anarchism, see *From Kavad to al-Ghazali: Religion, Law, and Political Thought in the Near East, 600–1100* (London, 2005), chap. 10; while the link between Mutazilite thought and Zoroastrianism is set forth in *The Cambridge History of Iran*, 4:555–59.

56. The best material on Tirmidhi is to be found in Wilferd Madelung, *Origins of the Controversy Over the Createdness of the Qu-ran* (Leiden, 1973).

57. See Josef van Ess's brilliant discussion in *The Flowering of Muslim Theology*, trans. Jane Maria Todd (Cambridge, 2006), chap. 3.

58. A. N. Nader's *Le système philosophique des mu'tazila* (Beirut, 1956) remains the clearest exposition of their doctrine; see also G. E. von Grunebaum, *Classical Islam: A History, 600–1258* (Chicago, 1970), 94ff.

59. O'Leary, *How Greek Science Passed to the Arabs*, 98.

60. Kennedy, *When Baghdad Ruled the Muslim World*, 250ff.

61. *The Civilizations of Central Asia*, vol. 4, pt. 2, 109.

62. Peter Adamson reviews this relationship in "Al-Kindi and the Mu'tazila: Divine Attributes, Creation and Freedom," *Arabic Sciences and Philosophy* 13 (2003): 45–77.

63. Nimrod Hurvitz, *The Formation of Hanbalism: Piety into Power* (London, 2002), 120.

64. This view is consistent with the research of John A. Nawas, "The Minha of 218 AH/833 AD Revisited: An Empirical Study," *Journal of the American Oriental Society* 116, 4 (1996): 698–708. On Mamun's absolutist designs, see Hodgson, *The Venture of Islam*, 1:280ff. and Gardet, *La cité muselmane, vie sociale et politique*, chap. 2.

65. For a further study, see John A. Nawas, "A Reexamination of Three Current Explanations for Al-Mamun's Introduction of the Mihna," *International Journal of Middle East Studies* 26 (1994): 615–29.

66. Translated by Walter Melville Patton, *Ahmad ib Hanbal and the Mihna* (Leiden, 1897), 56ff.

67. Biographical information on Hanbal can be found in Muhammad Abu Zahra, *The Four Imams* (Dar al Taqwa, 2001), 391ff.; and Christopher Melchert, *Ahmad ibn Hanbal* (Oxford, 2006), 1–22.

68. N. J. Coulson, *A History of Islamic Law* (Edinburgh, 1964), 91; Susan A. Spectorsky, "Ahmad Ibn Hanbal's Fiqh," *Journal of the American Oriental Society* 102, 3 (1982): 461–65.

69. Wesley Williams, "Aspects of the Creed of Imam Ahmad ibn Hanbal: A Study of Anthropomorphism in Early Islamic Discourse," *International Journal of Middle East Studies* 34 (2002): 441ff.

70. Nimrod Hurvitz, "Schools of Law and Historical Context: Reexamining the Hanbali *Madhhab*," *Islamic Law and Society* 7, 1 (2000): 47.

71. The title of Patricia Crone and Martin Hind's excellent study (Cambridge, 1986).

72. Matthew Gordon discusses these forces in detail in *The Breaking of a Thousand Swords: A History of the Turkish Military of Samarra* (New York, 2001).

73. Akhundova, *Tiurki v sisteme gosudarstvennogo upravleniia arabskogo khalifata*, 161ff.; Daniel Pipes, *Slave Soldiers and Islam* (New Haven, 1981), 145ff., stresses the importance of the forces' unfree status, which Kennedy minimizes in *When Baghdad Ruled the Muslim World*, 214.

74. Le Strange, *Baghdad during the Abbasid Caliphate*, 82.

75. Majid Fakhry, *A History of Islamic Philosophy* (New York, 1983), chap. 7.

CHAPTER 6

1. N. N. Negmatov, "States in North-Western Central Asia," in *History of Civilizations of Central Asia*, 2:443.

2. See Helen Wadell's classic, *The Wandering Scholars of the Middle Ages* (London, 1926).

3. *Khorazm Mamun Akademiiasi 1000 yil* (Tashkent, 2006), 2.

4. Kraemer, *Humanism in the Renaissance of Islam*, 57.

5. Heinrich Suter, *Die Mathematiker und Astronomen der Araber* (Leipzig, 1900), 65ff.

6. Bakhrom Abdukhalimov, *"Bait al-Khikma"va urta osie olimplarining bagdoddagi ilmii faoliiati* (Tashkent, 2004), chap. 5.

7. R. Lorch, "Al Saghani's Treatise on Projecting the Sphere," in *From Deferent to Equant: A Volume of Studies in Honour of E. S. Kennedy*, ed. D. A. King and G. Salibam (New York, 1987), 237ff.

8. Richard Lorch, *Al-Farghani on the Astrolabe* (Wiesbaden, 2005).

9. A. K Tagi-Zade and S. A. Vakhalov, "Astroliabi srednevekovogo Vostoka," *Istoriko-astronomicheskie issledovaniia* 12 (1975): 169–225; A. K. Tagi-Zade, "Kvadranty srednevekovogo Vostoka," *Istoriko-astronomicheskie issledovaniia* 13 (1977): 183–200.

10. Franz Rosenthal, "Al-Asturlabi and as-Samaw'al on Scientific Progress," *Osiris* 9 (1950): 555–64.

11. Abu Hayyan al-Tawhidi's (c. 930–1023) two volumes of discussions include sessions held at the Buyid court in Rayy, Iran, as well as in Baghdad. Neither is available in any Western language.

12. See the excellent essays on Sijistani in *Istoriia tadzhikskoi filosofii*, 119–22; and Kraemer, *Humanism in the Renaissance of Islam*.

13. Kraemer, *Humanism in the Renaissance of Islam*, 68.

14. Later Arab writers claimed Jabir and many others for themselves, arguing that he was in fact an Arab and not from Khurasan. But it is certain that he had close ties to the Barmak family, that his father spent many years in Khurasan, and that the earliest sources trace his origins to Tus, which had few Arab immigrants. For a contrary view, see William Newman, "New Light on the Identity of Geber," *Sudhoffs Archiv* 69 (1985): 76ff.

15. Seyyed Hossein Nasr, *An Introduction to Islamic Cosmological Doctrines* (Albany, 1993), 37.

16. A convenient overview is E. J. Holmyard, *Makers of Chemistry* (Oxford, 1931); also E. J. Holmyard, ed., *The Arabic Works of Jabir ibn Hayyan* (New York, 1928).

17. The best source on this extraordinary figure is by the late Prague scholar Paul Kraus, *Jâbir ibn Hayyân—Contribution à l'histoire des idées scientifiques dans l'Islam—Jâbir et la science grecque* (Paris, 1986).

18. Max Meyerhof, "Ali at-Tabari's 'Paradise of Wisdom,' One of the Oldest Arabic Compendiums of Medicine," *Isis* 6, 1 (1931): 7–12.

19. M. Z. Siddiqi, *Firdausy'l-Hikmat, or "Paradise of Wisdom," by Ali b. Rabban at-Tabari*, 8 vols. (Berlin, 1928).

20. Y. Tzvi Langermann, "The Book of Bodies and Distances of Habash al-Hasib," *Centaurus* 28 (1985): 108–13. Langermann's translation of Habash's *Book of Bodies and Distances* is on 122–27.

21. E. S. Kennedy and Richard Lorch, "Habash al-Hasib on the Melon Astrolabe," in *Astronomy and Astrology in the Medieval Islamic World*, ed. Edward S. Kennedy (Aldershot, 1998), 1–13.

22. J. L. Berggren, "The Mathematical Sciences," in *History of Civilizations of Central Asia*, 4:189.

23. I. Iu. Krachkovskii, *Arabskaia geograficheckaia literatura; izbrannye proizvedeniia*, 4 vols. (Moscow, 1957), 4:86.

24. Farghani's work is as yet unavailable in English. The best short study of it is Bahrom Abdukhalimov, "Ahmad al-Farghani and His Compendium of Astronomy," *Journal of Islamic Studies* 10, 2 (1999): 142–58; see also O. Buriev, *Al-Fargonii va uning namii merosi* (Tashkent, 1998). Cf. also A. I. Sabra, "Al Farghani," in *Dictionary of Scientific Biography*, 4:541–45.

25. George Sarton, *Introduction to the History of Science*, 3 vols. (Baltimore, 1927–1948), 1:667.

26. For an overview of Adelard's relation to Khwarazmi, see Lyons, *The House of Wisdom*, 121, etc.

27. L.C. Karpinski, *Robert of Chester's Latin Translation of the Algebra of Al-Khowarizmi: With an Introduction, Critical Notes and an English Version* (London, 1915).

28. Barnabas Hughes, "Gerard of Cremona's Translation of al-Khwārizmī's al-Jabr: A Critical Edition," *Medieval Studies* 48 (1986): 211–63.

29. Muhammad ben Musa, *The Algebra of Mohammed ben Musa*, trans. Friedrich August Rosen and Charles Theophilus Metcalfe (London, 1831), 3–4.

30. John Derbyshire, *Unkn(o)wn Quantity: A Real and Imaginary History of Algebra* (New York, 2006), 49.

31. *The Algebra of Mohammed ben Musa*, 150ff.

32. John Stillwell, *Mathematics and Its History* (New York, 1989), 48–49.

33. John J. O'Connor and Edmund F. Robertson, "Muḥammad ibn Mūsā al-Khwārizmī," MacTutor History of Mathematics Archive, University of St. Andrews.

34. Roshdi Rashed, "Algebra," in *Encyclopedia of the History of Arabic Science*, 2:352–53.

35. Roshdi Rashed, "Where Geometry and Algebra Intersect," in *From Five Fingers to Infinity*, ed. Frank J. Swetz (Chicago, 1994), 275. Needham, *Science and Civilization in China*, 3:147, argues that Chinese mathematicians had worked this out centuries earlier.

36. Derbyshire, *Unkn(o)wn Quantity*, 48.

37. The best review of this process is A. Allar, "Al-Khorezmi i proiskhozhdenie latinskogo algorizma," in *Mukhamad ibn Musa al-Khwarazmi*, ed. Iu. Iushkevich (Moscow, 1983), 53–67; The only European translation appears to be that of Menso Folkerts, *Die älteste lateinische Schrift über das indische Rechnen nach al-Hwārizmī* (Munich, 1997).

38. Boris A. Rozenfeld, "Sfiricheskaia geometriia Al-Khorezmi'a," in *Istoricheskie materialyi issledovaniia* (Moscow, 1990), 32–33:325–39.

39. Needham, *Science and Civilization in China*, 3:81–82, 112ff., 146–55; 4:48–52.

40. See A. Akhmedov, B. A. Rozenfeld, and N. D. Sergeeva, "Astronomicheskie I geograficheskie trudy al-Khorezmi, in Iushkevich, in *Mukhamad ibn Musa al-Khwarazmi*, 141–91; and M. N. Rozanskaia, "O znachenii 'Zidzha' al-Khorezmi v istorii, astronomii," in ibid., 192–212.

41. See G. J. Toomer's generally dismissive review, "Al-Khwarazmi, Abu Ja'far Muhammad Ibn Musa," in *Dictionary of Scientific Biography*, 361.

42. David A. King, "Al-Khwarazmi and New Trends in Mathematical Astronomy in the Ninth Century," Hagop Kevorkian Center, *Occasional Paper* no. 2 (New York, 1983).

43. Akhmedov, Rozenfeld, and Sergeeva, "Astronomicheskie I geografichestie trudy al-Khorezmi," 160–90.

44. Huburt Daunicht, "Der Osten nach der Erdkarte al-Huwārizmīs: Beiträge zur historischen Geographie und Geschichte Asiens," *Bonner orientalistische Studien* 19 (1968–1970).

45. N. N. Negmatov, ed., *Khorezm i Mukhamad al-Khorezmi v mirovoi istorii i culture* (Dushanbe, 1983), 19.

46. R. Bakhadirov reached this conclusion independently but provides no evidence for it in his *Iz istorii klassifikatsii nauk na srednevekovom musulmanskom vostoke* (Tashkent, 2000), 21.

47. A. R. Mukhamedjanov, "Economy and Social System in Central Asia in the Kushan Age," in *History of Civilizations of Central Asia*, 2:268.

48. Rtveladze, *Civilizations, States ,and Cultures of Central Asia*, 166–69.

49. A. Akhmedov, "The Persian and Indian Origins of Islamic Astronomy," in *History of Civilizations of Central Asia*, vol. 4, pt. 2, 195, 204.

50. O'Leary, *How Greek Science Passed to the Arabs*, 106.

51. There are solid grounds for believing that the purpose of the expedition was less diplomatic than to find the wall that Alexander the Great was said to have erected against Gog and Magog: Emeri van Donzel and Andrea Schmidt, *Gog and Magog in Early Christian and Islamic Sources: Sallam's Quest for Alexander's Wall* (Leiden, 2010). A careful account of this expedition is in Travis Zadeh, *Mapping Frontiers across Medieval Islam: Translation Geography and the Abbasid Empire* (London, 2011).

52. Negmatov, *Khorezm i Mukhamad al-Khorezmi v mirovoi istorii i culture*, 19–20.

53. On Khujandi, see Sevim Tekeli, "Al-Khujandi," in *Dictionary of Scientific Biography*, 7:352–54; also A. Akhmedov, "Astronomy, Astrology, Observatories and Calendars," in *History of Civilizations of Central Asia*, 4:202–3. On his sextant, see J. Frank, "Ueber zwei astronomische arabische Instrumente," *Zeitschrift für Instrumentenkunde* 41 (1921):193–200.

54. Joel L. Kramer, "Humanism in the Renaissance of Islam: A Preliminary Study," *Journal of the American Oriental Society* 104, 1 (January–March, 1984), especially 138ff.

55. The most authoritative study of Kindi is Peter Adamson, *Al-Kindi* (London, 2006); see also Amos Bertolacci, "From al-Kindi to Al-Farabi: Avicenna's Progressive Knowledge of Aristotle's Metaphysics According to His Autobiography," in *Arabic Sciences and Philosophy* 2 (2001): 289ff.; Corbin, *History of Islamic Philosophy*, 154–58.

56. Available in French, with excellent commentary, in R. Rashed and J. Jolivet, *Oeuvres philosophiques et scientifiques d al-Kindi*, 2 vols. (Leiden, 1998), 2:1–117; and in English, Al-Kindi, *Al-Kindi's Metaphysics: A Translation of Ya'qub ibn Ishaq al-Kindi's Treatise "On First Philosophy,"* ed. Alfred L. Ivry (New York, 1974).

57. Adamson, *Al-Kindi*, 45–48.

58. Richard Lemay, *Abu Ma'shar and Latin Aristotelianism in the 12th Century*, American University of Beirut, Publications of the Faculty of Arts and Sciences. Oriental Series, no. 38, 1962; also Pingree, "Abu Mashar al-Balkhi, Ja'far ibn Muhammad," 32–39.

59. Matti Moosa, "A New Source on Ahmed ibn al-Tayyib al-Sarakhsi: Florentine MS Arabic 299," *Journal of the American Oriental Society* 92, 1 (January–March 1972): 19ff. The best study of Sarakhsi is that of Franz Rosenthal, *Ahmad b. at-Tayyib as-Sarahsi* (New Haven, 1943).

60. Rosenthal, *Ahmad b. at-Tayyib as-Sarahsi*, 334–36.

61. See chapter 7.

62. M. Steinschneider, *Die arabische Literatur der Juden* (Frankfurt, 1902), 23ff.

63. A concise and balanced appreciation of Razi is Richard Walzer, *Greek into Arabic: Essays on Islamic Philosophy* (Cambridge, MA, 1962), 15–17.

64. Jennifer Michael Hecht, *Doubt: A History: The Great Doubters and Their Legacy of Innovation from Socrates and Jesus to Thomas Jefferson and Emily Dickinson* (San Francisco, 2003), 227–30.

65. Ibid.

66. Quoted in Muhammad Abdus Salam and H.R. Dalafi, *The Renaissance of Science in Islamic Countries* (Singapore, 1994), 252.

67. Notably Abu al-Rabban al-Balkhi, the most prominent Mutazilite in Baghdad.

68. Young, Latham, and Serjeant, *Religion, Learning, and Science in the Abbasid Period*, 371.

69. Seyyed Hossein Nasr, *Science and Civilization in Islam* (Cambridge, 1987), 47.

70. Gad Freudenthal, "Ketav Na-da'at or Sefer na Sekhel We-Ha-Muskalot: The Medieval Hebrew Translations of l-Farabi's Risalah FIL-AQL," *Jewish Quarterly Review* 93, 1–2 (2002): esp. 30–66.

71. G. Quadri, *La philosophie Arabe dans l'Europe medievale* (Paris, 1969), chap. 4.

72. On Farabi's biography, see Dimitri Gutas, "Farabi I, Biography," in *Encyclopedia Iranica*, http://www.iranica.com/articles/farabi-i.

73. On Farabi and music, see Askarali Radzhabov, "K istokam muzykalno-teoreticheskoi myslin epokhi samanidov," in *Samanidy:epokha i istoki kultury*, 234–46; and Saida Daukeeva's more exhaustive *Filosofiia muzyki Abu Nasra Mukhammada al-Farabi* (Almaty, 2002).

74. The Turkic thesis raised by Ibn Khallikan is rebutted by Goshtasp Lohraspi, "Farabi's Sogdian Origins," http://sites.google.com/site/ancientpersonalitiesofkhorasan/soghdian-farabi/farabi-soghdian-origin.

75. Gutas, "Farabi I, Biography"; Lohraspi, "Farabi's Sogdian Origins." Interesting and provocative alternative views on Farabi's Sogdian origins are to be found in *Istoriia tadzhikskoi filosofii*, 2:396ff.

76. From a fragment of Farabi's lost *On the Rise of Philosophy*, Majid Fakhry, *Al-Farabi, Founder of Islamic Neoplatonism* (Oxford, 2002), 7.

77. Muhsin Mahdi, "Al Farabi, Abu Nasr Muhammad ibn Muhammad ivn Tarkhan ibn wzalagh," in *Dictionary of Scientific Biography*, 4:523.

78. Like everything else about Farabi's life, this, too, is disputed, but the evidence for his birthplace at Faryab in northern Afghanistan is thin and contradicted by early sources which otherwise disagree with each other.

79. The most useful source on Otrar and its trade is K. M. Baibakov, *Srednevekovye goroda Kazakhstana na Velikom Shelkovom puti* (Almaty, 1998), 47–61.

80. H. Daiber makes a convincing case for this in "The Isma'ili Background of Farabi's Political Philosophy: Abu Ḥatim al Razi as a Forerunner of al-Farabi," in *Gottes ist der Orient, Gottes ist der Okzident*, ed. U. Tworuschka (Cologne, 1991), 143–50.

81. These are capably analyzed by many scholars, including Muhsin Mahdi, *Alfrabi's Philosophy of Plato and Aristotle* (Glencoe, 1962); and F. W. Zimmerman, ed., *Alfarabi's Commentary and Short Treatise on Aristotle's "De Interpretatione"* (London, 1981).

82. See Zimmerman, *Alfarabi's Commentary*, lxii–lxxiii.

83. Christopher A. Colmo, *Breaking with Athens: Alfarabi as Founder* (Lanham, 2004), chap. 1.

84. Walzer, *Greek into Arabic*, 208.

85. Richard Walzer, "Al-Farabi's Theory of Prophesy and Divination," in *Greek into Arabic*, 216.

86. Ian Richard Netton, *Alfarabi and His School* (Richmond, 1999), 28.

87. Joshua Parens, *An Islamic Philosophy of Virtuous Religions* (Albany, 2006), introduction, chap. 5; and by the same author, *Metaphysics as Rhetoric: Alfarabi's "Summary of Plato's Laws"* (Albany, 1995).

88. D. M. Dunlop, "Al-Farabi's Introductory Sections on Logic," *Islamic Quarterly* 2 (1966): 264–82; Nicholas Rescher, "Al-Farabi on Logical Tradition," *Journal of the History of Ideas* 24, 1 (January–March 1963): 127ff.

89. Abu Nasr al-Farabi, *On the Perfect State*, trans. Richard Walzer (Chicago, 1998). For a solid overview of Farabi's politics, see also Miriam Galston, *Politics and Excellence: The Political Philosophy of Alfarabi* (Princeton, 1990).

90. Alfarabi, *The Political Writings: Selected Aphorisms and Other Texts*, trans. Charles E. Butterworth (Ithaca, 2001).

91. Farabi, *On the Perfect State*, 8.

92. Ibid., 5.

93. Ibid., chap. 15.

94. Ibid., 281, 285, 289.

95. Ibid., 291.

96. Ibid., 299.

97. Ibid., 305.

98. Ibid., 15.

99. Parens advances this argument in *An Islamic Philosophy of Virtuous Religions*, 5–6, chap. 2.

100. Rémi Brague, "Note sur la traduction arabe de la Politique d'Aristote. Derechef, qu'elle n'existe pas," in *Aristote politique*, ed. Pierre Aubenque (Paris, 1993), 423–33.

101. The following paragraphs draw on Fred D. Miller, Jr. *Nature, Justice, and Rights in Aristotle's Politics* (Oxford, 1995); and Richard Kraut, *Aristotle: Political Philosophy* (Oxford, 2002).

102. A reserved version of this thesis is in Larbi Sadiki, *The Search for Arab Democracy* (New York, 2004), 208–18.

CHAPTER 7

1. Gafurov, ed., *Istoriia tadzhikskogo naroda* 2:129–32; Barthold, *An Historical Geography of Iran*, trans. Svat Souchek (Princeton, 1984), chap. 5; Bosworth, *The Medieval History of Iran, Afghanistan, and Central Asia*, 51–56; Richard N. Frye, *The Golden Age of Persia* (London, 1975), 188–94.

2. On the Tahirid dynasty's search for legitimacy, see Bosworth, *The Medieval History of Iran, Afghanistan and Central Asia*, chap. 7.

3. *The Cambridge History of Iran*, 4:420.

4. The author of both was Abu'l-Qasim Abdallah b. Ahmad al-Balkhi al-Ka'bi, on whom see Barthold, *Turkestan Down to the Mongol Invasion*, 11.

5. Richard N. Frye, "The Period from the Arab Invasion to the Saljuqs," in *The Cambridge History of Iran*, 4:98–99.

6. François De Blois, "A Persian Poem, Lamenting the Arab Conquest," in *Studies in Honour of Clifford Edmund Bosworth*, ed. Carole Hillenbrand (Leiden, 2000), 2:82–95.

7. Richard W. Bulliet, *The Patricians of Nishapur* (Cambridge, MA, 1972), 9–12.

8. Wilkinson, *Nishapur: Some Early Islamic Buildings and Their Decoration*, 82, etc.; also M. S. Dimand et al., "The Iranian Expedition, 1937: The Museum's Excavations at Nishapur," *Metropolitan Museum of Art Bulletin* 33, 11, pt. 2: 1–23.

9. M. S. Dimand, "Samanid Stucco Decoration from Nishapur," *Journal of the American Oriental Society* 58, 2 (June 1938): 258–61.

10. Charles K. Wilkinson, "The Glazed Pottery of Nishapur and Samarkand," *Metropolitan Museum of Art Bulletin*, New Series, 20, 3 (October 1961): 102–15.

11. James W. Allan, *Nishapur: Metalwork of the Early Islamic Period* (New York, 1982), 17–20.

12. Jan Rypka, *History of Iranian Literature* (Dordrecht, 1968), 33.

13. Jens Kroeger, *Nishapur: Glass of the Early Islamic Period*, Metropolitan Museum of Art (New York, 1995), 176–88.

14. Charles K. Wilkinson, "Life in Early Nishapur," *Metropolitan Museum of Art Bulletin*, New Series, 9, 2 (October 1950): 64.

15. Barthold, *Turkestan Down to the Mongol Invasion*, 213.

16. The following is drawn from Bulliet, *The Patricians of Nishapur*.

17. Ibid., 30ff.

18. Adam Mez, *Die Renaissance des Islams* (Hildesheim, 1968) (reprint of 1922 edition), 247.

19. Bulliet, *The Patricians of Nishapur*, 22.

20. Ibid., 81.

21. The Karramiyya of Nishapur were a later version of the Kharijitis who had first appeared in the earliest days of Islam in Basra, Iraq.

22. Barthold, *Turkestan Down to the Mongol Invasion*, 16. Manuscripts of this history exist in versions of six, eight, or twelve volumes. See Richard N. Frye, ed., *The Histories of Nishapur* (Cambridge, MA, 1965), 10ff.

23. Gafurov, *Istoriia tadzhikskogo naroda*, 2:50.

24. On Abu Mashar al-Balkhi (787–886), see below.

25. Frye, "The Period from the Arab Invasion to the Saljuqs," 420.

26. Shamolov, ed., *Istoriia tadzhikskoi filosofii*, 2:39ff.

27. Pingree, "Abū Maʿshar al-Balkhī, Jaʿfar ibn Muhammad," 36–37.

28. David Pingree, *The Thousands of Abu Maʿshar* (London, 1968), introduction.

29. The *Cambridge History of Iran*, 4:427.

30. S. Schechter, "The Oldest Collection of Bible Difficulties, by a Jew," *Jewish Quarterly Review* 13 (1901): 345–74.

31. Rawandi's *nisba* would suggest a birthplace in Iran, but several early sources trace his roots to Lesser Merv. The large and sophisticated Jewish community in nearby Balkh lends support to this possibility. His birth year is variously given as 815 and 827. The exact location of Marv al-Rud is disputed. What is certain is that it was at the point where the main east-west route from Sarakhs to Balkh crossed the Murgab River. See Paul Wheatley, *The Places Where Men Pray Together: Cities in Islamic Lands, Seventh through Tenth Centuries* (Chicago, 2000), 175–76.

32. A convenient introduction to this contentious figure is *Encyclopaedia of Islam*, 3:905. See also Sarah Stroumsa, *Freethinkers of Medieval Islam: Ibn al-Rawandi, Abu*

Bakr al-Razi, and Their Impact on Islamic Thought, Islamic Philosophy, Theology, and Science (Leiden, 1999), vol. 35.

33. See the discussion of Rawandi in *Istoriia tadzhikskoi filosofii*, 2:82–101; also Hecht, *Doubt: A History*, 223–27.

34. C. E. Bosworth, "An Alleged Embassy from the Emperor of China to the Amir Nasr. B. Ahmad," in *The Medieval History of Iran, Afghanistan and Central Asia*, 2.

35. Mez, *Die Renaissance des Islams*, 264–65.

36. Nurdeen Deuraseh, Mansor Abu Talib, "Mental Health in Islamic Medical Tradition," *International Medical Journal* 4, 2:76ff.

37. This passage on Naysaburi is based on Shereen El Ezabi's fine article, "Al-Naysaburi's *Wise Madmen*: An Introduction," *Alif: Journal of Comparative Poetics* 14 (1994): 192–205.

38. Ibid., 196.

39. *The Cambridge History of Iran*, 4:107.

40. Barthold, *An Historical Geography of Iran*, 68.

41. George N. Curzon, *Persia and the Persian Question*, 2 vols. (London, 1892), 1:227.

42. Fred H. Andrews, *Catalogue of Wall Paintings from Ancient Shrines in Central Asia and Seistan* (Delhi, 1933), 57–59; Trudy S. Kawami, "Kuh-e Khwaja, Iran, and Its Wall Paintings: The Records of Ernst Herzfeld," *Metropolitan Museum Journal* 22 (1987): 13–41.

43. For a good description based on the early sources, see G. Le Strange, *The Lands of the Eastern Caliphate* (New York, 1973), 333–51. For all aspects of Sistan and the Saffarids, see Clifford Edmund Bosworth's pioneering work, *The History of the Saffarids of Sistan and the Maliks of Numruz (247/861 to 949/1542-3)*, Columbia Lectures on Iranian Studies 8 (1994): 30–65.

44. Barthold, *Turkestan Down to the Mongol Invasion*, 216–20.

45. Bosworth, *The History of the Saffarids of Sistan and the Maliks of Numruz*, 119.

46. This history is briefly summarized in *The Cambridge History of Iran*, 4:107ff.

47. Note the similarity in treatment between Frye, *The Golden Age of Persia*, 194–99, and the thoughtful Soviet defense of the Saffarids in Gafurov, *Istoriia tadzhikskogo naroda*, 2:132ff.

48. Barthold, *An Historical Geography of Iran*, 64.

49. Bosworth, *The History of the Saffarids of Sistan and the Maliks of Numruz*, 13.

50. Bosworth, *The Medieval History of Iran, Afghanistan and Central Asia*, 59–60.

51. Barthold, *Turkestan Down to the Mongol Conquest*, 220.

52. Rypka, *History of Iranian Literature*, 129.

53. Ibid., 136.

54. Translation by Jackson, *Early Persian Poetry*, 26.

55. Le Strange, *The Lands of the Eastern Caliphate*, 388–90.

56. For this fact and all other insights on Tus, I am indebted to the fine PhD dissertation of Parvanch Pourshariati, "Iranian Traditions and the Arab Presence in Khurasan."

57. On the Bihafar Zoroastrian movement, see Gholam Hossein Sadighi, *Les mouvements religieux Iraniens au IIe et au III siècle de l'Hégire* (Paris, 1938), 116ff.

58. Pourshariati, "Iranian Tradition in Tus," 177.

59. Rypka, *History of Iranian Literature*, 151–53; also Pourshariati, "Iranian Tradition in Tus," 23.

60. Pourshariati, "Iranian Tradition in Tus," 107.

61. From the translation by Jackson, slightly revised by the author, *Early Persian Poetry*, 60–61. An alternative translation is by Edward Sachau, in Biruni, *The Chronology of Ancient Nations*, trans. and ed. C. Edward Sachau (London, 1879), xiii.

62. On Daqiqi Balkhi, see G. Lazard, *Les premiers poètes persans (IXe–Xe siècles)* (Tehran, 1964), 94–126; also Rypka, *History of Iranian Literature*, 153–54.

63. Rypka, *History of Iranian Literature*, 154.

64. Abolqasem Ferdowsi, *Shahnameh: The Persian Book of Kings*, trans. Dick Davis (Washington, DC, 1997), 853.

65. Edward G. Browne, *Literary History of Persia from Firdawsi to Sa'di* (New York, 1906), 142ff.

66. Davis, in Ferdowsi, *Shahnameh*, xiii.

67. The most penetrating analysis of these dimensions of *Shahnameh* is by Ferdowsi's translator, Dick Davis, *Epic and Sedition: The Case of Ferdowsi's "Shahnameh"* (Fayetteville, 1992), 65.

68. Ibid., 21.

69. Ibid., 20.

70. This is the central insight of Davis's brilliant analysis, ibid., 23.

71. See Davis's discussion of paternal authority in ibid., chap. 3.

72. Ferdowsi, *Shahnameh*, 845.

73. Ibid., 849.

74. Rypka, *History of Iranian Literature*, 161.

75. Davis, *Epic and Sedition*, 193.

76. Ferdowsi, *Shahnameh*, quoted by Davis, *Epic and Sedition*, 19.

77. Davis, *Epic and Sedition*, 179.

78. Ibid., 187.

79. This sentence has been slightly rearranged from ibid., 171.

80. Ferdowsi, *Shahnameh*, 854.

CHAPTER 8

1. These translations are by the poet Sassan Tabatabai, *Father of Persian Poetry: Rudaki and His Poetry* (Leiden, 2010), 58. The slight modifications entered by the author were made in full recognition of Tabatabai's invaluable and pioneering work. See also his PhD dissertation, "Rudaki, the Father of Persian Poetry," Boston University, 2000.

2. Tabatabai, *Father of Persian Poetry*, 84.

3. Ibid., 100.

4. Ibid., 98.

5. Mohammed H. Razi, "Rabiah Balkhi: Medieval Afgani Poet," in *Middle Eastern Muslim Women Speak*, ed. Elizabeth Warnock Fernea and Basima Qattan Bezirgen (Austin, 1977), 81.

6. Tabatabai, *Father of Persian Poetry*, 108.

7. Ata-Malik Al Juwaini, *Genghis Khan: The History of the World-Conqueror*, trans. John Andrew Boyle (Manchester, 1958), 97–98.

8. "He multiplies the wisdom of wise men. He enriches the powerful with knowledge." I. S. Bragitskii, *Abu Abdullakh Dzhafar Rudaki* (Moscow, 1989), 26.

9. Marufii of Balkh, a fellow poet at the Samani court, left the following verse in which the Fatamids refer to the Ismailis who ruled Egypt (translation by Hakim Elnazarov, who pointed out these lines to the author):

I heard from Rudaki, the master of the poets:
"Do not adhere to anyone, except the Fatimids."

Rudaki, *Divan*, ed. Qodiri Rustam (Almaty, 2007), 79.

10. Ibid., 91.

11. Tabatabai, *Father of Persian Poetry*, 56.

12. Browne, *A Literary History of Persia from the Earliest Times until Firdawsi*, 343–44.

13. Ibrahim b. Makhla al-Qadl Ismail b. Ali al Khuitabi, quoted in "Baghdad in the Tenth Century," http://www.eduplace.com/ss/hmss/7/unit/act2.1blm.html.

14. S. Kamoliddin, "On the Religion of the Samanid Ancestors," *Transoxonia* 11 (July 2006), http:/www.transoxonia.com.ar/11/kamoliddin-samanids.htm.

15. Klimburg-Salter, "Buddhist Painting in the Hindu Kush c. VIIth to Xth Centuries."

16. Barthold, *Turkestan Down to the Mongol Invasion*, 88, suggests the high figure of 500,000.

17. On Jayhani and his book on roads and kingdoms, see V. Minorsky, "A False Jayhani," *Bulletin of the School of Oriental and African Studies* 12 (1947–1948): 889–96; and Barthold, *Turkestan Down to the Mongol Invasion*, 12.

18. Barthold, *Turkestan Down to the Mongol Invasion*, 227–32.

19. C. E. Bosworth, "A Pioneer Arabic Encyclopedia of the Sciences: Al-Khwarazmi's *Keys to the Sciences*," *Isis* 54, 1 (1963): 100.

20. Z. Vesel, "Encyclopedias," in *History of Civilizations of Central Asia*, vol. 4, pt. 2, 363–65.

21. Gafurov, *Istoriia tadzhikskogo naroda*, 2:173.

22. Gafurov, *Central Asia: Pre-Historic to Pre-Modern Times* (Kolkata, 2005), 2:75.

23. E. A. Davidovich, "Coinage and the Monetary System," in *History of Civilizations of Central Asia*, vol. 4, pt. 1, 392–93.

24. James Owen, "Huge Viking Horde Found in Sweden," *National Geographic News*, October 28, 2010.

25. D. Austin and L. Alcock, eds., *From the Baltic to the Black Sea: Studies in Medieval Archaeology* (London, 1990), chap. 9.

26. On the horde of dirhams found in Gotland, see Stenberger, *Die Schatzfunde Gotlands der Wikingerzeit* (Stockholm, 1958). A collection of medieval "Arab" coins used in Sri Lanka can be found in the Central Bank of Sri Lanka's Currency Museum in Kotte.

27. Narshakhi, *The History of Bukhara*, 54.

28. Gafurov, *Central Asia*, 2:98.

29. Golden, *Central Asia in World History*, 64–66.

30. An authoritative account of the Samanids is by N. N. Negmatov, *Gosudarstvo Samanidov: Maverannakhr i Khorasan* (Dushanbe, 1977), vols. 9–10; summarized in N. N. Negmatov, "The Samanid State," in *History of Civilizations of Central Asia*, vol. 4, pt. 1, 83.

31. Zavadovskii, *Abu Ali Ibn Sina*, 197.

32. For a convenient summary of Samanid history, see *The Cambridge History of Iran*, 4:136–61.

33. Bragistskii, *Abu Abdullakh Dzhafar Rudaki*, 4.

34. Gafurov, *Central Asia*, 2:115.

35. Ibid.

36. A.C.S. Peacock, *Medieval Islamic Historiography and Political Legitimacy* (London, 2007), 37.

37. Dzhalilov, *Iz istorii kulturnoi zhizni predkov tadzhikskogo naroda i tadzhikov v rannem srednevekove*, 79.

38. Davlat Davudi, "Zoroastrizm i Islam posle arabskogo zavoevaniia v epokhu samanidov," in *Samanidy: epokha i istoki kultury*, 35–43; also Narshakhi, *The History of Bukhara*, 31.

39. Richard N. Frye, "Notes on the History of Transoxonia," *Harvard Journal of Asiatic Studies* 19, 1–2 (June 1956): 106–19.

40. For a description of the canal system, see Barthold, *Turkestan Down to the Mongol Invasion*, 103–7.

41. Richard N. Frye, *Bukhara: The Medieval Achievement* (Norman, 1964), 10.

42. Ibid., 93.

43. Ettinghausen, Grabar, and Jenkins-Madina, *Islamic Art and Architecture, 650–1250*, 112.

44. For example, the caves of Bamiyan. See Klimburg-Salter, *The Kingdom of Bāmiyān*, fig. 62, 72.

45. Shamsiddin Kamoliddin, "On the Origins of the Place-Name "Bukhara," http://www.transoxiana.org/12/kamoliddin-buxara.php.

46. Ibid.

47. This remarkable structure was barely known before G. A. Pugachenkova's research, "Mazar Arab-Ata v Time," *Sovetskaia arkheologiia* 4 (1961): 198–211.

48. Luke Treadwell, "Ibn Aafir Al-Azdi's Account of the Murder of Ahmad B. Ismail, Al-Salamani and the Succession of His Son Nasr," in *Studies in Honour of Clifford Edmund Bosworth*, 2:411.

49. Peacock, *Medieval Islamic Historiography and Political Legitimacy*.

50. William E. Gohlman, *The Life of Ibn Sina: A Critical Edition and Annotated Translation* (Albany, 1974), 33.

51. Ibid., 37.

52. Negmatov, "The Samanid State," 88–89.

53. Barthold, *Turkestan Down to the Mongol Invasion*, 13.

54. Text, with archaisms deleted by author, from Frye, *Bukhara*, 59–60.

55. On the Buyids, a valuable source is John J. Donohue, *The Buwayhid Dynasty in Iraq, 334H./945 to 403 H./1012. Shaping Institutions for the Future* (Leiden, 2003); also Kennedy, *The Prophet and the Age of the Caliphates*, 212–49.

56. On this branch of Shiism in Persia, see Farhad Daftary, "The Medieval Ismailis of the Iranian Lands," in *Studies in Honour of Clifford Edmund Bosworth*, 2:43ff.

57. Marshall G. S. Hodgson, *The Venture of Islam*, 3 vols. (Chicago, 1974), 2:36.

58. Gafurov, *Istoriia tadzhikskogo naroda*, 2:179; Hugh Kennedy, "Sicily and Al-Andalus under Muslim Rule," in *The New Cambridge Medieval History*, 3:614.

59. Paul Ernest Walker, *Early Philosophical Shiism: The Ismaili Neoplatonism of Abu Ya'qub al-Sijistani* (Cambridge, 1993).

60. Fakhry, *A History of Islamic Philosophy*, 166; also Peacock, *Medieval Islamic Historiography and Political Legitimacy*, 25ff; Corbin, *History of Islamic Philosophy*, 79ff.

61. Alessandro Bausani, "Scientific Elements in Isma'ili Thought: The Epistles of the Brethren of Purity," in *Isma'ili Contributions to Islamic Culture*, ed. Seyyed Hossein Nasr (Tehran, 1977), 123–37.

62. A. Paket-Chy and C. Gilliot, "Works on Hadith and Its Codification, on Exegesis and Theology," in *History of Civilizations of Central Asia*, vol. 4, pt. 2, 91.

63. Patricia Crone, *Medieval Islamic Political Thought* (Edinburgh, 2004), 130–41.

64. The phrase "Dome of Islam" comes from the Bukharan historian Abu Bakr Narshakhi (899–959), *History of Bukhara*, 56; see also Jonathan Brown, *The Canonization of Al-Bukhari and Muslim: The Formation and Function of the Sunni Hadith Canon* (Leiden, 2007).

65. C. Gilliot, "Theology," in *History of Civilizations of Central Asia*, vol. 4, pt. 2, 123; see the colorful example in Kraemer, *Humanism in the Renaissance of Islam*, 35–36.

66. Joseph Schacht, *Origins of Muhammedan Jurisprudence* (New York, 1950).

67. A. Paket-Chym, "The Contribution of Eastern Iranian and Central Asian Scholars to the Compilation of Hadiths," in *History of Civilizations of Central Asia*, vol. 4, pt. 2, 91.

68. M. M. Khairullaev, ed., *Iz istorii obshchestvenno-filosofskoi mysli i volnodumiia v Srednei Asii* (Tashkent, 1991), 26. An Afghan savant had heard of over 1,200 Hadith scholars: Adam Mez, *The Renaissance of Islam* (New Delhi, 1995), 247.

69. Alzhir Ahmet, "Amulyn Hadyscy Alymlary," *Miras* 35, 3 (2009): 87–92.

70. Ghassan Abdul-Jabbar, *Bukhari* (London, 2007), 10, 109.

71. Abdul Husain Muslim of Nishapur; Abu Isa Muhammad Tirmidhi from Tirmidh (Termez); Abu Dawud from Sistan; Ahmad al-Nasa'i from Nasa in Khurasan, and Bukhari.

72. Faisal Shafeek, *The Biography of Imam Bukhaaree* (Dar es Salaam, 2005), 48.

73. Ibid., 142–44, 178ff.; also Aikhairualiev, ed., *Iz istorii obshchestvenno filosofskoi mysli i volnodumiia v srednoi Azii* (Tashkent, 1991), 28–29.

74. Jonathan A. C. Brown details all the twists and turns of this process in his excellent *Canonization of Al-Bukhari and Muslim*.

75. Abdul-Jabbar, *Bukhari*, 16; an English translation of Abdul Husain Muslim's *Salih Muslim* is accessible at http://www.usc.edu/schools/college/crcc/engagement/resources/texts/muslim/hadith/muslim/.

76. On Tirmidhi, see U. Uvatov, "At-Termezi i ego tvorcheskoe nasledie," in *Iz istorii obshchestvenno-filosofskoi mysli i volnodumiia v Srednei Asii*, 31–33; on Abu Husain Muslim, see Brown, *The Canonization of al-Bukhari and Muslim*.

77. Brown, *The Canonization of al-Bukhari and Muslim*, 67.

78. Paul L. Heck, "The Epistemological Problems of Writing in Islamic Civilization," *Studia Islamica* 94 (2002): 90.

79. See Tarif Khalidi, *Arabic Historical Thought in the Classical Period* (Cambridge, 1996), 134–36.

80. Ignaz Goldziher raises these issues repeatedly in his *Muslim Studies*, 2 vols. (London, 1967–1971).

81. Tawhidi, cited in Kraemer, *Humanism in the Renaissance of Islam*, 43.

82. Heck, "The Epistemological Problems of Writing in Islamic Civilization," 89.

83. Ibid., 88n12.

84. Brown develops this thesis in *The Canonization of al-Bukhari and Muslim*.

85. Goodman, *Islamic Humanism*, 90–91.

86. "Certainly, deeds are by intentions, and each man will earn according to his intention." Abdul-Jabbar, *Bukhari*, 25.

87. Goodman, *Islamic Humanism*, 89.

88. Ibid., 86.

89. A. Ali, "Maturidism," in *A History of Muslim Philosophy*, ed. M. M. Sharif (Wiesbaden, 1963), 1:259–74.

90. A. Konyrbaev, V. Mokrynin, and V. Ploskikh, *Goroda veligogo puteshestvennika* (Bishkek, 1994), 50–51; Vladimir Ploskikh, "The Central Asian 'Atlantis': The Mystery of the Great Silk Way," n.d., 4–5.

91. Shaid Balkhi, in Browne, *Literary History of Persia*, 454.

92. The following section draws on Ibn Sina's autobiography (Gohlman, *The Life of Ibn Sina*); on Gotthard Strohmaier, *Avicenna* (Munich, 2006); and especially on the meticulous and authoritative analyses by Zavadovskii, *Abu Ali Ibn Sina*.

93. Zavadovskii, *Abu Ali Ibn Sina*, 49.

94. Ibid., 53.

95. Ibid., 59.

96. Ibn Sina's statement that "I listened to their conversation but my soul did not accept it" appears in many scholarly works (cf., for example, Gohlman, *The Life of Ibn Sina*, 19), but the latest edition of his autobiography, issued in the compendium *Osori Muntakhab* (Dushanbe, 2005), rejected it as a later interpolation. I am indebted to Hakim Elnazarov for this insight.

97. Thus in his last work he juxtaposes the "Worshiper," who "persists in exercising worship by prostration, fasting, etc.," to the "Knower," who disposes [his] thought through the sanctity of divine power, seeking the perpetual illumination of the light of the Truth into [his] innermost thought." Shams Inati, *Ibn Sina and Mysticism* (London, 1996), 81.

98. Zavadovskii, *Abu Ali Ibn Sina*, 66.

99. Ibid., 67.

100. Dimitri Gutas, *Avicenna and the Aristotelian Tradition* (Leiden, 1988), 181.

101. M. Achena, "Avicenna's Persian Poems," in *Encyclopedia Iranica*, http:/www.iranica.com/articles/Avicenna-xi.

102. T. N. Mardoni, "K istorii izuchenii poeticheskogo naslediia Abuali ibn Sina na arabskom iazyke," in *Ibn Sin-ova farhangi zamoniu* (Dushanbe, 2005), 220–22.

103. Zavadovskii, *Abu Ali Ibn Sina*, 66.

104. Ibid.

105. Gohlman, *The Life of Ibn Sina*, 25.

106. Zavadovskii, *Abu Ali Ibn Sina*, 82.

107. Ibid., 128; also Gohlman, *The Life of Ibn Sina*, introduction.

108. Gohlman, *The Life of Ibn Sina*, 43.

109. Ibid., 84–85.

110. Zavadovskii concurs with Biruni's translator, the German scholar Sachau, that the surviving letters are only a fragment of a much larger correspondence. Zavadovskii, *Abu Ali Ibn Sina*, 89–91; see also A. Sharipov, "Maloizvestnye stranitsy perepiski mezhdu Biruni i Ibn Sina," *Nauki v Uzbekistane* 11 (Tashkent, 1955): 598–668. For an excellent summary, see Strohmaier, "Ein programmatischer Briefwechsel," in *Avicenna*, 43–56.

111. The exchange has been partially translated into German by Gerhard Strohmaier, *Al Biruni, In den Gärten der Wissenschaft. Ausgewählte Texte aus den Werken des*

Muslimischen Universalgelehrten (Leipzig, 1991), 49–65; completely into Russian: *Biruni i Ibn Sina: perepiskia*, trans. Iu. N. Zavodovkii, ed. I. M. Mumin (Tashkent, 1973); and, less adequately, into English: "Ibn Sina-Al-Buiruni Correspondence," trans. Mazaffar Iqbal and Rafik Berjak, *Islam and Science* 1 (Summer and Winter 2003); 2 (Summer and Winter 2004); 3 (Summer and Winter 2005); 4 (Winter 2006); 5 (Summer 2007).

112. Seyed Hossein Nasr makes this case, with examples of later exchanges, in his translation of the exchange, *Al-Biruni and Ibn Sina*, ed. Seyed Hossein Nasr and Mahdi Mohaghegh (Tehran, 1972), 1. Nasr's introduction to this authoritative Arabic edition provides a concise overview of the exchange.

113. Ahmad Dallal, *Islam, Science, and the Challenge of History* (New Haven, 2010), 79.

114. For an exposition of Biruni's cosmological views from an Islamic perspective, see Nasr, *An Introduction to Islamic Cosmological Doctrines*, pt. 2.

115. Zavodavskii, *Abu Ali Ibn Sina*, 90.

116. Ibid.

117. Ibid., 91.

118. Originally published in *The New Statesman*, October 6, 1956; expanded to book form, *The Two Cultures* (Cambridge, 1960).

119. Zavadovskii, *Abu Ali Ibn Sina*, 93.

120. Jackson, *Early Persian Poetry*, 55.

121. The eleventh-century Khurasani historian Beyhaqi reported that Ibn Sina took an administrative post "with the *sultan*" (Zavadovskii, *Abu Ali Ibn Sina*, 95). Since Mahmud of Ghazni had only three years earlier created the title "sultan" for himself, this suggests that Ibn Sina was actually employed by his admirer and great protagonist, Mahmud. But by Beyhaqi's day (995–1077), the title "sultan" had been widely used by the Seljuks, among others, and the historian more probably misapplied the title to the Karakhanid rulers, who by then were in control of Bukhara.

122. Hakim Elnazarov notes (personal communication) the following Tajik and Iranian authors who support this explanation of Rudaki's blindness and exclusion from Bukhara: Abusaid Shokhumorov, *Somoniyon va Junbishi Ismoiliya: Falsafa dar ahdi Somoniyon* (Dushanbe, 1999); Said Nafisi, *Muhiti zindagi va ahvolu ash'ori Rudaki* (Tehran, 1922); and Hokim Qalandarov, *Rudaki va Ismoiliya* (Dushanbe, 2012).

123. Tabatabai, *Father of Persian Poetry*, 66–70, with slight revisions by the author.

CHAPTER 9

1. Bartold's assertion that the Amu Darya and Caspian were connected via the Aral Sea are to be found in "Svedeniia ob Aralskom more i nizovykh Amu-Dari s drevnei vremia do xvii-ogo veka," in *Sochineniia*, 2:109–45.

2. For an account of the northern route, see *Ibn Fadlan's Journey to Russia*, trans. Richard N. Frye (Princeton, 2005), 36ff.

3. V. Altman, "Ancient Khorezmian Civilization in the Light of the Latest Archaeological Discoveries (1937–1945)," *Journal of the American Oriental Society* 67, 2 (April–June 1947): 81.

4. The best sources on Khwarazmian cities are Tolstov, *Drevnii Khoresm* (Moscow, 1948), and his *Po sledam drevnekhorezmiiskoi tsivilizatsii* (Moscow, 1948), pt. 2; see also

Masson, *Strana tysiachi gorodov* (Moscow, 1966), 123–44; Barthold, *Turkestan Down to the Mongol Invasion*, 149; and N. N. Negmatov, "States in North-Western Central Asia," in *History of Civilizations of Central Asia*, 2:441, 446, 455.

5. Edgar Knobloch, *Archaeology, Art and Architecture of Central Asia* (London, 1972), 80.

6. These techniques continued into the Islamic era; cf. N. M. Bachinskii, *Antiseismika v arkhitekturnykh pamiatnikikakh srednei Azii* (Moscow, 1949), 6ff.

7. D. R. Hill, "Physics and Mechanics, Civil and Hydraulic Engineering," in *History of Civilizations of Central Asia*, vol. 4, pt. 2, 270.

8. Briefly discussed in E. V. Rtveladaze, ed., *Istoriia gosudarstevnnosti Uzbekistana*, 3 vols. (Tashkent, 2009), 1:210.

9. On Indian influence on Khwarazmian art, see Knobloch, *Beyond the Oxus*, 77; and "Toprak-Kala: dvorets," in *Trudy khorozemskoi arkheologo-etnograficheskoi ekspeditsii*, ed. Iu. A. Rapoport and E. E. Nerazik (Moscow, 1984), vol. 14; for a summary in English, see Iu. A. Rapoport, "The Palaces of Topraq-Qala," *Bulletin of the Asia Institute* 8 (1994): 161–85.

10. On the calendar, see G. Bulgakov, "Al-Biruni on Khwarazm," in *History of Civilizations of Central Asia*, 3:222–24.

11. G. Snezarev, *Remnants of Pre-Islamic Beliefs and Rituals among the Khwarazm Uzbeks* (from Russian original) (Berlin, 2003); C. E. Bosworth, "Al-Khwarazmi on Various Faiths and Sects, Chiefly Iranian," in *Iranica Varia: Papers in Honor of Professor Ehsan Yarshater*, Acta Iranica 30, 3rd ser., *Textes et memoires* (Leiden, 1990), 16:10ff.; G. Bulgakov, "Al- Biruni on Khwarazm," in *History of Civilizations of Central Asia*, 3:225–27.

12. Alison V. G. Betts and Vadim N. Yagodin, "The Fire Temple at Tash-k'irman Tepe, Chorasmia," in *After Alexander*, 435–54.

13. Henning, "The Khwarezmian Language," in *W. B. Henning Selected Papers*, 6:485–500; D. N. Mackenzie, "Khwrazmian Language and Literature," in *The Cambridge History of Iran*, vol. 3, pt. 2, 1244ff.; A. Taffazoli, "Iranian Languages," in *History of Civilizations of Central Asia*, 4:326.

14. Rtveladze, *Civilizations, States, and Cultures of Central Asia*, 101, 166.

15. Ibid., 101. Little is known of overall management under the Khwarazmshahs; see "Khorezmiiskoe gosudarstvo," in *Istoriia gosudarstvennosti Uzbekistana*, 1:193–210; E. E. Nerazik, "History and Culture of Khwarazm," in *History of Civilizations of Central Asia*, 3:207–10.

16. On currency, see Joe Cribb, "Money as a Marker of Cultural Continuity and Change in Central Asia," *Proceedings of the British Academy* 133 (2007): 372–73; Bulgakov, *Zhizn i trudy Beruni* (Tashkent, 1972), 1:20.

17. Biruni, *The Chronology of Ancient Nations*, 42.

18. Unfortunately the sole surviving later history of Khwarazm does not deal with the Mamun era: Barthold, *Turkestan Down to the Mongol Invasion*, 17; Shir Muhammad Mirab Munis and Muhammad Reza Mirab Agahi, *Firdaws al-Iqbal: History of Khorezm*, trans. Yuri Bregel (Leiden, 1999).

19. Al-Nadim, *The Fihrist of al-Nadim*, 2:632–33.

20. Al-Mugaddasi, *The Best Divisions for Knowledge of the Regions*, trans. Basil Anthony Collins (Reading, 1994), 255.

21. On Buzjani, see Ali Moussa, "Mathematical Methods in Abū al-Wafā's Almagest and the Qibla Determinations," *Arabic Sciences and Philosophy* 21, 1 (2011): 1–56; and John J. O'Connor and Edmund F. Robertson, "Mohammad Abu'l-Wafa Al-Buzjani," MacTutor History of Mathematics Archive, University of St. Andrews, http://www.gap-system.org/~history/Biographies/Abu%27l-Wafa.html.

22. G. Matvievskaia and Kh. Talashev, "Sochineniia Abu Masra Ibn Iraka o sverike," in *Iz istorii srednevekovoi vostochnoi matematiki i astronomii*, ed. S. Kh. Sprazhdnikov (Tashkent, 1993), 82–151; and Claus Jensen, "Abu Nasr Mansur's Approach to Spherical Astronomy as Developed in His Treatise *The Table of Minutes*," *Centaurus* 16, 1 (1971): 1–19; also Julio Samso, "Manṣūr Ibn ʿAlī Ibn ʿIrāq, Abū Nasr," in *Dictionary of Scientific Biography*, http://www.encyclopedia.com/doc/1G2-2830902803.html.

23. G. Matvievskaia and Kh. Talashev, "O nauchnom nasledii astronoma X–XI vv. Abu Nasra ibn Iraka," *Istoriko-astronomicheskie issledovaniia* 13 (1977): 219–32.

24. Barthold, *Turkestan Down to the Mongol Invasion*, 233–34.

25. The date of Iraq's birth is significant. If he was born in 948, as Bulgakov contends in *Zhizn i trudy Beruni* (Tashkent, 1972), 1:29, then it was Iraq himself, not his father, who adopted Biruni. But if Iraq was born either in 970 or 960, both of which have been suggested, then it would have been Iraq's father who adopted Biruni. The lifelong fraternal ties between Biruni and Iraq suggest one of the latter dates as the more probable.

26. Behaim's "Earth Apple" can be seen at the American Geographical Society's Digital Map Collection, http://collections.lib.uwm.edu/cdm4/document.php?CISOROOT=/agdm&CISOPTR=1228&CISOSHOW=1224.

27. Bulgakov, *Zhizn i trudy Beruni*, 1:33.

28. C. M. Bowra, *The Greek Experience* (Cleveland, 1957), 166,

29. Bulgakov, *Zhizn i trudy Beruni*, 1:20.

30. Ibid., 1:44, 54.

31. Ibid., 1:60–61.

32. Biruni, *The Chronology of Ancient Nations*, 214.

33. Ibid., 326–29.

34. Ibid., 7ff.

35. Ibid., 79.

36. Ibid., 178.

37. Ibid., 14, 76.

38. Ibid., 36.

39. Ibid., 84ff., 115.

40. Ibid., 61.

41. Ibid., 132.

42. Ibid., 147.

43. Ibid., 84.

44. Ibid., 73.

45. On temporal concepts in early Iran, see E. Bickerman, "Time-Reckoning," in *The Cambridge History of Iran*, vol. 3, pt. 2, 778–91.

46. Biruni, *The Chronology of Ancient Nations*, 15.

47. Sayyed Hossein Nasr, "Al-Biruni as Philosopher," in *Millenary of Abu Raihan Muhammad Ibn Ahmad al-Biruni*, Conference Proceedings, November 26–December 12 (Karachi, 1973), 7.

48. With further parenthetical clarifications, from Fuad I. Haddad, David Pingree, and Edward S. Kennedy, "Al-Biruni's Treatise on Astrological Lots," in *Astronomy and Astrology in the Medieval Islamic World*, 12.

49. Bulgakov, *Zhizn i trudy Beruni*, 1:85.

50. Ibid., 1:76.

51. A valuable collection of medieval sources on Gurganj is M. Aidfogdieva, ed., *Srednevekovye pismennye istochniki o drevnem Urgenche* (Ashgabat, 2000). See also Khemra Iusupov, *Serdtse drevnego khorezma* (Ashgabat, 1993), 26–50.

52. This chronology is based on Muhammad Nazim, *Life and Times of Sultan Mahmud of Ghazna* (Cambridge, 1931), app. F.

53. Not to be confused with the slightly later minaret nearby that Mahmud constructed after his conquest of the city. On Mamun's minaret, see Mukhammed Mamedov and Ruslan Muradov, *Gurganj* (Padua, n.d.), 46–48.

54. Statement of As-Sa'alibi, quoted by Zavadovskii, *Abu Ali Ibn Sina*, 101.

55. B. A. Abdukhalimov, ed., *Khorazm Mamun Akademiiasi* (Tashkent, 2005), 104; G. Matvievskaya, "History of Medieval Islamic Mathematics: Research in Uzbekistan," *Historia Mathematica* 20 (1993): 241.

56. Zavadovskii, *Abu Ali Ibn Sina*, 102.

57. This section is indebted to the fine studies of specific disciplines in *Khorazm Mamun Akademiiasi*.

58. Ghada Karmi, "A Medieval Compendium of Arabic Medicine: Abu Sahl al-Masuihi's *Book of the Hundred*," *Journal of the History of Arabic Science* 2, 2 (1978): 270–90.

59. R. Bakhadirov, *Iz istorii klassifikatsii nauk na srednekovom musulmanskom vostoke* (Tashkent, 2000), 37.

60. On the poets and prose writers see I. Elmurodov, "Adabiet," in *Khorazm Mamun Akademiiasi*, 243.

61. Bakhadinov, *Iz istorii klassifikatsii nauk na srednevekovom musulmanskom vostoke*, 74–83.

62. See article "Mu'tazila" in *Encyclopedia of Islam* (2nd ed.), http://referenceworks. brillonline.com/entries/encyclopaedia-of-islam-2/mutazila-COM_0822.

63. Kennedy, "Al-Biruni," in *Dictionary of Scientific Biography*, 2:149.

64. Bulgakov, *Zhizn i trudy Beruni*, 1:128.

65. Zavadovskii, *Abu Ali Ibn Sina*, 103.

66. Bulgakov, *Zhizn i trudy Beruni*, 1:122; Zavadovskii, *Abu Ali Ibn Sina*, 106.

67. Barthold, *Turkestan Down to the Mongol Invasion*, 275; also Nazim, *The Life and Times of Sultan Mahmud of Ghazna*, 56–60; Browne, *A Literary History of Persia from Firdawsi to Sa'di*, 96.

68. Nidhamu-i-Arudi-i-Samarqandi, *Chahar Maqala (The Four Discourses)*, trans. Edward G. Browne (London, 1921), 119; Zavadovskii, *Abu Ali Ibn Sina*, 97.

69. Nizami (Nidhami) Arudi, quoted by Dzhalilov, *Iz istorii kulturnoi zhizni predkov tadzhikskogo naroda i tadzhikov v rannem srednevekove*, 53.

70. Zavadovskii, *Abu Ali Ibn Sina*, 118.

71. Dzhalilov, *Is istorii kulturnoi zhizni predkov tadzhikskogo naroda i tadzhikof v rannem srednevekove*, 53.

72. Lenn Evan Goodman, "The Epicurean Ethic of Muhammad Ibn Zakariya ar-Razi," *Studia Islamica* 34 (1971): 5.

73. Mehdi Mohaghegh, "Note on the *Spiritual Physic* of Al-Razi," *Studia Islamica* 26 (1967): 5ff.

74. Young, Latham, and Serjeant, *Learning, Religion and Science in the Abbasid Period*, 373–74.

75. Ibid., 377.

76. Ibid., 373–74.

77. Emelie Savage-Smith, "Medicine," in *Encyclopedia of the History of Arabic Science*, 917.

78. Iu. Nuraliev, *Meditsinskaia sistema Ibn Siny (Avitsenny)* (Dushanbe, 2005), 236–39, 255ff.

79. O. Cameron Gruner, *A Treatise on The Canon of Medicine of Avicenna* (London, 1930), 205ff.

80. Emilie Savage-Smith, "Attitudes toward Dissection in Medieval Islam," *Journal of the History of Medicine and Allied Sciences* 50, 1 (1995): 67ff.

81. M. Nasser and A Tibi, "Ibn Hindu and the Science of Medicine," *Journal of the Royal Society of Medicine* 100, 1 (2007): 55–56.

82. Savage-Smith, "Medicine," 925.

83. Albert Z. Iskandar, "Ibn al-Nafis," in *Dictionary of Scientific Biography*, 9:602–6.

84. Dimitri Gutas, "Medical Theory and Scientific Method in the Age of Avicenna," in *Before and After Avicenna*, ed. David C. Reisman (Leiden, 2003), 148.

85. For example, Gruner, *A Treatise on The Canon of Medicine of Avicenna*, 96–97.

86. Gutas, "Medical Theory and Scientific Method in the Age of Avicenna," 157.

87. Zavadovskii, *Abu Ali Ibn Sina*, 104.

88. Because of the new materials he presents, on several points the account by Bulgakov supersedes those of Bartold and others. The following passage therefore draws on Bulgakov, *Zhizn i trudy Beruni*, 1:122–26.

89. Barthold, *Turkestan Down to the Mongol Invasion*, 275–76.

90. Hakim Muhammad Said and Ansar Zahid Khan, *Al Biruni: His Times, Life, and Works* (Delhi, 1990), 71. The claim that the line stretched to Lahore traces to Beyhaqi; the figure of five thousand is from Al-Utbi, *The Kitab-i-Yamani*, trans. James Reynolds (London, 1858), 447–49.

91. Zavadovskii, *Abu Ali Ibn Sina*, 68, 145, and chap. 8.

92. Avicenna, *The Metaphysics of the Healing*, trans. Michael E. Marmura (Provo, 2005).

93. Stephen Toulmin and June Goodfield, *The Ancestry of Science: The Discovery of Time* (Chicago, 1965), 64ff.

94. Quoted by Soheil N. Afnan, *Avicenna: His Life and Works* (London, 1958), 109. The following exposition draws on Afnan's brilliant chapters 3 and 4 on Ibn Sina's logic, metaphysics, and religious thought.

95. Ibid., 108. Ibn Sina's final thoughts on the soul are crystallized in his enigmatic and intriguing "Epistle of the Birds," on which see Peter Heath, "Disorientation and Reorientation in Ibn Sina's *Epistle of the Birds*: A Reading," in *Intellectual Studies in Islam. Essays in Honor of Martin B. Dickson* (Salt Lake City, 1990), chap. 10.

96. Afnan, *Avicenna*, 130–34. On Ibn Sina's crucial doctrine of emanation, see Herbert A. Davidson, *Alfarabi, Avicenna, and Averroes, on Intellect* (New York, 1992), 44ff.

97. On Ibn Sina's view on prophesy, see Walzer, *Greek into Arabic*, 218–19. On political suspects of Ibn Sina's views on prophesy, see James W. Morris, "The Philosopher-

Prophet in Avicenna's Political Philosophy," in *The Political Aspects of Islamic Philosophy*, ed. Charles E. Butterworth (Cambridge, 1992), chap. 4.

98. Ibn Sina, "The Healing," in *An Anthology of Philosophy in Persia*, 1:226–37.

99. On Ibn Sina and the West, see G. Quadri, *La philosophie arabe dans l'Europe médiévale, des origines à Averroès* (Paris, 1960), 95–121; G. M. Wickens, ed., *Avicenna: Scientist and Philosopher* (London, 1952), chaps. 5, 6.

100. Zavadovskii, *Abu Ali Ibn Sina*, chap. 8.

CHAPTER 10

1. Akhundova, *Tiurki v sisteme gosudarstvennogo upravleniia arabskogo khalifata*, 112, 128.

2. Ibid., 165–68, speaks of the Kara Bugry family under Caliph Mamun (r. 813–833) and thereafter.

3. The Seljuk vizier Nizam al-Mulk's description of this ritualized process is recounted by Barthold, *Turkestan Down to the Mongol Invasion*, 227.

4. The observation of Svat Souchek, *A History of Inner Asia* (Cambridge, 2000), 89.

5. J. H. Kramers, "La question Balḫī-Iṣṭaḫrī—Ibn Ḥawkal et l'atlas de l'Islam," *Acta Orientalia* 10 (1932): 9–30.

6. Needham, on "The Role of the Arabs," in *Science and Civilization in China*, 3:562–64.

7. Andreas Kaplony, "Comparing al-Kashgari's Map to His Text: On the Visual Language, Purpose, and Transmission of Arabic-Islamic Maps," in *The Journey of Maps and Images on the Silk Road*, 145.

8. Robert Dankoff and James Kelly, ed. and trans., *Mahmud al Kasgari, Compendium of the Turkic Dialects (Dīwān Lughat at-Turk)*, Sources of Oriental Languages and Literatures, vol. 7, pt. 1 (1982), 70–71. A. Rustamov is also translating the work into Russian: *Makhmud al-Kashgari, Divan Lugat-at-Turk*, ed. I. V. Kormushin, *Pismennosti Vostoka*, vol. 128 (Moscow, 2010). The first of three volumes has been issued.

9. On Barskhan, see V. Minorsky, " Tamim ibn Bahr's 'Journey to the Uyghurs,'" *Bulletin of the School of Oriental and African Studies* 12, 2 (1948): 290–91, 297.

10. Skrine and Ross in 1899 credited the Russian historian Grigoriev with this term: *The Heart of Asia* (London, 1899), 114n1.

11. For a concise review of the various theories on Karakhanid origins, see Peter B. Golden, *An Introduction to the History of the Turkic Peoples* (Wiesbaden, 1992), 214–16. Two recent studies are N. Necef, *Kaarahanlilar* (Selenge, 2005); and Ömer Soner Hunkan, *Türk Hakanligi (Karahanlilar) (766–1212)* (Istanbul, 2007).

12. Abu'l-Hasan Kalamati. See Barthold, *Turkestan Down to the Mongol Invasion*, 255.

13. The first Turkic khan to accept Islam was Satuq from Kashgar. Ibid., 76ff. For a more detailed analysis, see Omeljan Pritsak, "Von der Karluk zu den Karachaniden," *Studies in Medieval Eurasian History* (London, 1981), 292–94.

14. Barthold, *Turkestan Down to the Mongol Invasion*, 268.

15. E. A. Davidovich, "The Karakhanids," in *History of Civilizations of Central Asia*, vol. 4, pt. 1, 120.

16. For a convenient view of the literature on Tash Rabat, see D. D. Imankulov, "Iz istorii izucheniia Tash-Rabata," in *Pamiatniki Kyrgyzstana* (Frunze, 1982), vol. 5; on the most recent excavations, see D. D. Imankulov, "Novoe o Tashe-Rabate," in *Drevnii i srednevekovyi Kyrgyzstan* (Bishkek, 1996), 160–71.

17. B. D. Kochnev, *Karakanidskie monety: Istornikovedcheskoe i istoricheskoe issledovanie* (Moscow, 1993).

18. Davidovich, "The Karakhanids," 129.

19. Ibid., 128–29.

20. Dankoff and Kelly, *Mahmud al Kasgari, Compendium of the Turkic Dialects*, vol. 7, pt. 2 (1982), 76ff.

21. Ibid., 274.

22. The latter date is proposed in ibid., 7.

23. Sevim Tekeli, "Map of Japan Drawn by a Turk, Mahmud of Kashgar," *Turk Kulturunden Goruntuler* 7 (Ankara, 1986): 3–10.

24. Ibid., 6.

25. Quoted by Souchek, *A History of Inner Asia*, 90.

26. Rong, "The Nature of the Dunhuang Library Cave and the Reasons for Its Sealing," 247–75.

27. Barthold, *Turkestan Down to the Mongol Invasion*, 18.

28. Saidov et al., "The Ferghana Valley," 23.

29. Ibid., 17.

30. C. E. Bosworth, "Ozkend," in *Encyclopaedia of Islam*, 2nd ed. (Brill online, 2013), reference 17, February 2013, http://www.encquran.brill.nl/entries/encyclopaedia-of-islam-2/ozkend-SIM_6053.

31. A. Iu. Iakubovskii,"Dve nadpisi na severnom mavzole 1152 g. v Uzgende," *Epigrafika vostoka* 1 (1947): 27–32.

32. D. Imankulov, *Monumentalnaia arkhitektura iuga Kyrgyzstana*, Bishkek, 2005, 80.

33. A convincing summary of the case for Burana, i.e., the Kyrgyz case, is to be found in V. D. Goriacheva, *Srednevekovye gorodskie tsentry i arkhitekturnye ansampli Kirgizii* (Frunze, 1983); and M. E. Masson and V. D. Goriacheva, *Burana* (Frunze, 1985). The most recent argument for Ak Beshim (or Aktobe) is U. Kh. Shelekenov, *V-VIII Ghasyrlardaghy Balasaghun Qalasy* (Almaty, 2006).

34. Gerard Clauson, "Ak-Beshim-Suyab," *Journal of the Royal Asiatic Society*, New Series, 93 (1961): 1–13.

35. Konurbaev, Mokrynin, and Ploskikh, *Gorod velikogo puteshestvennika*, 177ff.; Ploskikh, "The Central Asian 'Atlantis.'"

36. Goriacheva, *Srednevekovye gorodskie tsentry*, 33ff.; Imankulov, *Monumentalnaia arkhitektura*, 154ff.

37. For a thorough discussion of this issue, see M. E. Masson, "Kratkaia istoricheskaia spravka o sredneaziatskikh minaretakh," *Materialy uzkostarosta* 2–3 (1933).

38. On Jam, see J. Sourdel-Thomine, *Le minaret Ghouride de Jam. Un chef d'oeuvre du XIIIe siècle* (Paris, 2004).

39. Imankulov, *Monumentalnaia arkhitektura*, 163–66.

40. Ettinghausen, Grabar, and Jenkins-Madina, *Islamic Art and Architecture, 650–1250*, 152.

41. Masson, "Kratkaia istoricheskaia spravka," as cited by Imankulov, *Monumental-naia arkhitektura*, 154n111.

42. Goriacheva, *Srednevekovye gorodskie tsentry*, 33.

43. Masson, "Kratkaia istoricheskaia spravka."

44. Ettinghausen, Grabar, and Jenkins-Madina, *Islamic Art and Architecture, 650–1250*, 152.

45. Muhammad Khadr, "Deux acteds de waqf d'un Qarahanide d'Asie Centrale," *Journal Asiatique* 255 (1967): 305–24; also Yaacov Lev, "Politics, Education, and Medicine in Eleventh Century Samarkand: A Waqf Study," *Wiener Zeitschrift für die Kunde des Morgenlandes* 93 (2003): 127–43.

46. Davidovich, "The Karakhanids," 130.

47. Ibid., 123.

48. Hajib, *Wisdom of Royal Glory*.

49. Ibid., 1.

50. Hajib, *Wisdom of Royal Glory*, 146.

51. Ibid., 159.

52. Robert Devereux presents a similar argument in his essay "Yusuf Khass Hajib and the Kudadgu Bilig," *The Muslim World* 4 (1961): 303.

53. Hajib, *Wisdom of Royal Glory*, 51–52.

54. Ibid., 256–57.

55. Barthold, *Turkestan Down to the Mongol Invasion*, 312.

56. On the Karakhitai, see Barthold, *Four Studies on the History of Central Asia*, 1:100–9; and D. Sinor, "The Kitan and the Kara Khitai," in *History of Civilizations of Central Asia*, vol. 4, pt. 1, 234–42.

57. Hajib, *Wisdom of Royal Glory*, 129.

CHAPTER 11

1. Major studies of Mahmud of Ghazni include C. E. Bosworth, *The Ghaznavids: Their Empire in Afghanistan and Eastern Iran 994–1040* (Edinburgh, 1963); Nazim, *The Life and Times of Sultan Mahmud of Ghazna*; and C. E. Bosworth, "The Early Ghaznavids," in *The Cambridge History of Iran*, vol. 4, chap. 5. Bartold's bibliography is on 18–24 of his *Turkestan Down to the Mongol Invasion* (London, 1928). For a somewhat dated bibliography, see Nazim, *Life and Times of Sultan Mahmud of Ghazna*, 1–15; as well as Bosworth's more thorough listing, "Early Sources for the History of the First Four Ghaznavid Sultans, 997–1041," *Islamic Quarterly* 7, 1–2 (1963): 3–22.

2. Kennedy, *The Prophet and the Age of the Caliphates*, 206ff.

3. C. E. Bosworth, "Ghaznavid Military Organization," *Islam* 36 (1960): 37–77; Bosworth, *The Ghaznavids*, 126–27.

4. C. E. Bosworth, "The Imperial Policy of the Early Ghaznawids," in *The Medieval History of Iran, Afghanistan, and Central Asia*, chap. 11, 49ff.

5. Nazim, *Life and Times of Sultan Mahmud of Gazna*, 140.

6. Frye, *Bukhara*, 87.

7. Barthold, *Turkestan Down to the Mongol Conquest*, 239.

8. Bosworth, *The Ghaznavids*, 71.

9. Karl August Wittfogel, *Oriental Despotism; a Comparative Study of Total Power* (New Haven, 1957); Nazim, *Life and Times of Sultan Mahmud of Ghazna*, 24–27.

10. Bosworth, "The Imperial Policy of the Early Ghaznawids," 49ff.

11. This argument parallels that offered for the other societies by Acemoglu and Robinson, *Why Nations Fail.*

12. On the principles of Buyid rule, see Herbert Busse, "The Revival of Persian Kingship under the Buyids," in *Islamic Civilization, 950–1150*, ed. D. S. Richards, vol. 3 of *Papers on Islamic History* (Philadelphia, 1973).

13. Al-Utbi, *The Kitab-i-Yamani*, 366; also C. E. Bosworth, "The Rise of the Karamiyyah in Khurasan," in *The Medieval History of Iran, Afghanistan, and Central Asia*, 10; Bosworth, *The Ghaznavids*, 145–205.

14. Nazim, *Life and Times of Sultan Mahmud of Ghazna*, chap. 5.

15. Barthold, *Turkestan Down to the Mongol Invasion*, 273.

16. Nazim, *Life and Times of Sultan Mahmud of Ghazna*, chap. 10, and especially Bosworth, *The Ghaznavids*, 48–96, 272–79.

17. Barthold, *Turkestan Down to the Mongol Invasion*, 293.

18. Salahuddin Khuda Bakhsh, *The Renaissance of Islam* (New Delhi, 1995), 175. On this and other titles, see C. E. Bosworth's learned analysis, "The Titulature of the Early Ghaznavids," in *The Medieval History of Iran, Afghanistan and Central Asia*, 210–32.

19. Bosworth, *The Ghaznavids*, 91ff.; Nazim, *Life and Times of Sultan Mahmud of Ghazna*, 145–47.

20. Browne, *A Literary History of Persia from Firdawsi to Sadi*, 104.

21. Nazim, *Life and Times of Sultan Mahmud of Ghazna*, 35ff.

22. This was the constant line of the poet Unsuri. See Browne, *A Literary History of Persia from Firdawsi to Sadi*, 120.

23. Crone, *Medieval Islamic Political Thought*, 377ff.

24. Heinz Halm, "Fatimiden und Ghaznawiden," in *Studies in Honour of Clifford Edmund Bosworth*, ed. Ian Richard Netton (Leiden, 2000), 1:209–20.

25. Nazim, *Life and Times of Sultan Mahmud of Ghazna*, 159–60; Barthold, *Turkestan Down to the Mongol Conquest*, 80–81.

26. R. W. Bulliet, "The Political-Religious History of Nishapur in the Eleventh Century," in *Islamic Civilization, 950–1150*, 3:74–77.

27. Ibid., 76.

28. Bosworth, "The Rise of the Karamiyyah in Khurasan," 10.

29. Nazim, *Life and Times of Sultan Mahmud of Ghazna*, 163.

30. Ibid., 165.

31. Ibid., 17, 29.

32. This figure appears for Gwalior and, later, for Somanatha; Catherine B. Asher and Cynthia Talbot, *India before Europe* (New York, 2006), 20; Nazim, *Life and Times of Sultan Mahmud of Ghazna*, 118.

33. For multiple perspectives on this campaign, see Romila Thapar, *Somanatha, the Many Voices of a History* (London, 2005).

34. Al-Utbi, *The Kitab-i-Yamani*, 462–63. See also C. E. Bosworth, "Mahmud of Ghazna in Contemporary Eyes and in Later Persian Literature," in *The Medieval History of Iran, Afghanistan, and Central Asia*, 28; Nazim, *Life and Times of Sultan Mahmud of Ghazna*, 3.

35. Browne, *A Literary History of Persia from Firdawsi to Sa'di*, 118–19.

36. Bosworth, *The Ghaznavids*, 90–92, 153; Alexander H. Krappe, "Shepherd and King in El Libro de Exemplos," *Hispanic Review* 14, 1 (January, 1946): 61–62.

37. A. A. Hakimov, "Arts and Crafts in Transoxonia and Khurasan," in *History of Civilizations of Central Asia*, vol. 4, pt. 2, 444–45.

38. This was noted by Bartold in his scathing portrait of Mahmud, *Central Asia Down to the Twelfth Century*, 271–93.

39. Bosworth, "The Titulature of the Early Ghaznavids," 215–23.

40. C. E. Bosworth, "An Embassy to Mahmud of Ghazna Recorded in Qadi ibn al Zubayr's Kitab adh-dhakha'ir wa't-tuhaf," *Journal of the American Oriental Society* 85, 3 (July–September 1965): 406.

41. Bosworth, *The Ghaznivids*, 211–12.

42. Nazim, *Life and Times of Sultan Mahmud of Ghazna*, 40.

43. Ibid., 24–27; Bosworth, *The Ghaznavids*, 129–45.

44. Soucek, "The Development of Calligraphy," in *History of Civilizations of Central Asia*, vol. 4, pt. 2, 491–92.

45. Gutas, *Avicenna and the Aristotelian Tradition*, 117.

46. Browne, *A Literary History of Persia from Firdawsi to Sa'di*, 96.

47. Biruni, *Alberuni's India*, 1:ix.

48. Ibid., 97. See also "Ibn Khammor," in *Istoriia tadzhikskoi filosofii*, 2:586–87.

49. Zavadkovskii, *Abu Ali Ibn Sina*, 98.

50. Abu'l Rayhan Muhammad ibn Ahmad al-Biruni, *The Book of Instruction in the Elements of the Art of Astrology*, ed. and trans. R. Ramsay Wright et al. (London, 1934), ii–vii.

51. Kennedy, "Al-Biruni," in *Dictionary of Scientific Biography*, 2:150; Said and Khan, *Al-Biruni*, 76.

52. Browne, *A Literary History of Persia from Firsdawsi to Sa'di*, 104–5.

53. A. H. Dani, "Southern Central Asia," in *History of Civilizations of Central Asia*, vol. 4, pt. 2, 559–63.

54. Pugachenkova, *Gurganj*, 78–80. On Mahmud's architecture, see Robert Hillenbrand, "The Architecture of the Ghaznavids and Ghurids," in *Studies in Honor of Clifford Edmund Bosworth*, 2124–86.

55. Barthold, *Turkestan Down to the Mongol Invasion*, 289.

56. Mauricio Taddei, "Some Structural Peculiarities in the Buildings at Tapi Sardar," in *Orient und Okzident im Spiegel der Kunst*, ed. Günter Burcher et al. (Graz, 1986), 389ff.; Mauricio Taddei, "Evidence of a Fire Cult at Tapa Sardar, Ghazni," in *South Asian Archaeology, 1981* (Cambridge, 1984), 263–69. For a pleasing recent description of these remains, see Peter Levi, *The Light Garden of the Angel King: Journeys in Afghanistan* (London, 1972), 115ff.

57. Nazim, *Life and Times of Sultan Mahmud of Ghazna*, 167. The second minaret at Ghazni was built by Mahmud's son Masud.

58. J. M. Rogers, "The 11th Century: A Turning Point in the Architecture of the Mashriq," in *Islamic Civilization, 950–1150*, 237–39.

59. Hillenbrand, "The Architecture of the Ghaznavids and Ghorids," 173.

60. Al-Utbi, *The Kitab-i-Yamani*, 465.

61. U. Scerato, "The First Two Excavation Campaigns at Ghazni,1957 1958," *East and West* 10, 1–2 (1959): 39ff.

62. Biruni, *Alberuni's India*, 2:103.

63. Al-Utbi, *The Kitab-i-Yamani*, 467.

64. All doubt about the ascription of this minaret to Mahmud has been removed by the discovery that the existing inscription dates to a later reconstruction, and by the detailed analysis of the brickwork, which closely parallels the exactly contemporary Karakhanid-built Kalyan minaret (1027) at Bukhara. See Pugachenkova, *Gurganj*, 78–85.

65. Daniel Schlumberger, Janine Sourdel-Thomine, and Jean Claude Gardin, *Lashkari Bazar: Une résidence royale ghaznévide et ghoride, Mémoires de la Délégation archéologique française en Afghanistan* (Paris, 1963–1978), vol. 18.

66. Hillenbrand, "The Architecture of the Ghaznavids and Gurids," 147.

67. Quoted in "Lashkari Bazar Palace Complex," *ArchNet*, http://archnet.org/library/sites/one.site.tel?site_id=11342.

68. Al-Utbi, *The Kitab-i-Yamani*, 461–62.

69. The full title of Utbi's book is *The Kitab-i-Yamini: Historical Memoirs of the Amir Sabaktagin and the Sultan Mahmud of Ghazna* (London, 1858); see also Bosworth, "The First Four Ghaznavid Sultans," 5–8.

70. Barthold, *Turkestan Down to the Mongol Invasion*, 19.

71. Among them was Gardizi, whose loyal yet pungent observations are equally noteworthy. See C. E. Bosworth, "Mahmud of Ghazni in Contemporary Eyes," in *The Medieval History of Iran, Afghanistan, and Central Asia*, 8ff; cf. Bosworth, "Arabic, Persian, and Turkish Historiography in the Eastern Iranian World," in *History of Civilizations of Central Asia*, vol. 4, pt. 2, 144.

72. Concise sources on Beyhaqi are C. E. Bosworth, "Early Sources for the History of the First Four Ghaznavid Sultans (977–1041)," *Islamic Quarterly* 7, 1–2 (1960): 10–14; and Nazim, *Life and Times of Sultan Mahmud of Ghazna*, 1–2. For secondary work on Beyhaqi, see Marilyn R. Waldman, *Towards a Theory of Historical Narrative: A Case Study in Perso-Islamicate Historiography* (Columbus, 1980), 77ff.; and Barthold, *Turkestan Down to the Mongol Invasion*, 22–23.

73. Barthold, *Turkestan Down to the Mongol Invasion*, 23.

74. The best modern edition is by C. E. Bosworth, *The 'History' of Beyhaqi (The History of Sultan Masud of Ghazna by Abu'l-l-Fazi Beyhaqi)*, commentary by Bosworth and Mohsen Ashtiany, 3 vols. (Cambridge, MA, 2011); for an excellent Russian edition of Beyhaqi, see A. K. Arends, trans., *"Abu-l-Fazl Baykhaki, Istoriia Masuda (1030–1041)*, 2nd ed. (Moscow, 1969).

75. C. Edmund Bosworth, "An Oriental Samuel Pepys? Abu'l Fadl Bayhaqi's Memoir of Court Life in Eastern Iran and Afghanistan, 1030–1041," *Journal of the Royal Asiatic Society* 14, 1 (April 2004): 15–16.

76. Ibid., 17.

77. Bosworth, "Early Sources for the History of the First Four Ghaznavid Sultans," 13.

78. Bosworth, "An Oriental Samuel Pepys?" 16.

79. Tabatabai, *Father of Persian Poetry*, 23. Arudi's very readable volume from 1155 to 1157 is available in translation, Nizami Arudi, *Chahar Maqala* (London, 1907).

80. Rybka, *History of Iranian Literature*, 174.

81. This late shift toward Arabic mainly affected the court chancellery. Rypka, *History of Iranian Literature*, 291.

82. Browne, *A Literary History of Persia from Firdawsi to Sa'di*, 125–26.

83. A. A. Seyed-Gohrab, "The Art of Riddling in Classical Persian Poetry," *Edebiyat* 12 (2001): 21–23.

84. Julie Scott Meisami, "The Persian Qasida to the End of the Twelfth Century," in *Qasida Poetry in Islamic Asia and Africa*, ed. Stefan Sperl and Christopher Shackle (Leiden, 1996), 1:146–62; and Rypka, *Persian Literature to the Beginning of the Twentieth Century*, 174–77.

85. Freely adapted by the author from the translation by Jackson, *Early Persian Poetry*, 47.

86. Freely adapted by the author from the translation by C. J. Pickering, quoted in ibid., 49. On Kisai as "a frivolous waster," see Ripka, *History of Iranian Literature*, 145.

87. Freely adapted from Peacock, *Medieval Islamic Historiography and Political Legitimacy*, 33.

88. Freely adapted from Browne, *A Literary History of Persia from Firdawsi to Sa'di*, 117–22.

89. Freely adapted from Jackson, *Early Persian Poetry*, 71.

90. Ibid., 80.

91. C. E. Bosworth, "The Development of Persian Culture under the Early Ghaznavids," *Iran* 6 (1968): 41.

92. Freely adapted by the author from Jackson, *Early Persian Poetry*, 80.

93. Rypka, *History of Iranian Literature*, 177–79.

94. Meisami, "The Persian Qasida to the End of the Twelfth Century," 150–56.

95. Rypka, *History of Iranian Literature*, 176.

96. This is suggested by Rybka in ibid., 155–57, on whose analysis of Ferdowsi's actions at this time this section is based.

97. Zavadovskii, *Abu Ali Ibn Sina*, 138.

98. "Ferdowsi," in *Encyclopedia Iranica*, http://www.iranica.com/articles/ferdowsi-i.

99. Nazim, *Life and Times of Sultan Mahmud of Ghazna*, 158.

100. Rypka, *History of Iranian Literature*, 162.

101. Nazim, *Life and Times of Mahmud of Sultan Ghazna*, 165; Bosworth, "The Development of Persian Culture under the Early Ghaznavids," 40.

102. Browne evaluates the various versions of these episodes in *A Literary History of Persia from Firdawsi to Sa'di*, 134–36.

103. Ferdowsi, *Shahnameh*, 851.

104. Notes to Russian edition of Biruni's *India*, quoted Ilias Nizamutdinov, *Iz istorii sredneaziatsko-indiiskikh otnoshenii, (ix–xviii vv)* (Tashkent, 1969), 12.

105. For a review of Biruni's Arabic and Indian sources, see Sachau's introduction to *Alberuni's India*, 1:xxi–xlii.

106. Nasr, *An Introduction to Islamic Cosmological Doctrines*, 109.

107. For a convincing argument that Biruni's relations with Mahmud were on the whole cordial, see Said and Khan, *Al-Biruni*, chap. 3.

108. See the review of this issue by M. S. Khan, "Al-Biruni and the Political History of India," *Orient* 25 (1976): 91n24.

109. Nasr, *An Introduction to Islamic Cosmological Doctrines*, 127.

110. Ibid.

111. J. L. Berggren, "The Mathematical Sciences," in *History of Civilizations of Central Asia*, vol. 4, pt. 2, 190.

112. Biruni, *Alberuni's India*, 1:198, 400.

113. Ibid., 1:57–70.

114. Ibid., 2:57–62; Khan, "Al-Biruni and the Political History of India," 110; E. S. Kennedy, "Al-Biruni," in *Dictionary of Scientific Biography*, 2:150.

115. R. Mercier, "Geodesy," in *The History of Cartography*, vol. 1, *Traditional Islamic and South Asian Societies*, ed. J. B. Harly and D. Woodward (Chicago, 1992), 175–87; James S. Aber, "Abu Rayhan al-Biruni," in *History of Geology*, http://academic.emporia.edu/aberjame/histgeol/biruni/biruni.htm. Nathan Camillo Sidoli points out (private correspondence) that Biruni on some points deferred to the earlier Mamun study on the grounds that it was the work of a large team of specialists with high-quality equipment.

116. Bulgakov, *Zhizn i trudy Beruni*, 1:143.

117. Ibid., 1:24. See also Kemal Ataman, "Re-Reading al-Biruni's *India*: A Case for Intercultural Understanding," *Islam and Islam-Christian Relations* 16, 2 (April 2005): 141–54.

118. It is not clear whether Biruni also translated into Hindi an Arabic compilation based on Ptolemy's *Almagest*, but this seems doubtful. Biruni, *Alberuni's India*, 1:137.

119. Ibid., 1:19–20, 179.

120. Ibid., 1:6; Khan, "Al-Biruni and the Political History of India," 87.

121. Translated and quoted by Khan, "Al-Biruni and the Political History of India," 112.

122. Ahmad Hasan Dani, "Al-Biruni's Indica: A Reevaluation," in *Al Biruni Commemorative Volume*, ed. Hakim Mohammad Said (Islamabad, 1973), 184.

123. Indeed, Biruni began his work with a detailed discussion on lying. Biruni, *Alberuni's India*, 1:2–6.

124. Ibid., 1:4.

125. Ibid., 1:147.

126. Ibid., 2:144.

127. Ibid., vol. 1, chap. 14.

128. Ibid., chap. 35.

129. Ibid., chap. 15. On time, see vol. 2, chaps. 1, 3.

130. Ibid., 2:10–11 (English modernized by author); Khan, "Al-Biruni and the Political History of India," 111.

131. Biruni, *Alberuni's India*, 1:111.

132. Ibid., 2:211ff., 234.

133. See discussion by Yohanan Friedmann, "Medieval Muslim Views of Indian Religions," *Journal of the American Oriental Society* 95, 2 (April–June 1975): 215.

134. Biruni, *Alberuni's India*, 1:195.

135. Ibid., chap. 11.

136. Ibid., 1:121.

137. Ibid., 1:32–39; W. Montgomery Watt, "Al Biruni and the Study of Non-Islamic Religions," in *Al-Biruni Commemorative Volume*, 416.

138. Biruni, *Alberuni's India*, 1:400–401. For this point on Malthus, see Arvind Sharma, *Studies in Alberuni's India* (Wiesbaden, 1983), 100–101.

139. Biruni, *Alberuni's India*, 1:400–401.

140. Quoted by Sharma, *Studies in "Alberuni's India,"* 7.

141. I am grateful to Dr. Nathan Camillo Sidoli of Waseda University for this insight.

142. Biruni, *Alberuni's India*, 2:167–68.

143. Ibid., 2:152. The text has been slightly condensed by the author.

144. Ibid., 2:149ff., 172ff.

145. Among those questioning the existence of such humanism are B. Spuler, "Hellenistisches Denken im Islam," *Saeculum* 5 (1954): 154ff.; and Gardet, *La cité musulmane, vie sociale et politique*, 273ff.; a more moderate voice is Joel L. Kraemer, "Humanism in the Renaissance of Islam: A Preliminary Study," *Journal of the American Oriental Society* 104, 1 (1984): 135.

146. Biruni, *Alberuni's India*, 2:104.

147. Arvind Sharma, "Alberuni on Hindu Xenophobia," in *Studies in "Alberuni's India,"* 127–28.

148. Biruni, *Alberuni's India*, 1:22.

149. Ibid., 1:xi–xii.

150. Ibid., 1:22.

151. Ibid., 2:12–13.

152. Khan, "Al-Biruni and the Political History of India," 113.

153. Biruni, *Alberuni's India*, 1:7.

154. Ibid., 1:77.

155. Frank Peters, quoted by Goodman, *Avicenna*, 23.

156. Adapted by the author from the translation by Jackson, *Early Persian Poetry*, 54.

157. Nazim, *Life and Times of Sultan Mahmud of Ghazna*, 124.

158. Browne, *A Literary History of Persia from Firdawsi to Sa'di*, 118.

159. David A. King, "Astronomy and Islamic Society: Qibla, Gnomonics and Timekeeping," in *Encyclopedia of the History of Arabic Science*, 1:141ff. Biruni and the other scientists who addressed these issues did so more as a way of exhibiting their skills and demonstrating their usefulness than out of deep piety, of which there is little evidence.

160. Bulgakov, *Zhizn i trudy Beruni*, 1:143–86.

161. Ibid., 1:155–57. This work remained unknown until the twentieth century, when E. S. Kennedy translated and edited it: *The Exhaustive Treatise on Shadows*, 2 vols. (Aleppo, 1976); also Mary-Louise Davidian, "Al-Biruni on the Time of Day from Shadow Lengths," *Journal of the American Oriental Society* 80, 4 (October–December 1960): 330–36; and van Brummelen, *The Mathematics of the Heavens and the Earth* (Princeton, 2009), 149–54.

162. Khan, "Al-Biruni and the Political History of India," 90.

163. Said and Khan, *Al-Biruni*, 95.

164. Biruni, *Alberuni's India*, 1:xliii. Saliba's fine history should be consulted on all aspects of medieval astronomy in Arabic: *A History of Arabic Astronomy* (New York, 1994).

165. Abu Rayhan Muhammad b. Ahmad al-Biruni, *Al-Qanunu'l-Masudi (Canon Masudicus)*, ed. Ministry of Education, Government of India, 3 vols. (Hyderabad, 1955), 1:3; Bulgakov, *Zhizn i trudy Beruni*, 2:244ff.

166. Ibid., 1:5.

167. On Sizji, see Glen van Brummelen, "Sizji," in *The Biographical Encyclopedia of Astronomers*, ed. Thomas Hockey et al. (New York, 2009), 1059.

168. Biruni, *Al-Qanunu'l-Masudi (Canon Masudicus)*, 1:xviii. The fresh translation from the archival original included here is by Dallal, *Islam, Science, and the Challenge of History*, 74.

169. Biruni, *Al-Qanunu'l-Masudi (Canon Masudicus)*, 1:xviii.

170. Ibid., 2:248ff.

171. Some argue that priority on this goes to another scientist from Khurasan, Abu Jafar Muhammad ibn Hasan Khazini (900–971).

172. Biruni, *Al-Qanunu'l-Masudi (Canon Masudicus)*, 2:xxxiv.

173. *Al-Biruni and Ibn Sina*, 8.

174. Said and Khan, *Al-Biruni*, 123–36.

175. An Egyptian scholar identified two hundred errors in the text and accompanying tables of this sole English edition, published in Hyderabad and cited above: Bulgakov, *Zhizn i trudy Beruni*, 2:277.

176. Aber, "Abu Rayhan al-Biruni."

177. Al-Biruni, *Sobranie svedenii dlia poznaniia dragotsennostei (Mineralogiia)*, trans. A. M. Belenitskii (Leningrad, 1963).

178. K. A. Nizami, "The Ghurids," in *History of the Civilizations of Central Asia*, vol. 4, pt. 1, 182.

179. Browne, *Literary History of Persia from Firdawsi to Sa'di*, 120.

CHAPTER 12

1. Al-Ghazali, *The Faith and Practice of al-Ghazali*, ed. William Montgomery Watt (London, 1998), 60; also Algazali, "Deliverance from Error," in *Philosophy in the Middle Ages: The Christian, Islamic, and Jewish Traditions*, ed. Arthur Hyman and James J. Walsh (New York, 1967), 275ff.

2. Farid Jabre, *La notion de certitude selon Ghazali: dans ses origins psychologiques et historiques* (Paris, 1958), esp. chap. 2.

3. Bulgakov, *Zhizn i trudy Beruni*, 1:161.

4. Cullen Murphy, *God's Jury: the Inquisition and the Making of the Modern World* (Boston, 2012), 9.

5. Barthold, "The Oghuz (Turkmans) before the Formation of the Seljuq Empire," in *Four Studies on the History of Central Asia*, 3:91–108.

6. A. Sevim, "The Origins of the Seljuqs and the Establishment of Seljuq Power in the Islamic Lands Up to 1055," in *History of Civilizations of Central Asia*, vol. 4, pt. 1, 146.

7. Bulliet, *Cotton, Climate and Camels in Early Islamic History*, chap. 3.

8. Ibid., 83, based on a report by Utbi.

9. Ibid.; important insights on these developments can be gleaned from A.C.S. Peacock, *Early Seljuq History. A New Interpretation* (London, 2010).

10. On Dandanqan, see the classic account in *Tabakat-i-Nasiri: A General History of the Muhammadan Dynasties of Asia*, trans. H. G. Raferty (London, 1970), 1:130–32; also S. G. Agadzhanov, *Gosudarstvo seldzhukidov i sredniaia Asia v XI–XII vv.* (Moscow, 1991), 47–57. Jurgen Paul points out that Nishapur did not put up a fight because it had no walls at the time, "The Seljuq Conquest(s) of Nishapur: A Reappraisal," *Iranian Studies*, vol. 38, no.4, December, 2005, 582–585.

11. Arminius Vambery, *History of Bukhara* (London, 1873), 94.

12. Skrine and Ross, *The Heart of Asia*, 130.

13. C. E. Bosworth, "The Political and Dynastic History of the Iranian World (AD 1000–1217)," in *The Cambridge History of Iran*, ed. J. A. Boyle (Cambridge, 1968)5:42–43.

14. This is the thesis of Paul Markham, "The *Battle of Manzikert*: Military Disaster or Political Failure?" (2005), www.deremilitari.org/resources/articles/markham.htm.

15. This view is contradicted by centuries of Turkish myth making. See Carole Hildebrand, *Turkish Myth and Muslim Symbol. The Battle of Manzikert* (Edinburgh, 2007).

16. Bosworth, "The Political and Dynastic History of the Iranian World (AD 1000–1217)," 60.

17. Ann K. S. Lambton, "Aspects of Saljuq-Ghuzz Settlement in Persia," in *Islamic Civilization, 950–1150*, 105–25.

18. Ibid., 55; a balanced and concise overview of Seljuk rule is Hodgson, *The Venture of Islam*, 2:42–55.

19. Rudolph Schnyder, "Political Centres and Artistic Powers in Saljuq Iran: Problems of Transition," in *Islamic Civilization, 950–1150*, 201–9; Gaston Wiet, *Baghdad, Metropolis of the Abbasid Caliphate* (Norman, 1971), 105.

20. On Nizam al-Mulk's background and upbringing, see Nizam al-Mulk, *Book of Government or Rules for Kings*, trans. Huburt Darke (London, 1978), ix–xi; also K. E. Schabinger-Schowingen's solidly researched "Zur Geschichte des Saldschuqen-Reichskanzlers Nisamu 'l-Mulk," in *Historische Jahrbücher* (1942–1949), 62–69:250ff.

21. Bosworth, "The Political and Dynastic History of the Iranian World (AD 1000–1217)," 58.

22. Davidovich, "Coinage and the Monetary System," 402; D. Dawudi, "Seljuklaryn mawerannahrdaky zikgehhanalry" *Miras* 55 (2009):66ff.

23. Agadzhanov, *Gosudarstvo seldzhukidov i sredniaia Asia v XI–XII vv.*, 107, 138.

24. Ibid., 151–57.

25. National Museum, Ashgabat, Turkmenistan.

26. Darke, in Nizam al-Mulk, *Book of Government or Rules for Kings*, xii, points out that Nizam al-Mulk's name for the work was *Manners of the Kings*. See also A.K.S. Lambton, "The Dilemma of Government in Islamic Persia: The Siyasat-Nama of Nizam al-Mulk," *Journal of the British Institute of Persian Studies* 22 (1984): 55–66. On Nizam al-Mulk's life, see S.A.A. Rizvi, *Nizam al-Mulk Tusi, His Contribution to Statecraft, Political Theory and the Art of Government* (Lahore, 1978).

27. Agadzhanov, *Gosudarstvo seldzhukidov i sredniaia Asia v XI–XII vv*, 99 ff.

28. Nizam al-Mulk, *Book of Government or Rules for Kings*, 78.

29. Barthold, *Turkestan Down to the Mongol Invasion*, 308

30. On continuity and change in Seljuk architecture, with emphasis on the latter, see Janine Sourdel-Thomine, "Renouvellement et tradition dans l'architecture saljuqide," in *Islamic Civilization, 950–1150*, 251–64.

31. Ettinghausen, and Grabar, Jenkins-Madina, *Islamic Art and Architecture, 650–1250*, 153–54.

32. Ibid., 139–44, 368.

33. Pope, *Persian Architecure*, 956.

34. Janine Sourdel-Thomine underscores elements of continuity from earlier eras in her "Renouvellement et tradition dans l'architecture saljuqide," 251ff.; by contrast, J. M. Rogers stresses the element of change in the Seljuk East, "The 11th Century—A Turning Point in the Architecture of the Mashriq?" in *Islamic Civilization, 950–1150*, 211–49.

35. G. A. Pugachenkova, "Puti razvitiia arkhitektury iuzhnogo Turkmenistana pory rabovladeniia i feodalizma," in *Trudy iuzhno-turkmenistanskoi arkheologicheskoi ekspeditsii* (Moscow, 1958), 6:314.

36. E. Baer, "Abd-al-Razzauq b. Masuud Naysabuure," in *Encyclopedia Iranica*, http://www.iranica.com/newsite/articles/v1f2/v1f2a071.html.

37. National Museum, Ashgabat.

38. Baer, "Abd-al-Razzauq Naysabuure."

39. Allan, *Nishapur: Metalwork of the Early Islamic Period*, 18– 20; Hakimov, "Arts and Crafts in Transoxonia and Khurasan," 423–30.

40. E. S. Kennedy, "The Exact Sciences in Iran under the Saljuqs and Mongols," in *The Cambridge History of Iran*, 5:671–72.

41. A. Abdurazakov and Ts. Khaidav, "Alchemy, Chemistry, Pharmacology and Pharmaceutics," in *Civilizations of Central Asia*, vol. 4, pt. 2, 230.

42. Charles Homer Haskins, *Renaissance of the Twelfth Century* (Cambridge, 1928), 64ff.

43. Bosworth, "The Political and Dynastic History of the Iranian World (AD 1000–1217)," 86.

44. Q. Mushtaq and J. L. Berggren, "Mathematical Sciences," in *Civilizations of Central Asia*, vol. 4, pt. 2, 181.

45. Bosworth, "The Rise of the Karamiyyah in Khurasan," 7–14.

46. The classic sources on Nishapur in this period are Richard W. Bulliet's *The Social History of Nishapur in the Eleventh Century* (Cambridge, 1967); and his "The Political-Religious History of Nishapur in the Eleventh Century," in *Islamic Civilization 950–1150*, 71–91.

47. The best Western translation is in French and was published in two parts: Asadi-Tusi, *Garšāsp-nāma*, ed. Cl. Huart (Paris, 1926); and Asadi-Tusi, *Garšāsp-nāma*, ed. H. Massé (Paris, 1951).

48. G. Patvievskaya, "History of Medieval Islamic Mathematics: Research in Uzbekistan," *Historia Mathematica* 20 (1993): 241.

49. Bosworth, "The Political and Dynastic History of the Iranian World (AD 1000–1217)," 86; and Bosworth, "Legal and Political Sciences in the Eastern Iranian World and Central Asia in the Pre-Mongol Period," 147.

50. Dzhalilov, *Iz istorii kulturnoi zhizni predkov tadzhikskogo naroda i tadzhikov v rannem srednevekove*, 64. In contrast, Charles Melville argues that the quantity and importance, if not the quality, of historical writing in Persian actually increased during the Seljuk period, "History: From the Saljuqs to the Aq Qoyunlu (ca. 1000–1500 C.E.)," *Iranian Studies* 31, 3–4 (Summer/Fall 1998): 474–76.

51. G. Hanmyradow, "Sarahs we gundogaryn medeni mirasy," *Miras* 3, 35 (2009): 60.

52. Robert E. Hall "Al-Khazini," in *Dictionary of Scientific Biography*, 7:335–51.

53. D. R. Hill, "Physics and Mechanics, Civil and Hydraulic Engineering, Industrial Processes and Manufacturing, and Craft Activities," in *Civilizations of Central Asia*, vol. 4, pt. 2, 254–61, on which the following paragraphs are based.

54. Waldren Grutz, "Oasis of Turquoise and Ravens," *Aramco World* 49, 4 (1998): 21.

55. Peter Avery and John Heath-Stubbs, *The Ruba'iyat of Omar Kayyam* (London, 1981), no. 16, 50.

56. Quoted by John J. O'Connor and Edmund F. Robertson, "Omar Khayyam," Mac-Tutor History of Mathematics Archive, http://www-history.mcs.st-andrews.ac.uk/Biographies/Khayyam.html.

57. The following exposition draws on Kennedy, "The Exact Sciences in Iran under the Saljuqs and Mongols," 665ff.; and O'Connor and Robertson, "Omar Khayyam."

58. O'Connor and Robertson, "Omar Khayyam."

59. Kennedy, "The Exact Sciences in Iran under the Saljuqs and Mongols," 666.

60. Bosworth, "The Political and Dynastic History of the Iranian World (AD 1000–1217)," 195.

61. Farhad Daftary, *The Assassin Legends: Myths of the Ismailis* (London, 1995).

62. For Ismaili perspectives on this conflict, see chapters in Farhad Daftary's *Medieval Ismaili History and Thought* (Cambridge, 1996), which includes Carole Hillenbrand's alternative view: "The Power Struggle between the Saljuqs and the Ismailis of Aklamut, 487–518/1094–1124: The Saljuq Perspective," 205–20.

63. W. M. Thackston, Jr., trans. *Naser-e Khosrow's Book of Travels* (New York, 1986); on Khusraw's life, see Alice C. Hunsberger, *Nasir Khusraw, the Ruby of Badakhshan* (London, 2000); also Daftary, "The Medieval Ismailis of the Iranian Lands," 59ff., 23ff.; and W. Ivanov's now dated *Problems in Nasir-i Khusraw's Biography* (Bombay, 1956).

64. Quoted in *Istoriia tadzhikskoi filosofii*, 2:378–79.

65. Peter Lamborn Wilson and Gholam Reza Aavani, *Nasir-i Khusraw, Forty Poems from the Divan* (Tehran, 1977), 72; noteworthy are the verse translations by Annemarie Schimmel, *Make a Shield from Wisdom: Selected Verses from Nasir-I Khuusraw's Divan* (London, 1993).

66. Schimmel, *Make a Shield from Wisdom*, 35.

67. Ibid., 96.

68. Nizam al-Mulk, *Book of Government*, **213–45.**

69. For a solid account of these shifting loyalties, see Bulliet, "The Political-Religious History of Nishapur in the Eleventh Century," esp. 74–84.

70. See Burhan al-Din al-Farghani al-Marghinani, *Al Hidayah: The Guidance* (Bristol, 2006); and Rosenthal, *Ahmad b. t-Tayyib as-Sarahsi.*

71. Sergej Chmelnizkij, "Architecture," in *Islam: Art and Architecture*, ed. Markus Hattstein and Peter Delius (Maspeth, 2008), 362–63.

72. These initiatives are discussed in detail by George Makdisi, "Muslim Institutions of Learning in Eleventh-Century Baghdad," in *Bulletin of the School of Oriental and African Studies* 24 (1961).

73. Le Strange, *Baghdad during the Abbasid Caliphate*, 298; on these institutions, see George Makdisi, "Muslim Institutions of Learning in Eleventh-Century Baghdad," in *Religion, Law and Learning in Classical Islam* (London, 1991), 1–55.

74. A. Bausani, "Religion in the Saljuq Period," in *The Cambridge History of Iran*, 5:283–87.

75. Asad Talas, *La Madrasa Nizamiyya et son histoire* (Paris, 1939).

76. Bosworth, "Legal and Political Sciences in the Eastern Iranian World and Central Asia in the Pre-Mongol Period," 136.

77. This section on Hayyam's philosophy is indebted to Seyyed Hossein Nasr's pioneering essay "The Poet-Scientist 'Umar Khayyam as Philosopher," in *Islamic Philosophy from its Origins to the Present* (Binghamton, 2006), 165–83; equally cogent is Mehdi Aminrazavi, *The Wine of Wisdom: The Life, Poetry, and Philosophy of Omar Khayyam* (Oxford, 2005), especially chaps. 5–6.

78. With slight modernizations of syntax by the author, Nasr, "The Poet-Scientist 'Umar Khayyam as Philosopher," 175.

79. While much valuable research ion Sufism has recently been issued, Franz Rosenthal's chapter "Knowledge Is Light" remains a concise and authoritative introduction, in *Knowledge Triumphant: The Concept of Knowledge in Medieval Islam* (Leiden, 1970), 155–82.

80. Mez, *The Renaissance of Islam*, 367.

81. On Balkhi, see Alexander Knysh's excellent *Islamic Mysticism: A Short History* (Leiden, 2000), 32.

82. It might also be noted that Tirmidhi ranked Christ higher than Muhammad and may have derived his notion of "friendship" with God from Christian doctrines. See Knysh, *Islamic Mysticism*, 104ff.

83. Two useful surveys of enthusiastic religion in the West are Gary Dickson, *Religious Enthusiasm in the Medieval West: Revivals, Crusades, Saints* (London, 2000); and David S. Lovejoy, *Religious Enthusiasm and the Great Awakening* (Upper Saddle River, 1969).

84. Seyyed Hossein Nasr provides a useful introduction in his concise *The Garden of Truth: The Vision and Promise of Sufism, Islam's Mystical Tradition* (San Francisco, 2007).

85. On Kharaqani, see ibid., 136.

86. For a listing of the early Khurasan Sufi groups, see Bausani, "Religion in the Saljuq Period," 297–98.

87. Sara Sviri, "The Early Mystical Schools of Baghdad and Nishapur: In Search of Ibn Munazil," *Jerusalem Studies in Arabic and Islam* 30 (2005): 457.

88. The Naqshbandiyya, Chishtiyya, and Kubrawiyya, founded in what is now Uzbekistan, Afghanistan, and Kazakhstan, respectively. For leading Sufi divines from the region, see *The Cambridge History of Iran*, 5:297–98.

89. Christopher Melchert, "Sufis and Competing Movements in Nishapur," *Journal of Persian Studies* 39 (2001): 237–47; Margaret Malamud, "Sufi Organizations and Structures of Authority in Medieval Nishapur," *International Journal of Middle East Studies* 26, 3 (August 1994): 427–42.

90. See Knysh, *Islamic Mysticism*, 125–26, 136ff.

91. Bosworth, "The Political and Dynastic History of the Iranian World (AD 1000–1217)," 138.

92. M. T. Houtsma, "The Death of Nizam al-Mulk and Its Consequences," *Journal of Indian History* 3 (1924): 147ff.

93. Bosworth, "The Political and Dynastic History of the Iranian World (AD 1000–1217)," 105–6.

94. Al-Ghazali, *The Faith and Practice of al-Ghazali*, 59, 60.

95. Aminrazavi, *The Wine of Wisdom*, 23.

96. Dunce Black MacDonald claimed that Ghazali withdrew from teaching because he had earned the displeasure of the new sultan, but in the same work he ascribes it to a purely religious and philosophical crisis: "The Life of al-Ghazali, with Especial Reference to His Religious Experiences and Opinions," *Journal of the American Oriental Society* 20 (1899): 82–92. See also Eric Ormsby's account in *Ghazali: The Revival of Islam* (Oxford, 2008), chap. 1.

97. Farouk Mitha, *Al Ghazali and the Ismailis* (London, 2001), 19ff.

98. Henry Corbin, "The Ismaili Response to the Polemic of Ghazali," in *Isma'ili Contributions to Islamic Culture*, 76.

99. Al-Ghazali, *The Faith and Practice of al-Ghazali*, 61.

100. Al-Ghazali, *The Incoherence of the Philosophers*, trans. Michael E. Marmura (Provo, 1997), 5–6.

101. This discussion draws extensively on Majid Fakhry's excellent exposition in *A History of Islamic Philosophy*, 228.

102. Ibid., 170.

103. Ilai Alon, "Al-Ghazali on Causality," *Journal of the American Oriental Society* 109, 4 (October–December 1980): 399.

104. Leor Halevi, "The Theologian's Doubts: Natural Philosophy and the Skeptical Games of Ghazali," *Journal of the History of Ideas* 63, 1 (January 2002): 26.

105. Richard M. Frank, *Creation and the Cosmic System: Al-Ghazali and Avicenna* (Heidelberg, 1992), 43. On Ghazali's metaphysics, see Sharif, *A History of Muslim Philosophy*, 1:581–641.

106. Frank Griffel, "Toleration and Exclusion: Al-Shafii and al-Ghazali on the Treatment of Apostates," *Bulletin of the School of African and Oriental Studies* 64, 3 (2001): 351.

107. Ibid., 354.

108. George Makdisi, "Al-Ghazali, disciple de Shafti'i en droit et en theologie," in *Religion, Law and Learning in Classical Islam*, 3:45–55.

109. Muhammad Al-Ghazzali, *The Alchemy of Happiness*, trans. Henry A. Homes (Albany, 1873).

110. Al-Ghazali, *The Faith and Practice of al-Ghazali*, 58.

111. Binyamin Abrahamov, *Divine Love in Islamic Mysticism: The Teachings of al-Ghazali and al-Dabbagh* (London, 2003), chap. 2.

112. S. M. Ghazanfar, "The Economic Thought of Abu Hamid Al-Ghazali and St. Thomas Aquinas: Some Comparative Parallels and Links," *History of Political Science* 32, 4 (2000): 869–79.

113. Ibn Rushd, *Tahafut al Tahafut: The Incoherence of the Incoherence*, trans. Simon van den Bergh, http://www.muslimphilosophy.com/ir/tt/; see also R. J. Kilcullen, "Al Ghazali and Averreos," http://www.humanities.mq.edu.au/Ockham/x52t07.html.

114. See the convenient history of these events in Agadzhanov, *Gosudarstvo seldzhukidov i srednaia Asia v XI–XII vv.*, 112ff.; and Bosworth, "The Political and Dynastic History of the Iranian World (AD 1000–1217)," 135.

115. Agadzhanov, *Gosudarstvo seldzhukidov i srednaia Asia v XI–XII vv.*, 154–56.

116. Barthold, *Turkestan Down to the Mongol Invasion*, 291.

117. Agadzhanov, *Gosudarstvo seldzhukidov i srednaia Asia v XI–XII vv.*, 170.

118. Yazberdiev, "The Ancient Merv and Its Libraries," 153.

119. Ibid., 185–94; *The Cambridge History of Iran*, 5:139.

120. Haskins, *The Renaissance of the Twelfth Century*.

121. Carla L. Klausner, *The Seljuk Vizerate: A Study of Civil Administration 1055–1194* (Cambridge, 1973), chap. 1; Ann K. S. Lampton, "The Administration of Sanjar's Empire as Illustrated in the 'Atabat al-kataba,'" *Bulletin of the School of Oriental and African Studies* 19 (1957): 367–88.

122. Agadzhanov, *Gosudarstvo seldzhukidov i srednaia Asia v XI–XII vv.*, 175.

123. Pope, *Persian Architecture*, 131.

124. Ibid., 154.

125. Typical of these impressive structures is the six-arched bridge still standing on the road from Mashad to Herat, pictured in Hill, "Physics and Mechanics," 263,

126. Tertius Chandler, *Four Thousand Years of Urban Growth: An Historical Census* (Lewiston, 1987), 337.

127. Chmelnizkij, "Architecture," 365, argues that Sanjar's tomb was built after his death by an admirer, but the state of confusion in those years makes this highly unlikely.

128. K.A.C. Creswell, "Persian Domes before 1400 AD," *Burlington Magazine for Connoisseurs* 26, 143 (February 1915): 208–213.

129. Chmelnizkij, "Architecture," 361.

130. Pugachenkova, "Puti razvitiia arkhitektury iuzhnogo Turkmenistana pory rabovladeniia i feodalizma," 320; Pope, *Persian Architecture*, 131.

131. For example, the great mausoleum of Oljeitu at Sultaniyya, Iran, 1315–1325.

132. Dzhamal al-Karshi, *Mulkhakat as-Surakh*, ed. and trans. Abdukakhkhora Saidova (Dushanbe, 2006), 31.

133. Abu Hamid al-Ghazali, "Letter to Sultan Sanjar Seljuki," http://en.wikisource .org/wiki/Al-Ghazali_letter_to_Sultan_Sanjar_Seljuki.

134. Anvari considered Shihab al-Din Tirmidhi from Termez to be far his superior as a poet. Rypka. "Poets and Prose Writers of the Late Saljuq and Mongol Periods," 563.

135. Modified from the translation of E. G. Browne, quoted by K. B. Nasim, *Hakim Auhad-ud-din Anwari* (Lahore, 1965), 41.

136. J. Rypka, "Poets and Prose Writers of the Late Saljuq and Mongol Periods," 563ff. The classic early work on Anvari is Valentin Zhukovskii, *Ali Avhadu-d Din Anvari: materialy k biografii* (St. Petersburg, 1883).

137. Browne, *A History of Persian Literature from Firdawsi to Sa'di*, 297.

138. Ibid., 298.

139. Agadzhanov, *Gosudarstvo seldzhukidov i srednaia Asia v XI–XII vv.*, 176ff.; Barthold, *Turkestan Down to the Mongol Invasion*, 323ff.

140. *Tabakat-i-Nasiri* 1:146ff.; Soucek, *A History of Inner Asia*, 99; Bosworth, "The Political and Dynastic History of the Iranian World (AD 1000–1217)," 147–49.

141. For an engaging account of this legend, see Charles E. Nowell, "The Historical Prestor John," *Speculum* 28, 3 (July 1953): 435–45; and L. N.Gumilev, *Searches for an Imaginary Kingdom: The Legend of the Kingdom of Prester John* (Cambridge, 1987).

142. Agadzhanov, *Gosudarstvo seldzhukidov i srednaia Asia v XI–XII vv.*, 179–81.

143. These passages, slightly modernized, are drawn from Browne's literal translation, *A Literary History of Persia from Firdawsi to Sa'di*, 384–90.

144. A. Youschkevitch and B. A. Rosenfeld, "Al-Khayyami," in *Dictionary of Scientific Biography*, 7:325. On Khayyam and Sanjar, see also J. A. Boyle, "Umar Khayyam: Astronomer, Mathematician, and Poet," in *The Cambridge History of Iran*, 4:660–61.

145. Avery and Heath-Stubbs, *The Ruba'iyat of Omar Khayyam*, no. 21, 51. All the quatrains cited herein are from the translation by Avery and Heath-Stubbs, on the grounds that theirs alone among available translations in English, French, or German were "intended to give as literal a . . . version of the Persian originals as readability and intelligibility permit," 42.

146. Ibid., no. 35, 55. Rypka's assessment of Khayyam's poetry can be found in *History of Iranian Literature*, 189–93.

147. Avery and Heath-Stubbs, *The Ruba'iyat of Omar Khayyam*, no. 74, 65.

148. Ibid., no. 10, 49.

149. Ibid., no. 6, 48.

150. Ibid., no. 87, 68.

151. Ibid., no. 85, 68.

152. Ibid., no. 86, 68.

153. Ibid., no. 104, 72.

154. Ibid., no. 84, 67.

155. Ibid., no. 91, 94.

156. Nasr, *Islamic Philosophy from Its Origins to the Present*, 168.

157. See also Swami Govinda Tirtha, *The Nectar of Grace: Omar Khayyam's Life and Works*, Allahabad, 1941, 173–288.

158. Avery and Heath-Stubbs, *The Ruba'iyat of Omar Khayyam*, no. 108, 73.

159. Nasr, *Islamic Philosophy from Its Origins to the Present*, 178.

160. Frank, *Creation and the Cosmic System*, 12–21.

161. Edward FitzGerald, *Rubáiyát of Omar Khayyám: A Critical Edition*, ed. Christopher Decker (Charlottesville, 1997), 69.

162. Avery and Heath-Stubbs, *The Ruba'iyat of Omar Khayyam*, no. 179, 91.

CHAPTER 13

1. Barthold, *Turkestan Down to the Mongol Invasion*, 404–5.

2. The classic source on the Mongol invasion is Ala-al-Din Ata-Malik Juvayni, *History of the World Conqueror*, trans. J. A. Boyle (Manchester, 1958). A three-volume version of this immense work was issued in London in 1913–1937. The Mongols themselves left no account of the campaign in Central Asia, but we do have an early source that some consider to have been written for them, and which became known as the *Secret History of the Mongols*, ed. and trans. Igor de Rachewiltz, 2nd ed. (Leiden, 2006). See also de Rachewiltz, "The Dating of the *Secret History of the Mongols*: A Reinterpretation," *Ural-Altaische Jahrbücher* 22 (2008): 150–84. A further contribution to the debate on authorship of this work is C. Atwood's clever "Informants and Sources for the *Secret History of the Mongols*," *Mongolian Studies* 24 (2007): 27–39.

3. Juvayni on the scholar Rukn al-Din Imam-Zadah, cited by Barthold, *Turkestan Down to the Mongol Invasion*, 410.

4. Rahula Sanktrityayana, *History of Central Asia* (Calcutta, 1964), 251; Barthold, *Turkestan Down to the Mongol Invasion*, 353–69.

5. Quoted by Edgar Knobloch, *Monuments of Central Asia* (London, 2001), 83.

6. *The Cambridge History of Iran*, 5:140ff.; , *Turkestan Down to the Mongol Invasion*, 379–80.

7. Arudia-i-Samarqandi, *Chahar Maqala*.

8. M. A. Gaipov, "Dzhurdzhani i ego trud "sokrovishche khorezmshakha," *Sovetskoe zdravookhranenie* 9 (1978): 71–74.

9. Georges C, Anawati, "Fakhr al Din al-Razi," in *Encyclopaedia of Islam*, 2:751–55.

10. Rypka, *History of Iranian Literature*, 200ff.

11. Sheila Blair, "The Madrasa at Zuzan: Islamic Architecture in Eastern Iran on the Eve of the Mongol Invasion," *Muqarnas* 3 (1985): 75–91.

12. Friedrich Heer, *The Medieval World, 1100–1350* (New York, 1962), 26.

13. These religious currents are the subject of Julian Baldick's *Animal and Shaman: Ancient Religions of Central Asia* (London, 2000); also Barthold, *Four Studies on the History of Central Asia*, 3:111–12.

14. I. Melikoff, "Ahmad Yesevi and Turkic Popular Islam," *Electronic Journal of Oriental Studies* 6, 8 (2003): 1–9. See also A. Kaymov, "Literature of the Turkic Peoples," in *Civilizations of Central Asia*, vol. 4, pt. 2, 381.

15. Barthold, *Four Studies on the History of Central Asia*, 1:54, 133.

16. Nile Green, "The Religious and Cultural Roles of Dreams and Visions in Islam," *Journal of the Royal Asiatic Society* 13, 3 (November 2003): 287–313.

17. Najm al-Dīn Dāya, "The Wetnurse," wrote *The Path of God's Bondsmen: From Origin to Return* (North Haledon, 1980).

18. Green, "The Religious and Cultural Roles of Dreams and Visions in Islam" 287ff.

19. By far the most useful sources on Attar are Hellmut Ritter, "Philologika X: Faradaddin Attar," *Islam* 25 (1919): 134–73, and F. Meier, "Der Geistmensch bei dem persischen Dichter Attar," *Eranos-Jahrbuch* 13 (1945): 286ff. The following exposition relies on these authors but passes over Ritter's insightful periodization of Attar's development.

20. An impressively graceful translation is Farid ud-Din Attar, *The Conference of the Birds*, trans. Afkham Darbandi and Dick Davis (London, 1984).

21. Browne, *A Literary History of Persia from Firdawsi to Saʿdi*, 513.

22. Attar, *The Conference of the Birds*, 206.

23. The key link has been identified as Ghazali's "Treatise on Direct Knowledge from God," *al-Risala al-Laduniyya*.

24. Devin A. de Weese, "Islamization in the Mongol Empire," in *The Cambridge History of Inner Asia: The Chinggisid Age*, ed. Nicola Di Cosmo, Allen J. Frank, and Peter B. Golden (Cambridge, 2009), 120–34; a dissenting view is offered by Jürgen Paul, who minimizes the pre-Mongol impact of Sufism: "Islamizing Sufis in Pre-Mongol Central Asia," in *Islamization de l'Asie Centrale*, ed. Étienne de La Vaissière (Paris, 2008), 297–317.

25. Barthold, *Four Studies on the History of Central Asia*, 1:114–18.

26. Barthold, *Turkestan Down to the Mongol Invasion*, 402.

27. J. J. Saunders, *The History of the Mongol Conquests* (Philadelphia, 1972) , 55.

28. Owen Lattimore, "The Geography of Ghingis Khan," *Geographical Journal* 129, 1 (1963): 6–7.

29. Morris Rossabi, *Khubilai Khan: His Life and Times* (Berkeley, 1988), 4.

30. H. Göckenjan, "Zur Stammesstruktur und Heeresorganization altaischer Völker. Das Dezimalsystem," in *Europa Slavica-Europa Orientalis. Festschrift für Herbert Ludat zum 70. Geburtstag*, ed. K-D Grothusen and K. Zernack (Berlin, 1980), 51–86.

31. The widespread view that Chinggis Khan also established a written legal code was attacked by D. O. Morgan, "The Great Yasa of Chingis Khan and Mongol Law in the Ilkhanate," *Bulletin of the School of Oriental and African Studies* 49, 1 (1986): 163–76, but Igor de Rachewiltz has confirmed that the "Jasagh" was a written law code. For a solid recent biography, see Ratchnevsky, *Genghis Khan, His Life and Legacy*, trans. T. N. Haining (Oxford, 1992).

32. Barthold, *Four Studies on the History of Central Asia*, 1:130.

33. Juvaini, *History of the World Conqueror*, 77–78; Sanktrityayana, *History of Central Asia*, 247; Barthold, *Turkestan Down to the Mongol Invasion*, 403–4.

34. Juvaini, *The History of the World-Conqueror*, 135.

35. Barthold, *Turkestan Down to the Mongol Invasion*, 398. The Russian text is somewhat fuller on this point, *Turkistan v epokhu mongolskogo nashestviia*, 465.

36. Juvaini, *The History of the World-Conqueror*, 127–29, 135; Vambery, *History of Bukhara*, 123.

37. Barthold, *Turkestan Down to the Mongol Invasion*, 439–40. On Jalal al-Din's tumultuous career, see *Shikhab ad-din Mukhammad an-Nasavi. Sirat as-Sultan Dzhalil ad-din Mankburny*, trans. A. N. Buniiatov (Moscow, 1996).

38. Douglas S. Benson, *The Mongol Campaign in Asia* (Chicago, 1991), 145; Saunders, *The History of the Mongol Conquests*, 140ff.

39. This section draws mainly on Juvaini, *The History of the World-Conqueror*, 120ff.; and Barthold, *Turkestan Down to the Mongol Invasion*, 429–33.

40. Juvaini, *The History of the World-Conqueror*, 127; Wittfogel, *Oriental Despotism*, 30.

41. Juvaini, *The History of the World-Conqueror*, 133–34.

42. Barthold, *Four Studies on the History of Central Asia*, 1:127ff.

43. Juvaini, *The History of the World-Conqueror*, 173. Some recent scholars have questioned Juvaini as a source and even suggest that the Mongols did not destroy either Merv or Nishapur. In the absence of firm proof of this hypothesis, this account follows the more traditional view. For a skeptical view of the tales of destruction, see George Lane, *Ghengis Khan and Mongol Rule* (Westport, 2004).

44. John Andrew Boyle, *The Mongol World Empire, 1206–1370* (London, 1977), 617.

45. Juvaini, *The History of the World-Conqueror*, 176, 127.

46. Barthold, *Turkestan Down to the Mongol Invasion*, 408, 341.

47. Juvaini, *The History of the World-Conqueror*, 98ff.

48. Boyle, *The Mongol World Empire, 1206–1370*, 618.

49. Barthold, *Four Studies on the History of Central Asia*, 1:122–27.

50. Vambery, *History of Bukhara*, 123.

51. Juvaini, *History of the World Conqueror*, 91–94.

52. Reuven Amitai has written cogently on the Mongol-Mamluk conflict in his *Mongols and Mamluks: The Mamluk-Ilkhanid War, 1260–1281* (Cambridge, 1995), especially 214–35.

53. Rossabi, *Khubilai Khan: His Life and Times,* 53.

54. Ibid., 131ff.

55. Stephen Turnbull, *The Mongol Invasions of Japan, 1274 and 1281* (Oxford, 2010); on the recent discovery of ships from Khubilai Khan's fleet, see 'Divers Find Thirteenth Century Wreck from Kublai Khan's Invasion Fleet That Was Destroyed by a 'Divine' Typhoon," *Daily Mail*, October 26, 2011.

56. Rossabi, *Khubilai Khan*, 142.

57. Ibid., 179–99.

58. Ibid., 145–51.

59. Peter Jackson and David Morgan, eds., *The Mission of Friar William of Rubruck*, trans. Peter Jackson (London, 1990), 240ff..

60. Thomas T. Allsen, *Culture and Conquest in Mongol Eurasia* (Cambridge, 1971), 161.

61. M. C. Johnson, "Greek, Moslem, and Chinese Instrument Design in the Surviving Mongol Equatorials of 1279 AD," *Isis* 32, 1 (July 1940): 27–43; and especially Needham, *Science and Civilization in China*, 5:352–74.

62. Allsen, *Culture and Conquest in Mongol Eurasia*, 166–68.

63. Ibid., 168–70; Kennedy, *Astronomy and Astrology in the Medieval Islamic World*, 59–74.

64. Grousset, *Empire of the Steppes*, chap. 9.

65. George Lane, *Early Mongol Rule in Thirteenth Century Iran: A Persian Renaissance* (London, 2003).

66. Andre Godard, "The Mausoleum of Öljeitü at Sultaniy," in *A Survey of Persian Art*, ed. Arthur Upham Pope and Phyllis Ackerman (Tehran, 1977), 1103–18.

67. Judith Kolbas, *The Mongols in Iran: Chinggis Khan to Uljaytu, 1220–39* (London, 2006), 375. It may fairly be argued that the problem arose equally from the lack of receptivity of Mediterranean Muslims to European civilization.

68. On Shams al-Din and other Central Asians in the Mongol administration, see Rossabi, *Khubilai Khan*, chaps. 5, 7.

69. Browne, *A Literary History of Persia from Firdawsi to Sa'di*, 486.

70. Barthold adjudicates between these in his *Turkestan Down to the Mongol Invasion*, 343, 355–59.

71. Rypka, "Poets and Prose Writers of the Late Saljuq and Mongol Periods," 5:576–77. On the negative impact of panegyrics, see A. Afsahzod, "Persian Literature," in *History of Civilizations of Central Asia*, vol. 4, pt. 2, 374.

72. On the events of 1208–1212 in Samarkand, see Barthold, *Turkestan Down to the Mongol Invasion*, 361–67. On Rumi's birthplace and other biographical details, see Franklin D. Lewis's authoritative *Rumi Past and Present, East and West* (Oxford, 2000), 42–56. The dispute over Rumi's traditional birthdate of 1207 can be found in Lewis and Annemarie Schimmel, *The Triumphal Sun: A Study of the Works of Jalaloddin Rumi* (New York, 1993), 12–14.

73. Shimmel, *The Triumphal Sun*, 9.

74. Reynold A. Nicholson, *Rumi, Poet and Mystic* (Oxford 1995), 32.

75. Kabir Helminksi, ed., *The Rumi Collection* (Boston, 2000), 45.

76. Ibid., 177.

77. Ibid., 87.

78. Hamid Dabashi refutes the oft-told tale that Tusi had been held at Alamut against his will: "Khwaja Nasir al-Din Tusi and the Ismailis," in *Medieval Ismaili History and Thought*, ed. Farhad Daftary (Cambridge, 1996), 231ff.

79. On recent excavations at the observatory and at Alamut as a whole, see "New Findings in Alamut Excavations," http://previous.presstv.ir/detail.aspx?id=76699; and "Medieval Observatory Unearthed in Northern Iran," http://www.medievalarchives.com/tag/iran.

80. In spite of his *nisba*, it is sometimes claimed that Tusi was born in Rayy, near modern Tehran.

81. Dabashi, "Khwaja Nasir al-Din Tusi and the Ismailis," 239.

82. Bakhtyar Husain Siddiqi, "Nasir al-Din Tusi," in *A History of Islamic Philosophy*, ed. M. M. Sharif (Wiesbaden, 1963), 565ff.

83. Farid Alakbarov, "A 13th-Century Darwin? Tusi's Views on Evolution," *Azerbaijan International* 9 (Summer 2001): 48ff.

84. Ibid.

85. Ahmad S. Dallal, *An Islamic Response to Greek Astronomy: Kitab Tadil Hayat Al-Aflak of Sadr Al-Sharia*, ed. D. Pingree (Leiden, 1997).

86. Van Brummelen, *The Mathematics of the Heavens and the Earth*, 186ff.

87. F. Jamil Ragep, "Freeing Astronomy from Philosophy: An Aspect of Islamic Influence on Science," *Osiris* 16 (2001): 60.

88. Kennedy, *Astronomy and Astrology in the Medieval Islamic World*, 287, 369–77.

89. E. S. Kennedy, "Late Medieval Planetary Theory," *Isis* 57, 3 (1966): 78. Also Kh. Talashev, "Ob arifmeticheskom tractate Nasir ad-Dina ad-Tusi," in *Iz istorii tochnykh nauk na sredenvekovom blizhnem i srednem vostoke* (Moscow, 1972), 210–19; and George Saliba, "The First Non-Ptolemaic Astronomy at the Maraghah School," *Isis* 70, 4 (1979): 571–76.

90. An excellent selection of Tusi's philosophical writings is in Nasr and Aminrazavi, eds., *An Anthology of Philosophy in Persia*, 2:344ff.

91. Jackson and Morgan, eds., *The Mission of Friar William of Rubruck*, 141.

92. Morris Rossabi, "The Legacy of the Mongols," in *Central Asia in Historical Perspective*, ed. Beatrice F. Mainz (Boulder, 1998), 31–32.

93. Browne, *A Literary History of Persia from Firdawsi to Sa'di*, 492ff.

94. Ibid., 470.

95. Ilias Nizamutdinov, *Iz istorii sredneaziatsko-indiiskikh otnoshenii (ix–xviii vv.)* (Tashkent, 1969), 17–23; Iqbal Ghani Khan, "Technical Links between India and Uzbekistan (ca. 1200–1650 AD)," in *Historical and Cultural Links between India and Uzbekistan* (Patma, 1996), 342.

96. Ibn Battuta, cited by Le Strange, *Lands of the Eastern Caliphate*, 463.

97. B. Akhmedov, "Central Asia under the Rule of Chingis Khan's Successors," in *History of Civilizations of Central Asia*, vol. 4, pt. 1, 265.

98. Grousset, *Empire of the Steppes*, 352–53.

99. Juvayni cites a Chinese visitor who claimed three-fourths had perished, in *History of the World-Conquerer*, 129–32. Gafurov estimates that half the population died: *Istoriia tadzhikskogo naroda*, 2:220.

100. Barthold, *Turkestan Down to the Mongol Invasion*, 461.

101. Juvaini, *History of the World-Conqueror*, 2:618ff.

102. John Masson Smith, Jr., "Demographic Considerations in Mongol Siege Warfare," *Archivum Ottomanicum* 13 (1994): 329.

103. Boyle, *The Mongol World Empire, 1206–1370*, 621–22.

104. David Morgan, *The Mongols* (London, 1986), 74.

105. J. Byrne, *The Black Death* (London, 2004), 58–62.

106. John Masson Smith, Jr., "Mongol Manpower and Persian Population," *Journal of Economic and Social History of the Orient* 18, 3 (October 1975): 291.

107. Gafurov, *Istoriia tadzhikskogo naroda*, 1:305.

108. The one historian to stress this important point is Morgan, *The Mongols*, 80ff.

109. Juvaini, *History of the World-Conqueror*, 127, 257. Barthold, *Turkestan Down to the Mongol Invasion*, 437, claims without evidence that the Amu Darya ceased to flow into the Caspian as a result of the Mongols' assault on hydraulic systems.

110. Juvaini, *History of the World-Conqueror*, 165; also Hafiz Abru, quoted by Le Strange, *The Lands of the Eastern Caliphate*, 402.

111. Barthold, *Turkestan Down to the Mongol Invasion*, 480.

112. Mez, *The Renaissance of Islam*, 634.

113. Barthold, *Turkestan Down to the Mongol Invasion*, 427.

114. Michal Biran, *Qaidu and the Rise of the Independent Mongol State in Central Asia* (London, 1997), 97ff.

115. Gafurov, *Istoriia tadzhikskogo naroda*, 2:298.

116. Biran, *Qaidu and the Rise of the Independent Mongol State in Central Asia*, chap. 1.

117. Souchek, *A History of Inner Asia*, 111.

118. Gafurov, *Istoriia tadzhikskogo naroda*, 2:302. On the criminalization of refusal to accept Mongol currency, see Davidovich, "Coinage and the Monetary System," 404.

119. On the Tarabi revolt of 1238, see Akhmedov, "Central Asia under the Rule of Chinggis Khan's Successors," 266–67.

120. Ibid., 99–110.

121. Biran, *Qaidu and the Rise of the Independent Mongol State in Central Asia*, 94–97.

122. Barthold, in *Four Studies on the History of Central Asia*, 1:121, asserts that no historical writing occurred under the Mongols, so the *History of Herat* by Sif ibn Yaqub, or Saifi, was a rare exception.

123. Rypka, *History of Iranian Literature*, 235, 254, 263.

124. Barthold, *Turkestan Down to the Mongol Invasion*, 461.

125. The sole biography, flawed by many errors, is Ch. G. Baiburdi, *Zhizn i tvorchestvo Nizari-persidskogo poeta* (Leningrad, 1963); also Rypka, *History of Iranian Literature*, 255ff.

126. Nizari's Ismaili affirmations are ably assessed by Nadia Eboo Jamal, *Surviving the Mongols: Nizami Quhistani and the Continuity of Ismaili Traditions in Persia* (London, 2002).

127. The classic study of this structure and its problematic inscription, which casts doubt on its popularly accepted link with Manas, is by G. A. Pugachenkova, *Gumbez Manassa* (Moscow, 1950); also S. Peregudova, "Mausoleum of Manas," Kyrgyzstan Freenet Website, http://freenet.bishkek.su/kyrgyzstan/gumbez.html.

128. A. T. Hatto has translated Valikhanov's text in *The Memorial Feast for Kokotoy-Khan* (Oxford, 1977); for a modern rendering, see Walter May, trans., *Manas* (Bishkek, 2004).

129. An exhaustive and useful introduction to these debates is M. I. Bogdanov, A. A. Petrosian, and V. M. Zhirmunskii, eds., *Kyrgyzskii geroicheskii epos Manas: Voprosy izucheniia eposa narodov SSR* (Moscow, 1961), and especially Zhirmunskii's overview, 85–196. The charge of forgery was made by Robert Auty in Robert S. Auty and A. T. Hatto, *Traditions of Heroic and Epic Poetry*, 2 vols. (London, 1980), 2:234.

130. This issue is treated exhaustively by Pugachenkova, *Gumbez Manassa*, who concludes that the relation to Manas, while indirect, was real.

131. V. V. Bartold, *Kyrgyzy: Istoricheskii ocherk* (Frunze, 1927).

132. R. Z. Kydyrbaeva, "The Kyrgyz Epic *Manas*," in *History of Civilizations of Central Asia*, vol. 4, pt. 2, 403.

133. Hodgson, *The Venture of Islam*, 2:426.

134. Souchek, *A History of Inner Asia*, 117–18.

135. Shaykh Muhammad Hisham Kabbani, *The Naqshbandi Sufi Way: History and Guidebook of the Saints of the Golden Chain* (Chicago, 1995), 164–65.

136. On Naqshband and his order, see Ikschak Weismann, *The Naqshbandiyya: Orthodoxy and Activism in a Worldwide Sufi Tradition* (London, 2008); and Hadrat

Sharkh Muhammad Hisham Kabbani's thoughtful *The Naqshbandi Sufi Way (History and Guidebook of the Saints of the Golden Chain)* (n.p., 1995).

137. On how direct occult transmission from the dead later became a feature of Central Asian Sufism, see Juliam Baldick, *Imaginary Muslims: The Uwaysi Sufis of Central Asia* (New York, 1993), 15–39.

138. Among many studies of Naqshband's influence abroad, see Arthur F. Buehler, *Sufi Heirs of the Prophet: The Indian Naqshbandiyya and the Rise of the Mediating Sufi Shaykh* (Columbia, 1998).

139. Rypka, *History of Iranian Literature*, 232–33.

140. Jacques Verger, "The Universities," in *The New Cambridge Medieval History*, ed. Michael Jones (Cambridge, 2000), 6:66–73.

CHAPTER 14

1. John E. Woods, "Timur's Genealogy," in *Essays Written in Honor of Martin B. Dickson* (Salt Lake City, 1990), 85ff.

2. Ruy Gonzales de Clavijo, *Narrative of the Embassy of Ruy Gonzalez de Clavijo to the Court of Timour at Samarcand, AD 1403–6*, trans. Clements R. Markham (London, 1859), 126.

3. L. B. Baimatov, "Kamal Khujandi and His Time," in *To the 675 Anniversary of the Great Kamal Hujandy* (Dushanbe, 1996), 126.

4. Relevant materials on this meeting are assembled by Walter J. Fischel, *Ibn Khaldun and Tamerlane: Their Historic Meeting in Damascus, 1401 AD (803 A.H.)* (Berkeley, 1959).

5. Peter Jackson, *The Mongols and the West, 1221–1410* (London, 2005), 243ff.

6. An analytic overview of these campaigns is provided by Jean Aubin, "Comment Tamerlan prenait les villes," *Studia Islamica* 19 (1963): 83–122.

7. Gibbon, *The Decline and Fall of the Roman Empire*, 5:665.

8. Dennis C. Twitchet and Herbert Franke, eds., *Alien Regimes and Border States, 907–1368, Cambridge History of China* (Cambridge, 1994), 6:622.

9. Quoted by R. Z. Kyrdyrbaeva, "The Kyrgyz Epic *Manas*," in *History of Civilizations of Central Asia*, vol. 4, pt. 2, 405.

10. R. G. Mukminova, "Craftsmen and Guild Life in Samarkand," in *Timurid Art and Culture: Iran and Central Asia in the Fifteenth Century*, ed. Lisa Golombek and Maria Subtelny (Leiden, 1992), 29–35.

11. For an overview of his architectural projects, see Pulat Zokhidov, *Temur Davrining Memorii Kakhkashoni/ Arkhitekturnye sozvezdie epokhi Temura* (Tashkent, 1996).

12. See the comprehensive description of Timur's buildings in Shahrisabz in M. E. Masson and G. A. Pugachenkova, "Shakhri Sabz pri Timure i Ulug-Beke," *Trudy Tsentroaziatskogo Gos. Universiteta* (1963): 17–96, trans. J. M. Rogers in *Iran* 18 (1978): 103–26; and 20 (1980): 121–43.

13. Lisa Golombek, "Tamerlane, Scourge of God," *Asian Art* 2, 2:34.

14. See chapter 10.

15. This passage is indebted to Thomas W. Lentz and Glenn D. Lowry, "Timur and the Image of Power," in *Timur and the Princely Vision: Persian Art and Culture in the*

Fifteenth Century (Washington, DC, 1989), chap. 1; see also Sheila S. Blair and Jonathan M. Bloom, *The Arts and Architecture of Islam, 1250–1800* (New Haven, 1994), 34–54.

16. From the Timur-era historian Abdul-Razzaq Samarqandi, quoted in *Timur and the Princely Vision*, epigraph; also see *Amir Timur in World History* (Tashkent, 2001), 123.

17. Clavijo, *Narrative of the Embassy of Ruy Gonzalez de Clavijo to the Court of Timour at Samarcand, AD 1403–6*, xlviii.

18. Ibid., 43.

19. An authoritative account of governance under Timur and his successors is Beatrice Forbes Manz, *The Rise and Rule of Tamerlane* (Cambridge, 1989), esp. chap. 6 and appendix, 167–75; recent writings on the political system are reviewed by D. S. Abidzhanova, "Izuchenie voprosov gosudarstvennogo ustroistva v maveraunnakhre v kontse xiv–nachale xv vekov v otechestvennoi i zarubezhnoi literature," in *Uzbekistan v srednie veka: istorii i kultura* (Tashkent, 2003), 60–69.

20. K. Z. Ashrafyan, "Central Asia under Timur From 1370 to the Early Fifteenth Century," in *History of Civilizations of Central Asia*, vol. 4, pt. 1, 337.

21. Maria Eva Subtelny and Anas B. Khalidov, "The Curriculum of Islamic Higher Learning in Timurid Iran in the Light of the Sunni Revival under Shah-Rukh," *Journal of the American Oriental Society* 115, 2 (April–June 1995): 210–36.

22. For a useful view of these and other sites today, see Allchin and Hammond, *The Archaeology of Afghanistan*, 379–89.

23. On the Kitabkhana, see Glen D. Lowry, "The Kitabkhana and the Dissemination of the Timurid Vision," in *Timur and the Princely Vision*, ed. Lentz and Lowry, 159–36.

24. O. Grabar, "The Visual Arts," in *The Cambridge History of Iran*, 5:648–51.

25. See Yolande Crowe, "Some Timurid Designs and Their Far Eastern Connections," in *Timurid Art and Culture*, ed. Golombek and Subtelny, 168–78.

26. Ahmad ibn Muhammad ibn Arabshah, *Tamerlane, or Timur the Great*, trans. J. H. Sanders (2008), 320.

27. Sarah Chapman, "Mathematics and Meaning in the Structure and Composition of Timurid Miniature Painting," *Persica* 29 (2003): 33–45.

28. L. Richter-Bernburg, "Medicine, Pharmacology and Veterinary Science in Islamic Eastern Iran and Central Asia," in *History of Civilizations of Central Asia*, vol. 4, pt. 2, 314.

29. A. A. Akhmedov, introduction to *Mukhammad Taragai Ulughbeg, Zidzh: Novye Guraganovy astronomicheskie tablitsy* (Tashkent, 1994), 10–11.

30. Bartold's biographical study remains the invaluable source on Ulughbeg: Barthold, "Ulugh-Beg," in *Four Studies on the History of Central Asia*, 2:1–183.

31. Mohammad Bagheri, "A Newly Found Letter of Al-Kashi on Scientific Life in Samarkand," *Historia Mathematica* 24 (1997): 243.

32. Tirmidhi, *Hadiths*, no. 218; also Ibn Majah, *Hadiths*, 1, 224.

33. Sergej Chmelnizkij, "Timurid Architecture," in *Islam: Art and Architecture*, 423.

34. Hodgson, *The Venture of Islam*, 2:439ff.

35. Bagheri, "A Newly Found Letter of Al-Kashi on Scientific Life in Samarkand," 250.

36. Edward S. Kennedy, "A Letter of Jamshid al-Kashi to His Father: Scientific Research and Personalities at a Fifteenth Century Court," *Orientalia* 29 (1960): 191–213; Bagheri, "A Newly Found Letter of Al-Kashi on Scientific Life in Samarkand," 241–56.

37. Bagheri, "A Newly Found Letter of Al-Kashi on Scientific Life in Samarkand," 249.

38. E. S. Kennedy, "The Exact Sciences in Timurid Iran," in *The Cambridge History of Iran*, ed. Peter Jackson (Cambridge, 1986), 6:575–78.

39. Edward S. Kennedy has analyzed these devices in a series of pioneering studies: "A Fifteenth-Century Planetary Computer: al-Kashi's 'Tabaq al-Manateq' I. Motion of the Sun and Moon in Longitude," *Isis* 41, 2 (1950): 180–83; "Al-Kashi's Plate of Conjunctions," *Isis* 38, 1–2 (1947): 56–59; "An Islamic Computer for Planetary Latitudes," *Journal of the American Oriental Society* 71, 1 (1951):. 13–21; and "A Fifteenth-Century Planetary Computer: al-Kashi's 'Tabaq al-Maneteq' II: Longitudes, Distances, and Equations of the Planets," *Isis* 43, 1 (1952): 42–50.

40. Kennedy, "The Exact Sciences in Timurid Iran," 578ff.

41. *Tabulae long. ac lat. stellarum fixarum ex observatione Ulugh Beighi*, trans. Thomas Hyde (London, 1665).

42. A. A. Akhmedov, introduction to *Ulughbeg Mukhammad Taragai (1394–1449: "Zidzh," Novye Guraganovy astronomicheskie tablitsy* (Tashkent, 1994), 15.

43. Barthold, *Four Studies on the History of Central Asia*, 2:125–26.

44. Based on A. A. Akhmedov's citations from the Turkish historian of science, in Zaki Salih and Akhmedov, *Mukhammad Taragai Ulughbeg*, 6.

45. F. Jamil Ragep, "Freeing Astronomy from Philosophy: An Aspect of Islamic Influence on Science," *Osiris*, 2nd Series, 16 (2001): 49–71.

46. R. M. Savory, "The Struggle for Supremacy in Persia after the Death of Timur," *Der Islam* 40 (1964): 35–65.

47. Barthold, "Mir Ali-Shir," in *Four Studies on the History of Central Asia*, 3:34. Some of Khoja Ahrar's letters have come down to us: see Jo-Ann Cross and Ason Urumbaev, ed. and trans., *The Letters of Khwaja 'Ubaydallah Ahrar and His Associates* (Leiden, 2002).

48. For an overview of the political history of these years, see H. R. Roemer, "The Successors of Timur," in *The Cambridge History of Iran*, 6:121–25.

49. Blair and Bloom, *The Arts and Architecture of Islam, 1250–1800*, chap. 3.

50. Maria Eva Subtelny, "Socioeconomic Bases of Cultural Patronage under the Later Timurids," *International Journal of Middle East Studies* 20, 4 (November 1988): 379–505.

51. Barthold, "Mir Ali-Shir," in *Four Studies on the History of Central Asia*, 3:17. The classic study of Navai's life remains E. E. Bertels, *Navai: opyt tvorcheskoi biografii* (Moscow, 1948); see also Bertels's *Navai i Dzhami* (Moscow, 1965).

52. Barthold, "Mir Ali-Shir," 22.

53. On the funding of such entities, see Maria Eva Subtelny, "A Timurid Educational and Charitable Foundation: The Ilkhlasiyya Complex of Al Shir Navai in 15th Century Herat and Its Endowment," *Journal of the American Oriental Society* 111, 1 (January–March 1991): 38–61.

54. On Bihzad (Behzod in Uzbek), see Ebadollah Bahari, *Bihzad: Master of Persian Painting* (London, 1997); and Eleanor Sims, Boris Il'ich Marshak, and Ernst J. Grube, *Peerless Images: Persian Painting and Its Sources* (New Haven, 2002), 60ff.

55. Subtelny, "A Timurid Educational and Charitable Foundation," 45.

56. Zahiru'd-din Muhammad Babur Padshah Ghazi, *Babur-Nama* (Delhi, n.d., after edition of 1921), 304–6.

57. Dawlatshah, *The Tadhkiratush-Shuara, "Memoirs of the Poets" of Dawlatschah*, trans. E. G. Browne (Leiden, 1901).

58. Rypka, *History of Iranian Literature*, 283.

59. Barthold, "Mir Ali-Shir," 3:30; also Ehsan Yarshater, "Persian Poetry in the Timurid and Safavid Periods," in *The Cambridge History of Iran*, 6:973–74.

60. Rypka, *History of Iranian Literature*, 286–88.

61. Mir Ali Shir, *Muhakamat Al-Lughatain*, trans. Robert Devereux (Leiden, 1966).

62. Richard C. Foltz, *Mughal India and Central Asia* (Karachi, 1998), 133ff.

63. John F. Richards, *The New Cambridge History of India* (Cambridge, 1993), 282.

64. Gabor Agoston, "Early Modern Ottoman and European Gunpowder Technology," in *Multicultural Science in the Ottoman Empire*, ed. Ekmeleddin Ihsanoglu et al. (Turnhout, 2002), 13ff.

65. John D. Hoag, "The Tomb of Ulugh Beg and Abdu Razzaq at Ghazni, a Model for the Taj Mahal," *Journal of the Society of Architectural Historians* 27, 4 (December 1968): 234–48.

66. Franz Babinger, *Mehmed the Conqueror and His Time*, trans. Ralph Manheim (Princeton, 1992), 466.

67. For an intriguing overview, see the essays in Attilio Petruccioli, ed., *Gardens in the Time of the Great Muslim Empires* (Leiden, 1997); see also Ebba Koch, "Mughal Palace Gardens from Babur to Shah Jahan, 1526–1648," in *Muqamas XIV*, ed. Guiru Necipoglu (Leiden, 1997), 143–65.

68. On Isfahan, see Stephen Blake, *Half the World: The Social Architecture of Safavid Isfahan, 1590–1722* (Costa Mesa, 1999); and Roger Savory, *Iran under the Safavids* (Cambridge, 1980), chap. 7.

69. Maurice S. Dimond, "Mughal Painting under Akbar the Great," *Metropolitan Museum of Art Bulletin*, New Series, 12, 2 (October 1953): 46–51. On Mughal painting, see Milo Cleveland Beach, *Early Mughal Painting* (Cambridge, 1987); and *The Imperial Image: Paintings for the Mughal Court* (Washington, DC, 1981); on overall links, see Richard Foltz, "Cultural Contacts between Central Asia and Mughal India," *Central Asiatic Journal* 42, 1 (1998): 44–65.

70. Curiously, it is generally thought that Persian poetry actually declined under the Safavids: Rypka, *History of Iranian Literature*, 292. On Persian in Mughal India, see Muzaffar Alam, "The Pursuit of Persian: Language Policy in Mughal Politics," *Modern Asian Studies* 32, 2 (May 1998): 317–49.

71. For example, Annemarie Schimmel and Stuart Cary Welch, *Anvari's Divan: A Pocket Book for Akbar* (New York, 1983).

72. Henry Corbin and Seyyed Hossein Nasr first proposed the existence of this school, and Nasr himself gives it a later evaluation in "The School of Isfahan Revisited," in *Islamic Philosophy from Its Origins to the Present*, 209ff.

73. Svat Soucek, *Piri Reis and Turkish Mapmaking after Columbus: The Khalili Portolan Atlas, Studies in the Khalili Collection* (London, 1992), vol. 2.

74. David A. King, "Two Iranian World Maps for Finding the Direction and Distance to Mecca," *Imago Mundi* 49 (1997): 62–82.

75. Irfan Habib, "Akbar and Technology," *Social Scientist* 20, 9/10 (September–October 1992): 3–15.

76. Ekmeleddin İhsanoğlu argues that observations began in 1573. *Science, Technology and Learning in the Ottoman Empire* (London, 2004), 19.

77. Cyril Elgood, *Safavid Medical Practice* (London, 1970), 70–88.

78. Roger Savory, *Iran under the Safavids* (Cambridge, 1980), 27ff.; see also Said Amir Arjomand, "The Clerical Estate and the Emergence of a Shi'ite Heirocracy in Safavid Iran: A Study in Historical Sociology," *Journal of the Economic and Social History of the Orient* 28, 2 (1985): 169–219.

79. See Rula Jurdi Abiusaad, *Converting Persia: Religion and Power in the Safavid Empire* (London, 2004), especially 61ff.; also Vera Moreen, "The Status of Religious Minorities in Safavid Iran from 1617–61," *Journal of Near Eastern Studies* 40, 2 (April 1981): 119–34.

80. For the views of one his Jesuit consultants, see John Correia-Afonso, S.J., ed., *Letters from the Mughal Court* (St. Louis, 1981), 4ff.

81. Sri Ram Sharma, *The Religious Policies of the Mughal Emperors* (Bombay, 1962), chap. 3; Hodgson, *The Venture of Islam*, 3:73.

82. Pran Nath Chopra, *Some Aspects of Society and Culture during the Mughal; Age, 1526–1707* (Agra, 1963), 155.

83. *Encyclopedia Iranica*, http://www.iranicaonline.org/articles/cap-print-printing -a-persian-word-probably-derived-from-hindi-chapna-to-print-.

84. William J. Watson, "Ibrahim Muteferrika and Turkish Incunabula," *Journal of the American Oriental Society* 88, 3 (1968): 435–41.

85. On the South Indian origins of printing in India, see B. S. Kesavan, *History of Printing and Publishing in India* (New Delhi, 1997).

CHAPTER 15

1. Bertrand Russell, *The Problems of Philosophy* (London, 1912).

2. Dimitri Gutas rightly defends early Islamic leaders against the widely repeated charge that they were hostile to Greek philosophy: *Greek Thought, Arabic Culture*, 167ff.

3. Steven Johnson, *Where Good Ideas Come From: The Natural History of Innovation* (New York, 2010), chap. 1.

4. Abdus Salam, *Ideals and Realities: Selected Essays of Abdus Salam* (Singapore, 1989), 147.

5. Mohamad Abdalla, "Ibn Khaldun on the Fate of Islamic Science After the 11th Century," *Islam and Science* 5, 1 (2007): 61–70.

6. Sonja Brentjes, "The Prison of Categories—'Decline' and Its Company," in *Islamic Philosophy, Science, Culture, and Religion: Studies in Honor of Dimitri Gutas*, ed. Felicitas Opwis and David Reisman, *Islamic Philosophy, Theology, and Science* (Leiden, 2012), 83:131–37.

7. Dallal, *Islam, Science, and the Challenge of History*, 151ff.

8. Sarah Stroumsa, *Freethinkers of Medieval Islam* (Leiden, 1999), 241.

9. Nigel Cliff's convincing case for Vasco da Gama's religious and civilizational motives does not change the broader reality that Europeans were seeking more secure and cheaper routes to the East: *Holy War: How Vasco da Gama's Epic Voyages Turned the Tide in a Centuries-Old Clash of Civilizations* (New York, 2011).

10. Iqbal Ghani Khan, *Historical and Cultural Links between India and Uzbekistan* (Patna, 1994), 337.

11. See Morris Rossabi's balanced treatment of political and economic factor in "The 'Decline' of the Central Asian Caravan Trade," in *The Rise of Merchant Empires*, ed. James D. Tracy (Cambridge, 1990), chap. 11.

12. Frye, "The Period from the Arab Invasion to the Saljuqs," 420. For an earlier expression of the same thesis, see Barthold, *Turkestan Down to the Mongol Invasion*, 180.

13. Timur Kuran develops this argument, as well as the case against Islamic partnership law, in *The Long Divergence: How Islamic Law Held Back the Middle East* (Princeton, 2011), chap. 5.

14. Biruni, *Alberuni's India*, 1:xv.

15. Dallal, *Islam, Science, and the Challenge of History*, 154.

16. Algazali, "Deliverance from Error," 273–74.

17. Savage-Smith, "Medicine," 927.

18. Golden, *Central Asia in World History*, 115.

19. Ernest Renan, *An Essay on the Age and Antiquity of the Book of Nabithean Agriculture, to Which Is Appended a Lecture on "The Position of the Shemitic Nations in the History of Civilization"* (London, 1863), 131ff.; Bartold, "Teokraticheskaia ideia i svetskaia vlast v musulmanskom gosudarstve," *Otchet S. Peterburgskogo Universiteta za 1902 god.* (St. Petersburg, 1903), 1–26; and J. J. Saunders, "The Problem of Islamic Decline and Decadence," in *Muslims and Mongols: Essays on Medieval Asia*, ed. G. W. Rice (Canterbury, 1977), 99–127.

20. Ira M. Lapidus, "The Separation of State and Religion in the Development of Early Islamic Society," *International Journal of Near East Studies* 6, 4 (October 1975): 363–85.

Index

❀

Page numbers in *italics* refer to maps, plates, and figures.

Abbas (Muhammad's youngest uncle), 119
Abbas (poet), 140
Abbas, Shah, 508
Abbas faction, 119–20
Abbasid banner, 142–43
Abbasid revolution, 121, 123–27, 154, 515
Abdallà, Mohamad, 522–23
Abd al-Razzaq, Abu Mansur, 215–17, 234, 244–45
Abidharmapitaka, 86
Abu Dulaf, 239
Abu Hanifa, 201, 557n129
Abu Hasan, xxxiii
Abu Hashim, xxxii, 119
Abu Ishak of Merv (Kisai), 351–52
Abu 'l-Fadl mausoleum, 426
Abu 'l-Faraj ibn al-Tayyib, 259, 264
Abu Mashar al-Balkhi, 21, 94, 179–80, 203
Abu Muslim, 119–25, 127, 560n65
Abu Said, 500
Abu Tahir, 398
An Abridged Collection of Authentic Hadiths with Connected Chains [of Transmission] Regarding Matters Pertaining to the Prophet, His Practices, and His Times (Bukhari), 246
academic institutions. *See* educational institutions
Achaemenids, 49
adaptation, 61, 519
Adelard of Bath, 168, 170–71
Aeschylus, 138
aesthetic cultures, 529
Afghanistan, *xxxix*, 6, 32, 520; Arab conquest of, 107–8; Buddhism in, 84; as source of luxury goods, 41; Taliban occupation of, 554n67
Afrasiab (Samarkand), 29–30, 49–50, 235; alphabet or script, 69; climate, 36; glassmaking, 65–66; irrigation system, 38–39
Afshin, 152–53

Age of Enlightenment, 4–20, 99, 515–17; Baghdad as center of, 157–58; cause of, 517–20; climate of, 35–36; decline of, 522–24, 533–39; intellectual class of, 16–20; as "intensive" civilization, 37–40; key players, xxi–xxix, 7–15; Khwarazm as center of, 271; lifetime of, 537–38; under Mahmud of Ghazni, 341–44, 357–60; Nishapur as center of, 196–207; relevance of, 538; starting point, 530–31; translations, 135
Agra, 508
agriculture, 37–40, 46, 57–58, 524–25
Ahmad, 451
Ahmad ibn Musa, 145–46, *147*
Ahrar, Khoja, 500, 606n47
Ahriman, 72
Ahuramazda, 72
Ai Khanoum, 57, 67, 76–81, *79–80*, 553n61
Airtam Frieze, *23*, 84–85
Ajina-Tepe (Witches Hill), 84–85
Ajmer, 335–36
Akbar, xxxvii, 336, 508–9, 511–12
Ak Beshim, 317–18
Akhsikent, 30, 46
aksakals (white beards), 522
Ak Serai or White Palace, 482–83, *483*, 483–84
Ala al-Din. *See* Juvayni
Alamut, 459
Albatenius. *See* Battani
Albigensians, 459
Albumasar. *See* Abu Mashar al-Balkhi
alchemy, 162–63, *163*
Alchemy of Happiness (Ghazali), 420
alcohol: poetry about drinking, 133; wine, 15, 51, 257, 296
Aleppo, 480
Alexander the Great, xxi, xxxi, 49–50, 69, 76, 176–77, 376
Alfarabius. *See* al-Farabi, Abu Nasr Muhammad